Underutilized Plant Resources of the Himalayas

THE AUTHORS

Dr. S.K. Sood is Professor and Dean Life Sciences, Himachal Pradesh University, Shimla. He obtained his M.Sc. (1974) and M.Phil. (1976) from Himachal Pradesh University and Ph.D. (1981) from Nagarjuna University, Guntur, specializing in Plant Systematics and Embryology. For his post-doctoral, he underwent specialized training in Ethnobotany. He has had a brilliant academic career (triple gold medallist) followed by over 30 years of teaching and conducting research in Botany at postgraduate as well as research levels. His current major interests are in the theory and practice of systematics, flora of North-West Himalaya, ethnic tribes and conservation research.

Dr. Anju Batta Sehgal received her Ph.D. in the year 2013 from Himachal Pradesh University, Shimla. She is gold medalist of HPU. At present, Dr. Sehgal is working as Principal of Govt. College in Himachal Pradesh. Recently Dr. Sehgal presented a paper in University of California, Santa Barbara, USA.

Dr. T.C. Bhalla is former director, Institute of Integrated Himalayan Studies, Himachal Pradesh University, Shimla. He received his Ph.D. from the Punjab University, Chandigarh. At present, he is the Chairman of the Department of Biotechnology.

Underutilized Plant Resources of the Himalayas

S.K. Sood
Anju Batta Sehgal
T.C. Bhalla

2016
Regency Publications
A Division of
Astral International Pvt. Ltd.
New Delhi 110 002

Cataloging in Publication Data—DK
Courtesy: D.K. Agencies (P) Ltd. <docinfo@dkagencies.com>

Sood, S. K. *(Life scientist)*, **author.**
Underutilized plant resources of the Himalayas/ S.K. Sood, Anju Batta Sehgal, T.C. Bhalla.
pages cm
Study conducted in Hamīrpur District of Himachal Pradesh, India.
Includes bibliographical references.
ISBN 978-93-86071-02-6 (International Edition)

1. Ethnobotany—India—Hamīrpur (Himachal Pradesh : District) 2. Plants, Useful—India—Hamīrpur (Himachal Pradesh : District) 3. Botany, Economic—India—Hamīrpur (Himachal Pradesh : District) 4. Plant diversity—India—Hamīrpur (Himachal Pradesh : District) I. Sehgal, Anju Batta, author. II. Bhalla, T. C., author. III. Title.

QK98.4.I4S66 2016 DDC 581.63095452 23

Published by : **Regency Publications**
A Division of
Astral International Pvt. Ltd.
– ISO 9001:2008 Certified Company –
4760-61/23, Ansari Road, Darya Ganj
New Delhi-110 002
Ph. 011-43549197, 23278134
E-mail: info@astralint.com
Website: www.astralint.com

Digitally Printed at : **Replika Press Pvt. Ltd.**

This copy is for sale outside Indian Sub-Continent.

Preface

Plant diversity is the basis of life for all living creatures. India with about 68 million people belonging to about 550 tribal communities and the plant wealth comprising of 17,500 species, is regarded as a veritable emporium of vast ethnobotanical knowledge. This has led to a keen interest by pharmaceutical companies to use this knowledge base as a reserve for research and development programme with the hope of producing novel drugs for amelioration of human suffering. Ironically, with the advent of modernization in tribal pockets and changes in sustenance economy, the rich traditional knowledge base in India as is true elsewhere, is fast eroding for which the need of the hour is to develop sound database on their ethnobotanical prudence before it is lost forever to posterity. As a matter of fact, tribal knowledge of plants is important not just for the tribal people themselves, but for the wider world. Viewed in this context, the preparation of this invaluable compendium has been necessitated, and is based on intensive ethnobotanical surveys undertaken by the authors in various rural pockets of the Hamirpur Himalayas during the years 2008-2012, a region of The Outer Himalayas significantly rich in ancient heritage, floristic and cultural diversity and ethnic variability. Main body of the textural matter comprised of an exhaustive account of 296 promising plants (medicinal 263, edible 133, fodder 108, ornamental 67, sacred 52, magico-religious 34, fibre 29, veterinary 123, miscellaneous 84) with multifaceted information on updated Nomenclature; Vernaculars, English, Hindi, Sanskrit and Regional Names; Distribution; Description; Flowering & Fruiting; Habitat Ecology; Part/s Used; Folk Uses; Active Constituents; Biological Activity and Uses in Literature to serve as a valuable ready-reckoner to different stakeholders like pharmacists, medical practitioners, chemists, biotechnologists, plant geneticists, new entrepreneurs and technocrats, extension workers, students, researchers, policy planners *etc.* Further, inclusion of information on economic valuation and availability of the underutilized Himalayan plant resources, nativity; 40 photographic plates, 63 tables, 39 figures, 3 useful appendices are the other added attractions that enhance the value of this treatise significantly. It is hoped that the information comprehended in the present manual offers tremendous scope for its bio-prospecting and as well can be utilized by policy planners in developing effective strategies for ensuring a meaningful sustainable rural development plan for hill communities of the Himalayan States. No institutional library in India and abroad should be without it.

We are indebted to Prof. A.D.N. Bajpayee, Vice-Chancellor, Himachal Pradesh University, Shimla, Prof. T.N. Lakhanpal and Prof. V.K. Mattu for their encouragement, support and well wishes; to Dr S.K. Srivastva (Director) and Dr Prashant (herbarium incharge, Botanical Survey of India) for their kind help in authentication of the plant specimens; to Dr Maninder Karan, Professor UIPS, Punjab University, Chandigarh for pharmaceutical analysis of some of the collected plants; and to Dr Kuldeep Dogra, Dr Dheeraj Rawat, Dr Suresh Kumar, Dr Romita Sharma, Mr Ved Prakash, Mr Rakesh, Ms Chandrika, Ms Parul, Ms Indu, Ms Poonam and Mr Bal Krishan for their ungrudging help in various ways. We extend our gratefulness to the publisher for bringing out the book in its present form. Thanks are also due to authors of books, papers and figures listed in the treatise. Finally, we thank the our families for bearing the opportunity cost.

The work does not pretend to be complete; suggestions for improving it, if any, are most welcome.

S.K. Sood
Anju Batta Sehgal
T.C. Bhalla

Abbreviations

AD	Anno domini
Alc.	Alcoholic
Alk.	Alkaline
Aq.	Aqueous
amsl	Above Mean Sea Level
Ass.	Assamese
Ayur.	Ayurvedic
B.	Bengal
BC	Before Christ
Bo.	Bombay
Bu	Butyl
C	Carbon
C.	Central
Ca	Calcium
CBD	Centre for Biological Diversity
Cm	Centimeter
CNS	Central Nervous System
d	Dextro
E.	East
EL	Ethnobotany Laboratory
Eng.	English
et al.	(et alia) and other authors
Etc.	(et cetera) and so on.
Extr. (s)	Extract (s)
Fig. (s)	Figure (s)
G	Gujrati
Gar.	Garhwal
HCN	Hydrogen Cynide
H.P.	Himachal Pradesh
HPU	Himachal Pradesh University
i.e.	(id est) that is
K	Potassium
Kan.	Kannada
Kash.	Kashmir
kg	Kilogram
Kum.	Kumaon
l	laevo
lat.	Lattitude
long.	Longitude
m	meter
Mal.	Malyalam
Mar.	Marathi
mg	milligram
Mg	Magnesium
M.P.	Madhya Pradesh
ml	millilitre
′	Minute

NIT	National Institute of Technology
(-) ve	Negative
N	Nitrogen
N.W.	North West
O	Oxygen
OAc	Acetoxy
OH	Hydroxy
OMe	Oxymethyl
Ori.	Oriya
P	Punjab
%	Per cent
Ph	Phenyl
Pl. (s)	Plate(s)
"	Second
S.	South
Sans.	Sanskrit
Sp.	Specie
Spp.	Species
Sq Km.	Square Kilometer
Syn.	Synonym
S.W.	South West
Tam.	Tamil
Tel.	Telgu
TIV	Total Importance Value
U.P.	Uttar Pradesh
Var.	Variety
Vern.	Vernacular
viz.	Vide licet (namely)
Vit.	Vitamin
Vet.	Veterinary
Vol. (s)	Volume (s)
W.	West
WHO	World Health Organisation
WTO	World Trade Organisation
&	And
o	Degree

List of Illustrations

Contents

Chapter 1
Introduction

Man and Plants

Prior to the coinage of the term 'Ethnobotany' by Dr John Harshberger in 1895, usage of plants by human beings found place in Sanskrit, Greek and Arabic literature, ethnographies travelogues, herbals, etc. (Mudgal, 1995). Since then, the scope, concepts and implications of ethnobotany have been expanding at a very fast rate (Schultes, 1960; Jain, 1967, 1986, 1987). Alcorn (1984) regarded ethnobotany as the study of contextualized plant use. Wickens (1990) defined ethnobotany as "The study of useful plants prior to their commercial exploitation and eventual domestication; it includes the use of plants by both tribal and non-tribal communities without any implication of primitive or developed societies". This plant-man relationship has been classified into two categories: 1) Abstract, and 2) Concrete (Jain, 1987). The abstract relationship deals with taboos, avoidances, sacred plants, worship, folklore, etc. while the latter includes mainly with the material use and the acts of domestication, conservation, improvement or destruction of plants. More importantly, this study of plants in relation to people includes both wild and domesticated plants (Heiser, 1995). Pushpangadan (1990) treated ethnobotany/ ethnobiology as a study of the knowledge system pertaining to the multidimensional perspective of life, culture, traditions as well as interaction of traditional or less advanced human communities like tribals with their local flora (ethnobotany) or fauna (ethnozoology). Pei (2001) regarded ethnobotany as the science which comprehends the relationship of a given society with its environment and, in particular, with the plant world; these relationships may be social, economic, ecological, symbolic, religious, commercial or artistic. Now-a-days, the subject of ethnobotany has been recognised as a rapidly expanding multidisciplinary natural science throughout the world, with many workers becoming involved in the practical application of its data in areas such as biodiversity prospecting and conservation biology.

Indigenous Societies

Indigenous societies/ tribals/ aborigines all over the world in different geographical regions are an invaluable bank of traditional knowledge. They are scattered over the face of earth in around 70 countries. Among them well over 150 million live in Asia; two thirds in China and India (Sinha, 1996). The strong basis of this traditional knowledge has been the necessity, instinct, curiosity and keen observation, constant trial and error, long experience and close association of indigenous people with nature (Jain, 2004). In fact, these societies are human conservatories which can't be duplicated by application of science and technology. The decade beginning from January 1, 1995 was observed as the International Decade for the world's Indigenous People. All over the world, the indigenous people have protected the biodiversity (forest and wildlife) because of their knowledge and traditional practices. The science of pharmacognosy also owes its development at least in part to the native medicine men who recognized and used the therapeutic qualities of herbs. Undeniably, tribal knowledge of plants is important not just for the tribal people themselves but for the wider world (Maheshwari, 1987a, b). In this regard Swaminathan (1995) rightly pointed: "Those who have conserved biodiversity tend to remain poor, while those who have converted such genetic diversity into commercial products through biological technology are rich". Unfortunately, there is no record of this knowledge. Janaki Ammal (1996) lamented that the tribal traditions are fast disappearing due to urbanisation, rapid industrialization and changes in sustenance economy. Of late, there has been a resurgence of interest all over the world in the study of primitive communities and tribals with an eye to potential future use for the ultimate welfare of humanity. One of the foremost and challenging tasks before the world community is to inventorise and record all ethnobiological information among the diverse ethnic communities before the traditional cultures are lost for ever (Rao, 1996).

India—A Veritable Emporium of Ethnic Variability and Floristic Diversity

India (lat. 8°-04′ N to 37°-06′ N and long. 68°-07′ E to 97°-25′ E; 329 million hectares of geographical area; 7,500 km long coast line), a federal constitutional republic with a parliamentary democracy consisting of 28 states and seven union territories, lies within the Indo-Malaya ecozone and endowed with a rich biological heritage. With more than 53 million natives or tribal communities under 227 ethnic groups covering 5,000 villages, the region is perhaps the largest country possessing a valuable heritage of accumulated indigenous/tribal knowledge (Anonymous, 1994; Pushpangadan, 1994). As many as 106 different languages and 227 subsidiary dialects are spoken by tribals in India (Arora, 1995). Phytogeographically, India can be classified into Western Himalayas, Eastern Himalayas, Assam, Indus Plain, Gangetic Plain, Deccan, Malabar and the Andamans. The wide variety in physical features and climatic conditions has resulted in a diversity of ecological habitats, which harbour more than 15,000 species out of the total 4,22,000 flowering plants reported from the world (Govaerts, 2001). It has been estimated that about 6,000 species belonging to over 141 genera of 47 families are endemic to India. Due to diverse agro-climatic conditions, uniqueness and richness of the flora, India is recognised as one of the World's top 12-mega diversity countries of the world (Zeven & Zhukorsky, 1975). The floristic richness has also rendered the North-eastern region and Western Ghats to be recognised as two out of 25 Hot spots of the world (Myers, 1990).

The Himalayas

The Great Himalayas (27°50′ - 27°06′ N lat. and 72°30′ - 97°25′E) represent the largest mountain chain covering about 5 million km^2 in surface area for which the first Prime minister of independent India, Shri Jawahar Lal Nehru once stated: "Himalayas are not only near to us, but also very dear; for they have always been a part of our history and tradition, our thinking and poetry, our worship and devotion. They are not only physically present in India, dominating the vast Indian plains, but for every Indian they convey a deeper message". Owing to enormous size and elevation, the Himalayas include parts of Trans, North-West, Central and Eastern Himalayas (Rodgers & Panwar, 1988) with as many as 8,000 species of flowering plants (about 50% of the total Indian flora) estimated to occur in this region. From time immemorial, the region has influenced the life and culture of the diverse ethnic communities living all along its length of the mountainous chain (300-8,000m amsl) with vegetation comprising of sub-tropical, temperate, sub-alpine and alpine types. Himachal Pradesh, one of the pioneer Himalayan states nestled in the lap of North-West Himalayas between 30°22′ - 33°12′ N and 75°3′ to 79°E. There are altogether 16,997 villages of different sizes covering a geographical extent of 55,673 sq kms area under its 12 districts having altitudinal range between 244 - 6,750m. The state boasts of a wide diversity of plant forms of Mediterranean, West Asian, Tibetan, Japanese and Burmese elements in its sub-tropical, temperate, alpine and cold desert regions. A number of diverse ethnic communities such as Gaddies, Gujjars, Kinnaurs or Kananrs, Jads, Lahaulis, Spitians, Pangwals and Swangalas with distinct differences in socio-economic and socio-cultural conditions inhabit the region. These communities have special contributions and significance in protection of ecosystems and bioresources of the region. Unfortunately, acculturation and technological changes are destroying much of this plant wealth, and their much valued traditional practices.

Current Status of Work on Ethnobotanical Spectrum of Indian Flora

Even prior to the coining of the term Ethnobotany, the uses of plants by human beings found its mention in ancient Sanskrit, Greek, Chinese and Arabic literature. In India, the earliest credit goes to the work of Watt (1889-1896, 1908) which provided first true reflection of ethnobotany and indigenous knowledge with an index of 3,000 vernacular names of plant products and various uses from its different parts (Shah, 2005). Later, credit for efforts in this direction to pioneered ethnobotanical research had been that of Janaki Ammal (1955) who studied subsistence on food plants of certain tribals of South India particularly to explore better prospects of Dioscoreas under the official programme in the Economic Botany Section of Botanical Survey of India (BSI). Since then uses of plants by the tribals are being recorded for variety of purposes (Jain, 1981, 1989). In the last few decades, the research and developmental activities in the area of ethnobotany has gained tremendous momentum all over the world with India not an exception to it. *Glimpses of Indian Ethnobotany* by Jain (1981) is the first book dealing with Indian ethnobotany. Subsequently, the subject matter had been reviewed in some of the important compilations like 'Bibliography of Ethnobotany' (Jain *et al.*, 1984), cumulative index to ethnobotany from 1989-2000 with the analysis of regions and the themes covered (Srivastava & Jain, 2000), Indian ethnobotanical literature during last two decades—a graphic view and the future directions (Jain & Srivastava, 2001), a recent detailed one listing 1,600 references of which over 1,250 relate strictly to the ethnobotanical work or very closely allied themes on India and the others on concept, global evolution of the discipline, and methodology of new areas from works of some outstanding foreign ethnobotanists (Jain, 2002) and that on ethnobotanical spectrum of Indian flora; an overview during the past 20 years in the journal ethnobotany (Goel & Tripathy, 2009) have brought together the ethnobotanical literature from disparate sources. In recent times, a number of useful treatises on the subject from Indian perspective have been those of Maheshwari (1992), Joshi (1995), Arora & Pandey (1996), Rama Rao & Henry (1996), Kaushik (1998), Pandey (1998), Singh & Pandey (1998), Dagar & Dagar (1999), Mudgal *et al.* (1999), Singh & Kumar (2000a), Sood *et al.* (2001, 2005a, b, 2009a, b, 2010, 2011a, b, 2012a, b), Singh *et al.* (2003a, b), Sood & Thakur (2004), Trivedi (2004, 2009, 2010), Sood & Kaushal (2008), Sood & Sudershna (2008), etc. and all these contributed significantly towards our understanding on the various aspects of ethnobotanical science. Barring these, other useful works describing linkages of ethnobotany with the other sciences and disciplines (Manilal, 1989), quality control requirements of medicinal plants used in traditional medicines (Mehrotra, 1990), efforts and concerns about ethnobotanical studies on the Indian plant genetic resources (Arora, 1995), role of ethnobiologists in the traditional knowledge and the sustainable development (Rao, 1996), need to compute the relative reliability of ethnobotanical claims and ethnobotany in the 21st century in India (Khan, 2001, 2005), ethnobotanical perspective of the Sanskrit plant names (Patil, 2000), Indian plant species used in Tibetan traditional medicines (Jain, 2003; Jain & Goel 2005), role of traditional and folklore herbals for the development of new drugs (Mehrotra & Mehrotra, 2005), CBD, WTO and Biodiversity Act of India in the ethnobotanical perspective (Pushpangadan & Kumar, 2005), ethnobotany in the new millennium—some thoughts on future direction in Indian ethnobotany (Jain, 2006), and prospects of ethnobotany in 2008 (Shah, 2008) have also been done in recent years.

Ever since the pioneering work of William Spencer in 1817, the state of Himachal Pradesh had been extensively explored floristically by several workers (Atkinson,

1882; Hooker, 1872-1897; Collett, 1902; Nair, 1964; Chowdhery & Wadhwa, 1984; Polunin & Stainton, 1987; Stainton, 1988; Aswal & Mehrotra, 1994; Dhaliwal & Sharma, 1999; Kaur & Sharma, 2004; Retnam & Martin, 2006; Singh & Sharma, 2006). Mention of uses of plants are also there in some of the above floras. A good number of studies on diversity of medicinal and aromatic plants in different physiographic regions of the state have also been done (Ahluwalia, 1952; Rastogi, 1960; Gupta, 1961, 1964, 1971; Rau, 1961; Shabnam, 1964; Sarin, 1967; Uniyal & Chauhan, 1971-1973; Dhiman, 1976; Arora *et al.*, 1980; Sarin & Chopra, 1985; Chauhan, 1974, 1983, 1989a, b, 1990, 1995, 1999, 2003; Singh & Aswal, 1992; Singh, 1993, 1999a, b; Dobriyal *et al.*, 1997; Sharma & Sood, 1997; Badola, 2001; Chaurasia *et al.*, 2001). Ethnobotanically, the available information in the context of Himachal Pradesh is scarce (Goel & Tripathi, 2009). So far 46 publications including 5 books have been published and the data is available only for Gaddies, Lahaulas and Spitians. In this regard, ethnobotanical data documented for some of its districts, viz. Chamba (Uniyal *et al.*, 2006: Traditional use of medicinal plants among the tribal communities of Chhota Bhangal), Hamirpur (Prakash *et al.*, 2003: Ethnobotanical studies on some ethnobotanical plants of district Hamirpur; Rawat *et al.*, 2009: Ethnobotanical studies on dental hygiene in district Hamirpur H.P., India), Kangra (Uniyal & Chauhan, 1973b: Traditionally important medicinal plants of Kangra Valley in Dharamshala forest circle; Kapur, 1993: Ethnomedico plants of Kangra Valley; Sharma & Chauhan, 2000: Ethnobotanical studies of Gaddies—a tribal community of district Kangra; Singh & Kumar, 2000b: Observations in ethnoveterinary medicine among the Gaddi tribe of Kangra Valley (H.P.) in Kangra region; Rawat & Kharwal, 2010: Indigenous traditional method for making liquid detergent "cchoa" from *Zea mays* L. in Shivalik Hills (H.P.), Kinnaur (Negi & Subramani, 2002: Ethnobotanical studies in the Chhitkul of Sangla Valley; Singh, 2004: Indigenous knowledge and conservation practices in tribal society of Western Himalaya: a case study of Sangla Valley), Kullu (Uniyal & Chauhan, 1973a: Traditionally important medicinal plants of Kullu forest division; Singh, 1999a: Ethnobotanical studies of useful plants of the Kullu district; Singh, 2004: Ethno-medicinal plants of Kullu Valley, Himachal Pradesh; Jain & Puri, 1994: An ethno-medico-botanical survey of Parvati valley; Singh *et al.*, 1998a: Ethno-cultural analysis of sacred species and ecosystems in Chhakinal watershed; Uniyal, 2003: Utilization of medicinal plants by the rural women of Kullu; Sharma *et al.*, 2004: Observations on the traditional phytotherapy among the inhabitants of Parvati valley; Sharma *et al.*, 2005: Studies on plant associated indigenous knowledge among the Malanis of Kullu district, Himachal Pradesh; Negi & Subramani, 2006: Ethnobotanical study in the Nagger area of Kullu Valley, Himachal Pradesh), Lahaul-Spiti (Koelz, 1979: Notes on ethnobotany of Lahaul—a province of the Punjab; Duhoon *et al.*, 1996: Seabuckthorn (*Hippophae* Spp)—a less known wonder plant of ethno-medico-botanical importance in cold desert of India; Kala, 2003: Ethno-medico-botany of Indian Trans-Himalayas—a case study from Spiti; Chauahn & Lakhanpal, 2000: Ecological and ethnobotanical studies on the flora of Spiti, Himachal Pradesh—a cold desert in the western Himalaya; Kala &

Manjrekar, 2000: Ethno-medico-botanical study of Indian Trans-Himalaya—a case study from Spiti; Sood *et al.*, 2001: *Ethnobotany of Cold Desert Tribes of Lahaul-Spiti (N.W. Himalaya)*; Chandra Sekar & Srivastava, 2003: Ethnobotanical studies in Pin Valley National Park; Srivastava & Chandra Sekar, 2004: Ethnomedicine of the Pin Valley National Park; Chandra Sekar & Srivastava, 2005: New reports on aphrodisiac plants from Pin Valley National Park; Singh & Chauhan, 2005: Traditional practices of herbal medicines in Lahaul valleys; Chaurasia *et al.*, 2007: Ethnobotany and plants of Trans-Himalaya, Uniyal *et al.*, 1973: Preliminary observations on medicinal plants of Lahaul Spiti division in Himachal Pradesh), Mandi (Gaur & Singh, 1993: Ethnomedicinal plants of Mandi district, Himachal Pradesh; Singh 1996b: Wild edible plants of Mandi district in North-West Himalaya; Sood & Thakur, 2004: *Ethnobotany of Rewalsar Himalaya*; Rajasekaran & Singh, 2009: Ethnobotany of Indian horse chestnut (*Aesculus indica*) in Mandi district; Sood *et al.*, 2009b: *Ethnobotanical Studies on Trees, Shurbs and Climbers of Himalaya)*, Shimla (Chauhan, 1996a: Ethno-medico-botany of Pabbar Valley; Sharma *et al.*, 2009: Some important supplementary food plants and wild edible fungi of upper region of district Shimla), and Solan (Verma & Chauhan, 2006: Ethno-medico-botany of Kunihar forest division; Verma & Chauhan, 2007: Indigenous medicinal plants knowledge of Kunihar forest division) and the entire State (Chauhan, 1988b: Ethnobotanical study of medicinal plants of Himachal Pradesh; Chauhan, 1996b: Plant resources of economic use in Himachal Pradesh; Lal *et al.*, 1996: Supplementary foods of Gaddies in H.P. and plants used as ethnomedicine and supplementary food by Gaddies of Himachal Pradesh; Sharma & Rana, 1999: Traditional medicinal uses of plants in Himachal hills; Singh & Kumar, 2000a: Ethnobotanical wisdom of Gaddie tribe of Western Himalayas; Aggarwal *et al.*, 2004: Ethnobotanical studies on some wild plants of lower hills of Himachal Pradesh; Rana, 2004: *Lesser Known Food Plants of Himachal Pradesh*; Bhalla & Bhalla, 2006: Traditional food and beverages of Himachal Pradesh) have enriched our knowledge on the subject. Besides, noteworthy contributions on the allied aspects providing useful information on the importance of *Physochaina praelta* in shaping Lahaul's economy (Sarin *et al.*, 1963), present status and future prospects of medicinal and aromatic resources of Himachal Pradesh (Sarin & Chopra, 1985; Gupta, 1988), role of forestry extension in exploitation of medicinal and aromatic plants of Himachal Pradesh (Sagwal & Gupta, 1988), endangered ayurvedic pharmacopoeial plant resources of H.P. (Chauhan, 1989c), medicinal plants of Himachal Pradesh used in Indian pharmaceutical industry (Singh & Aswal, 1992), medicinal plants of ayurvedic importance from Mandi district of H.P. (1993), occurrence, distribution and diversity of soft fruits in N.W. and W. Himalaya and prospects of their conservation and utilization (Sharma & Chandel, 1996), endangered medicinal plant species in Himachal Pradesh (Badola & Pal, 2002), collection and marketing strategies of important medicinal and aromatic plants in H.P. (Negi & Bhalla, 2002), conservation assessment and management prioritization for the medicinal plants of Himachal Pradesh (Ved *et al.*, 2003), baseline data on medicinal plants conservation and sustainable utilization (Himachal Pradesh)

(Mishra & Gokhale, 2003) and potential application of biotechnology in medicinal plants of Himachal Pradesh (Sethi, 2003) have also appeared during the last few decades.

From the scrutiny of above literature it is evident that though some work has been done on ethnobotanical aspects of plants in many parts of Himachal Pradesh, no attempt has been made so far to document information on this aspect from district Hamirpur, Himachal Pradesh (India). To bridge this gap in our understanding on the subject, the present work has, therefore, been envisaged to explore the empirical knowledge of the local inhabitants on the multifacet uses of plants from the region with a view not only to conserve it from being lost irreversibly to varied biotic pressures but also for using them as valuable clues for social forestry endeavours, therapeutic agents, supplementary foods, sustainable management of species as well as their habitats, local and regional problems, and socio-economic upliftment of the mountain people and the society at large.

The Study Area

Location and Extent

The district Hamirpur was created as a result of trifurcation of district Kangra on September 1, 1972. The region is situated between 76°-17'-50□ to 76°-43'-42″east longitudes and 31°-24′-48″ to 31°-53′-35″ north latitudes in the southwestern part of Himachal Pradesh with elevation ranging from 400- 1,232m. The region is bounded in the north by river Beas which separates it from Kangra, in the east by Mandi, in the south by Bilaspur and in the west by Una (Figs. 1.1-1.3). Entire population of the area is 4,54,293 people (male - 2,16,742 and female - 2,87,551) with highest literacy rate of 89.01% (male - 95.28% and female - 83.44%) as per 2011 census (Pl. 1A-D). Under forest, the total geographical area in the district amounts to 19,278 hectares. Main hill ranges of the district are Jakh Dhar and Sola Singhi. Infact, the Jakh Dhar is in continuation of Kali Dhar range of district Kangra and it enters district Hamirpur near Nadaun and traverses further into southeastern direction. Town of Hamirpur lies to the east of this range in the north. The Sola Singhi Dhar is the longest range of the tract and known under various names, such as Chintpurni and Jaswan Dhar.

Etymology and Historical Perspective

During Mahabharta period Hamirpur was a part of the old Jallandhar- Trigarta empire. Panini referred to the people of this kingdom as great warriors and fighters and that tradition continues even today as large number of people are in defence forces. It is believed that in the ancient period, the rulers of Gupta dynasty had set up their sovereignty over this part of the land. During middle ages, presumably the area fell under the control of Mohammed Gazani, Timurlang and later ruled by Ranas (Feudal hill chiefs). The Katoch dynasty put these Ranas under its control and the dynasty became prominent during the period of Raja Hamir Chand who ruled the area from 1700-1740 A.D. Present town derived its name from this

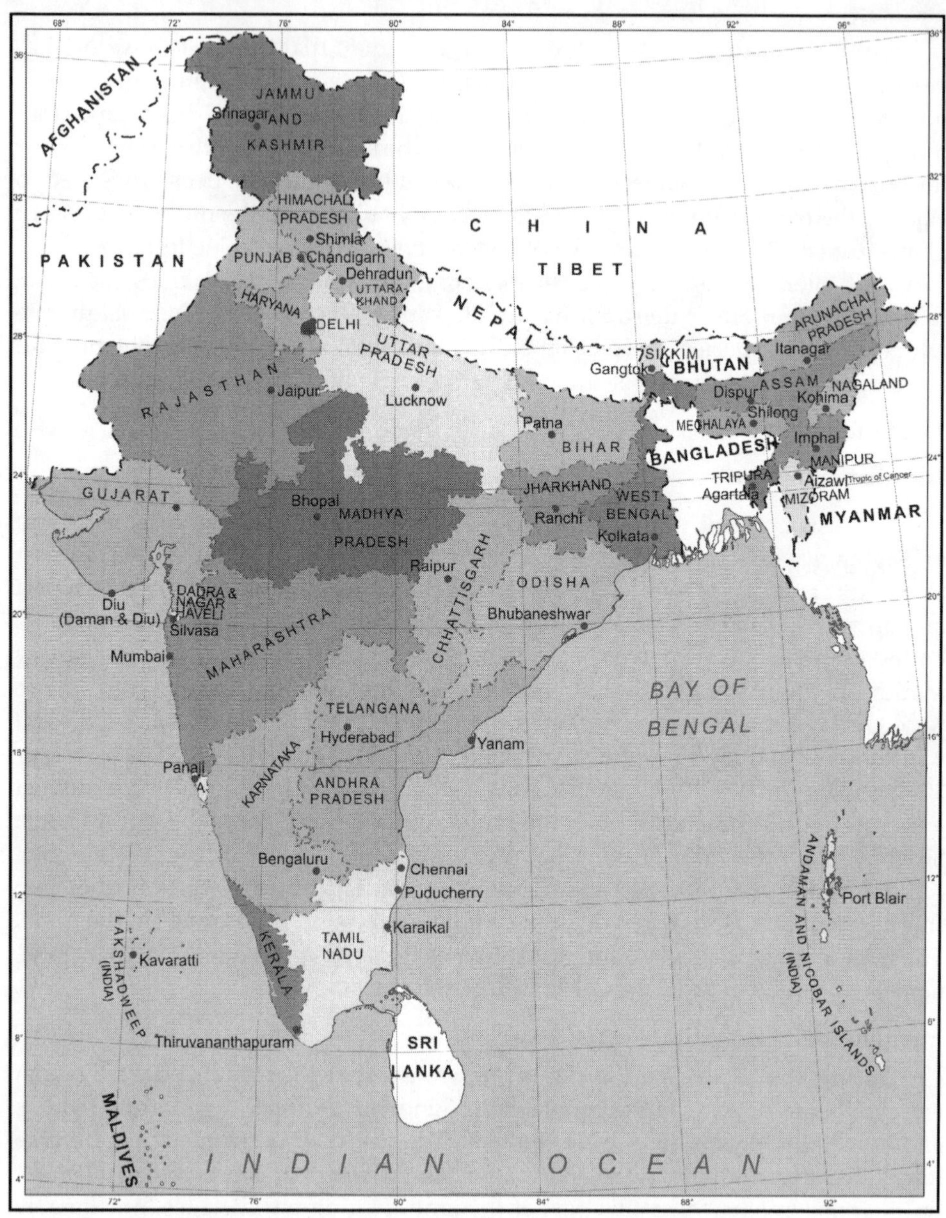

Fig. 1.1. Map Showing the location of Himachal Pradesh in India

Fig. 1.2: Map of Himachal Pradesh enlarged to show location of district Hamirpur

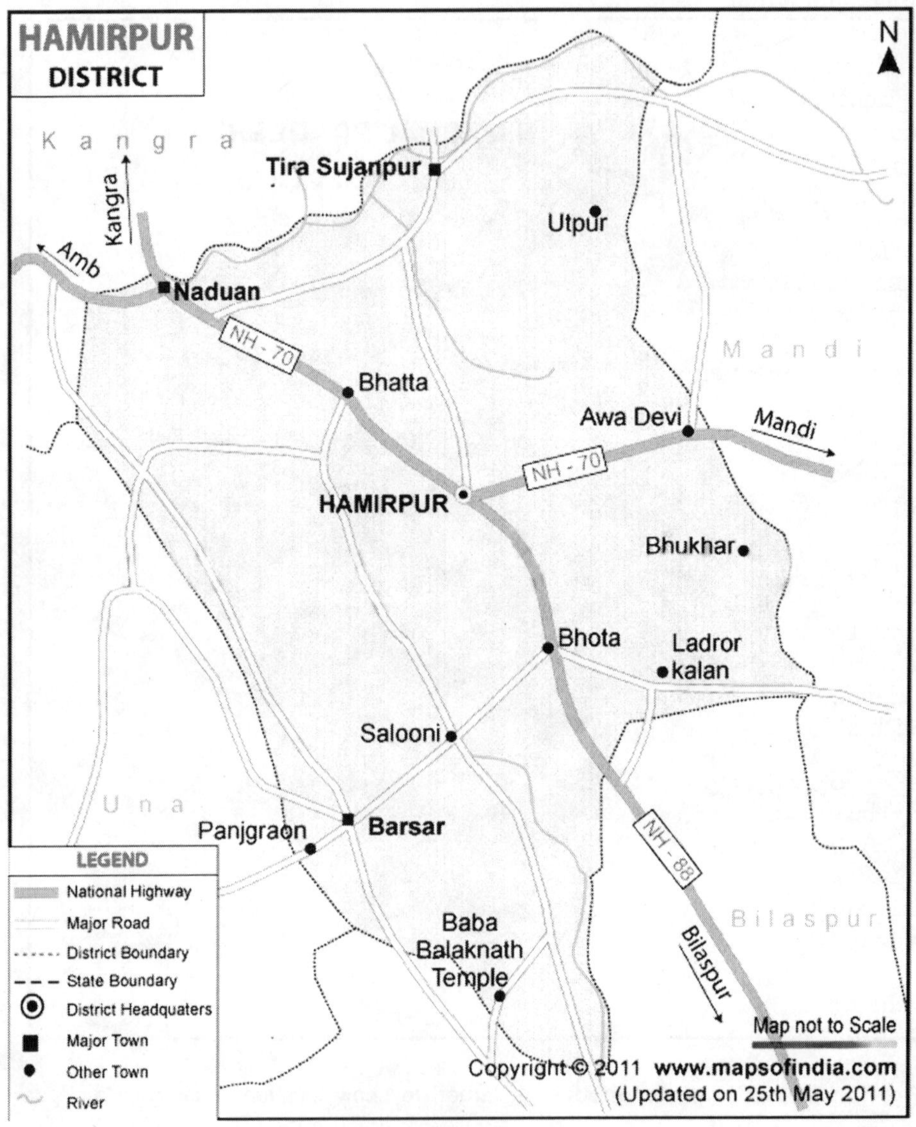

Fig. 1.3: Enlarged map of districit Hamirpur showing its boundary with adjoining districits

ruler (Pl. 1A). The district came to lime light in the period of Raja Sansar Chand-II, who built Sujanpur fort and various temples and ruled the region from 1775-1823 A.D. In 1806 A.D., Gurkhas ruthlessly looted the area between the fort of Kangra and Mahal Morion and virtually destroyed its various villages. The area was later ruled by Sikhs in 1809 A.D. and British Empire in 1846 A.D. Untill 1966 Hamirpur remained the part of Panjab province, and it was later merged in Himachal Pradesh on September 1, 1972. To start with, two tehsils Hamirpur and Barsar were created whereas Tira Sujanpur, Nadaun and Bhoranj were added to it in 1980.

Land, Climate and People

Total area of the district Hamirpur amounts to 1,118 sq kms. The middle and upper Shivaliks and the comparatively recent deposits constitute the main geological formations, comprising mainly of sand, silt and clay stones. Forests are of deciduous type and lie in the subtropical zone. Monsoon starts in the end of June and lasts till third week of September. October and November constitute the post monsoon season with an average rainfall amounting to 152 cms. The maximun temperature reaches to 40°C during summers (April-June) and has an average minimum of 3-5°C during winters (December-February). Humidity is high during south-west monsoon season whereas the air is drier in the remaining part of the year. Predominantly, the population comprises of Hindus with Brahmins, Rajputs, Kanaits, Rathis and Jhinwars whereas Muslims, Sikhs, Christians and Buddhists are in minority. Caste system prevails. The people are hardworking (Pl. 2A-D), culturally bound, god fearing and superstitious. Simplicity and honesty are their main attributes. People speak 'western Pahari' with dialect akin to adjoining areas of Mandi, Bilaspur and Kangra districts. Commonly, the joint family system prevails, which of late appears to be vanishing due to impact of modernisation.

Dwellings

The typical houses of peasantry are of mud mixed with hay (unburnt bricks) (Pl. 6A, B), and two storied with doors and windows made up of 'tunni' (*Cedrela toona* Roxb.), 'ber' (*Zizyphus jujuba* Mill.) or 'simbal' wood (*Bombax ceiba* L.) and roofs usually of slates (Pl. 5B). Inmates reside in the lower storey and the upper being used as kitchen (Pl. 2D, 6C) and store room for food grains, fuel and fodder (Pl. 6D). Thatched huts (Pl. 6A) and concrete buildings (Pl. 4D) are also common in the district. For describing the various units of traditional dwellings in the district, associated terms followed in the region are described in Table 1.1. Unlike cities, the village schools are small-sized (Pl. 7A).

There exists magnificent palaces (Pl. 3A-C), 'havelies' (Pl. 5A), mansions and historical buildings, built in the year 1700 (Hamirpur fort, Sujanpur fort, Nadaun fort) (Pl. 3B, C), 1888 (tehsil building in Hamirpur town) (Pl. 4A, B), and 1875 (Thana of Hamirpur, (Pl. 4C), old historical temples (Pl. 3D) in the district. Due to simple living 'charpai' (cot) (Pl. 8A), mats of paddy straw and/or leaves of date palm trees and 'patra' (low height wooden stool) 'binna' of 'khajoor' and bamboos are used for sitting. Brooms of 'khajoor'(Pl. 8B) and grass are employed for cleaning the houses. Moreover, baskets, mats, 'chhaj', and 'binnas' are sold by villagers in the nearby markets to earn livelihood.

Table 1.1: Various Units of Traditional Dwellings in District Hamirpur

Dwelling Units	Name in English
Baranda	Courtyard
Dwar	Door
Ghar or Makan	House
Tabella	Stable
Waan	Ground floor living room
Kamra	Room
Bor	Room on first floor
Ladda	Roof top
Baithak	Sitting room
Kand	Wall
Dwari	Window
Sirrhi	Stairs

Household utensils made of brass or bronze are used whereas furniture of the houses is of wood. The fuel wood oftenly comprises of dried sticks of 'beol' (*Grewia oppositifolia*), mango (*Mangifera indica)*, 'garna' (*Carissa inermis*), stalk of maize (*Zea mays*). In addition, 'chalakhadian' (female cones of Pine trees), 'gohttu' (dried cow dung mixed with hay or needles of pine tree) (Pl. 6C-D, 7C) are used as supplements. For drying purposes, grains are usually spread on roof top whereas 'changer' (bamboo baskets) is used for spices and other food items. Like most of village houses in India, 'sil-bata' (Pl. 8C) is employed for grinding purposes.

Table 1.2: Ornaments Worn by Womenfolk of Hamirpur

Ornament	Material	Body Part
Anguthari	Silver or Gold	Toe
Balis	Silver or Gold	Ear
Bala	Silver or Gold	Ear
Balu	Silver or Gold	Nose
Baluk	Silver or Gold	Nose
Bang	Silver or Gold	Wrist
Churi	Silver or Gold	Arms
Chak	Silver or Gold	Head
Chiri (Dora)	Silver or Gold	Hair
Mang Tikka	Silver or Gold	Head
Clip	Silver or Gold	Hair
Dandi	Silver or Gold	Ear
Dhodun	Silver or Gold	Ear
Dhol	Silver or Gold	Neck
Jumkas (Kanta)	Silver or Gold	Ear
Jhanjar	Silver or Gold	Feet
Kanbali	Silver or Gold	Ear
Koka	Silver or Gold	Nose
Kantha (Kanthi)	Silver or Gold	Neck
Pher	Silver or Gold	Ear
Tili (Koka)	Silver or Gold	Nose
Pajeb	Silver or Gold	Anklet
Singi	Silver or Gold	Neck
Mundris (Anguthi)	Silver or Gold	Finger
Chaini	Silver or Gold	Neck

Dress and Ornaments

Oftenly, men wear 'kameez' ('kurta') and 'pyjama' (Pl. 2A), 'pahari jootas' (shoes) with no cap ('topi'). Women are fond of wearing 'kameez', 'salwar', coloured embroidered 'chadaroo' (dupatta) and 'jootis' (local shoes) (Pl. 7B). Silver bengals, 'kadas', 'necklace', 'chak', 'karanphool', 'nath', 'chuda', ear hangings and 'pajeb' (Pl. 9A-C; Table 1.2) are commonly adorned by womenfolk. In winter, people prefer woollen clothes like coat without arms and sweaters. At the time of marriage bride bears 'ghagri' (a type of 'lahenga') made up of embroidered silk.

Food and Beverages

Basically, 92% of the populace are agriculturist and live in villages. They grow wheat, barley, gram, lentils in rabi season and maize, paddy, black gram, 'kulth' (horse gram), etc during kharif season. People prefer wheat, rice as well as maize 'rotis' with butter milk ('lassi') and 'sarson-ka-saag', curries and local vegetables. Percentage of nonvegetarians is less with mutton, chicken and fish curries greatly relished. Alcoholic beverages are largely consumed by men. Milk and milk products are also used in large quantity. For storage and cooking, earthen or metallic pots of different sizes are used. Food is taken four times a day; morning meal is called 'datyalu' (breakfast), mid- day as 'dupehri' (lunch), evening as 'tukuru' and the dinner as 'sanjyalu'. On special occasions like marriages and festivals, 'dham' (community lunch) is prepared in large-sized 'beltoi' (Pl. 10B) and 'sagla' in open kitchen ('rasoee'). Food items cooked in 'dham' are rice, 'madara' of grams or 'rajmash'(red grams), 'mah'(black gram), 'khatta', 'kadi' and 'meetha' (Pl. 10C). Festival foods are 'til choli', 'khichari' (Lohri and Makar Sakranti), 'ankli' (after harvesting of rice), 'babroo' in Nolle festival after Diwali, 'kheer' on (Baishakhi), and 'seviyan' on Sair. Food taboos are observed on certain occasions such as death, fast, diseases like measles, chicken pox, etc.

Social Customs and Beliefs

Birth Rituals

The delivery is attended by midwife and/or elderly women of the household. The umbilical cord is cut by midwife holding on to a silver coin or silver ornament. After delivery, the infant is bathed and wrapped in a clean cloth. The first thirteen days are called 'sootak' and the family observes abstention from all religious rituals. During this period no one accepts food and drink from newly become mother excepting people from the lower caste. The first feeding to the infant is known as 'ghutti' and it is believed that the child imbibes the temperament of the person who gives him/her its first dose. After 'sootak' is over, the cleansing up ceremony 'goontar' ('shudhi') is performed by a Brahmin priest. A mixture of curd, sugar, cow's urine, milk and Ganga-jal (holy water from Ganges) is sprinkled all over the house and is also taken by all the members of the family. Other associated ceremonies, namely 'namkaran' (naming of the infant), 'annaprashan' (first tasting of cereal by the baby), 'jamalu'/ 'mundane' (first hair cut) are also performed with great joy and hoisting of lunch for all the local people.

Marriage System

'Roka' or 'sagai' (betrothel) is the first step in marriages in which boy's family sends clothes, ornaments and money ('shagun') for the girl. Ceremonies like 'saand', 'samuhat', 'mehndi', 'tael', 'sehrabandi', 'milani' and welcome to 'barat' (marriage party) are performed by singing of local marriage songs related to the occasion. Dances are performed on the beat of 'dholak' and 'nagaras'(Pl. 10A). At bride's place ceremonies like 'lagan-bed' and 'kanayadan' are performed. Bride and groom circumbulate sacred fire amongst chanting of mantras by the Brahmin priest. At grooms place the newly wed couple worship 'peepal' (*Ficus religiosa*) and 'daru' (*Punica granatum*) tree for their long and happy married life. 'Muklawa', welcome of the daughter along with her husband to the parent's house after marriage, takes place the next day after marriage.

Divorce

It is uncommon in the villages. For this, the matter is first taken to the Panchayat for settlement; if it is not resolved then legal separation is got done through court.

Death Rituals

Death rituals are performed by a 'Acharj' as per Sanatan Dharma. As death approaches, the sick man's body is lowered on the ground with head pointing to the North and smeared with fresh cow-dung. After death of the person, conch shells are blown and mustard oil lamp is lit near the corpse. Before the funeral rites, the body is bathed and covered with a colourful cloth and taken in bier made of bamboos to cremation ground. Prior to cremation, a 'pind-dan' (ritual feeding of balls made of cooked rice, sesame seeds and curd to the birds in the name of person died) ceremony is performed. On the third day, bones and ashes ('astu') are collected for its immersion in holy Ganges. Mourning is observed for thirteen days. On the tenth day, clothes are washed ('kapad-dhulai') and a priest recites the holy 'Garud Purana'. At the end of mourning period, a ceremony 'kriya-karma' (a religious rite for the peace of the departed soul) is performed. The annual death rite is called 'barkhi' whereas the last death rite, 'chabarkh' is performed in the fourth year.

Faith and Worship

Hamirpur district is one of the popular religious spots of Himachal Pradesh. Narvdeshwar temple in Sujanpur, built by queen of Raja Sansar Chand (Pl. 11A), is a unique architectural marvel of its time. In the month of Baishakh, thousands of devotees visit this place to take a glimpse of holy Shivling of Bil-Kaleshwar temple (400 yrs old), which is situated at a distance of 5 km on Nadaun-Sujanpur road at confluence of river Beas and Kunah Khad. Murli-Manohar mandir (constructed in 'Shikar Style') located in historical chowgan of Sujanpur (Pl. 11B), Gasota Mahadev temple (Pl. 11C; over 400yrs old and 8 km from Hamirpur town), Tauni Devi temple of Goddess Durga's sister (12 km away and 200 yrs old; Pl. 12A), Awah-Devi temple of Goddess Jalapa (Pl. 12B; 24 km from Hamirpur), Jhaniyari Devi temple, Kalanjari Devi, Gurudwara Sahib Nadaun, Peer-Saheb grave, Baba Balak Nath temple (Pl. 12C) and Jawalaji temple are the other famous religious spots in and around the district.

River System

Beas, a major river of this area, is drained by a number of perennial streams.

Bakar Khad, Kunah Khad, and Maan Khad drain into river Beas while Shukar Khad and Mundkhar Khad drain into Seer Khad which ultimately merges into Satluj. Various natural water reservoirs such as Tal, Telkar Kangoo, Dimmi, Bilkaleswar spring, a water fall near Sujanpur, 'Khatri' (rain water harvesting reservoir dug into hill) and 'Bauli' are also exploited for domestic and agricultural purposes (Pl. 13A-C).

Occupation and Economy

Besides defence services, the main occupation of the people in the district is agriculture, animal husbandry and horticulture. Cultivated land comprises of terraced fields (Pl. 14A) which is usually irrigated by 'kuhls' (irrigation channels) and rainfall. Various implements used for agricultural and household operations are listed in Table 1.3. Main crops are maize, wheat and paddy (Pl. 14B, C). Vegetables are grown as cash crop in some of the isolated pockets. Amongst pulses, new high yielding varieties are being introduced. Sugarcane, potatoes, 'toria' and mixed cropping of maize with gram and soyabeans are extensively cultivated.

For horticulture operations, demonstration orchards at Bhumpal near Nadaun and Badiana at Bhoranj have been set up for raising sand peas, mango, 'galgal' (*Citrus limon*), lime, 'kinnow', etc for distribution. The district has seven veterinary hospitals and 24 dispensaries, ten artificial insemination centres, sheep breeding farm at Tal and milk chilling plant at Jalari. An industrial state 2 km from Hamirpur has been set up where in a training institute, Polytechnic College and Mahilla Gramadhyog (Pl. 15A-C) are rendering good services by imparting training in carpentry, weaving, electrical fitting, shawl production and carpet making.

Table 1.3: Agricultural Implements and Household Accessories Used by Local Folks of Hamirpur

Local Name/s of Implements and Accessories	English Name	Used For
Tools		
Kulhari	Axe	Cutting wood
Tokri	Basket	Carrying manure and cow dung
Kera		Dropping seeds in plough furrows
Chabbal	Digger	Digging
Thora	Hammer	Breaking big stones
Hal	Plough	Ploughing
Drat or drati	Sickle	Cutting branches of trees
Jungla	Yoke	Yoking bullocks for ploughing
Mayee		Levelling the fields
Pottery		
Chillam		Smoking
Jhawa (Chaun)		Scrubbing dead skin from feet.

contd...

Table 1.3: *contd...*

Local Name/s of Implements and Accessories	English Name	Used For
Karuae		Keeping 'suhagi' on 'Karva Chauth' (fast praying for long life of husband)
Ghara	Pitcher	Carrying water
Handko	Small pitcher	Storing milk, ghee or lassi (butter milk)
Basketry		
Chatai	Mat	Wooven from palm leaves and paddy husk-Pral for sitting purposes
Chhabra/Chhabri	Basket	Keeping vegetables and fruits
Chhikri	Small basket	Covering mouth of cattle during ploughing
Pakhi	Fan	A hand held device for creating a current of cool air
Kardu or Karandu	Basket with handle and lid	Keeping vegetables, eatables or wool
Peru	Huge basket	Storing grains
Semi circular basket	Tokri	Horticulture and agricultural operations
Utensil		
Belan		Rolling chappaties
Chakla		Base for rolling chappaties
Ghara		Storing water
Madani		Churning milk or curd
Tawa		Baking chappaties on fire
Furniture		
Baksa	Wooden box	Keeping clothes and valuables
Choki	Low height square table	Worshipping
Patra	Low height table	Sitting on ground
Mattresses		
Khinde	Cushion, Materices	Bed cushion stuffed with small pieces of waste clothes
Pand	Mat	Mat woven with palm leaves and bamboos for sitting purposes
Tapar	Mat	Mat stuffed with straw for sitting purposes
Leph	Quilt	Thin quilt filled with cotton

Fairs, Festivals and Folklores

The colourful fairs and festivals of Hamirpur provide healthy recreation for its people. Special foods are cooked on the day of festivals. People dress on new and colourful attires, worship at temples before celebrations. Some of the festivals like 'Gasota- Mahadev' fair, 'Hamir Utsav' (State fair), 'Holi fair' of Sujanpur Tira (State fair, Pl. 18A, B), Baba- 'Balaknath fair', Deotsidh, Ghasian fair, Awah-Devi

Table 1.4: Fairs and Festivals of District Hamirpur

Fairs and Festivals	Occasion	Remarks
Festivals		
Chet (Dholru)	First month of the lunar year	For happiness and prosperity in family
Navratras	Durga-Ashtmi in April and October	Appeasing goddess Durga for her blessings
Haryali	16th July	Signifies advent of the rainy season
Rakhri	Full moon day in Shrawan	Sisters tie sacred thread on the wrist of their brothers and prey for long healthy life
Sair	September	Worshipping of local deity for prosperous future
Dashera	September/ October	Signifies victory of good over evil and celebrated by burning effigies of Ravana, Kumbakarane and Meghnath on 'Dashmi', i.e. 10th day of celebration.
Diwali	November	Symbolising celebration of returning of Lord Rama's to his kingdom from exile after killing Ravana
Shivratri	February	Occasion for keeping fast to appease Lord Shiva to invoke his blessings
Fairs		
Bisowa	13th April	Agarian fair when people celebrated it by taking holy dip at Kaleswar Mahadev
Gugga	August/September	Gugga worshipping constitutes a form of snake cult
Lohri	1st Magh (mid January)	Celebrated by eating and dancing around bonhomie of fire
Gasota-Mahadev	1st Monday of Jeshtha	Celebrated religiously by singing songs, eating, dancing and buying
Holi Mela at Sujanpur	In March	A commercial cum religious fair
Hamir Utsav	1st week of September	Colourful cultural programmes and exhibition are its main attraction
Baba-Balaknath fair	March	A religious cum commercial fair
Markanda fair	April	A holy dip taken in natural water spring believed to cure all skin diseases

fair, Tauni-Devi fair, Markanda fair and Googa fair of snakes are celebrated with great enthusiasm (Pl. 18A-C; Table 1.4). A few festivals, cattle fair of Jahu, Cha kmoh and Dhirar are held for trading animals.

The rich folk culture of region is usually reflected at the time of marriage, fairs and festivals. The group folk songs of girls are called 'jatras' and that of young men as 'jheras'; the most common 'jhehra' being 'gugga chauha', Raja Sansar Chand, Raja Bhangal and Raja Sirmour. Folk dances of the region worth mentioning are 'gidha' of womenfolk and 'chanderwali' dance of menfolk. As this district was once the part of Kangra, many of its folk songs/ cultural activites are similar to it. In marriages traditional musical instruments like 'nagara', 'shahnai', 'dholak', 'damroo', 'dhafli', 'thali', 'tabla', 'flute', 'harmonium', etc (Pl. 17A-D) are commonly employed.

Important Institutions in Hamirpur

The district is considered as an educational hub of Himachal Pradesh. Several educational institutions like National Institute of Technology (a national level technical university offering undergraduate, postgraduate, and doctoral programmes in engineering, science and technology) (Pl. 19A), Netaji Subhash Chandra Bose memorial government postgraduate college (Pl. 19C), Institute of Biotechnology and Environmental Sciences-Neri, MIT college of Engineering and Management-Bani, Sainik School at Sujanpur and 9 colleges of education are contributing for the cause. Besides, a Radha Swami charitable trust hospital at Bhota (Pl. 19B) and Subordinate Service Selection Board at Mattani are also serving the local populace.

Recreational Sports

Sujanpur Palace and Chowgan are known for organising adventure sports like paragliding, angling, rafting and trekking. Sainik School of Sujanpur hoists many of these activites every year (Pl. 20A-C).

Places of Interest

Palace of Sujanpur Tihra founded by Raja Abhay Chand—the king of ruling Katoch dynasty of Kangra in 1748 A.D., tomb of Nawab Gulam Mohammed of Rampur, the abode of Markandya Rishi at Markanda (6 km from Dera Parole on the bank of Kunah Khad) known for its ancient water spring in possessing miraculous healing properties against skin ailments, famous Amtar fort of Nadaun where Raja Sansar Chand used to held his court in princely days, Bilkaleshwar Mahadev temple (believed to be founded by the Pandavas), Gurudwara Sahib Nadaun (built in 1929), Deotsidh temple of Baba Balak Nath (45 km from Hamirpur), Narvdeshwar temple (Pl. 11A; built by the chief queen of Maharaja Sansar Chand), Murli Manohar Mandir of historical Chowgan of Sujanpur Tira (Pl. 11B), 200 years old Tauni Devi temple (Pl. 12A), Awah Devi temple (the highest point of district) located at 24 km from Hamirpur (Pl. 12B), 400 years old Gasota Mahadev temple (Pl. 11C; distance of 8 km from Hamirpur), Jhaniyari Devi temple (6 km from Hamirpur), 200 years old Sai Fazal Shah Peer Saheb's grave (1½ km from Nadaun) are some famous places of the district worth visiting.

Collection of Ethnobotanical Data

For documenting first hand information pertaining to ethnobotanical exploration of Hamirpur District, Himachal Pradesh (India), intensive and extensive surveys were undertaken in all subdivisions, blocks and tehsils of the district over the period of four years (2008-2012). Field tours were planned in such a way so as to collect ethnobotanically important species either in flowering or in fruiting stage (Table 1.5). Participatory rural appraisal (Etkin, 1993; Lipp, 1989; Hedberg, 1993; Pei & Long, 1998), observations and key informant (family heads, healers, old experienced and knowledgeable informants, medicinemen) (Pl. 16A-C) were interviewed to collect information for the field survey. A questionnaire based on the proforma designed by Jain & Goel (1995) was prepared to gather data for this purpose; and the resultant information was recorded in the ethnobotanical field notebook. Uses of plants narrated by a person were cross-checked and considered notable only if the author observed actual application, or similar application was reported by at least three informants. For the medicinal uses, information pertaining to method of preparation (powder, aqueous extract, infusion or decoction), dosage and combination with other plant, if any was also noted.

Table 1.5: Collection Sites with Respective Altitude and Flowering and Fruiting Season of Ethnobotanically Important Plants Collected from the Study Area

Name of the Plant	Place/s	Altitute (m)	Flowering & Fruiting Season
Abelmoschus esculentus (L.) Moench.	Hamirpur Sujanpur,	787 567	July-September
Abelmoschus moschatus Medic.	Khaggal	920	July-September
Abroma augusta Jacq.	Khaggal	920	February-May
Abrus precatorius L.	Anu Khurd	858	May-August
Acacia catechu (L. f.) Willd.	Dhoh, Kanjiyan	1034	May-July
Acacia concinna (Willd.) DC.	Neri	381	January-April
Acacia nilotica (L.) Delile	Patta	1020	July-September
Achras zapota L.	Anu Khurd	858	December-January
Achyranthes aspera L.	Krishana Nagar	858	May-October
Achyranthes bidentata Blume.	Tikker Didwin Tal	800 861	June-September
Acorus calamus L.	Khas graoan	858	June-July
Actinidia chinensis var. *deliciosa* A. Chev.	Bhots	765	February-November
Adhatoda vasica Nees	Tauni Devi	1050	April-June
Adiantum capillus-veneris L.	Hiranagar	900	July-October

contd...

Table 1.5: *contd...*

Name of the Plant	Place/s	Altitute (m)	Flowering & Fruiting Season
Adiantum lunulatum Burm.	Sujanpur Fort	567	July-October
Aegle marmelos (L.) Correa *ex* Roxb.	Tal	861	April-July
Aerva lanata (L.) Juss. *ex* Schult.	Pucca Bharo	840	May- August
Agave americana L.	Tauni Devi	1050	March-May
Ageratum conyzoides L.	Awah Devi	1236	July-August
Albizia lebbeck Benth.	Daneta	596	April-May
Albizia stipulata Boiv.	Nadaun Anu	483 858	May-June
Allium cepa L.	Jhaniari	787	March-April
Allium sativum L.	Jhaniari	787	March-April
Aloe barbedensis Mill.	Khaggal	920	June-January
Aloe vera Tourn. *ex* L.	Chabutra	797	March-April
Alstonia scholaris R.Br.	Anu	900	May-June
Amaranthus paniculatus L.	Hamirpur	787	May-June
Amaranthus spinosus L.	Dimmi	1006	September-November
Amaranthus viridis L.	Lambloo	850	May-June
Amorphophallus campanulatus (Roxb.) Blume *ex* Decne.	Bhira, Bharmoti	780 531	April-June
Andrographis paniculata Wall. *ex* Nees	Neri	381	July-October
Anetham sowa Roxb. *ex* Flem.	Bajroh, Bhoranj	950	August-September
Annona squamosa L.	Nadaun	483	May-August
Arachis hypogaea L.	Bharmoti	483	October-December
Argemone mexicana L.	Dhoh	1034	April-May
Arisaema tortuosum (Wall.) Schott.	Bandroh	850	July-October
Artemisia japonica Thunb.	Rangas	765	September-December
Artemisia vulgaris L.	Bassi	987	August-October
Artocarpus integrifolia L.	Bari Mandir	1060	May-June
Artocarpus lakoocha Roxb.	Khas graon	858	March-June
Asclepias curassavica L.	Tarkwari Bassi	980- 987	Throughout the Year
Asparagus adscendens	Dungi	906	October-November

contd...

contd...

Name of the Plant	Place/s	Altitute (m)	Flowering & Fruiting Season
Asparagus racemosus Willd.	Khaggal	920	July-October
Avena sativa L.	Jhaniari	900	November-January
Azadirachta indica A. Juss.	Anu	900	June-September
Bacopa monnieri (L.) Penn.	Neri	381	June-September
Bambusa arundinaceae (Retz.) Roxb.	Dungi	906	Sporadically
Bambusa nutans Wall.	Tikker Khatrian	908	Flowering sporadically
Bauhinia vahlii Wight & Arn.	Dhaneta	596	April-June
Bauhinia variegata L.	Tauni Devi	1050	March-August
Berberis asiatica Roxb. *ex* DC.	Bhaleth	506	May-September
Boerhavia diffusa L.	Anu Khurd	858	July-September
	Brahmni	858	
Bombax ceiba L.	Jhanikker	1050	February-May
Bougainvillea spectabilis Willd.	Bhira	780	January-June
Brassica campestris L.	Kuthera	900	March-September
Brassica juncea (L.) Czern. & Coss.	Ladraur	978	March-September
Brassica nigra (L.) Koch.	Hamirpur	787	March-September
Bryophyllum calycinum Salisb.	Krishana Nagar	858	March-May
Butea monosperma (Lam.) Kuntze.	Jhaniari, Anu	900	March-April
Callistemon lanceolatus DC.	Bhota	765	March-April
Calotropis procera (Ait.) R.Br.	Salooni	1514	February-May
Canna indica L.	Barru	800	January-February
Cannabis sativa L.	Krishana Nagar	858	June-September
Canscora diffusa (Vahl.) R.Br.	Khaggal	920	October-November
Capsicum annuum L.	Bhareri	920	Greater part of the year
Cardiospermum halicacabum L.	Bhadhani	950	August- September
Carduus nutans L.	Dhaneta	596	Greater part of the year
Carica papaya L.	Kalinjadi	900	Greater part of the year
Carissa carandas L. var. *congesta* (Wt.) Bedd.	Garane-ke-Gallu	900	April-July

contd...

Table 1.5: *contd...*

Name of the Plant	Place/s	Altitute (m)	Flowering & Fruiting Season
Carissa spinarum auct. non L.	Garane -ke - Gallu	900	May-June
Carthamus oxycantha Bieb.	Tauni Devi	1050	December-April
Carum copticum Hiern.	Bassi	987	April-May
Cassia fistula L.	Nadaun	483	March-July
Cassia occidentalis L.	Bari Mandir	1060	May-September
Catharanthus roseus (L.) G. Don	Krishana Nagar	858	Throughout Year
Cedrela toona Roxb. *ex* Rottl.	Bassi	987	May-June
Celastrus paniculatus Willd.	Bhota, NIT	765 900	April-June
Celosia argentea L.	Rangas	765	August-December
Centella asiatica (L.) Urb.	Bhira, Bassi	780 , 987	April-May
Centipeda minima (L.) A. Br. & Aschers.	Bassi	987	September-November
Cestrum nocturnum L.	Hira Nagar, Sujanpur	900 567	August-October
Chenopodium album L.	Kanjiyan	910	January-March
Chenopodium ambrosioides L.	Kanjiyan	910	July-August/October-November
Cicer arietinum L.	Bijari, Nadaun	823, 483	March-April
Cinnamomum camphora (L.) Nees & Eberm.	NIT, Neri	900, 381	April-August
Cinnamomum tamala Nees & Eberm.	Nadaun, Neri	483, 381.	June-August
Cissampelos pareira L.	Patta	1020	May-November
Citrus aurantium L.	Bijari Chakmoh	823	October -March
Citrus grandis (L.) Osbeck.	Kangoo, Anukhurd	649, 900	November-March
Citrus jambhiri Lush.	Dimmi	1006	October-March
Citrus limon (L.) Burm. f.	Nalti Galore	787	November-February
Citrus medica L.	Barsar	815	April-September
Clematis grata Wall.	Ladraur	978	July-September
Colebrookia oppositifolia Sm.	Bhoranj	987	April-November
Colocasia esculenta (L.) Schott	Sameerpur	1196	July-October

contd...

Table 1.5: *contd...*

Name of the Plant	Place/s	Altitute (m)	Flowering & Fruiting Season
Cordia dichotoma Forst. f.	Tal	861	April-July
Coriandrum sativum L.	Sujanpur	567	October-December
Cucumis sativus L.	Dhaneta	596	May-September
Cucurbita hispida Wall.	Naduan	483	June-October
Cucurbita pepo L.	Ladraur	978	June-September
Curcuma domestica (Medik.) Valeton.	Chakmoh	736	September-October
Cuscuta reflexa Roxb.	Thana Drogan	950	September-October
Cycas revoluta L.	Bhota	765	February-May
Cymbopogon citratus (DC.) Stapf	Khaggal	920	September-January
Cynodon dactylon (L.) Pers.	Kot	950	August-November
Dalbergia sissoo Roxb. *ex* DC.	Nalti	851	May-July
Datura metel auct. non L.	Krishana Nagar	900	June-August
Datura stramonium L.	Rangas	765	June-October
Daucus carota L.	Jhaniari	900	November-March
Dendrocalamus hamiltonii Nees & Arn.	Tikker Khatrian	908	May-August
Dioscorea bulbifera **L.**	Mair	840	June-August
Dodonaea viscosa L. Jacq.	Tikker Khatrian	908	February-April
Duranta repens L.	Saur	840	April-August
Eclipta prostrata Roxb.	Khaggal	920	August-November
Elaeocarpus ganitrus Roxb. *ex* G. Don	Neri	381	December-April
Elettaria cardamomum (L.) Maton	Hamirpur	787	April-August
Emblica officinalis Gaertn.	Gallu Amb Tarkwari	980	April-September
Equisetum debile Roxb. *ex* Vauch.	Dhoh	1034	September-October
Eriobotrya japonica Lindl.	Bhota	765	July-September
Eruca sativa Lam.	Barsar	815	February-April
Eucalyptus umbellata Domin	Pucca Bharo	840	March-April

contd...

Table 1.5: *contd...*

Name of the Plant	Place/s	Altitute (m)	Flowering & Fruiting Season
Eugenia cerasoides Roxb.	Anu Khurd, Balooni	900, 958	April-July
Eugenia jambolana Lam.	Brahmni	858	May-July
Eupatorium adenophorum Spreng.	Lamloo	850	February- May
Euphorbia pulcherrima Willd.	Lohardi	858	June-July
Euphorbia royleana Boiss.	Tikker Khatrian	908	June-July
Ficus benghalensis L.	Tauni Devi	1050	February-June
Ficus carica L.	Sujanpur	567	March-August
Ficus glomerata Roxb.	Bhoranj	950	April-July
Ficus hispida L.f.	Kanjiyan	910	June-August
Ficus pamata Forsk.	Kalinjhari	900	April-September
Ficus religiosa L.	Didwin Tikker	800	March-October
Ficus roxburghii Wall. *ex* Miq.	Brada	1120	April-August
Flacourtia ramontchi L' Herit	Sujanpur	567	March-June
Foeniculum vulgare Mill.	Gasota	850	August-September
Fragaria chiloensis (L.) Duchesne	Bhota	765	May-June
Fumaria indica (Haussk.) Pugsley.	Jhanikker	1050	February-March
Glycine max Merr.	Dhaneta	596	September-December
Glycyrrhiza glabra L.	Chabutra	797	May-September
Gossypium arboreum L.	Chakmoh	736	July-September
Grevillea robusta A. Cunn. *ex* R. Br.	Bhota	765	March-July
Grewia oppositifolia Roxb.	Tikker Khatrian	1000	May-June
Habenaria pectinata Don	Mattani	920	July-September
Helianthus annuus L.	Rangas	765	April-July
Helicteres isora L.	Bhota	765	May-August
Hibiscus rosa-sinensis L.	Salooni	1514	February-October
Holarrhena pubescens (Buch.-Ham.) Wall. *ex* G. Don	Anu	858	June-February
Hordeum vulgare L.	Barsar	815	November-February
Hypericum cernum Roxb.	Kangoo	649	April-October
Impatiens balsamina L.	Hamirpur	787	July-September

contd...

Table 1.5: *contd...*

Name of the Plant	Place/s	Altitute (m)	Flowering & Fruiting Season
Ipomoea cairica (L.) Sweet	Awahdevi	1236	June-November
Ipomoea carnea Jacq.	Bhota	765	May-July
Jackranda acutifolia auct. non- Humb. & Bonpl.	Sujanpur	567	April-June
Jasminum humile L.	Chakmoh	736	May-September
Jasminum multiflorum Roth.	Ladraur	978	February-April
Jasminum sambac (L.) Ait.	Bhota	765	February-April
Jatropha curcus L.	Jhanikker	1050	February-May
Juglans regia L.	Anukhurd	858	April-October
Lagenaria siceraria (Molina) Standl.	Kanjiyan	910	July-September
Lagerstroemia indica L.	Nadaun	483	June-October
Lannea coromandelica (Houtt.) Merr.	Bhota, Nadaun	765, 483	July-September
Lantana camara L.	Bhaleth	506	June-October
Lavandula angustifolia Mill.	Khaggal	920	February-June
Lawsonia alba Lam.	Neri	381	June-September
Lens culinaris Medic.	Joh	980	October-December
Lepidium sativum L.	Lambloo	850	November-January
Leucaena leucocephala (Lam.) de Wit.	NIT	900	September-November
Linum usitatissimum L.	Dhaneta	596	September-October
Litchi chinensis (Gaertn.) Sonn.	Salasi	805	April-June
Luffa aegyptiaca Mill.	Patta	1020	May-September
Lycopersicon esculentum Mill.	Sakand gram	1000	May-June
Madhuca indica J. F. Gmel.	Nadaun	483	November-January
Mallotus philippinensis (Lam.) Muell.-Arg.	Anu	900	November-April
Malvaviscus conzatii Greenm.	Thankari	1000	Major part of the year
Mangifera indica L.	Ukhali	840	March-June
Martynia annua L.	Neri	381	June-October
Melia azedarach L.	Kohli	851	March-May
Melissa officinalis L.	Khaggal	920	August-October
Mentha piperata L.	Bhota	765	July-September
Mentha spicata L.	Sujanpur	567	March-July
Mesua ferrea L.	Neri	381	June-October

contd...

Table 1.5: *contd...*

Name of the Plant	Place/s	Altitute (m)	Flowering & Fruiting Season
Millettia extensa Benth. *ex* Baker	Bhoranj	950	October-December
Mimosa pudica L.	Salasi	805	July–September
Mirabilis jalapa L.	Sujanpur	567	June-September
Momordica charantia L.	Patta	1020	June-October
Moringa oleifera Lam.	Neri	381	February-April
Morus alba L.	Jhaniari	900	March-May
Murraya koenigii (L.) Spreng.	Mattani	920	March-June
Murraya paniculata (L.) Jack.	Nadaun	483	June-December
Musa sapientum L.	Galore	787	Throughout year
Nasturtium officinale R. Br.	Dimmi	1006	April-September
Nerium indicum Mill.	Rangas	765	April-November
Nicotiana tabacum L.	Khaya	787	April-August
Ocimum basilicum L.	Hamirpur	787	June-September
Ocimum sanctum L.	Sujanpur	567	July-November
Opuntia dillenii (Ker-Gawl.) Haw.	Tikker Khatrian	908	April-May
Oroxylum indicum Vent.	Tal	947	June-August
Oryza sativa L.	Bharari	920	September-December
Oxalis corniculata L.	Bhota	765	April-November
Papaver rhoeas L.	Bhira	780	May-October
Parthenium hysterophorus L.	Chauri	858	August-September
Paspalum scrobiculatum L.	Lambloo	850	July- September
Pedilanthus tithymaloides (L.) Poir.	Anu	900	August-September
Pelargonium graveolens L'Hert.	Khaggal	920	March-July
Pennisetum americanum L.	Tauni Devi	1050	August-October
Pennisetum purpureum Schum.	Dera Parol	910	September-November
Phoenix sylvestris Roxb.	Bhaleth	506	May-June
Phragmites karka Trin. *ex* Steud.	Khya	787	July -September
Phyllanthus urinaria L.	Khaya	787	August-September
Physalis peruviana L.	Jhaniari	900	April- September
Pinus roxburghii Sarg.	Hiranagar	900	February -May
Piper longum L.	Bijhari	823	July-January

contd...

Table 1.5: *contd...*

Name of the Plant	Place/s	Altitute (m)	Flowering & Fruiting Season
Pistacia integerrima Stewart *ex* Brandis	Tal Mahal	947	March-May
Plumbago zeylanica L.	Bhira	816	April-November
Pogostemon benghalensis Burm. f.	Nalti	787	February-April
Polyalthia longifolia (Sonn.) Thw.	Bhota	765	February-April
Pongamia pinnata Pierre.	Awahdevi	1236	April-June
Populus nigra L.	Dimmi	1006	April-July
Premna latifolia Roxb.	Masayana	900	July-September
Prunus amygdalus Baill.	Sujanpur	567	April-August
Prunus domestica L.	Kuthera	900	April-August
Prunus persica Benth. & Hook. f.	Sujanpur	506	February-May
Psidium guajava L.	Nadaun	583	July-September
Punica granatum L.	Brahmani	858	May-September
Putranjiva roxburghii Wall.	Krishana Nagar	858	March-April
Pyrus communis L.	Tauni Devi	1050	June-August
Pyrus pashia Ham.	Tauni Devi	1050	March-May
Randia dumetorum (Retz.) Lam.	Khaya	787	April-June
Ranunculus muricatus L.	Kuthera	900	February-April
Raphanus sativus L.	Kohli	858	November-March
Rauwolfia serpentina Benth. *ex* Kurz.	Brahmni, Kaggal	858, 1000	May-September
Reinwardtia indica Dum.	Badhani	908	November-May
Rhynchostylis retusa Blume	Anu	900	May-June
Ricinus communis L.	Jhaniari	900	March-August
Rosa damascena Mill.	Muthan	900	Greater part of the year
Rosmarinus officinalis L.	Khaggal	920	February-June
Roylea cinerea (D. Don) Baill.	Patnaun	1196	March-November
Rubus ellipticus Sm.	Hiranagar	900	January-May
Rumex hastatus D. Don	Krishana Nagar	858	November-April
Ruta graveolens L.	Neri	381	July-September
Saccharum arundinaceum Retz.	Tikker	800	October-December

contd...

Table 1.5: *contd...*

Name of the Plant	Place/s	Altitute (m)	Flowering & Fruiting Season
Saccharum officinarum L.	Chakmoh	736	January-April
Saccharum spontaneum L.	Nalti	787	Greater part of the year
Salvia coccinea L.	Tarkwari	980	June-August
Santalum album L.	Brahmani	858	April- August
Sapindus mukorossi Gaertn.	Tal	947	May-August
Saraca asoca (Roxb.) De Wild.	Khaggal	920	March-September
Sechium edule Sw.	Muthan	900	June-September
Sesamum indicum DC.	Nadaun	483	July- September
Setaria viridis Beauv.	Kakkar	850	September-December
Solanum erianthum D. Don	Tal	947	July-October
Solanum melongena L.	Ghodiwala Pind	930	June-September
Solanum nigrum L.	Dhoh	1034	January-July
Solanum surattense Burm. f.	Tal	947	April-October
Solanum tuberosum L.	Lambloo	850	May-September
Sonchus asper Hill	Jhaniker	1050	March-April
Soymida febrifuga A. Juss.	Panjot, Anu	1020, 900	August-December
Spilanthes acmella auct. non (L.) Murr.	Sujanpur	567	May-August
Spinacia oleracea L.	Bhira	780	January-April
Stephania glabra (Roxb.) Miers	Balooni	858	June-August
Tabernaemontana divaricata (L.) R. Br.	Nadaun	483	July-October
Tagetes erecta L.	Tikker Didwin	800	October-January
Tamarindus indica L.	Khaya Village	787	May-August
Taraxacum officinale Weber ex Wigg.	Kanjian	910	March-November
Tecoma grandiflora Loisel.	Loharari	858	April-August
Tecoma stans (L.) H.B. & K.	Pucca Bharo	840	April-August
Terminalia arjuna (Roxb.) Wt. & Arn.	Hamirpur	900	May-September
Terminalia bellirica (Gaertn.) Roxb.	Jhanikar	1050	May-September

contd...

Table 1.5: *contd...*

Name of the Plant	Place/s	Altitute (m)	Flowering & Fruiting Season
Terminalia chebula Retz.	Sukarkhad	381	April-June
Thevetia neriifolia Juss. *ex* Steud.	Dimmi	1006	July-September
Thuja orientalis L.	Sameerpur	1196	July-October
Tinospora cordifolia (Willd.) Miers *ex* Hook. f. & Thoms.	Mattani	920	May-July
Trichodesma indicum R.Br.	Kehdru	842	September-December
Tridax procumbens L.	Kuthera	900	July-August
Trifolium alexandrinum L.	Kangoo	649	March-June
Trigonella foenum-graecum L.	Sujanpur	567	May-August
Triticum aestivum L.	Chakmoh	736	November-April
Tylophora indica (Burm.f.) Merr.	Khaggal	920	May-October
Urtica dioica L.	Dhaneta Dhar	1070	June-October
Vanda roxburghii R.Br.	Sujanpur	567	May- June
Verbascum thapsus L.	Kehdru	842	January-April
Vernonia anthelmintica Willd.	Patnaun	1000	July-August
Vicia pallida Facquem.	Bhareri	920	December-April
Viola canescens Wall.	Dhoh	1034	April-November
Vitex negundo L.	Bhota	765	March-June
Vitis jacquemontii Parker	Chauri	858	June-August
Vitis vinifera L.	Anu Khurd	858	March-June
Wendlandia heynei (Roem. & Schult.) Sant. & Merch.	Man khad	483	April-June
Withania somnifera (L.) Dunal	Brahmni	858	June-November
Woodfordia floribunda Salisb.	Pucca Bharo	840	March-June
Xanthium strumarium L.	Chhoti Kashi Nadaun	483	September-November
Yucca aloifolia L.	NIT Anu	900	June-September
Zanthoxylum armatum DC.	Anu Khurd	900	April-June
Zea mays L.	Bhira	780	July-August
Zingiber officinale Rosc.	Khaya	787	July-October
Zizyphus jujuba (Lam.) Gaertn. non-Mill.	Tikker Khatrian	908	June-August

contd...

Table 1.5: *contd...*

Name of the Plant	Place/s	Altitute (m)	Flowering & Fruiting Season
Terminalia chebula Retz.	Sukarkhad	381	April-June
Thevetia neriifolia Juss. *ex* Steud.	Dimmi	1006	July-September
Thuja orientalis L.	Sameerpur	1196	July-October
Tinospora cordifolia (Willd.) Miers *ex* Hook. f. & Thoms.	Mattani	920	May-July
Trichodesma indicum R.Br.	Kehdru	842	September-December
Tridax procumbens L.	Kuthera	900	July- August
Trifolium alexandrinum L.	Kangoo	649	March-June
Trigonella foenum-graecum L.	Sujanpur	567	May-August
Triticum aestivum L.	Chakmoh	736	November-April
Tylophora indica (Burm.f.) Merr.	Khaggal	920	May-October
Urtica dioica L.	Dhaneta Dhar	1070	June-October
Vanda roxburghii R.Br.	Sujanpur	567	May-June
Verbascum thapsus L.	Kehdru	842	January-April
Vernonia anthelmintica Willd.	Patnaun	1000	July-August
Vicia pallida Facquem.	Bhareri	920	December-April
Viola canescens Wall.	Dhoh	1034	April- November
Vitex negundo L.	Bhota	765	March-June
Vitis jacquemontii Parker	Chauri	858	June-August
Vitis vinifera L.	Anu Khurd	858	March-June
Wendlandia heynei (Roem. & Schult.) Sant. & Merch.	Man khad	483	April-June
Withania somnifera (L.) Dunal	Brahmni	858	June-November
Woodfordia floribunda Salisb.	Pucca Bharo	840	March-June
Xanthium strumarium L.	Chhoti Kashi Nadaun	483	September-November
Yucca aloifolia L.	NIT Anu	900	June-September
Zanthoxylum armatum DC.	Anu Khurd	900	April-June
Zea mays L.	Bhira	780	July-August
Zingiber officinale Rosc.	Khaya	787	July-October
Zizyphus jujuba (Lam.) Gaertn. non-Mill.	Tikker Khatrian	908	June-August

Enumeration:

The collected plant specimens were dried, preserved and mounted as per the known herborizing practices outlined by Jain & Rao (1977). The specimens were identified using regional floras and various revisionary and monographic works (Chauhan, 1999; Chowdhery & Wadhwa, 1984; Collett, 1902; Dhiman, 1976; Hooker, 1872-1897; Nair, 1977; Polunin & Stainton, 1984; Stainton, 1988) and later got authenticated by carefully matching with the specimens at the herbaria of Botanical Survey of India (BSI), Northern Circle and Forest Research Institute (FRI), Dehradun. Nomenclature of these taxa was confirmed from Bennet (1986) and Wielgorskaya (1995). Excepting for the minor modifications to conform to the present day circumscription, families and genera are delimited mainly after The Flora of British India (Hooker, 1872-1897). Total importance values (TIV) of all the presently studied ethnobotanically important species were worked out as per the methodology given by Belal & Springuel (1996). The nativity of the species was determined on the basis of Mathews (1969), Maheshwari & Paul (1975), Nayar (1977), Sharma (1984), Hajra & Das (1982), Saxena (1991), Pandey & Parmar (1994), Reddy et al., (2000), Reddy & Reddy (2004), Murthy et al., (2003), Negi & Hajra (2007) and Reddy (2008). Festivals and other ceremonial occasions were also attended and relevant information pertaining to the uses of plants was documented. The novelty of the claims was confirmed from The Wealth of India (Anonymous, 1948-1976) and the publications of the earlier workers (Agarwal, 1986; Ambasta, 1986; Jain, 1991; Jain & De Filipps, 1991; Kirtikar & Basu, 1994; Maheshwari, 1996; Nayar et al., 1989). Attempt has also been made presently to collect data on nutritional value of ethnobotanically important plants of the region (Bennet et al., 1991; Dwivedi, 1993; Pruthi, 1976). Likewise, phytochemical analysis of Mallotus philippinensis (fruits with ripe and unripe seeds) and Zizyphus jujuba leaves was got done at Chowksi laboratory Indore by following the standard methodology (Pearson's chemical analysis of foods, 1987), whereas analysis for glycosides (Molisch's test), tannins (Braymer's test), saponins (Liebermann-Burchard (LB) and foam test), terpenoids (CHCl3+conc. H2SO4 and LB test), alkaloids (Meyer's and Dragendroff test) and flavonoids (Shinoda and Lead acetate tests) (Arulpriya et al., 2011; Harborne, 1973; Raaman, 2008; Trease, 1983; Kokate, 1994) was done in pharmacognosy laboratory of University Institute of Pharmaceutical Sciences Punjab University Chandigarh. Photographs of ethnically important plants in natural habitats with inset showing fruits/ inflorescence were also clicked. The voucher specimens of all the presently studied taxa (HPU-EL-SKS-ABS-15001-15296) have been deposited in the herbarium section of Ethnobotanical Laboratory, Department of Biosciences, Himachal Pradesh University, Shimla.

Methdology:

Systematic enumeration of the plants is in alphabetical order of their botanical names, synonyms of the species followed by their respective families. Besides information on Ayurvedic, English, Sanskrit and Regional Names, distribution, description, reproductive cycle, habitat ecology, material examined, part/s used, folk use/s, active constituents, biological activity and uses in literature has also been furnished and presented, if available. The data gathered were screened with the help of available literature (Ambasta, 1986; Anonymous, 1948-76; Chopra et al., 1956, 1969; Jain, 1991; Kirtikar & Basu, 1994; Maheshwari & Singh, 1984; Singh, 1988; Singh et al., 1987) besides many other books and articles published in different journals. Appendices include informations on, index to ethnobotanically important families of district Hamirpur (Appendix I), index to vernacular names and families of ethnobotanically important plants of district Hamirpur (Appendix II) and glossary of words commonly used by the local populace of Hamirpur (Appendix III).

Chapter 2

Underutilized Himalayan Plant Wealth of Economic Significance

***Abelmoschus esculentus* (L.) *Moench* (Pl. 21A).**

Syn.: *Hibiscus esculentus* L.

Family: Malvaceae.

Vern.: Bhindi.

Ayurvedic, English, Sanskrit and Regional Names:

Ayur.: Bhaandi, Bhendaa, Bhindaka;

Eng.: Gumbo, Lady Finger, Okra;

Sans.: Bhenda, Darivka,Tindisa;

B.: Dheras;

Bo.: Bhendi;

Folk: Bhindi, Raamturai.

G.: Binda;

H.: Bhindi, Bhindi tori;

Kan.: Bendo;

Mal.: Bendai, Venda;

Mar.: Bhendi;

Oriya: Bhendi;

P.: Bhinda tori, Bhindi, Ramturai;

Tam.: Vendai;

Tel.: Kukkutamu, Sinnaguruginja.

Distribution: Cultivated throughout India up to 1,350m.

Description: Pubescent or hairy herbs with erect simple stems and usually deeply lobed leaves. Flowers axillary solitary, large and showy. Calyx bell-shaped, 5-lobed. Ovary five-celled. Capsules many-seeded and enclosed in the persistent calyx and splitting from the top in five valves.

Flowering & Fruiting: Twice in a year: February-April; July-September.

Habitat Ecology: Cultivated throughout the district.

Material Examined: HPU-EL-SKS-ABS-15001; 05.05.2011.

Parts Used: Leaves. Stem, Roots, Flowers, Pods, Seeds.

Folk Uses: Pods used as vegetable and for pickling. Decoction (20 ml, once daily for 5-7 days) of immature pods considered good emollient*, demulcent and diuretic. Seeds consumed as antispasmodic. Leaves used as fodder. Fibres obtained by retting of its stems for 20 days, used for making ropes. Powdered roots and crushed stems added for purifying jaggery*.

Active Constituents: These are quercetin, hyperin, hydrolysate of precipitated mucilage, proanthocyanidins, D-glucose, D-glucuronic and galacturonic acids (ripe *fruits*) and flavonol glycosides and anthocyanins (fresh *flowers*).

Biological Activity: The plant exhibits +ive emollient, demulcent and diuretic activities.

Uses in Literature: Known to be used earlier as anticancerous, antispasmodic, antitumorous, demulcent, diuretic, fodder, pickle, vegetable; and for catarrhal affection, cordages, dysentery, dysuria (Ambasta, 1986; Bhatt *et al.*, 2002; Chauhan, 1999; Dager & Dager, 1996; Datt, 1996; Gill & Nyawuame, 1994; Girach *et al.*, 1998; Goud & Pullaiah, 1996a,b; Jain, 1991, 1996; Katewa *et al.*, 2001; Khanna, 2002; Khare, 2007; Kothary & Moorthy, 1996; Kshirsagar *et al.*, 2003; Kumar & Chauhan, 2005; Kumar & Pullaiah, 1998; Kumar *et al.*, 2004; Kurian, 1995; Malik, 1996; Mandal & Basu, 1996; Mishra *et al.*, 1996b; Nazarudeen *et al.*, 1996; Painuli & Maheswari, 1996; Parrotta, 2001; Pawar & Patil, 2004; Prasad *et al.*, 1996; Pullaiah & Kumar, 1996; Ram *et al.*, 2002; Rama Rao & Henry, 1996; Rao & Pullaiah, 2001; Ravisankar & Henry, 1992; Reddy *et al.*, 1998; Retnam & Martin, 2006; Saini, 1996b; Satapathy & Brahmam, 1996; Sen & Behera, 2003; Sharma, 1996; Singh & Pandey, 1996; Singh & Prakash, 1994; Singh *et al.*, 1996b; Siwakoti & Verma, 1996; Sood *et al.*, 2005b; Verma & Chauhan, 2006).

Abelmoschus moschatus Medic. (Pl. 21B).

Syn.: *Hibiscus abelmoschus* L.

Family: Malvaceae.

Vern.: Kasturi Bhindi.

Ayurvedic, English, Sanskrit and Regional Names:

Ayur.: Lataakasturi, Lataakasturika, Kattaphala, Kayuka;

Eng.: Ambrette Plant, Musk Mallow;

Sans.: Latakasturikam;

B.: Mushakdana;

Bo.: Kasturi-bhenda;

G.: Muskdana;

H.: Muakdana;

Kan.: Kadukasturi;

Mal.: Kattukasthuri;

Mar.: Kasturi-bhenda;

Tam.: Varttilai-kasthuri, Katukasturi;

Tel.: Kasturibendavittu.

Distribution: Tropical areas throughout India.

Description: Aromatic, hirsute, erect herbs having palmately lobed leaves and yellow coloured flowers with crimson centre. Pods with reniform and compressed seeds. Seeds have a sweet and heavy fragrance similar to that of musk.

Flowering & Fruiting: July-September.

Habitat Ecology: Cultivated in lower hills of Himachal Pradesh; Khaggal, 920m.

Material Examined: HPU-EL-SKS-ABS-15002; 03.07.2008.

Parts Used: Leaves. Shoots. Roots. Fruits. Seed.

Folk Uses: Shoots cooked as vegetable*. Dried leaves kept for perfuming clothes*. Whole plant used as fodder* and dried ones used as fuel*. 5g of its powdered seeds along with 'harar', *Terminalia chebula* Retz. 'bhera', *Terminalia bellirica* (Gaertn.) Roxb., 'amla', *Emblica officinalis* Gaertn. and 'allia', *Cassia fistula* L. prescribed twice daily with water for curing constipation*, cold* and cough*. Seeds used as tonic against hysteria, nervous disorders and after delivery to ladies for strength and body heat (20 mg powder with a glass of milk). Also seeds are sucked for checking mouth ulcers* and pain whereas inhaling of its puff checks nasal congestion*.

Active Constituents: Seeds rich in marcocyclic musks 5(Z)-tetradecen-14-olide and 7(Z)-hexadecen-16-olide also called ambrettolide and 18.9% linoleic acid, phosphatidylcholine, phosphatidylserine.

Biological Activity: Insecticidal and spasmolytic activites confirmed.

Uses in Literature: Reported to be used as antihalitosis, antiseptic, antispasmodic, aphrodisiac, aromatherapic, carminative, cooling, demulcent, digestive, diuretic, insecticidal, nervine, stimulant, stomachic, tonic; and for aching joints, anxiety, cordages for the leather hoops of musical instruments, cramps, cuts, external massage, flat baskets, flavouring tobacco and coffee, fibers, poor circulation, sizing paper, sprains, straps for tying on the neck of the beast and wounds (Ambasta, 1986; Bhatt *et al.*, 2002; Brown, 1995; Chauhan, 1999; Chopra *et al.*, 1986; Dhiman, 2003; Facciola, 1990; Grieve, 1984; Jain & Srivastava, 2001; Manandhar, 2002; Mittre, 1997; Parrotta, 2001; Singh & Kumar, 2000a; Singh & Pandey, 1998; Varghese, 1996).

Abroma augusta Jacq. (Pl. 21C).

Syn.: *Ambroma augusta* L. f., *A. angulata* Lam., *A. wheeleri* Retz., *A. fastuosum* Gaertn.

Family: Sterculiaceae.

Vern.: Ulat Kamal, Olat kambal.

Ayurvedic, English, Unani, Sanskrit and Regional Names:

Ayur.: Pishaacha Kaarpaasa, Pivari;

Eng.: Devil's cotton, Perernnial Indian hemp;

Unani: Ulat Kambal;

Sans.: Pisacakarpasa, Pivari,Yosini;

B.: Ulatkambal;

Folk: Kamal, Snukapaasi;

H.: Ulatkamal;

Kan.: Divvahatti;

Mar.: Olatkambal;

Tam.: Sivappututti.

Distribution: Throughout hotter and moister parts of India from Punjab and Uttar Pradesh east-wards to Arunachal Pradesh, Assam, Meghalaya, Tripura; ascending to 1,200m southwards in Peninsular India.

Description: Evergreen shrubs possessing cordate, ovate-oblong, serrulate sometimes angled leaves and a few-flowered peduncles opposite the leaves. Purplish or dark marrun petals scarcely exceeding the sepals. Capsules obpyramidal, glabrous, five-valved. Seeds numerous, small, blackish, covered with silky hair.

Flowering & Fruiting: February-May.

Habitat Ecology: Rich and well drained soil; Khaggal, 920m.

Material Examined: HPU-EL-SKS-ABS-15003; 25.04.2010.

Parts Used: Root. Bark. Stem. Leaves.

Folk Uses: Bark used as substitute for hemp*. Juice of fresh leaves good against disorders of uterus (10-15 ml, once daily for a week). 50 g juice of leaves taken for 14 days for sexual diseases. Bark, stem, roots and leaves used as medicine for irregular menstruation, gonorrhoea and boils*.

Active Constituents: *Roots* possess abromine, friedelin, abromasterol, abromasterol A, choline, beta-sitosterol, stigmasterol and octacosanol, taraxerol, its acetate and lupeol.

Biological Activity: Amenorrhoeal, abortifacient, galactotrophic and dysmenorrhoeal activites found to be +ive.

Uses in Literature: Known in India as antibacterial, antifungal, aphrodisiac, arbortifacient, emmenagogue, galactotrophic, hypolipidemic, insecticidal, uterine tonic; and for amenorrhoea, diabetes, dermatitis, disorder of genital organs, dysmenorrhoea, gonorrhoea, stomachache, treatment of sores (Ahuja, 1965;

Agarwal, 2003; Ambasta, 1986; Asolkar *et al.*, 1992; Chauhan, 1999; Chhetri *et al.*, 2005; Dastur, 1964; Gupta *et al.*, 2011; Halim, 2002; Jain, 1991; Jain *et al.*, 1991; Jha & Basak, 1994; Katewa *et al.*, 2001; Khan *et al.*, 2003; Khare, 2007; Maheshwari & Singh, 1987; Masih, 2003; Mishra *et al.*, 1996a; Mittre, 1997; Nahar *et al.*, 2010; Prajapati *et al.*, 2003; Rahamtullah *et al.*, 2010; Retnam & Martin, 2006; Shukla & Verma, 1996; Siwakoti & Verma, 1996; Watt, 1972).

***Abrus precatorius* L. (Pl. 21D).**

Syn.: *A. minor* Desv., *A. pauciflorus* Desv.

Family: Fabaceae.

Vern.: Rati.

Ayurvedic, English, Sanskrit and Regional Names:

Ayur.: Chirihintikaa, Chirmiti, Gunjaa, Gunjaka, Kabjaka, Kakanti, Raktikaa, Tiktikaa;

Eng.: Crab's Eye, Indian-liquorice, Jequirity, Papernoster pea, Rosary pea, Weather plant;

Sans.: Angaravallari, Aruna, Bhilabhushana, Chatak, Gunja, Kanchi, Rakta, Raktala, Raktika;

Ass.: Gunja, Krishanala, Kakachinchi;

B.: Ghunhati, Gunj, Kawet;

Bo.: Ghungchi Gunja;

G.: Chanoti, Gunja;

H.: Chirmiti, Ghumghachi;

Kan.: Guengungi;

Kash. : Shangir;

Mal.: Irattimadhuram, Shekkunni;

Mar.: Chanoti, Kunch;

Oriya: Kainsho, Kotibepolo;

P.: Labri, Rattak;

Tam.: Adisamiyai, Attighurie-ghenza;

Tel.: Kukkutamu, Sinnaguruginja;

U.P.: Ghumchi, Labri, Rattak, Ujarki.

Distribution: Throughout India; ascending to 1,050m in outer Himalayas.

Description: Copiously branched climbers with slender, greenish yellow branches, obovate or oblong leaflets and subsessile pale purple-yellowish flowers in crowded racemes. Pods having bright scarlet seeds with a black spot on the hilum.

Flowering & Fruiting: May-August.

Habitat Ecology: Open dry slopes, shrubberies, wastelands; Anu Khurd, 858m.

Material Examined: HPU-EL-SKS-ABS-15004; 12.05.2010.

Parts Used: Leaves. Roots. Seeds.

Folk Uses: 25ml decoction of the plant considered beneficial for hair (once daily, 15-25 days). Poultice of leaves applied for wound healing in animals. ½ teaspoon of powdered roots given with honey once daily for 5 days to cure whooping cough* and sore throat. Smoke of burnt seeds given to animals to ward of evil spirits; its paste mixed with root paste of *Plumbago zeylanica* L. good for applying locally on skin diseases. Milk of unripe seeds used as substitute for 'shilajeet' for treating fractured bone. Also, seeds used as measure of weight by goldsmiths as a seed considered equal to 1.75g or 1 carat. Seed extract (few drops daily) used in treatment of ulcers. Locally, seeds also used for making necklaces*, bracelets, rosaries* and decorating baskets.

Active Constituents: Found to contain abrin (a toxalbum similar to ricin from castor seeds), a glucoside abralin an alkaloid abrine and abrine A & B, abricin, abridin, abric acid, hypaphorine, alainine, serine, abralin, abrin, abramine (*seeds*), precol, abrol precasin, isoflavoquinones I, II, III (*roots*), xyloglucosyldelphinidin and p-coumaryl galloyl-glucosldephnidin (*seed coat*).

Biological Activity: Anthelmintic, antispermatogenic, antitumorous and spasmogenic activities +ve.

Uses in Literature: Known to be used as an abortifacient, alexiteric, antibacterial, antidiarrhoeal, antidote to poison, antifertility, antiinflammatory, antiperiodic in preventing malaria, antiseptic, antitumorous, aphrodisiac, blood purifier, CNS depressant, colic, contraceptive, diuretic, emetic, febrifuge, insecticidal, laxative (vet.), masticatory, measure of weight for goldsmiths, molluscicidal, purgative, refrigerant, symbolic (to ward off evil spirits), tonic; and for affections of hair and nervous system, anthrax (vet.), boils, bone fracture (vet.), bronchitis, convulsions, costipation (vet.), cold, coma, congestion, cough, cuts, decoration of baskets and bracelets, diarrhoea, dropsy, dysentery (vet.), expulsion of placenta (vet.), eye complaints, facilitating delivery, flatulence, gall neck (vet.), gonorrhoea, haemorrhoids, headache, hemicrania, indigestion, inflammation, insect bites (vet.), liver disorders, liver disorders (vet.), menorrhoea, menstrual pains, muscular pains, nervous disorders (vet.), paralysis, parturition, peptic ulcers, pneumonia, prolapsed uterus (vet.), retained placenta (vet.), retention of urine, rheumatism, ringworms, scabies, sciatica, scorpion bite, skin eruptions, snake bite (vet.), sore throat, stiffness of the shoulder joints, swelling, syphilis, ulcer (vet) and wounds (vet.) (Ambasta, 1986; Bhatt *et al.*, 2002; Chauhan, 1999; Dager & Dager, 1996; Datt, 1996; Gill & Nyawuame, 1994; Goud & Pullaiah, 1996a, b; Jain, 1991, Jain & Tarafder, 1970; Katewa *et al.*, 2001; Khanna, 2002; Khare, 2007; Kiruba *et al.*, 2006; Kothary & Moorthy, 1996; Kshirsagar *et al.*, 2003; Kumar & Chauhan, 2005; Kumar & Pullaiah, 1998; Kumar *et al.*, 2004; Kurian, 1995; Malik, 1996; Mandal & Basu, 1996; Mishra *et al.*, 1996b; Nazarudeen *et al.*, 1996; Painuli & Maheswari, 1996; Pawar & Patil, 2004; Prasad *et al.*, 1996; Pullaiah & Kumar, 1996; Ram *et al.*, 2002; Rama Rao & Henry, 1996; Rao & Pullaiah, 2001; Ravisankar & Henry, 1992; Reddy *et al.*, 1998; Retnam & Martin, 2006; Saini, 1996b; Satapathy & Brahmam, 1996; Sen & Behera, 2003; Shah *et al.*, 1981; Sharma, 1996; Singh & Pandey, 1996; Singh & Prakash, 1994, 1996; Singh *et al.*, 1996b; Siwakoti & Verma, 1996; Sudarsanam *et al.*, 1995; Tarafder & Chaudhuri, 1981; Varghese, 1996; Verma & Chauhan, 2006).

Major Ayurvedic Formulations: Gunjadi taila, Gunja-pralepa, Mrtasanjivani-gutika, Nili-bhringadi-taila, Uchatta.

Trade: Sold in market as 'Gunchi' and 'Rati.

Safety Profile: A fatal poisoning in children has been reported after chewing its seeds due to aborin-a highly toxic protein similar to ricin.

Substitute or Adulterant: Roots used as adulterant or substitute for 'mulhati' (*Glycyrrhiza glabra* L.).

Acacia catechu Willd. (Pl. 22A).

Syn.: *A. polyacantha* Willd., *A. wallichiana* DC., *A. catechuoides* DC.,
A. sundra Bedd.,M. *catechu* Roxb.,M. *catechuoides* Roxb.

Family: Fabaceae.

Vern.: Khair.

Ayurvedic, English, Sanskrit and Regional Names:

Ayur.: Dantdhaavan, Gaayatri, Kadara, Khadira, Kantaki, Raktasaara, Sonavalka;

Eng.: Black catechu, Catechu, Cutuch, Nigraum, Pegu;

Sans.: Dantadhavana, Jivumangalaya, Vakrakantaka;

Ass.: Kat, Koir;

B.: Kahyer, Kuth;

Bo.: Khaderi, Lalkhair;

G.: Kher, Kheribayal;

H.: Babul, Katha, Khair;

Mal.: Kadaram, Senkainnali;

Mar.: Khaderi, Lalkheir;

Oriya: Bimbu, Khodira;

Tam.: Karangalli, Kasukkalti;

Tel.: Khadiramu, Mallasandra.

Distribution: Throughout the drier region of India, Peninsula and from Indus eastwards to Assam & N. India: 500-1,350m.

Description: A moderate-sized tree upto 13m, 1.5m girth, recurved prickles, rough dark coloured bark, small white or pale yellow flowers and bipinnate leaves. Pods stalked, strap-shaped, glabrous-oblong with a triangular beak.

Flowering & Fruiting: May-July.

Habitat Ecology: Roadsides, drier areas, wastelands, Hilly slopes; Dhoh (near Kanjiyan), 1,034m.

Material Examined: HPU-EL-SKS-ABS-15005; 08.07.2008.

Parts Used: Twigs. Leaves. Heartwood ('katha').

Folk Uses: Twigs used as 'datun' (tooth brush). Leaves given as fodder* for goats. 'Katha' obtained from its heartwood used for curing spongy gums, tonsils, ulceration of mouth and throat diseases. Wood used for construction of poles*, doors, water resistant panels* and furniture.

Active Constituents: These are catechin, catechu, catechutanic acid and tannic acid, tannin (*bark*), α-β-ã catechin and 1-epicatechin (*wood*).

Biological Activity: Antiviral, antiinflammatory, hepatoprotective, hypoglycaemic and spasmolytic activities confirmed.

Uses in Literature: Reported earlier as abortifacient, adhesive for plywood, anticancerous, aphrodisiac, astringent, dentrifice, digestive, dye, dying canvas, expectorant, fibre, fishing nets, gum arabic, host for lac, insect flies, tonic; and for affection of gums, asthma, bleeding from nose, blood dysentery, bronchial affection, burns, chest pain, child birth, colic pain, conjection of faeces, conjuctivitis, constipation (vet.), cough, curing impotency in excess, diarrhoea with pyrosis, eczema, gonorrhoea, gravel pthisis, haemoptysis, hoarseness, increasing the secretion of milk after delivery, inflammation, intermittent fevers, itching, leprosy, leucorrhoea, menorrhagia, menstrual complaints, Mithila paintings, muscular growth (vet.), otorrhoea, pain in chest, piles, printing, rheumatism, scurvy, skin disease (vet.), suppressing sexual desire, swelling (vet.), syphilis, tanning, tonsilis, toothache, ulcer and wounds (vet.) (Ahuja, 1965; Ambasta, 1986; Asolkar *et al.*, 1992; Chauhan, 1999; Hooker, 1878; Jain, 1991; Jain *et al.*, 1991; Jha & Basak, 1994; Katewa *et al.*, 2001; Khare, 2007; Maheshwari & Singh, 1987; Masih, 2003; Mishra *et al.*, 1996b; Retnam & Martin, 2006; Saini, 1996a; Shukla & Verma, 1996; Siwakoti & Verma, 1996; Watt, 1972).

Major Ayurvedic Formulations: Arimedadi-taila, Chandanadi-vati, Dasnar churna, Khadiraadi-kwatha, Khadiraastaka, Khadiradi-gutika, Khadiradivati, Khadirishta, Khadir-sar, Kutajarishta, Lavangadi-vati.

Trade: Kathaa, Katthaa, Paakhraa.

Acacia concinna (Willd.) DC. (Pl. 22B).

Syn.: *A. sinuata* (Lour.) Merr., *A. rugata* (Lamk.) Ham.,
A. hooperiana Zippel, *Mimosa concinna* Willd.

Family: Fabaceae.

Vern.: Shikaa-kaai, Ban Reetha.

Ayurvedic, Sanskrit and Regional Names:

Ayur.: Kantvalli, Saatalaa, Saptalaa, Shitalaa, Shrivalli;

Sans.: Aila, Rassaul;

B.: Banritha, Toldang;

Bo.: Sikekdi;

G.: Chikakai;

H.: Kocji, Ritha;

Kan.: Sige;

Mal.: Chikaka;

Mar.: Shikakai;

Tam.: Shikai;

Tel.: Shikaya.

Distribution: Tropical jungles throughout India especially in Deccan, Eastern Himalayas.

Description: Climbers or trees with copius scattered prickles, pale green leaflets and flowers in copiously panicled globose heads. Pinnae 12-16, leaflets 30-50 membranous. Stipules and bracts cordate-ovate. Pods strap-shaped, thick succulent with slightly waved sutures.

Flowering & Fruiting: January-April.

Habitat Ecology: Usually in dry and wet deciduous forests, planted; Neri, 381m.

Material Examined: HPU-EL-SKS-ABS-15006; 07.04.2010.

Parts Used: Bark, Leaves, Flowers, Pods.

Folk Uses: Bark added in preparation of local wine* 'sur' and one tsp of its burnt ash with cow's milk, recommended for 7-10 days, for cure of asthma and lungs dysinfection*. Leaves used as fodder*. Pods used as shampoo for hair and conditioner. Also, the powdered pods of the plant along with 'amla' (*Emblica officinalis* Gaertn.), henna (*Lawsonia alba* Lam.) and curd applied as hair dye. Wood used as timber for construction*.

Active Constituents: Principle ingredients are lupeol, α-spinasterol, hexacosanol, α-spinasterone, acacic acid, lactone (*bark*); and a nortiterpene-acacidal, acactic acid lactone 3-acetate, sonunin-I and sonunin-II (*pods*).

Biological Activity: Acacia lactone inhibits mitogenic activity and used as immunosuppressant.

Uses in Literature: Known so far as antidiarrhoeal, antimicrobial, deobstruent, diuretic, detergent, dye, emetic, expectorant, fish poison, haemolytic, mild cathartic, nauseant, purgative, spasmolytic, spermicidal, tonic; and for bilious affections, chutney, dandruff, flatulence, jaundice, malarial fever, skin diseases, tanning, and washing hair (Ambasta, 1986; Chopra *et al.*, 1956; Chunekar & Pondel, 1999; Khare, 2007; Kirtikar & Basu, 1994).

Acacia nilotica L. (Pl. 22C).

Syn.: *A. arabica* (Lam.) Willd.

Family: Fabaceae.

Vern.: Babul, Kikkar, Shammi.

Ayurvedic, English, Sanskrit and Regional Names:

Ayur.: Aabhaa, Baavari, Babbuula, Babbuuri, Kinkiraata, Pitapushpaka, Shitaka, Shuulikaa, Sukshamapatra, Yugmakantaka;

Eng.: Arabic Tree, Black Babool, Indian Babul;

Sans.: Babbuka;

H.: Babul, Kikkar;

P.: Kikkar;

Tam.: Karuvelei;

Tel.: Nallatumma.

Distribution: Throughout drier parts of India.

Description: Trees or shrubs with straight finally grey-downy, prickly branchlets and white or yellow flowers in round heads. Leaflets membranous, green, glabrous. Flowers sessile, actinomorphic, hypogynous, polyandrous stamens. Pods indehiscent, rigidly coriaceous.

Flowering & Fruiting: July-September.

Habitat Ecology: Wild as well as cultivated in dry places, wastelands, hilly slopes; Patta, 1,020m.

Material Examined: HPU-EL-SKS-ABS-15007; 10.07.2010.

Parts Used: Stem. Bark. Gum. Heartwood. Pods. Whole Plant.

Folk Uses: Plant considered religious and worshipped in Navgraha Vatika*. Branches used as tooth brush* for scouring teeth. Stem used as fuel and for making agricultural implements, boats, tent pegs. Green pods and young leaves make excellent fodder for goats. Fresh leaves chewed twice a day for five days to cure mouth sores. Paste of leaves applied for skin diseases*.

Active Constituents: Isolated constituents are steroids, glycosides, flavonoids, tannins, gums, amino acids, fatty acids, organic acids, polyphenols; tannin (av.c.12%), + catechin-5-gallate, gallic acid its methyl ester naringenin, catechin, catechol, pyrogallol (*bark)*, stearic acid, kaemferol glucoside, isoquercitin, leucocynidin (*flowers*).

Biological Activity: Stem bark hypoglycaemic, antiprotozoal, spasmolytic, CVC active and analgesic.

Uses in Literature: Known in India for agricultural implements, body swelling, burning sensation in eyes, cholera, cold, cooling effect (vet.), cough, dental caries, diarrhoea, dysentery, earache, ear discharge, fencing, fodder, indigestion (vet.), leprosy, leucorrhoea, mouth sores, musical instruments, premature ejaculation, pyorrhoea, rinderpest, spermatorrhoea, syphilitic ulcer, toothache, whooping cough; and as an astringent and galactagogue (Ambasta, 1986; Bhatt *et al.*, 2002; Kalpesh *et al.*, 2004; Khanna, 2002; Khanna & Kumar, 2000; Khare, 2007; Kshirsagar *et al.*, 2003; Kumar & Chauhan, 2005, 2006; Kumar *et al.*, 2004; Mohanty *et al.*, 1998; Pandey & Verma, 2002; Patel *et al.*, 2003; Punjani, 1998; Ram *et al.*, 2002; Singh, 2003; Sen & Batra, 1997; Sen & Behera, 2003; Sen *et al.*, 2002).

Major Ayurvedic Formulations: Khadiradi-vati, Khadira-rishta, Khadie-Sar, Lavangadi-Vati, Babool-toothpaste, Babool-arista, Chandaradi-vati, Kutaja-rista.

Adulterant or Substitute: Seeds used as adulterant or substitute for 'kaladana' (*Ipomoea hederacea* (L.) Jacq.).

Achras zapota L. (Pl. 22D).

Syn.: *A. sapota* L., *Mimusops manilkara* G. Don,

Manilkara zapota (L.) P. Van Royan,

Manilkara achras (Mill.) Fosberg, *Sapota achras* Mill.

Family: Sapotaceae.

Vern.: Chiku.

English and Regional Names:

Eng.: Sapodilla Plum, Bully Tree;

H.: Chiku;

Tam.: Shimai , Eluppai.

Distribution: Native to central America, cultivated chiefly in Maharashtra Tamilnadu, West Bengal and rest of India.

Description: Evergreen small trees with pubescent branches, shining, oblong-elliptic leaves crowded at apex and solitary-axillary white flowers. Corolla funnel-shaped. Berries globose, indehiscent with black, polished seeds embedded in a light-brown pulp of fruit.

Flowering & Fruiting: December-January.

Habitat Ecology: Cultivated; Anu Khurd, 858m.

Material Examined: HPU-EL-SKS-ABS-15008; 22.05.2011.

Parts Used: Leaves. Bark. Fruit. Seeds. Wood.

Folk Uses: Leaves used as fodder* for goats. Fruits edible. Bark considered febrifugal*. Seeds prescribed as diuretic*. Wood used as timber*.

Active Constituents: *Bark* contains latex 20-25% (gutta-percha), tannin (11.8%), α and β—amyrin acetate, taraxerol and ursolic acid.

Uses in Literature: Used so far as edible, chewing gum; and for colouring sails, dental surgery and fishing tackles (Ambasta, 1986; Khare, 2007).

Achyranthes aspera L. (Pl. 23A).

Syn.: *Digera muricata* Mart.

Family: Amaranthaceae.

Vern.: Puthkanda.

Ayurvedic, English, Sanskrit and Regional Names:

Ayur.: Aaghaat, Adahshalya, Apaamaarga, Chirchitaa, Kanihi, Kharamanjari, Kharapushpaa, Mayura, Mayuraka, Pratyakpushpaa, Shaikharika, Shikhari, Vashira;

Eng.: Prickly Chaff Flower;

Sans.: Adhoganta, Adhvashalya, Aghamargava, Aghata, Apangaka, Dhamargava, Malakanta, Mayuraka, Sthalamangari;

Ass.: Apang;

B.: Apang;

Bo.: Aghada;

G.: Agheda, Aghedi;

H.: Apang, Chichra, Chirchira, Latjira;

Kan.: Uttarane;

M.P.: Apamara;

Mal.: Katalati;

Mar.: Aghada, Aghara;

P.: Kutri;

Tam.: Nayuviri;

Tel.: Antisha, Apamaragamu;

U.P.: Latzira, Lahchichra.

Distribution: Tropical and subtropical regions throughout India up to 2,100m.

Description: Undershrubs with orbicular, obovate, pubescent tomentose or valvety leaves and 30-45 cms long spikes. Seeds inverted.

Flowering & Fruiting: May-October.

Habitat Ecology: Wastelands, roadsides, forest edges, drier places; Krishna Nagar, 858m.

Material Examined: HPU-EL-SKS-ABS-15009; 12.07.2008.

Parts Used: Whole Plant. Roots. Seeds.

Folk Uses: Plant used as 'datun' (tooth brush). 2-3g ash of roots used in asthma (twice daily, for 15-20 days); its paste applied as antidote to snake bite and for tooth infection. A gargle of root juice useful for throat infection* and fever. Roots tied with 'bere' (*Zizyphus jujuba* Mill.) for religious purposes. Also, 3-6 cm long pieces of fresh roots tied to the tail of animals considered good for facilitating expulsion of placenta*. Seeds considered good for suppressing hunger*. Plant worshipped in 'chananchhati fast', Navgraha Vatika and a must in 'hawan' ceremony.

Active Constituents: Main constituents are achyranthine, betaine (*whole plant*) and ecdysterone (polypodine A), acdysone, oleanolic acid, (*roots*), linoleic, oleic, palmitic, stearic, dehenic, arachidic, myristic and lauric acids, saponin A and B, saponin α-methylester (*seeds*).

Biological Activity: Antibacterial, anthelmintic, anti-inflammatory, cytotoxic, cardiotonic activites confirmed.

Uses in Literature: Reported so far as abortifacient, antidote to snake bite, antiseptic, appetiser (vet.), astringent, cardiotonic, cereals in times of scarcity, colic, diuretic, dye, emetic, febrifugal, galactagogue (vet.), germicidal, laxative, magico-therapeutic in labour, manure, oligogalactia, psychoactive, spinach, tonic after delivery, tooth brush, tooth powder; and for asthma, blisters in mouth, boils (vet.), bronchitis (vet.), burns (vet.), cataract, cholera, cramps (vet.), dental caries, diarrhoea, edible purposes (leaves, seeds), eye complaints (vet.), facilitating delivery, fracture (vet.), hemicrania, hydrophobia, indigestion, injuries (vet.), jaundice, malarial fever, migraine, pyorrhoea, removing dead foetus (vet.), renal complaints, scorpion stings, scrofula, skin diseases, swelling, toothache, ulcer, whooping cough and wounds (vet.) (Ambasta, 1986; Aminnudin & Girach, 1991, 1996; Bhandary *et al.*, 1996; Borthakur, 1993; Chauhan, 1999; Chowdhery, 1996; Datt, 1996; Dvarakan & Ansari, 1996; Gill *et al.*, 1997; Girach, 1992; Girach & Aminnudin, 1995; Goud & Pullaiah, 1996b; Hembrom, 1996; Henry *et al.*, 1996;

Hosagoudar & Henry, 1996c; Jain *et al.*, 1994; Jha *et al.*, 1991, 1996; Joshi, 1995; Kiruba, *et al.*, 2006; Khanna *et al.*, 1996b; Krishna & Singh, 1987; Lal *et al.*, 1996; Malik *et al.*, 2009; Manadhar, 1989a, b, 1996; Mishra *et al.*, 1996b; Mohanty & Padhy, 1996; Pandey & Verma, 2003, Prasad *et al.*, 1996; Pullaiah & Kumar, 1996; Raghupathy & Mahadaven, 1991; Rajenderan *et al.*, 1996; Ramadas *et al.*, 2000; Rama Rao & Henry, 1996; Rao *et al.*, 1996; Reddy *et al.*, 1996; Saini 1996b; Saklani & Jain, 1994; Samwatsar & Diwanji, 1996b; Sharma & Borthakur, 2003; Sharma & Sood, 1997; Shome *et al.*, 1996; Shukla & Verma, 1996; Sikarwar, 1996; Singh & Maheshwari, 1992; Singh & Pandey, 1996; Singh & Parkash, 1996; Singh & Rao, 2003; Srivastava *et al.*, 2003; Sudarsanam *et al.*, 1995; Sundaravelu *et al.*, 2003; Swamy & Gupta, 1996; Tirkey *et al.*, 2003b; Verma, 1998; Vihari, 1995).

Major Ayurvedic Formulations: Abhaya-lavana, Agastihairtaki-leha, Agnimukha, Apamargaksara, Apamargaksara-taila, Gudapippali, Jyotismati-taila.

Achyranthes bidentata Blume (Pl. 23B).

Syn.: *A. wightiana* Wall., *A. lanceolata* Wall., *A. aspera* Wall.

Family: Amaranthaceae.

Vern.: Puthkanda.

Ayurvedic and Regional Names:

Ayur.: Shveta Apaamaarga;

Ass.: Apamarga, Bankhat;

H.: Adhajhara, Oga;

Kash.: Puthkandiari;

U.P.: Latzeera.

Distribution: Temperate and subtropical Himalayas: Kishtwar to Sikkim at 1,200-3,200m; Khasi hills; H.P. upto 1,500m elevation.

Description: Undershrubs with stiff stem, ovate-lanceolate opposite, greenish white or pinkish leaves and stiffy flowers crowded in spikes having spinous bracteoles. Utricles enclosed in the hardened perianth. Seeds inverted truncate at apex and rounded at the base.

Flowering & Fruiting: April-September.

Habitat Ecology: Wastelands, grasslands, drier places; Tikker (Didwin), 800m, Tal, 861m.

Material Examined: HPU-EL-SKS-ABS-15010; 29.05.2011.

Parts Used: Whole Plant. Roots. Seeds.

Folk Uses: 5-10ml decoction of the plant and spikes given twice daily to the patients bitten by mad dogs and against cough. 2-3g of powdered plant with honey consumed twice a day as diuretic; plant also used in making local wine. ½ teaspoon of powdered roots prescribed twice a day for 5 days with black pepper (*Piper nigrum* L.) and honey to cure cough. Gargles of its juice good for throat infection*. Also, its amulet tied around waist of pregnant women to facilitate delivery. Freshly picked roots tied to the tail of animals for removal of placenta. Paste of seeds applied as poultice against snake bite and scorpian bite. Also, eaten raw for suppressing hunger*.

Active Constituents: Isolated constituents are two saponins, saponin A and saponin B, saponin α-methyl ester, β-methyl ester (*seeds*), ecdysterone, ecdysone, emodins and physcion (*roots*).

Biological Activity: Antimicrobial (*root* extr.), anti-inflammatory, cardiotonic (*seeds*), diuretic and abortifacient (*stem bark*) activities confirmed.

Uses in Literature: Reported so far as an astringent, diuretic, galactagogue (vet.), tooth brush, tooth powder; and for asthma, blisters in mouth, cholera, dental caries, diarrhoea, dropsy, edible purposes (leaves, seeds), facilitating delivery, hemicrania, hydrophobia, indigestion, jaundice, ophthalmia, pyorrhoea, rabies, scorpion stings, scrofula, skin diseases, swelling, toothache and whooping cough (Chhetri, 2005; Chowdhery, 1996; Jain, 1991; Jain *et al.*, 1991; Kapur & Singh, 1996; Khare, 2007; Lalramnghinglova, 1996; Manandhar, 1990, 1996; Molla & Roy, 1996; Pawar & Patil, 2004; Retnam & Martin, 2006; Singh *et al.*, 1980; Sudarsanam *et al.*, 1995; Verma & Chauhan, 2006).

Major Ayurvedic Formulations: Abhaya, Agastihairtaki, Agnimukha, Apamargaksara, Lavana, Gudapippali, Jyotismati-taila.

Acorus calamus L. (Pl. 23C).

Family: Araceae.

Vern.: Barae, Barre, Barya.

Ayurvedic, English, Sanskrit and Regional Names:

Ayur.: Ghorbach, Golomi, Kshudra-patra, Maangalyaa, Shadgranthaa, Shataparvaa, Tikshnagandhaa, Ugragandhaa, Ugraa Vachaa;

Eng.: Calamus, Sweet flag;

Sans.: Bhadra, Galani, Golomi, Jalaja, Jatila, Kunga, Shadagrantha, Ugra, Ugragandhaha, Vacha;

B.: Bach;

G.: Gandhilovaj, Godavaj, Vekhand;

H.: Bach, Ghor, Gorbach, Ghorbach;

P.: Bariboj, Wach;

Tam.: Vadaja, Vasa, Vashambutel, Wasa.

Distribution: Throughout India; Kinnaur, Kullu, Mandi, Kangra, Shimla, Hamirpur, Sirmour, Chamba near river beds in H.P.

Description: Perennial aromatic herbs, with branched rhizomes creeping in mud and stout joints. Leaves bright green, thickened in the middle and wavy margins. Spadix sessile, dense flowered, yellow-green. Berries turbinate, few seeded which are oblong.

Flowering & Fruiting: June-July.

Habitat Ecology: Frequent along water banks and marshy places; Khas Graon, 858m.

Material Examined: HPU-EL-SKS-ABS-15011; 07.07.2008.

Parts Used: Whole Plant. Leaves. Rhizomes.

Folk Uses: Paste of rhizomes and leaves in mustard oil make good lotion for joint pains and body massage* of kids. Cut and dried pieces of rhizome roasted in ghee chewed against flatulence* and cough. 1g of powdered rhizome prescribed with cow's milk twice daily till cure for nasal congestion*, internal injuries* and wounds. Crushed rhizome mixed with mustard oil (*Brassica campestris* L.) applied for relief in joint pains in animals. Poultice of the whole plant considered good against stomachache.

Active Constituents: Main ingredients are vit. C, luteolin-6, 7-diglucoside (*leaves, rhizome*), oxalic acid 0.078 and calcium 0.006% (*leaves*), asarone, α pinene, d-camphene, calamene, calamenol and sesquiterpenes (oil extracted from *rhizome*), acorin, calamene, tannin, mucilage starch, vit. C, fatty acid, and calcium oxalate (*roots*).

Biological Activities: Rhizome exhibits CNS depressant and spasmogenic activities. Roots vermifugel, stomachic, carminative in small doses.

Uses in Literature: Known earlier for the treatment of abdominal pains, aching joints, arthritis, asthma, bodyache, bronchial asthma, catarrh, cold, cough, cuts, chronic diarrhoea, dysentery, dyspepsia, dyspepsia (vet.), epilepsy, fever, gastric (vet.), general horse care (vet.), headache, heart and lung trouble, hysteria, indigestion, indigestion (vet.), inflammation, insect bites, insecticidal (vet.), internal parasite removal (vet.), leucorrhoea, lice (vet.), malaria, pain in neck, psychoneurosis, pthisis, rheumatism, ringworms (vet.), septicaemia (vet.), skin disease, snake bites, teething in children, throat pain, wounds (vet.); and as analgesic, emetic, insecticidal sedative, stimulant (vet.) and tonic (Agarwal, 1997; Ambasta, 1986; Borthakur & Sarma, 1996; Borthakur *et al.*, 2004; Chak & Sharma, 1965; Chhetri, 2005; Dandiya & Sharma, 1961, 1962; Geetha *et al.*, 1996b; Gogoi & Borthakur, 2001; Jain, 1991; Joshi, 1993; Khare, 2007; Kumar & Jain, 1998; Maheshwari *et al.*, 1980, 1981; Mohanty *et al.*, 1998; Nath & Borodoloi, 1989; Pal & Jain, 1998; Ramdas *et al.*, 2000; Ranjan & Sethuraman, 1997; Retnam & Martin, 2006; Shah, 1987; Sudarsanam *et al.*, 1995; Tiwari & Pande, 2006; Uniyal, 1968; Upadhye & Kulkarni, 2004; Verma, 1998; Verma & Chauhan, 2006; Vir Jee *et al.*, 1984).

Major Ayurvedic Formulations: Abana tablets, Ashwagandharista, Bacha churn Chandraprabha-vati, Laghu, Mahashankhavati, Sanjivani-vati, Saraswat churn.

Trade: Decorated and transversely sliced rhizomes sold in market.

Remarks: Rhizome emetic in large doses. Also, it may cause disturbed digestion and gastroenteritis.

Adulterant or Substitute: Rhizomes used as an adulterant or substitute for that of 'galangal' (*Alpinia galangal* L.) and 'bish' (*Aconitum* spp).

Actinidia chinensis var. *deliciosa* A. Chev. (Pl.23D).

Family: Actinidiaceae.

Vern.: Kiwi.

English Name:

Eng.: Chinese Goose Berry.

Distribution: Cultivated up to 2,100m.

Description: Woody, twining, climbing shrubs reaching upto 9m. Apices of leaves at the end of branches sharp pointed. Young leaves coated with red hair whereas mature ones dark green and hairless on the upper surface. Fruits oblong, rusty brown with hairy surface.

Flowering & Fruiting: February-November.

Habitat Ecology: Planted; Bhota, 765m.

Material Examined: HPU-EL-SKS-ABS-15012; 15.07.2010.

Part Used: Fruit.

Folk Uses: Fruits edible as source of vit. C* and considered as cardiotonic*; jams and jellies also prepared from it.

Uses in Literature: Reported to be used for decorating ice creams, marmalade, pencil making, preparing jams, paper, tube, wine; and as edible, and insecticide (Ambasta, 1986).

Trade: Kiwi.

Adhatoda vasica Nees. (Pl. 24A).

Syn.: *A. zeylanica* Medic., *Justicia adhatoda* L.

Family: Acanthaceae.

Vern.: Safed basuti.

Ayurvedic, English, Sanskrit and Regional Names:

Ayur.: Aataruushaka, Simhaasya, Simhaparni, Sinhavadanaa, Vaajidanta, Vaasaa, Vaasaka, Vaasikaa, Vrisha;

Eng.: Malabarnut leaf, Nut tree;

Sans.: Arus, Atarusha, Bhisangmata, Raktappittaghni, Vajidantakaha, Vasaka;

Ass.: Herb uksha;

B.: Bakas, Basaka patra, Vasaka;

Bo.: Adalsa, Vasuka;

G.: Adalsa, Adsoge, Adulaso, Arduosa, Arusa, Arusha, Paandru;

H.: Adalsa, Adarsa, Arusa, Banasa, Vasaka;

Kan.: Adusoge, Attalotakam;

Mal.: Atalotakam;

Mar.: Adulsa, Baksha;

Oriya: Rottoomuli;

P.: Bhekar;

Tam.: Adadoli, Kattumurunghai;

Tel.: Adasara, Addasaramu;

U.P.: Adusa, Arusa, Roos, Rusa.

Distribution: Panjab-Assam: 500-1,600m.

Description: Evergreen glabrous dense shrubs with dark green, elliptic-lanceolate leaves and white flowers in dense axillary pedunculate spikes. Seeds orbicular-oblong, tubercular-verrucose and glabrous.

Flowering & Fruiting: March-May.

Habitat Ecology: Open slopes, damp places, wastelands, roadsides; Tauni Devi, 1,050m.

Material Examined: HPU-EL-SKS-ABS-15013; 13.04.2008.

Parts Used: Whole Plant. Leaves. Flowers. Roots.

Folk Uses: Decoction of leaves along with that of *Punica granatum* L. and *Acorus calamus* L. given twice daily for a week for cough, cold and asthma. Slightly warmed leaves applied as band aid on pimples* and acne*. Leaves used for hot water fomentation*, and given to livestock for cudding*. Poultice of 50g of its dried leaves along with 25g 'ashwagandha' (*Withania somnifera* (L.) Dunal), 5g 'jayaphal'(*Myristica fragrans* Houtt.) and 10g 'cloves' (*Syzygium aromaticum* (L.) Merr. & Perry.) prepared in mustard oil considered good for massage* in rheumatic pains and arthritis. Tender leaves given to cattle for 4-5 days for curing loss of appetite and as fodder. Hot decoction of its leaves (2-3 litres) given twice daily to animals to check cough. 1g powdered flowers with honey given on empty stomach for 15-30 days to check asthma. 10 ml decoction of roots given twice a day for 2-4 days to cure high fever. Plant used as a packing material* for unripe fruits.

Active Constituents: Active principles associated to vascinin, vascinine, vascicone vasinone, vascine, β-sitosterol, tritriacontane, vasicinine, vascicinol, vascicoline, adnatodine, vasciolinone, antistione (*leaves*), vasicinol, 6-hydroxyvasicine (*root*), vasicoline, adhatodine, anisotine, betaine, 6-hydroxypaganine (*plant*).

Biological Activities: Alkaloid vasicine acts as uterine stimulant, uterotonic, abortifacient, bronchio-dilator, respiratory stimulant and hypotensive whereas root extr. exhibits positive hypoglycaemic activities.

Uses in Literature: Known to be used as an anthelmintic, antipyretic, antiseptic, blood purifier, ectoparasite attachment (vet.), edible, expectorant, expectorant (vet.), febrifuge, febrifuge (vet.), fuel for brick burning, green manure, insect repellent for stored grains, insecticidal, psychoactive; and for abortion, anthrax (vet.), asthma, boils, bronchial problems (vet.), bronchitis, child birth, cold, cough, diarrhoea (vet.), dislocation of joints, dysentery, dysentery (vet.), eczema, eye complaints, fencing, foetus removal (vet.), fever (vet.), gastric problems, gonorrhoea, headache, heart troubles, ingredient of 'chyawanprash', jaundice, loss of memory, making beads of rosaries, malaria, mastitis (vet.), panting (vet.), preparation of rice bear, pthisis, pulmonary affections, rheumatism, rheumatism (vet.), scabies (vet.), skin diseases (vet.), stomach disorder (vet.), swelling, toothache, uterine tonic, uterine tonic (vet.), wounds (vet.) and wrapping paddy (Ambasta, 1986; Aswal, 1996; Borthakur, 1993; Bhattarai, 1992; Borthakur & Sarma, 1996; Borthakur *et al.*, 2004; Chopra *et al.*, 1956; Chowdhury *et al.*, 1996; Dixit & Pal, 1996; Girach, 1992; Girach *et al.*, 1996; Gogoi *et al.*, 2003; Huidrom, 1996; Jain *et al.*, 1991; Jha *et al.*, 1991; Kapur & Singh, 1996; Khanna & Kumar, 2000; Khare, 2007; Kiruba *et al.*, 2006; Kurian, 1995; Lalramnghinglova, 1996; Mehrotra *et al.*, 1995; Mokat & Deokule, 2004; Negi *et al.*, 1993; Pal & Jain, 1998; Prasad *et al.*, 1996; Punjani, 1998; Ranade, 1998; Raghupathy & Mahadevan, 1991; Rama Rao & Henry, 1996; Ramdas *et al.*, 2000; Reddy *et al.*, 1998; Reddy & Raju, 2000; Rekha *et al.*, 1998; Retnam & Martin, 2006; Sharma &

Borthakur, 2003; Sharma & Sood, 1997; Sikarwar, 1996; Sikarwar *et al.*, 1994; Srivastava *et al.*, 2003; Sudarsanam *et al.*, 1995; Swami & Gupta, 1996; Tirkey, 2004; Tirkey *et al.*, 2003 a,b; Varma, 1997; Watt, 1972).

Major Ayurvedic Formulations: Asokarista, Chandanadi-taila, Chyawanprasha, Kanakasava, Kasamitra, Kasni, Punarnavarishta Pradarai-rasa, Vasaaca-Churna, Vasakamadhu, Vasakasava Vasarista, Vasaputaphaka, Vasavaleha, Viaspanaka.

Trade: Adusa, Baansapatti, Vasakapatti.

Adulterant or Substitute: Roots used as an adulterant or substitute for snake root or 'senega' (*Polygala senega* L.).

Adiantum capillus-veneris L. (Pl. 24B).

Family: Adiantaceae.

Vern.: Hansaraj.

Ayurvedic, English and Regional Names:

Ayur.: Hansaraaja, Hansapadi;

Eng.: American Maiden hair Fern, Venus hair, Rock Fern;

Tam.: Seruppadai;

Folk.: Mubaaraka.

Distribution: Widely distributed in temperate and tropical Himalayas; 1,800-2,700m.

Description: Ferns possessing underground rhizome covered with flat, multiseriate, lanceolate palea (scales) and circinately coiled young leaves. Fronds bipinnate; petioles dark shining and black with median groove. Sporangia borne on reflexed outgrowths of fertile pinnules, leptosporangiate with tetrahedral, triangular, smooth walled spores.

Flowering & Fruiting: July-October.

Habitat Ecology: Damp shady places on hill slopes and roadsides; Hira Nagar, 900m.

Material Examined: HPU-EL-SKS-ABS-15014; 04.06.2010.

Parts Used: Rhizomes. Leaves.

Folk Uses: Poultice of grounded rhizomes in mustard or sesame oil (*Sesamum indicum* L.) massaged for joint pains* and rheumatism*. Also, rhizomes cooked as vegetables* and pickled* by native people. Leaves used as fodder* and eaten raw for cough.

Active Constituents: Flavonoid glucosides including rutin, isoquercetin, astragalin, kaempferol, hydroxycinnamic acid esters and terpenoids, adiantone (*whole plant*) have been isolated.

Uses in Literature: Used so far as antibacterial, anticancerous, antifungal, antitussive, antiviral, aphrodisiac, astringent, demulcent, expectorant, emmenagogue, emollient, hypoglycaemic, purgative, stimulant, tonic; and for bronchial disorders, chronic catarrh, cold, cough, fever, hard swelling and hard tumours of spleen, imposthumes of uterus, lever and other viscera (Chopra *et al.*, 1956; Chunekar & Hota, 1998; Chunekar & Pondel, 1999; Dixit, 1974; Khare, 2007; Kirtikar & Basu, 1994; Kumar *et al.*, 2003; Parihar & Parihar, 2006; Vyas, 1987).

Adiantum lunulatum Burm. (Pl. 24C).

Syn.: *A. philippense* L.

Family: Adiantaceae.

Vern.: Hansraj.

Ayurvedic, English and Regional Names:

Ayur.: Hansapaadi, Hansapadi, Hansaraaja, Kitamaata, Raktapaadi, Tripaadikaa;

Eng.: Black Maiden hair, Walking Maiden hair Fern;

Tam.: Seruppadai.

Distribution: Throughout the greater part of India up to 1,200m.

Description: Ferns having underground, ascending rhizomatous stem densely covered with fibrous roots, scales and leaf bases and unipinnate compound, circinately coiled leaves and black rachis. Sori 1-3, large roundish placed in a distinct hollow on the upper edge.

Flowering & Fruiting: July-October.

Habitat Ecology: Shady and moist slopes, edges of ravines; Sujanpur Fort, 567m.

Material Examined: HPU-EL-SKS-ABS-15015; 04.09.2010.

Parts Used: Whole Plant. Leaves. Rhizomes.

Folk Uses: Dried plant used in fear psychosis*. Leaves given as fodder*. 5-10ml decoction of rhizomes given twice a day to cure fever, whereas its poultice prepared in mustard or 'til' oil massaged to check joint pains*.

Active Constituents: Hopane and isohopane type terpenoids, lupane and norlupane triterpenoids (*whole plant*) have been isolated.

Uses in Literature: Known to be used as alexiteric, antidysentric, emetic, febrifugal, soothing agent in erysipelas; and for asthma, atrophy, bleeding from nose, cough, diseases of blood, dysentery, emaciation, fever, hair fall, indigestion, leprosy, muscular pains, strangury, ulcers and urinary diseases (Caius, 1935; Chopra *et al.*, 1956; Chunekar & Hota, 1998; Chunekar & Pondel, 1999; Dixit, 1974; Khare, 2007; Kirtikar & Basu, 1994; Parihar & Parihar, 2006; Sharma & Vyas, 1985; Vyas, 1987).

Major Ayurvedic Formulations: Maanasamitra Vataka prescribed for mental disorders.

Aegle marmelos (L.) Correa *ex* Roxb. (Pl. 24D).

Syn.: *Feronia pellucida* Roth., *Crateava marmelos* L., *C. religiosa* Ainslie.

Family: Rutaceae.

Vern.: Bel.

Ayurvedic, English, Sanskrit and Regional Names:

Ayur.: Bilva, Mahaakapitha, Shaandilya, Sadaaphala, Shailuusha Shalya, Shriphala;

Eng.: Bael fruit tree, Bengal Quince, Holy fruit, Indian Quince, Stone Apple;

Sans.: Asholam, Adhavaruha, Atimangaliya, Bilva, Duraruha, Mahakapithakya, Malura, Shandilya, Satyadharma, Tripatra;

Ass.: Bel;

B.: Bela, Vilva;

Bo.: Bila;

G.: Bil, Billy;

H.: Bel, Bili, Siriphal, Sirphal;

Kan.: Bilvapatra;

M.P.: Bael;

Mal.: Kuvulam, Mavilạmu;

Mar.: Bael, Bel;

Oriya: Biliva, Sripholo;

Tam.: Alvuvigam, Iyalbudi;

Tel.: Bilvamu, Sailushamu.

Distribution: Throughout India.

Description: Medium-sized glabrous trees armed with sharp straight axillary thorns, 3 glabrous, ovate-lanceolate leaflets and greenish-white, sweet scented flowers in short axillary panicles. Berries globose, yellowish and having woody rind.

Flowering & Fruiting: April-July.

Habitat Ecology: Wastelands, open slopes, drier areas, temple courtyards; Tal, 861m.

Material Examined: HPU-EL-SKS-ABS-15016; 21.04.2008.

Parts Used: Leaves. Fruits. Gum. Wood. Stem Bark.

Folk Uses: 1-3g powdered leaves taken on empty stomach with luke warm water for 15-30 days to cure diabetes; also these ('bilpatri') are used in death rituals* and offered to appease Lord Shiva. Fruit juice consumed to check constipation. Ripe fruits edible and considered refrigerant; also used for making 'murraba' (preserve) and pickles Fruit gum used as an adhesive; 10g of its gum in combination with equal quantity of 'gur' (jaggery) taken twice a day against dysentery. 2-5g of powdered stem bark with honey given once daily for asthma* and jaundice. Wood used in 'hawan'*.

Active Constituents: Main constituents are aegeline, β-sitosterol (*leaves*), umbelliferone, dictamnine, xanthotoxol, xanthotoxin, scoparone, isopimpinellin, isoimperation, N-2-methoxy 2 (4 methoxy phenyl), ethyl cinnamamide, marmelin, bergapten, marmesin, osthol and auraptin (*pericarp*).

Biological Activity: Hypoglycaemic activity +ve.

Uses in Literature: Used earlier as abortifacient, adhesive, appetiser, astringent, beads, candy, cardiac depressant, cementing mixture, fish poison, fuel, ingredient of 'chywanprash', laxative, psychoactive, refrigerant, sacred, scorbutic, soaps, soothing agent, substitute for tea, sweet drinks, tannin and varnish, tree motif of Shiva temples, wish granting tree; and for abdominal accummulation of cellular tissue accompained with constipation and jaundice, abdominal pain for abscess, allaying urinary troubles, anthrax (vet.), boils (vet.), bowel complaints, breast pains, bronchitis, burning sensation of soles of the feet, catarrhs, chest pains, cholera, cold, convulsions, diabetes, diarrhoea (vet.), dyeing, dysentery, edible purposes, eye complaints, gastric trouble, headache, heart palpitation, hyperacidity, indigestion, injuries, irritation of alimentary canal, jaundice, joint pains (vet.), keeping evil spirits away, menstrual disorders, mouth sores, nausea, night fever, paintings (Mithila), skin infections, spermatorrhoea, sunstroke, toys, tympany, vomiting and wounds (Ambasta, 1986; Aminuddin & Girach, 1993; Bajpayee & Dixit, 1996; Bhatt *et al.*, 2001; Chowdhury *et al.*, 1996; Datt, 1996; Dwarakan & Ansari, 1996; Goud & Pullaiah, 1996b; Huidrom, 1996; Jain *et al.*, 1994; Jha & Basak, 1994; Kadel & Jain, 2006; Katewa & Arora, 1997; Khare, 2007; Kothari & Moorthy, 1996; Kshirsagar & Singh, 2000; Kurian, 1995; Lalramnghinglova, 1996; Manandhar, 1990; Masih, 2003; Mehrotra *et al.*, 1995; Mishra *et al.*, 1996a; Mohanty *et al.*, 1998; Negi & Gaur, 1991; Painuli & Maheswari, 1996; Patole & Jain, 2002; Pullaiah & Kumar, 1996; Rama Rao & Henry, 1996; Reddy *et al.*, 1998; Retnam & Martin, 2006; Sen & Behera, 2003; Sharma & Singh, 2001; Singh, 2003; Singh & Rao, 2003; Singh *et al.*, 1997; Srivastava *et al.*, 2003; Tirkey *et al.*, 2003a,b; Venkataratnam & Venkata Raju, 2004; Vihari, 1995; Watt, 1972).

Major Ayurvedic Formulations: Amritarista, Bilva-Churna, Bilvadi-Kwatha, Bilvadi-leha, Bilvadi-majadi-yog, Bilwamooltwak, Brhatgangadhra-churna, Chywanprasha, Dasmula-taila, Dhanyapanchka-kvatha, Gangadhara-churna, Maha-narayan-taila, Pushyanuga-churna, Vatsakadi-kvatha.

Trade: Dried pulp roots and 'belgiri' fruits sold in the drug markets of India.

Aerva lanata (L.) Juss. *ex* Schult. (Pl. 25A).

Syn.: *Aerva floribunda* Wight, *Achyranthes villosa* Forsk., *Illecebrum lanatum* L.

Family: Amaranthaceae.

Vern.: Bhadra, Chaya.

Ayurvedic and Regional Names

Ayur.: Aadddanpaaki, Gorakshaganjaa, Paashaanabheda, Shatkabhedi;

Tam.: Sirupeelai;

Folk.: Paashaanabheda.

Distribution: Warmer parts of India, up to 1,000m.

Description: Hoary-tomentose undershrubs reaching upto 90cm and emitting camphor like aroma from roots. Spikes long, densely woolly forming subglobose clusters. Flowers dioecious enveloped in long wool. Utricles indehiscent; seed inverse, testa coriaceous.

Flowering & Fruiting: May-August.

Habitat Ecology: Common in dry and wasteplaces; Pucca Bharo, 840m.

Material Examined: HPU-EL-SKS-ABS-15017; 28.05.2008.

Parts Used: Whole Plant. Leaves. Roots.

Folk Uses: 1-3g of powdered plant given with luke warm water (twice daily for 5 days) for diarrhoea. Plant also used for performing 'hawan' and in religious ceremonies*. Leaves cooked as spinach and also added to flavour soups. Gargles of decoction of leaves prescribed for cough* and sore throat*. Leaf sap used in eye complaints. Poultice of flowers and roots checks headache, whereas that of roots good against snake bite. 5-10ml decoction of its flowers and of roots given twice daily for 15 days for kidney stones*. Branches used for hanging at the entrance to ward off evil spirits.

Active Constituents: β-sitosterol, α-amyrin and palmitic acid (*whole plant*).

Biological Activity: Plant considered lithontriptic and diuretic.

Uses in Literature: Known so far as anthelmintic, antidiarrhoeal, anticholerin, bechic, diuretic, demulcent; and for backache, catarrh of bladder, child birth, diarrhoea, eye complaints, gonorrhoea, headaches, hepatitis, lithiasis, sores, soups and vegetable, smoke inhalation, snake bite, stones in stomach, strangury, urethral discharges, wash babies that have become unconscious from malaria and warding off evil spirits (Chopra *et al.*, 1956; Chunekar & Hota, 1998; Chunekar & Pondel, 1999; Freedman, 1998; Khare, 2007; Kirtikar & Basu, 1994).

Agave americana L. (Pl. 25B).

Syn.: *A. cantula* Roxb.

Family: Agavaceae.

Vern.: Jangali keora, Kuer Chatyan.

Ayurvedic, English, Sanskrit and Regional Names:

Ayur.: Kaantala;

Eng.: American Aloe, Carata, Century plant;

Sans.: Kantala;

B.: Banskeora, Bilati-ananash, Bilatipat, Carata, Koyam;

G.: Fangli-kunvara;

H.: Banskeora, Barakanwar, Hatisengar, Kantala, Rakas-pattah, Ramkanta;

Kan.: Bhuttale, Budukattalenaru;

Mal.: Panam, Katrazha;

Mar.: Parkand;

Tam.: Anaik-kat-razhai, Pithakalabuntha;

Tel.: Rakshimatalu.

Distribution: Cultivated as hedge plant throughout India.

Description: Shrubby rhizomatous stout plants having variegated, narrow lanceolate, spine tipped leaves with spongy margins and in lux rosette having curved tip and marginal hooked teeth. Inflorescence huge, terminal and spicate with dull greenish-yellow flowers in large branched cluster. Fruits loculicidal capsule or berry; seeds black, fleshy and endospermic.

Flowering & Fruiting: March-May.

Habitat Ecology: Wastelands, roadsides, drier areas; Tauni Devi, 1,050m.

Material Examined: HPU-EL-SKS-ABS-15018; 15.05.2009.

Parts Used: Leaves. Shoots.

Folk Uses: Planted as an hedge and the plant also used for thatching roof. Leaves used to tie bundle of grasses and as fodder for animals. Paste of its crushed rhizome, turmeric and rock salt tied with bamboo sticks to set fractured bones. Mucilage of leaves applied to heal external wounds*. Tender shoots ('mannu') edible. Fibres extracted from leaves by keeping in water for 7 days used for ropes and cordages.

Active Constituents: *Leaves* contain hecogenin, cholrogenin, 10 steroidal saponins, agavoside A, B, C, D, E, F, G, H, I, glucose, galactose, xylose and rhamnose.

Biological Activity: Antibacterial and anticancerous activities found to be +ve.

Uses in Literature: Known so far as anthelmintic, antisyphilitic, diuretic, fire wood, fish poison, fodder, vegetable; and for abortion, boils, burns, constipation, cuts, dropsy, dysentery, earache, exoparasite (vet.), expelling worms, fencing, fever, goitre, mats and cordage, paralysis, skin diseases, skin diseases (vet.), syphilis, thatching roofs, venereal sores, veterinary diseases, waist pain, warts, warts (vet.) (Ambasta, 1986; Chopra *et al.*, 1956; Goel *et al.*, 1984; Jain, 1991; Jain *et al.*, 1991; Kapur, 1996; Khare, 2007; Kumar & Jain, 1998; Lalramnghinglova, 1996; Maliya & Singh, 2003; Mittre, 1981; Punjani, 1998; Rana *et al.*, 1996; Rao, 1996; Reddy *et al.*, 1998; Rekha *et al.*, 1998; Saxena & Vyas, 1981, 1983; Singh, 1996b; Watt, 1972).

Adulterant or Substitute: Extr. of roots and leaves used as an adulterant or substitute for Sarsaparilla (*Smilax* spp.).

Ageratum conyzoides L. (Pl. 25C).

Syn.: *A. cordifolium* Roxb.

Family: Asteraceae.

Vern.: Phulanu.

Ayurvedic, English and Regional Names:

Ayur.: Dochunty, Sahadevi, Uchunti;

Eng.: Goat weed, White weed;

B.: Dochunty;

Bo.: Osari, Sahadevi;

G.: Ajgandha, Gandharisedardi, Mankdamari;

M.P.: Koobhi;

Mal.: Appa, Muryanpacha;

Mar.: Ghaneraosadi;

Oriya: Boksunga, Poksunga.

Distribution: Throughout India.

Description: Erect hairy annual herbs having opposite, petaloid, ovate, crenate leaves and small purple flower heads in dense terminal corymbs. Bracts striate, acute. Ray flowers many. Achenes black, pappus awned.

Flowering & Fruiting: July-August.

Habitat Ecology: Edges of cultivated areas, wasteplaces; Awah Devi, 1,236m.

Material Examined: HPU-EL-SKS-ABS-15019; 22.07.2009.

Parts Used: Whole Plant. Leaves.

Folk Uses: Paste of leaves applied as poultice on honey bee bites*; also its juice applied to check bleeding due to cut injuries. Plant before flowering used as fodder*; its extr. also used for repelling insects.

Active Constituents: These are phenols (eugenol) 5% and phenol ester (*leaves, flowers*), friedelin, β-sitosterol, stigmasterol, α-sphinasterol, coumarin, flavonoids and anti-gonadotropic hormone (*plant*).

Biological Activity: Antibacterial, anthelmintic and anticancerous activities confirmed.

Uses in Literature: Reported earlier as antilithic, antiseptic, haemostatic, insecticidal, nerve tonic, shampoo, symbolic; and for acclerating delivery, body swellings, boils, burns, cancer, cuts, diarrhoea, dysentery, fever, flatulence, headache, healing, kidney stone, leprosy, muscular pain, piles, prolapsed anus, ringworm, scabies, sores, skin diseases, tetanus, uterine disorders, wounds and edible purposes (Ambasta, 1986; Augustine & Sivadasan, 2004; Baruah & Sarma, 1984; Chhetri, 2005; Jain, 1991; Khare, 2007; Kumar & Jain, 1998; Samwatsar & Diwanji, 1996a; Sikarwar, 2002; Sinha, 1986; Upadhyay & Singh, 2005).

Adulterant or Substitute: Leaves used as substitute for that of 'ayapana' (*Eupatorium triplinerve* Vahl.).

Albizia lebbeck Benth. (Pl. 25D).

Syn.: *A. latifolia* Boivin., *Acacia speciosa* Willd., *A. sirisa* Ham., *A. lebbeck* Willd., *Mimosa speciosa* Jacq.,M. *sirisa* Roxb.,M. *lebbeck* L.

Family: Fabaceae.

Vern.: Sirin.

Ayurvedic, English, Sanskrit and Regional Names:

Ayur.: Bhandi, Bhandila, Kapitana, Mridupushpa Shirisha, Shitapushpa;

Eng.: Parrot tree, Siris tree, Sizzling tree;

Sans.: Barhapuspha, Bhandi, Bhandika, Kapitaka, Kapitana, Krishnasiris, Lomashapushpaka, Murdhapushpa, Plavaga, Shankiniphala, Vishavghati, Vishahanta;

B.: Siris, Sirisha;

Bo.: Garso, Harreri, Siris;

G.: Kaliosavas, Kalokansakio, Kalosadasadu, Kaloshirish, Pilosarshio;

H.: Garso, Kalshish, Kalsir, Lasrin, Mathirshi, Shiris, Shirish, Sirai, Sirin, Tantia;

Kan.: Baggera;

M.: Vagu;

Mal.: Kattuvaka, Nenmani, Vaku;

Mar.: Chichola, Kalashiras, Mothasiras, Shirish;

Oriya: Vaghe, Kotvaghe;

Tam.: Adukkavajai, Kalindi, Kattuvagai, Pandil, Perunvagai, Piram, Undugam;

Tel.: Dirisan.-amu, Girisamu, Kapitanamu, Paddadirisan-amu, Sirishamu, Siriso.

Distribution: Ascending throughout India upto 1,650m.

Description: Unarmed deciduous trees having bipinnate leaves, white, fragrant flowers in globose umbellate heads and leaves with a large gland near the base of the main petiole. Pinnae 4-8, with or without a gland between the lowest. Pods strap-shaped, pale-yellow, smooth shining and 6-10 seeded.

Flowering & Fruiting: April-May.

Habitat Ecology: Wasteplaces, edges of cultivated fields; (frequently planted); Dhaneta, 596m.

Material Examined: HPU-EL-SKS-ABS-15020; 06.04.2009.

Parts Used: Leaves. Wood. Gum. Twig.

Folk Uses: Wood employed for making furniture, body of trucks*, house floors* and agricultural implements (plough, 'jungla'). Gum of the plant added in 'halwa'* (sweet pudding) preparation and considered good for ladies after delivery for strength*. Leaves used as fodder and twigs as fire wood.

Active Constituents: Flavandiols, melacacidin, lebbeacacidin (*Heartwood*), d-catechin, leucocyanidin, condensed tannins (*bark*), vicenin-2, ã-cynoutrin, rutin, myricitin, robibin and albegenin (*leaves*) are the main ingredients.

Biological Activity: Antiasthmatic and antianaphylactic activities found to be +ve.

Uses in Literature: Reported in literature as an alternative, antidote to poison, aphrodisiac, astringent, camel fodder, detergent, expectorant, fire wood, psychoactive plant, restorative, symbolic (god of love uses its flowers on arrow of desires), tannin, tonic; and for agricultural implements, arthritis, asthma, boats, boils, carriages, cataract, cold, carbuncle, cooling, cough, diarrhoea, dyeing, dysentery, edible purposes, eruptions, eye troubles, fever (vet.), furniture, gonorrhoea, gum ulcers, house building, infected teeth, interior decoration, leather, leprosy, leucoderma, night blindness, oil mills, picture frame, piles, railway carriage works, reducing enlargement of cervical glands, scorpion sting, snake bite, sugar cane crushers, swelling, toy manufacturing, tympany (vet.) and ulcer (Ambasta, 1986; Chauhan, 1999; Gill & Nyawwame, 1994; Goud & Pullaiah, 1996b; Jain, 1991; Jain & Patole, 2001; Khanna *et al.*, 1996b; Khare, 2007; Kurian, 1995; Raghupathy & Mahadevan, 1991; Ram *et al.*, 2002; Rama Rao & Henry, 1996; Siwakoti & Verma, 1996; Srivastava *et al.*, 2003; Watt, 1972).

Major Ayurvedic Formulations: Deshanglepa, Mahasirisagad, Sirisarishata, Sirisavlena.

Albizia stipulata Boiv. (Pl. 26A).

Syn.: *A. chinensis* (Osbeck) Merr., *Acacia stipulata* DC., *A. smithiana* Wall., *Mimosa stipulata* Roxb.,M. *smithiana* Roxb.,M. *chinensis* (Osbeck) Merr., *Arthrosprion stipulatum* Hassk.

Family: Fabaceae.

Vern.: Oyaee.

Hindi and Regional Names:

Ass.: Sau, Sow;

B.: Chakua;

Bo.: Udala;

H.: Sirin;

Kan.: Hotta bage, Kal-boge;

Mal.: Potta vaga;

Mar.: Laeli;

Oriya: Reyi, Gudanaudia;

Pb.: Oe, Oi, Shirsha, Sirin, Kasir;

Tam.: Katturanji, Pili vagei;

Tel.: Chindaga, Chiragu, Konda.

Distribution: Throughout India.

Description: Trees with pubescent or hairy leaves, pink and large-sized stipules, and yellow flowers in panicled heads. Pods glabrous, indehiscent, 8-10 seeded.

Flowering & Fruiting: May-June.

Habitat Ecology: Frequent along roadsides, field borders, cultivated areas; Nadaun, 483m, Anu, 858m.

Material Examined: HPU-EL-SKS-ABS-15021; 03.05.2011.

Parts Used: Leaves. Branches. Root. Bark. Wood.

Folk Uses: Leaves used as fodder. Root bark mixed with hot mustard oil applied as poultice for checking baldness*. 2-4g mixture of powdered bark in combination with 25g 'mulathi' (*Glycyrrhiza glabra* L.) given twice daily for 10 days to restore normal functioning of liver*. Also, its bark and roots serve as important constituents of local wine* 'sur'. Wood used as timber for construction* and agricultural implements. Dried branches used as fuel.

Active Constituents: These are saponin, beta-sitosterol, tannins.

Biological Activity: 50% EtOH extr. of aerial parts spasmogenic and diuretic. Seeds showed marked hypoglycaemic activity in normal rats. Aq. extr. of bark acts as uterine constrictor.

Uses in Literature: So far reported in India for agricultural implements, boxes, building, cart wheels, cuts, furniture, scabies, sizing daphne paper, skin diseases, tea boxes, turnery articles, wooden bells; and as cattle fodder, fuel and gum (Agarwal, 2003; Ambasta, 1986; Khare, 2007; Kirtikar & Basu, 1994; Sharma & Rana, 2005; Sood & Thakur, 2004; Watt, 1972).

Allium cepa L. (Pl. 26B).

Family: Lilaceae.

Vern.: Piaz.

Ayurvedic, English, Sanskrit and Regional Names:

Ayur.: Durgandh, Palaandu;

Eng.: Onion;

Sans.: Basl, Dirghapatra, Mahakanda, Palandu, Rajapriya;

Ass.: Piaz;

B.: Piyas;

Bo.: Kanda;

H.: Piaj, Palandu;

Kan.: Bawang;

Mal.: Lund;

Mar.: Kanda;

Tam.: Vulli-gaddalu, Nirulli;

Tel.: Vengayam, Nirulli, Kunballi.

Distribution: Cultivated throughout India.

Description: Perennials with thick globular bulb having membranous tunic. Leaves subdistichous fistular shorter than the tall scape. Head globose, many and dense fid. Flowers white, many. Seed 1-2 in each cell, black.

Flowering & Fruiting: March-April.

Habitat Ecology: Cultivated; Jhaniari, 787m.

Material Examined: HPU-EL-SKS-ABS-15022; 05.04.2008.

Parts Used: Bulb. Leaves. Seeds.

Folk Uses: One tsp juice of the bulb along with *Zingiber officinale* Rosc. given twice a day for vomiting. Poultice of leaves or roasted bulb applied to cure boils, pimples and acne*, also leaves burnt by natives to expel mosquitoes* and snakes*. 5-7 drops of bulb extr. used as ear drops to check ache. One tsp infusion prepared from its bulbs given thrice daily for fifteen days for expelling intestinal worms, digestive disorders* and as diuretic; also eaten raw for curing malaria* and flavouring food stuff. Seeds used in pickles.

Active Constituents: Diosgenin (0.18%), (+) S-methyl-L-cystein-sulphoxide and (+) S-propyl-L-cystein sulphoxide, propenyl sulphonic acid and lipooxygenase metabolites of linoleic acid, isomeric 9, 10, 13 trihydroxy-11-octadecenoic acid (*bulb*) have been isolated.

Biological Activity: Antiperiodic, antibacterial, aphrodisiac, and antitumour activities exhibited.

Uses in Literature: So far known as anthelmintic, appetiser, bactericidal, diuretic, emetic, expectorant, heart stimulant, pungent, spice, tonic, vegetable; and for asthma, bad taste of mouth, bleeding piles, body pain, boils, cold, diarrhoea, dysentery, earache, flatulence, flavouring, foot sores, impotency, indigestion, opthalmia, pickling, pimples, promoting bile production, pyorrhoea, scabies, soup, urinary discharge and unconsciousness (Ambasta, 1986; Kadel & Jain, 2006; Katewa *et al.*, 2001; Khanna & Kumar 2000; Khare, 2007; Mohanty *et al.*, 1998; Pandey & Verma, 2002; Retnam & Martin, 2006; Samati, 2004; Sharma & Singh, 2001; Viswanathan *et al.*, 2001).

Allium sativum L., Roxb. (Pl. 26C).

Family: Liliaceae.

Vern.: Lasun.

Ayurvedic, English, Sanskrit and Regional Names:

Ayur.: Lashuna, Rasona, Yavaneshta, Ugragandha, Mahaushadh, Arishta;

Eng.: Garlic;

Sans.: Sum, Sir;

Ass.: Rasun;

B.: Naharu;

G.: Shunam;

H.: Rasun, Lasun, Lashan;

Kan.: Gokpas;

Mar.: Lasan;

Tam.: Velluli-fella-gadda;

Tel.: Belluli.

Distribution: Cultivated all over India.

Description: Erect perennials with flat leaves, slender scape, and terminal umbel. Flowers small, perianth bell-shaped, 6-parted. Spathes long-beaked, heads bearing bulbils and flowers. Sepals lanceolate and acuminate. Seeds 1-2 in each cell, compressed, testa black.

Flowering & Fruiting: March-April.

Habitat Ecology: Cultivated; Jhaniari, 900m.

Material Examined: HPU-EL-SKS-ABS-15023; 17.04.2010.

Parts Used: Bulb.

Folk Uses: Bulb of garlic burnt in mustard oil used as drops to cure earache. Its cloves consumed raw in the morning or ½ spoon of its juice diluted with 4 spoons of water and a pinch of salt given to cure stomachache and stomach disorders*. Gargles of ½ spoon of its powdered bulb check tonsilitis*. Paste of cloves applied locally for joint pains*, head scalds*, dandruff* and hairfall*.

Active Constituents: These are allicin (2-propenyl propane thiosulphate), allin, allinase, scorclinins, selenium, vit. A, B, C & E (fresh *garlic* oil), allitin, sativin I & II, seordinine, scordine, anthocyanins, glycosides of kaempferol and querectin and essential oil.

Biological Activity: Aq. garlic extr. (2ml/ 100mg), significantly increases level of blood serum urea (6.27 vs 4.1 m mol/I) and activity of blood serum aspartate amino transferase (130 vs 120 units) in rats as compared to control oil.

Uses in Literature: Used earlier for arthritis, ascites, brain diseases, duodenal ulcer, earache, epilepsy, fattening, gangrene of lung, hepatitis, laryngeal tuberculosis, lumps, malaria, psychic disorders, pthisis, ringworms, respiratory disorders, scorpion bite, sexual potency, skin trouble, whooping cough, wounds; and as anthelmintic, antibiotic, antithrombic, aphrodisiac, colic, condiments, digestive, diuretic, spice and tonic (Ambasta, 1986; Borthakur *et al.*, 2004; Gogoi & Borthakur, 2001; Kadel & Jain, 2006; Khare, 2007; Retnam & Martin, 2006; Samati, 2004; Sen *et al.*, 2002; Sen & Behera, 2003; Sharma & Singh, 2001; Sharma *et al.*, 2001).

Major Ayurvedic Formulations: Rasan-vati, Rason-pind, Rasonashtka, Lahsunadyaghrita, Garlic capsule, Garlic pearls.

Aloe barbadensis Mill. (Pl. 26D).

Syn: *Aloe abyssinica* Lam., *A. chinensis* Baker, *A. indica* Royle, *A. perfoliata* L. *A. vulgaris* Lam.

Family: Liliaceae.

Vern.: Ghi-kawer, Kuarian.

Ayurvedic, English and Regional Names:

Ayur.: Eleyaka, Ghritkumaarika, Grihkanyaa, Kanyaa, Kanyaasaara, Kumaari, Kumaarikaa;

Eng.: Broad-leaved aloe, Soap aloe;

H.: Ghee Kunwar.

Distribution: Kerala, Tamilnadu & other warmer regions of India below 1,500m.

Description: Erect succulent herbaceous shrubs having lanceolate-ovate leaves and upper surface pale-dark green with white blotches in comparison to lower ones. Capsule oblong. Seeds blackish-brown.

Flowering & Fruiting: June-January.

Habitat Ecology: Escape from cultivation, dry wastelands; Khaggal, 920m.

Material Examined: HPU-EL-SKS-ABS -15024; 03.07.2008.

Part Used: Fleshy Leaves.

Folk Uses: Leaves used as rope* to tie bundles of grasses. Leaf pulp or juice applied externally in burns and white paches on skin. 10 ml extr. of leaves given once a day for 3 days in pain during menstruation*. Fleshy leaves utilised for making preserve ('murraba'*) and 'gulkand'. Deeply cut leaves filled with salt given to animals to check dysentery and its decoction (1-2 litre) given once daily to check constipation* and dehydration*. Pakoras of its flowers or its 1tsp powder twice a day, given for curing diabetes*.

Uses in Literature: Reported earlier as an anti-inflammatory, antimicrobial, digestive (vet.), emollient emmenagogue, purgative; and for burn (vet.), delivery (vet.), face packs, healing wounds, indigestion (vet.), insect bite (vet.), making cosmetics, mastitis (vet.), miscarriage (vet.), pickles, skin diseases (vet.), sores (vet.), swelling (vet.) and wound (vet.) (Bhatt *et al.*, 2001; Borthakur & Sarma, 1996; Chopra *et al.*, 1956; Geetha *et al.*, 1996b; Kadel & Jain, 2006; Khare, 2007; Kirtikar & Basu, 1994; Kiruba *et al.*, 2006; Mishra *et al.*, 1996b; Reddy & Raju, 2000; Sudarsanam *et al.*, 1995; Varshney, 1998; Watt, 1972).

Aloe vera (L.) Webb & Benth. (Pl. 27A).

Syn.: *A. barbadensis* Mill., *A. perfoliata* var. *vera* L.

Family: Liliaceae.

Vern.: Ghi-kawer, Kuarian.

Ayurvedic, English, Sanskrit and Regional Names:

Ayur.: Ghritkumaarika, Kanyaasaara;

Eng.: Aloe, Barbado, Alo, Common Indian, Aloe, Curcago Aloe, Jaffarabad Aloe;

Sans.: Adala, Ajara, Amara, Ambudhisrava, Aphala, Atipichhila, Bahupatri, Brahmaghni, Grihakanya, Kantakapravrita, Sthuladala, Sukantaka, Tarani, Vipulsrava;

Ass.: Chaikuwari;

B.: Ghirtakumari;

G.: Kadvikunwar, Kunvar, Kuvara;

H.: Ghiguvara, Ghikumari, Gvarpatha, Kumari, Kuvarapatha;

Kan.: Loli-Sara;

Mal.: Cheninayakam, Cherukattala, Ghirtakumari, Kaltala, Kumari;

Mar.: Koraphada, Koraphanta, Korkand;

Oriya: Ghikumari, Ghritokumari, Kumari;

Tam.: Angani, Angini, Ettakalavanta, Kalavandu, Musamvaramu, Sottukkattala.

Distribution: Warmer regions of India below 1,500m; planted in Indian gardens.

Description: Stoloniferous, perennial shrubs having rosette of leaves arising from the ground, and large succulent, spinose leaves. Flowers vermillion coloured arising in racemes. Fruits a loculicidal capsule. Seeds albuminous.

Flowering & Fruiting: March-April.

Habitat Ecology: Wasteplaces, cultivated along edges of field borders; Chabutra, 797m.

Material Examined: HPU-EL-SKS-ABS-15025; 21.07.2010.

Parts Used: Leaves. Plant.

Folk Uses: Plant regarded sacred and its garland hung at the entrance of newly constructed house to ward off evil spirits. 500-600ml decoction prepared from its leaves, 'ajwain' (*Trachyspermum ammi* L.) and black pepper (*Piper nigrum* L.) given to cattle for 7 days to cure stomach disorders. Poultice of leaves considered good for joint pains; and its pulp applied to heal boils, burns and skin patches. Pulp of leaves consumed raw for effective treatment of piles and rectal fissures*. Also, the pulp roasted in 'desi' ghee given to cure 'pitta' disease (liver ailments) and considered refrigerant. Decoction of leaves given to animals to check constipation and dehydration* and leaves given raw to cure dysentery.

Active Constituents: Chief constituents are barbaloin, aloesin, aloesone, and β-sitosterol (*leaves*) and 4 partially acetylated glucomannans, β-barboline and aloeanodin (*jelly*).

Biological Activity: Barbaloin inhibits the growth of *Mycobacterium tuberculosis*. Plant extr. (100mg/kg) prevents implantation in 50 and 100% rats at oral dose of 100 and 200mg/kg. Supernatant of gel cytotoxic against C3 HIOTI/2 cells at intiation stage of transformation.

Uses in Literature: Reported to be used as an aperient in apoplexy, aromatic, carminative, cathartic, colic, coolant in fever, diaphoretic, diuretic, emmenagogue, fibre, flavouring agent, mastitis, refrigerant, stimulant, stomachic, tonic, vegetable, vermicidal; and for acute indigestion, arthritis, asthma, blisters, blunt injuries, boils (vet.), bone sprain, breast tissue hardening, bronchitis, burns (vet.), burnt part, checking growth of tumour, chest pain, cold, constipation, cough, dermatitis, digestion (vet.), dyeing, easy shedding of placenta during delivery (vet.), enlarged lymphatic glands, eye flu, falling hair, fencing, flatulence, foot swelling, fresh cuts and injuries, giddiness, gonorrhoea, headache, healing wounds, itching (vets.), jaundice, joint diseases, leprosy, leucorrhoea, liver disorders, menstrual complaints, miscarriage (vet.), obesity, ostisis in horses, piles, preparation of gin, relieving muscle pains, sexual vigour and vitality, skin diseases, spleen ailments, ulcer, washing hair (vet.), wounds and X-ray burns (Ambasta, 1986; Bhatt *et al.*, 2001, 2002; Borthakur & Sarma, 1996; Borthakur *et al.*, 2004; Chauhan, 1999; Datt, 1996; Dwivedi *et al.*, 2005; Geetha *et al.*, 1996a, b; Girach & Aminuddin, 1995; Gogoi & Borthakur, 2001; Gogoi *et al.*, 2003; Hosagoudar & Henry, 1996a; Jain, 1991; Jain & Patole, 2001; Jadhav, 2006; Kadel & Jain, 2006; Katewa *et al.*, 2001; Khanna & Kumar, 2000; Kiruba *et al.*, 2006; Kurian, 1995; Mishra *et al.*, 1996b; Punjani, 1998; Rama Rao & Henry, 1996; Reddy & Raju, 2000; Retnam & Martin, 2006; Saini, 1996b; Samwatsar & Diwanji, 1996b; Satapathy & Brahmam, 1996; Sen *et al.*, 2002; Sharma & Borthakur, 2003; Shome *et al.*, 1996; Silori & Rana, 2000; Singh & Pandey, 1996; Sudarsanam *et al.*, 1995; Varshney, 1998; Verma & Chauhan, 2006; Watt, 1972).

Major Ayurvedic Formulations: Aloes compound, Bringamaladi, Cukkumtipalyadi gutika, Kumaryadi-vati, Kumaryasava, Loharasayana, Rajarpravartini-vati, Rajpravartain-vati.

Trade: Fresh Leaves, Aloe Vera Gel, Aloe Vera Powder.

Alstonia scholaris R.Br. (Pl. 27B).

Syn.: *Echites scholaris* L.

Family: Apocynaceae.

Vern.: Dita bark, Devil tree, Shaitan ka per.

Ayurvedic, English, Sanskrit and Regional Names:

Ayur.: Saptaahvaa, Saptachhada, Saptaparna, Saptaparni Shaarada, Vishaaltvak, Vishamchhada;

Eng.: Ditabark, Devil's tree;

Sans.: Deva-Vriksha, Saptaahva, Saptaparna;

Ass.: Chatian, Satiana;

B.: Chatittin;

Bo.: Satvin;

Kan.: Chatiun;

Mal.: Pala;

Mar.: Satvin;

Oriya: Chhotina, Kumbaro, Soptorposi;

Tam.: Aclilampali, Pala;

Tel.: Aedakularite-chettu, Palgaruda.

Distribution: Throughout the moist regions of India and tropical parts of western Himalayas.

Description: Evergreen trees with tall trunk having buttressed base and whorled branches. Milky juice bitter. Leaves coracious, whitish beneath. Cymes umbellately branched. Follicles upto 70 cm.

Flowering & Fruiting: May-June.

Habitat Ecology: Wastelands in moist regions, cultivated as pollution absorber along the roadsides; Anu, 900m.

Material Examined: HPU-EL-SKS-ABS-15026; 15.05.2011.

Parts Used: Bark. Wood.

Folk Uses: 1-3g powdered bark given twice a day for 3 days against malaria, diarrhoea and dysentery, also as astringent. Wood used for making frames in construction*.

Active Constituents: Plant contains alkaloids echtamime, ditamine, ditaine and echitenines and several fatty and resinous substances.

Biological Activity: Bark antiperiodic, spasmolytic, antidysenteric. Echitamine inhibits *Plasmodium berghei.*

Uses in Literature: Considered useful as alternative to quinine, anticholeric, antidysenteric, antiperiodic, astringent, emmenagogue, febrifugal, hypotensive, spasmolytic, uterine stimulant; and for boards, beri beri, coffin making, corks, household utensils, internal fevers, malaria, net floats, slates, trenchers and vulnerary diseases (Ambasta, 1986; Jain, 1968; Khan & Khanum, 2005; Khare, 2007; Kirtikar & Basu, 1994).

Amaranthus paniculatus L. (Pl. 27C).

Syn.: *A. speciosus* Sims, *A. strictus* Willd., *A. sanquineus* L., *A. frumentaceus* Ham., *A. farinaceus* Herb., *A. anacardana* Ham., *A. flavus* var. *bracteatus* L.

Family: Amaranthaceae.

Vern.: Cholayee.

English, Sanskrit and Regional Names:

Eng.: Prickly Amaranth;

Sans.: Rahadri, Rajagiri, Rajashakini;

B.: Bathu, Chuko, Natya;

G.: Chuko, Rajaro;

H.: Chaulai, Chuamarsa, Ganhar;

Kan.: Kire soppu;

Kas.: Bustanafroz;

Mar.: Rajagira;

Tam.: Pungikirai.

Distribution: Throughout India.

Description: Robust tall herbs having striated stem, long-petioled elliptic or ovate-lanceolate finely acuminate leaves and red or yellow flowered dense spikes. Bracts long recurved. Seeds black.

Flowering & Fruiting: September-November.

Habitat Ecology: Cultivated fields, roadsides; Hamirpur, 787m.

Material Examined: HPU-EL-SKS-ABS-15027; 08.06.2010.

Parts Used: Whole Plant. Leaves. Seeds.

Folk Uses: Plant cooked as pot herb, considered good appetiser*, laxative and source of iron*. Also, its poultice applied against joint pains*. Seeds edible, eaten especially on fasts*. Powdered seeds (6g) in divided doses given with cold water thrice a day for 5 days to check diarrhoea.

Active Constituents: Reported to contain β-carotenes, ascorbic acid, folate, saponins (*plant*), hentriacontane and α-spinasterol (*leaves, stem*), oxalic acid, palmitic acid, stearic acid, oleic acid, linoleic and docosaenoic acid.

Uses in Literature: Considered useful in allergic asthma, allergic rhinitis, biliousness, colic pain, diarrhoea, diathesis, eczema, fever, gonorrhoea, leucorrhoea, menorrhoea, piles, scorpion and snake bites, sores, spermatorrhoea, warmth; and as antiseptic, astringent, blood purifier, diuretic, emollient, galactogenic, hemorrhagic, lactagogue, laxative, spasmolytic and vermifuge (Ambasta, 1986; Bondya *et al.*, 2006; Chandra Sekar & Srivastava, 2003; Chopra *et al.*, 1956; Gogoi & Borthakur, 2001; Khare, 2007; Kirtikar & Basu, 1994; Kshirsagar *et al.*, 2003; Kumar & Chauhan, 2005; Kumar & Goel, 1998; Maliya, 2004; Maya *et al.*, 2003; Mohanty *et al.*, 1998; Patil, 2004; Retnam & Martin, 2006; Upadhyay & Singh, 2005).

Amaranthus spinosus L. (Pl. 27D).

Family: Amaranthaceae.

Vern.: Chelri, Cholyai.

Ayurvedic, English, Sanskrit and Regional Names:

Ayur.: Alpamaarish, Meghnaad, Megharava, Tandulaka, Tanduliya, Vishaghn;

Eng.: Prickly Amaranth;

Sans.: Alpamarisha, Bahuvirya, Bhandira, Granthila, Kandakamarisha, Kandera, Meghananda, Sushaka, Tandula;

B.: Kanta notya;

G.: Kantanu-dant;

H.: Kataili chauli;

Kan.: Mulla harive soppu;

Mal.: Kattu-mullen-keera;

Mar.: Kante math;

Tam.: Mulluk-kirai;

Tel.: Mullatota-kura.

Distribution: Throughout India.

Description: Robust spinous annual herbs possessing straight spines, ovate-lanceolate leaves and greenish flowers in axillary, panicled slender spikes. Utricles rugose, acute, indehiscent. Seeds shining, oblong, black and compressed.

Flowering & Fruiting: May-June.

Habitat Ecology: Wastelands, cultivated fields, roadsides; Dimmi, 1,006m.

Material Examined: HPU-EL-SKS-ABS-15028; 08.07.2010.

Parts Used: Whole Plant. Leaves.

Folk Uses: Plant cooked as pot herb*, its poultice applied to check joint pains* and considered good appetiser*, laxative and source of iron*. Infusion of plant recommended in urinary troubles. Leaves used as fodder*.

Active Constituents: Reported to contain saponins (*plant*), hentriacontane and α-spinasterol (*leaves, stem*).

Biological Activity: Petroleum ether extr. juvenomimetic +ive.

Uses in Literature: Considered useful in abscesses, allergic asthma, allergic rhinitis, bruises, burns, colic pain, diarrhoea, eczema, fever, gonorrhoea, inflammatory swelling, leucorrhoea, menorrhagia, piles, scorpion and snake bites, sores, spermatorrhoea, warmth, wounds; and as antiseptic, diuretic, emollient, febrifuge, galactogenic, lactagogue, laxative, spasmolytic and vermifuge (Ambasta, 1986; Bondya *et al.*, 2006; Chandra Sekar & Srivastava, 2003; Gogoi & Borthakur, 2001; Khare, 2007; Kirtikar & Basu, 1994; Kshirsagar *et al.*, 2003; Kumar & Chauhan, 2005; Kumar & Goel, 1998; Maliya, 2004; Maya *et al.*, 2003; Mohanty *et al.*, 1998; Patil, 2004; Retnam & Martin, 2006; Upadhyay & Singh, 2005).

Amaranthus viridis L. (Pl. 28A).

Syn.: *A. caudatus* L., *Chenopodium caudatum* Jacq., *Euxolus viridis* Moq. *A. polystachyus* Ham., *A. fasciatus* Roxb.

Family: Amaranthaceae.

Vern.: Chelri, Cholyai.

Ayurvedic, English, Sanskrit and Regional Names:

Ayur.: Alpamaarish;

Eng.: Green or Wild Amaranth, Green Pigweed;

Sans.: Tanduliya,Vishaghna;

B.: Ban note;

G.: Dhindo;

Mal.: Bayam munyit;

Mar.: Lhanamat;

Tam.: Kuppai-keerai, Sinna-keerai;

Tel.: Chailaka thotakura.

Distribution: Throughout India.

Description: Erect glabrous herbs possessing grooved stems, long, petiolated ovate leaves and pale green flowers in axillary and panicled slender spikes. Sepals 3. Utricles rugose, acute, indehiscent. Seeds shining black.

Flowering & Fruiting: May-June.

Habitat Ecology: Wastelands, cultivated fields, roadsides; Lambroo, 850m.

Material Examined: HPU-EL-SKS-ABS-15029; 01.05.2011.

Parts Used: Whole Plant. Leaves. Root.

Folk Uses: Plant cooked as pot herb as a good source of iron and appetiser*. Leaves used as fodder. 10 ml root juice taken with 'amla' (*Emblica officinalis* Gaertn.) twice a day for 3-4 days to control excessive bleeding during menstruation*.

Active Constituents: Plant contains water, proteins, fat, carbohydrates, fibres, ash, minerals, calcium, phosphorus, magnesium, sodium, potassium, zinc, vitamins and saponins.

Biological Activity: Petroleum ether extr. juvenomimetic +ive.

Uses in Literature: Considered useful for acne, dysentery, edible purposes, inflammation. scorpion sting, snake bites, swelling, urinary trouble; and as cattle fodder, emollient, laxative, pot herb and vermifuge (Ambasta, 1986; Chopra *et al.*, 1956; Girach *et al.*, 1998; Khare, 2007; Kirtikar & Basu, 1994; Jain, 1991; Patole & Jain, 2002; Retnam & Martin, 2006).

Adulterant or Substitute: Young shoots used as an adulterant or substitute for *Asparagus* and *Spinach.*

Amorphophallus campanulatus (Roxb.) Blume *ex* Decne. (Pl. 28B).

Syn.: *Amorphophallus paeoniifolius* (Densst.) Nicolson, *Arum campanulatum* Roxb.

Family: Araceae.

Vern.: Zamikand.

English, Sanskrit and Regional Names:

Eng.: Elephant Food Yam;

Sans.: Arsaghana;

B.: Ban note;

G.: Suran;

H.: Arsaghana, Balukand, Zamikand;

Kan.: Surarna gerdde;

Mal.: Chenakarona-kizhana;

Mar.: Jangli suran;

Tam.: Kurruma kalungh;

Tel.: Kunda-muncha-kunda.

Distribution: Humid parts of E. & W. India; also cultivated in some parts of India.

Description: Cormous herbs flowering before leafing. Leaves lobed 3-partite, segments pinnatisect. Spathe various; limb campanulate, infundibular convolute or open, marcescent. Spadix hardly longer than spathe. Flowers unisexual. Fruits clustered, obovoid, red.

Flowering & Fruiting: April-June.

Habitat Ecology: Cultivated; Bhira, 780m; Bharmoti, 531m.

Material Examined: HPU-EL-SKS-ABS-15030; 25.06.2010.

Parts Used: Leaves. Tuber.

Folk Uses: Leaf juice used as antiseptic to clean wounds. Tuber used as vegetable* and also pickled*. 2-5g powdered tubers with honey taken once a day for 15-30 days to cure joint pains*, piles, abdominal pains, asthma and rheumatism*.

Active Constituents: Reported to contain tyrosinase, laccase betulinic acid, β-sitosterol, stigmasterol, lupeol, triacontane, glucose, galactose and xylose (*corm*).

Biological Activity: Tyrosinase exhibits mono-and di-phenolase activities specific for L-DOPA as substrate.

Uses in Literature: Known to be used as anthelmintic, antibacterial, antifungal, appetizer, carminative, cytotoxic, expectorant, stomachic, tonic; and for bruises, dysentery, liver and spleen enlargement, piles, stomach disorders, toothache and glandular enlargement (Ambasta, 1986; Ghani, 1998; Kirtikar & Basu, 1994; Khare, 2007; Khan *et al.*, 2007; Retnam & Martin, 2006).

Andrographis paniculata **Wall.** ***ex*** **Nees (Pl.28C).**

Syn.: *Justicia paniculata* Burm.

Family: Acanthaceae.

Vern.: Kalmegh.

Ayurvedic, English and Regional Names:

Ayur.: Bhuunimba, Bhuuminimbaka, Kaalmegha, Kaltanaatha, Kiraata-tikta, Vishwambharaa, Yavtikta,

Eng.: Creat;

Tam.: Nilavembu.

Distribution: Throughout India from Lucknow -Assam.

Description: Erect annuals having quadrangular stems and hairy, white, spotted, rose purple 2-liped corolla for atleast half its length. Capsule glandular, hairy glabrous.

Flowering & Fruiting: July-October.

Habitat Ecology: Cultivated; Neri, 381m.

Material Examined: HPU-EL-SKS-ABS-15031; 13.03.2008.

Parts Used: Whole Plant. Leaves. Shoots. Rhizomes.

Folk Uses: 15-30ml decoction of the plant given in fever, typhoid, jaundice and stomach trouble (twice daily, 3-10 days). Leaves and shoots eaten raw for worms in stomach*, cough*, cold*, diarrhoea* and respiratory troubles*.

Active Constituents: Andrographolide, deoxyandrographolide, diterpenes, Panicoline (*leaf, rhizomes*). 14-Deoxy-11-dehydroandrographolide, 14-Deoxy-11-oxoandrographolide, paniculide-A, B & C (*plant*).

Biological Activity: Alcoholic extr. of the plant exhibited antidiarrhoeal activity against *E. coli.*

Uses in Literature: Earlier reported as antiallergic antibiotic, antihepatitic, antiinflammatory, antimalarial, antipyretic, antisnake venom antithrombogenic, antityphoidal, bitter tonic, hepatoprotective, and immunostimulant agent (Ambasta, 1986; Calabrese *et al.*, 2000; Chopra *et al.*, 1956; Coon & Ernst, 2004; Khare, 2007; Kirtikar & Basu, 1994).

Anetham sowa Roxb. *ex* Flem. (Pl. 28D).

Syn.: *A. graveolens* L. var. *sowa* Roxb., *A. graveolens* DC., *Peucedanum sowa* Roxb., *Peucedanum graveolens* Benth.

Family: Apiaceae.

Vern.: Kaudi saunf.

Ayurvedic, English, Sanskrit and Regional Names:

Ayur.: Shataahvaa, Shatapushpa;

Eng.: Indian Dill, Sowa, Garden Dill, Anet;

Sans.: Madhura, Misroya, Satapushpa, Satapuspi, Shatapushpa;

H.: Anithi, Sotapa, Sowa, Soya;

Tam.: Sadakuppai.

Distribution: Throughout tropical and subtropical parts of India.

Description: Aromatic annual, glabrous herbs with hollow, finely grooved stem, pinnately compound leaves and yellow flowers in compound umbels. Fruits narrowly winged, 2-3 times as broad as thick.

Flowering & Fruiting: August-September.

Habitat Ecology: Cultivated; Bajroh fields, Bhoranj, 950m.

Material Examined: HPU-EL-SKS-ABS-15032; 10.08.2009.

Parts Used: Leaves. Seeds.

Folk Uses: Seeds used as condiment. 15-30ml decoction of the same given twice a day till cure for diarrhoea*. Powdered seeds mixed with 'ajwain' (*Carum copticum* Hiern.), 'methi' (*Trigonella foenum-graecum* L.) 'hing' (*Ferula asafoetida* L.) and dry ginger given to animals for stomach disorders.

Active Constituents: 5% volatile oil (carvone), flavonoids, coumarins, xanthones and triterpenes (*seeds*).

Biological Activity: Carvone exhibited antimicrobial and antifungal properties.

Uses in Literature: Known so far as antifungal, antimicrobial, antispasmodic, carminative, stomachic; and for dilating blood vessels, crack palate (vet.), flatulence, gastrointestinal tract, kidney, lowering blood pressure, making gripe water, sleep disorders, slow heart rate (vet.), stimulating respiration and urinary tract spasms (Ambasta, 1986; Chopra *et al.*, 1956; Khare, 2007; Kirtikar & Basu, 1994; Mishra *et al.*, 1996b).

Annona squamosa L. (Pl. 29A).

Syn.: *Annona asiatica* L., *A. cinerea* Dunal, *Guanabanus squamosus* (L.)M. Gomez, *Xylopia glabra* L., *Annona biflora* Moc. & Sesse, *A. forskahlii* DC.

Family: Annonaceae.

Vern.: Sitaphal.

Ayurvedic, English, Sanskrit and Regional Names:

Ayur.: Gandagaatra, Sitaaphala;

Eng.: Custard Apple, Sweet Sop, Sugar Apple;

Sans.: Agrimakhya, Atripya, Bahubijika, Gandagatra, Sitaphala, Subha, Suda;

Ass.: Mandargon;

B.: Ata, Katal;

H.: At, Atasitaphal, Sitaphal;

Mal.: Sita-Palam, Sita Pasham;

Oriya: Ato, Sitapholo;

Tam.: Sitapandu;

Tel.: Atta.

Distribution: Naturalised in Western Region.

Description: Small trees possessing oblong, membranous leaves with acute base. Flowers solitary, greenish-yellow. Petals 3, narrow-oblong. Fruits fleshy, areolate and yellow green.

Flowering & Fruiting: May-August.

Habitat Ecology: Cultivated; Nadaun, 483m.

Material Examined: HPU-EL-SKS-ABS-15033; 18.06.2009.

Parts Used: Fruits, Seeds, Leaves.

Folk Uses: Fruits edible and considered good in anaemia* and burning sensation. Dried leaves burnt to repel insects. Powdered seeds mixed with its juice utilized for killing livestock bugs and lice in hair of human beings.

Active Constituents: Reported to contain folic acid (*fruit*), polyphenol (*fruit pulp*).

Biological Activity: Anthelminitc activity against *Caenorhabditis elegans* +ve.

Uses in Literature: Reported as an abortifacient, astringent, antiscorbutic, cooling, emetic, febrifugal, pectoral, purgative, tonic, tooth brush, wormicide; and for carbuncle, diabetes, diarrhoea, edible purposes, excess hairfall, indigestion, dysentery, killing lice, licks, pyorrhoea, tumours and worms (Ambasta, 1986;

Asolkar *et al.*, 1992; Bhatt *et al.*, 2002; Chatterjee & Pakrashi, 1997; Chopra *et al.*,1956, 1983; Girach *et al.*, 1997; Kadel & Jain, 2006; Katewa *et al.*, 2001; Khare, 2007; Kirtikar & Basu, 1994; Kumar & Goel, 1998; Maliya & Singh, 2003; Mohanty *et al.*, 1998; Mokat & Deokule, 2004; Patole & Jain, 2002; Rao & Pullaiah, 2001; Retnam & Martin, 2006; Sen & Behera, 2003; Wunderlin & Hansen, 2008).

Remarks: Seeds strong irritant to conjuctiva and cause blindness.

Trade: Annona seeds, leaves and oil.

Arachis hypogaea L. (Pl. 29B).

Family: Fabaceae.

Vern.: Moongphali.

Ayurvedic, English and Regional Names:

Ayur.: Bhuumimudga; Mandapi, Tailamudga;

Eng.: Groundnut, Peanut;

H.: Moongphali;

Tam.: Nelakadalai, Verkadalai.

Distribution: Widely grown for pods in southern India, Maharashtra and Gujarat.

Description: Annual herbs, with pinnate leaves, yellow flowers in racemose clusters, which after fertilization bend down. The stock elongates and the young pods burried under the ground due to elongation of stock. Seeds non-endospermic.

Flowering & Fruiting: October-December.

Habitat Ecology: Cultivated for seeds, oil and fodder; Bharmoti, Nadaun, 483m.

Material Examined: HPU-EL-SKS-ABS-15034; 10.11.2010.

Parts Used: Leaves. Seeds.

Folk Uses: Seeds edible, also the oil extracted from it employed for cooking, massaging and as an adulterant* for 'desi' ghee. Peanuts also given to tuberculotic patients* for early recovery. Peanut flour prescribed as tonic for sick people. Leaves and oil cake ('khal') used as fodder*. Outer brown covering of seeds used as fuel*. Peanuts also make a good snacks by deep frying them in mustard oil and mixing with salt.

Active Constituents: Bioflavonoids, vitamin P, tannins, lipooxidase, p-coumaric acid, coenzyme Q10, niacin, folate, vitamin E, polyphenols, p-coumaric acid characterized from *seeds.*

Biological Activity: Antifungal, capric acid obtained from red skin variety showed activity against *Aspergillus niger.*

Uses in Literature: Reported to be used for baking, cardiovascular diseases, chutneys, cooking, frying, haemophiliacs, lowering cancer risk, massage oil, source of vitamin Q10 and niacin, snacks; and as antiaging, antibacterial, antifungal, adulterant, antioxidant and dryfruits (Ambasta, 1986; Chopra *et al.*, 1956; Chunekar & Pondel, 1999; Khare, 2007; Kirtikar & Basu, 1994).

Argemone mexicana L. (Pl. 29C).

Family: Papaveraceae.

Vern.: Jangali Post.

Ayurvedic, English, Sanskrit and Regional Names:

Ayur.: Hemaahvaa, Hemavati, Himaavati, Kaanchan-kshiri, Katuparni, Pitadugdhaa, Svarnkshiri;

Eng.: Mexican Poppy, Prickly poppy;

Sans.: Brahmadandi, Hemashikha, Kshirini, Srigalakanta;

B.: Baroshial-kanta, Sialkanta;

H.: Bharbhand, Satiyanashi;

Kan.: Datturi;

Mal.: Brahmadanti;

Mar.: Daruri, Kontedhotra;

P.: Bhatkateya, Bherband, Kandiari, Sialkanta;

Tam.: Bramadandu, Kurukkum;

Tel.: Brahmadandi, Chettu;

Urdu: Baramdandi.

Distribution: Naturalised Throughout India.

Description: Annual herbs with yellow latex, thistle-like leaves, white spinous veins, clasping stem, yellow prickly flowers which are poppy-like. Stamens indefinite. Ovary 1-celled. Capsule oblong-ovoid, opening by 4-6 valves. Seeds numerous, globose.

Flowering & Fruiting: April-May.

Habitat Ecology: Rocky slopes, drier areas, along nallahs; Dhoh, 1,034m.

Material Examined: HPU-EL-SKS-ABS-15035; 29.03.2008.

Parts Used: Whole Plant. Stem. Leaves. Seeds.

Folk Uses: Plant beaten with sticks and given to milching animals before flowering to increase milk content. Also, planted as hedge* plant. One tsp of pulverized twigs and seeds given twice a day with honey for 5 days to cure cough. Smoke of seeds and 'ajwain' (*Trachyspermum ammi* L.) considered effective for tooth problems and spongy gums. 2-3g of powdered leaves taken with a pinch of black pepper in the morning with cow's milk to cure diabetes*. About 25g powdered seeds mixed with wheat flour prescribed in animals for placental evagination*.

Active Constituents: Protopine (0.36%), berberine (0.74%), resin and toxic principle (*roots, stems*), norsanguinarine, norchelerythrine, cryptopine, berberine, benzo-phenanthridine, coptisine, sanguinarine and chelerythrine (*plant*), sanguinarine and allocryptopine (*seed oil*) have been isolated.

Biological Activity: Antifungal, anthelmintic, antileprotic, antidiarrhoeal, antiviral activities confirmed. Berberine exhibits significant myocardial depressant activity and total alkaloid fractions reported to cause fall in blood pressure.

Uses in Literature: Known earlier as an anodyne, antiallergic, antidote against snake bite, blood purifier, diuretic, edible, mouth wash; and for arthritis, asthma, boil, cataract, chronic cough, constipation, copious urination, dog bites, dropsy, eczema, eye complaints, fertility, gum trouble, headache, indigestion, inflammation, itching, jaundice, joint diseases, leucorrhoea, leprosy, ophthalmia, piles, psoriasis, pyorrhoea, regulating fertility, ringworm, scabies, scorpion bites, skin diseases, sun stroke, syphilis, toothache, typhoid, ulcers, and whooping cough, treatment of malaria, venereal sores, scorpion bite, photophobia, rheumatism (Ambasta, 1986; Bhatt & Sabnis, 1987; Chopra *et al.*, 1956; Dey, 1973; Dey *et al.*, 1973; Dwivedi *et al.*, 2005; Jain, 1991; Jain & Patole, 2001; Jain *et al.*, 1991; Kemparaja *et al.*, 1998; Khanna & Kumar, 2000; Kumar & Chauhan, 2006; Kumar & Jain, 1998; Kumar & Pullaiah, 1998; Maliya, 2004; Pal & Jain, 1998; Patil, 2004; Reddy *et al.*, 1998; Retnam & Martin, 2006; Sarangi & Sahu, 2004; Sen *et al.*, 2002; Sikarwar, 2002; Silori & Rana, 2000; Upadhyay & Singh, 2005; Verma & Chauhan, 2006; Willcox *et al.*, 2007).

Arisaema tortuosum (Wall.) Schott *(Pl. 29D).*

Syn.: *Arum tortuosum* Wall., *Arisaema helleboriefolium* Schott, *A. curvatum* Kunth. sensu Hook. f.

Family: Araceae.

Vern.: Kedae-Ri-Chhalli.

English and Regional Names:

Eng.: Cobra flower;

P.: Don, Gurin, Jangosh, Kirakal, Kirkichalu.

Distribution: Temperate and subtropical Himalayas.

Description: Erect herbs, possessing tuberous roots, pedately compound, leaves and 9-15 unequal leaflets, Fruiting spike naked. Spathe pale-green. Berries 1-seeded.

Flowering & Fruiting: July-October.

Habitat Ecology: Open rocky slopes, roadsides, dry areas; Bandroh grasslands, 850m.

Material Examined: HPU-EL-SKS-ABS-15036; 10. 08. 2010.

Part Used: Roots.

Folk Uses: Root paste applied externally to reduce swelling* in cattle. Oil from roots considered good against prolapsed uterus* (vet.). 50-100g crushed roots given orally to animals suffering from 'dandu' (prolapsed uterus) or in combination with black pepper powder (10g) mixed with ball-sized wheat dough given for 'basaihar' (mastitis)*.

Active Constituents: Petroleum ether extr. of corm yields n-alkanes, n-alkanols, stigmasterol, sitosterol, campesterol and cholesterol.

Biological Activity: Plant accredited with antifertility, insecticidal and insect repellent properties

Uses in Literature: Used so far as nutritive, pot herb, stomachic; and for throat swelling and wound (Anonymous, 1985; Asolkar *et al.*, 1992; Bhatt, 1990; Bhattarai, 1992; Jain, 1991; Khare, 2007; Rama Rao & Henry, 1996; Sikarwar *et al.*, 1994).

Artemisia japonica Thunb. (Pl. 30A).

Syn.: *A. parviflora* Don

Family: Asteraceae.

Vern.: Chamber.

Regional Names:

H.: Pati;

P.: Kanyurts.

Distribution: Temperate Himalayas.

Description: Tall aromatic undershrubs with lower wedge-shaped leaves. Heads small, greenish-yellow. Pappus none. Achenes black.

Flowering & Fruiting: September-December.

Habitat Ecology: Meadows, open slopes, wastelands; Rangas, 765m.

Material Examined: HPU-EL-SKS-ABS-15037; 26.08.2010.

Part Used: Leaves.

Folk Uses: Powdered aerial parts mixed with 'ashwagandha' (*Withania somnifera* (L.) Dunal) in ratio of 2:1 given for 5-7 days to cure rheumatism*. Poultice of leaf paste applied in skin affections. Also, its paste along with that of *Vitex negundo* L. applied externally for sprains and swellings*.

Active Constituents: Active constituents are coumarins 6-7-di-methyl-aesculetin, aesculetin and scopoletin (*leaves*).

Biological Activity: Antiviral and diuretic activity +ve.

Uses in Literature: Used so far for convulsion in children, nervous affections, skin diseases, sprains, throat infections; and as antibacterial, antifungal, antilithic and spasmodic (Jain, 1991; Khare, 2007; Megoneitso & Rao, 1983).

Artemisia vulgaris L. (Pl. 30B).

Syn.: *A. indica* Willd., *A. paniculatum* Roxb.,.*A. nilagirica* (Clarke) Pamp.

Family: Asteraceae.

Vern.: Charmad.

Ayurvedic, English and Regional Names:

Ayur.: Da manaka, Gandhotkata, Pushpachaamara;

Eng.: Fleabane, Indian Wormwood, Dungwort, Mugwort, Wild Wormwood;

H.: Nagdona;

Sans.: Danta, Nagadamani.

Distribution: Throughout the mountain region of India upto 3,500m.

Description: Tall, aromatic, much branched, tomentose shrubs, having pinnately lobed leaves of which ultimate segments are narrow. Heads small, solitary or fascicled. Achenes very minute, ellipsoid-oblong with no pappus. Seeds exalbuminous.

Flowering & Fruiting: August-October.

Habitat Ecology: Meadows, open slopes, wastelands; Bassi, 987m.

Material Examined: HPU-EL-SKS-ABS-15038; 22.07.2009.

Parts Used: Aerial parts. Leaves. Flowers.

Folk Uses: Paste of aerial parts applied over burn injuries* (vet.). Powdered leaves with cold water relieve headache* if taken twice daily for 3 days. Decoction of flower tops taken on empty stomach as an antilithic substance.

Active Constituents: Active constituents are cineol, thujone, thujyl, citral and substitute for cinchona (*plant*).

Biological Activity: Antibacterial, antifungal properties confirmed.

Uses in Literature: Used so far as anthelmintic, antibacterial, antifungal, antilithic, choleretic, diaphoretic, emmenagogue, menstrual regulator, nervine, stomachic, substitute for cinchona in fevers; and for cure of nervous and spasmodic affections (Ambasta, 1986; Jain, 1991; Khare, 2007; Kirtikar & Basu 1994; Megoneitso & Rao, 1983).

Artocarpus integrifolia L. (Pl. 30C).

Syn.: *Artocarpus heterophyllus* Lam., *A. integra* (Thunb.) Merr.

Family: Moraceae.

Vern.: Katal.

English and Regional Names:

Eng.: Jack Fruit, Jack tree, Kathal;
Ass.: Dawa;
G.: Phanas,Vanas;
B.: Dahu;
H.: Kathal, Lovi;
Kan.: Daho;
Kol. : Chama, Chamba, Dewa;
Mar.: Vonte;
Mal.: Chakka;
Ory. : Kantokalo, Panasa, Panoso;
Tel.: Lakucha, Verupanasa.

Distribution: Cultivated throughout India.

Description: Evergreen trees. Leaves entire, coriaceous, base rounded and obtuse apex. Solitary flowers in axillary receptacles. Fruits tuberculated. Achenes deeply sunk in fleshy mass.

Flowering & Fruiting: May-June.

Habitat Ecology: Cultivated; Bari-Mandir, 1,060m.

Material Examined: HPU-EL-SKS-ABS-15039; 15.04.2011.

Parts Used: Leaves. Fruit.

Folk Uses: Ripe fruits edible and good for dysentery*. Sorosis also used as vegetable and pickled. Leaves and fruits of the plant used as fodder* for cattle.

Active Constituents: Artaocarpetin, artocarpanone, artocarpin, cycloartenone, cycloartenol, β-sitosterol, isoartocarpin (*leaves*) flavonoids artocarpesin and norartocarpetin isolated (*fruits*).

Biological Activity: Antiameobic, aphrodisiac (unripened fruits), acholine (seeds) and esterogenic (leaves), bacteriolytic (latex), activities confirmed.

Uses in Literature: Known to be used as coolant, diuretic, gum, pickle, laxative, vegetable; and for abscess, asthma, carpentry, diarrhoea, furniture, musical instruments, promoting healing, skin disease, and turnery (Ambasta, 1986; Chatterjee & Pakrashi, 1997; Chopra *et al.*, 1969; Diwakar & Ansari, 2003; Dwivedi, 1993; Drury, 1978; Khare, 2007; Parrotta, 2001; Prajapati *et al.*, 2003; Sahni, 1998; Saklani & Jain, 1994; Samati, 2004; Varghese, 1996.).

Artocarpus lakoocha Roxb. (Pl. 30D).

Syn.: *Artocarpus lacucha* Buch.-Ham.

Family: Moraceae.

Vern.: Dheu.

Ayurvedic, English, Sanskrit and Regional Names:

Ayur.: Granthiphala, Kshudra, Lakucha, Panas, Pitanaasha;

Eng.: The Jack Fruit tree, Monkey Jack;

Sans.: Peingnai, Pienne;

Ass.: Kanthar;

B.: Knathal;

H.: Kanthal, Kathal;

Kan.: Teprong;

Kol. : Panasa;

Mar.: Pila, Pilapasham;

Tam.: Panasa-pandu, Pansa, Veru-panasa;

Tel.: Halsu, Heb-helsu, Halsina.

Distribution: Sub-Himalayan track and Outer Hills

Description: A large deciduous tree possessing, rough, grey bark, coriaceous, elliptic, entire leaves and flowers in axillary globose pedunculate heads. Fruits lobulate, smooth, yellow. Seeds oblong, a few broad.

Flowering & Fruiting: March-June.

Habitat Ecology: Cultivated; Khas Groan, 858m.

Material Examined: HPU-EL-SKS-ABS-15040; 15.04.2011.

Parts Used: Leaves. Fruit. Bark.

Folk Uses: Fruits cooked as vegetable and pickled, also its milk applied for suppression of boils. Heated leaves applied as bandage to wounds*. Leaves used as fodder for increasing milk content* in cattle. Poultice of bark good against pimples*; also its fibres used for cordages.

Active Constituents: Some water soluble bases (*fruits*), cycloartenyl acetate, cycloartenol and cycloartnone, oxyresveratrol (*bark*) have been confirmed.

Uses in Literature: Used as edible, gum, pickle, purgative, rafter, vegetable, vermifuge; and for beams, boat building, cracked skin, boils, dental caries, diarrhoea, furniture, posts, and toothache (Ambasta, 1986; Khare, 2007; Kumar & Goel, 1998; Mohanty *et al.*, 1998; Samati, 2004).

Asclepias curassavica L. (Pl. 31A).

Family: Asclepiadaceae.

Vern.: Ghamera, Parvini.

Ayurvedic, English, Sanskrit and Regional Names:

Ayur.: Kaakanaasikaa;

Eng.: Curassavian, Swallow-Wort, West Indian Ipeca-cuanha, Wild Ipecacuanha;

Sans.: Kakatundi;

Bo.: Kakatundi, Kuraki;

Mar.: Kurki;

P.: Kaktundi.

Distribution: Throughout India up to 1,500m.

Description: Undershrubs with opposite, oblong, lanceolate leaves and orange-red corolla. Follicles long, smooth, swollen, pointed.

Flowering & Fruiting: Flowers throughout the year.

Habitat Ecology: Common in wasteplaces, roadsides, forest edges, fields; Tarkwari, Bassi, 980-987m.

Material Examined: HPU-EL-SKS-ABS-15041; 12.09.2009.

Parts Used: Flowers. Plant.

Folk Uses: Plant grown as hedges* and its flowers considered sacred* for offering to deities*. Also, plant exploited for making fibers*.

Active Constituents: Hesperidin, cardenolides, calotropagenin, coroglaucigenin, uzarigenin, asclepin, its glucosides and uzarin (*plant*).

Uses in Literature: Known in Indian literature as adulterant, anthelmintic, antibacterial, astringent, antidysenteric, cardiotonic, cathartic, cytotoxic, emetic, fish poison, spasmogenic, styptic, sudorific; and for abdominal pain, cancer, corns, gonorrhoea, piles, sores, stuffing, ulcer, warts and wounds (Ambasta, 1986; Jain, 1991; Jain *et al.*, 1997; Khare, 2007; Kurian, 1995; Watt, 1972).

Asparagus adscendens Roxb. (Pl. 31B).

Syn.: *A. volubilis* Wall., *Asparagus adscendens* Kunth.

Family: Liliaceae.

Vern.: Sansban.

Ayurvedic, English, Sanskrit and Regional Names:

Ayur.: Mahaashtaavari, Mushali;

Eng.: Asparagus;

Sans.: Bon, Musli;

Bo.: Dholimusali, Saphedmusali;

G.: Dholimusali, Saphedmusali, Ujlimusli;

H.: Satavar, Safedmusali;

Mar.: Dholimusali, Safedmusali.

Distribution: W. Himalayas, Punjab-Kumaon. H.P. upto 1,500m.

Description: Herbs with terete grooved stem, stout, suberect branches, long stout spines and densely crowded cladodes in tufts. Flowers pure white, fragrant. Berries globose, green, small, 1-seeded.

Flowering & Fruiting: October-November.

Habitat Ecology: Damp places, edges of cultivations, wastelands; Dungi, 906m.

Material Examined: HPU-EL-SKS-ABS-15042; 03.04.2009.

Parts Used: Whole Plant. Leaves. Root.

Folk Uses: Plant regarded as sacred* and worshipped; 'Maternal milani' performed during marriages using its twigs. Half tsp powdered leaves and flowers given with luke warm water to relieve headaches (2 times daily). Paste of leaves mixed with black pepper given for nausea and vomiting. Poultice of roots, twigs and leaves applied on stomach to cure menorrhagia*. Roots considered emetic and coolant (vet.). 2g dry powdered roots given to pregnant ladies 1-2 times a day as coolant. Poultice of roots given before delivery in labour*. Mixture of its root with 'gur' (jaggery) given to animals for stomach disorders.

Active Constituents: Asparagine and saponin steroidal (*roots*), β-sitosterol, disogenin and sarsasapogenin (*fruits*) are the principle constituents.

Biological Activity: Antimicrobial activity found to be +ve.

Uses in Literature: Recorded properties in literature are aphrodisiac, appetiser, astringent, coolant, demulcent in veterinary medicine, diaphoretic, diuretic, edible, galactagogue, substitute for 'salep misree', tonic; and for anthrax, debility, diarrhoea, dysentery, headaches, inducing heat, joint diseases, increasing lactation in animals, leucorrhoea, liver disorders, nausea, miscarriage, premature ejaculation, snake bites, spermatorrhoea, stomach pain, throat complaints and vomiting (Ambasta, 1986; Borthakur *et al.*, 2004; Dhiman, 1976; Dwivedi *et al.*, 2005; Girach *et al.*, 1998; Hembrom & Goel, 2005; Kadel & Jain, 2006; Khare, 2007; Kirtikar & Basu, 1994; Kumar & Pullaiah, 1998; Manandhar, 1980; Mohanty *et al.*, 1998; Mokat & Deokule, 2004; Rao & Pullaiah, 2001; Retnam & Martin, 2006; Reddy *et al.*, 1998; Sarangi & Sahu, 2004; Sen & Behera, 2003; Singh, 2003).

Major Ayurvedic Formulations: Musali Pak, Aswagandharista, Geriforte tablets, Maharishi, Amrit Kalash-5, Shatavarighrita, Narayana taila, Vishnu taila.

Trade: Dried decorticated harvested roots sold in the market.

Asparagus racemosus Willd. (Pl. 31C).

Family: Liliaceae.

Vern.: Satmusli, Satavari, Shakakul.

Ayurvedic, English, Sanskrit and Regional Names:

Ayur.: Atirasaa, Bahusutaa, Indivari, Shataavari, Shatmuuli, Shatpadi, Shatviryaa, Vari, Vhiru;

Eng.: Willd Asparagus;

Sans.: Satamuli, Satavari, Svetmuli;

B.: Satamuli;

G.: Satawari;

H.: Shatamuli;

Mar.: Satawarmul;

Mal.: Shatavali;

Oriya: Chotaru, Mohajolo, Satabari, Satamuli, Sotabori;

Pub.: Bozidon, Satawar;

Tam.: Ammai-kodi, Inli-chedi, Nili-chedi;

Tel.: Chalagadu, Pili-pegalu.

Distribution: Throughout subtropical and tropical regions of India upto 1,500m.

Description: Climbing undershrubs with curved prickles. Stems tall climbing, cladodes 1-3cm., curved terete, spreading in tufts of 2-6. Flowers fragrant and in racemes.

Flowering & Fruiting: July-October.

Habitat Ecology: Roadsides, wild in hill slopes in loose soils; Khaggal, 920m.

Material Examined: HPU-EL-SKS-ABS-15043; 03.07.2008.

Parts Used: Root. Young Shoots.

Folk Uses: Roots boiled in milk given as medicine in dysentery. Poultice of boiled leaves mixed with ghee applied for suppuration of boils*. Used as one of the constituents of 'hawan'* and a must in marriages as religious plant*. Plant relished as fodder* by sheep and goats.

Active Constituents: Flavonoids, rutin and quercetin-3-glucuronide, saponins-shatavarins I-IV (*whole plant*), shatavarin IV, sitosterol; 4, 6-dihydroxy-2-O-benzaldehyde (*root*).

Biological Activity: Antimicrobial activity found to be +ve.

Uses in Literature: So far used as an astringent, galactagogue, intestinal disinfectant, nervine tonic, styptic, ulcer healing agent; and for bleeding disorders, diarrhoea, genitourinary tract troubles, haematuria, hyper acidity, lactic disorders and sexual debility (Ambasta, 1986; Borthakur *et al.*, 2004; Khare, 2007; Kirtikar & Basu, 1994).

Avena sativa L. (Pl. 31D).

Family: Poaceae.

Vern.: Javi, Joee.

Ayurvedic, English, Sanskrit and Regional Names:

Ayur.: Yavikaa;

Eng.: The Oat;

Sans.: Ganer, Kulijud;

H.: Jai, Javi;

Pub.: Javi.

Distribution: Cultivated in north India from Bengal-Indus and in Himalayas upto 3,700m.

Description: Cultivated herbs possessing adventitious fibrous roots, erect stem with distinct nodes and internodes and cauline, alternate leaves having sheathing leaf bases which are sessile and multicostate. Flowers in panicle of spikelets, each spikelet with 2 or 3 florets.

Flowering & Fruiting: November-January.

Habitat Ecology: Wastelands, cultivated as fodder crop; Jhaniari, 900m.

Material Examined: HPU-EL-SKS-ABS-15044; 27.03.2010.

Parts Used: Plant. Seeds.

Folk Uses: Plant considered a good fodder for sheep and goats; 'toori' (straw) made from its dried stem and leaves used as fodder. Grains used as low quality food ('satu' meal); these if cooked in milk are prescribed as good laxative*. Paste of its seeds fed to animals for enhancing lactation*.

Active Constituents: Fats, proteins 87%, 26 flavonoids of apigenin, luteolin and tricin isolated (*seeds*).

Biological Activity: Avenacosides exhibit strong antifungal activity *in vitro*.

Uses in Literature: Used earlier as anti-depressant, antispasmodic, cardiac tonic, emollient, fodder, nervine tonic, stimulant, thymoleptic; and for colitis, debility, diarrhoea, dysentery, meals, menopausal phase, palpitation, sleeplessness, spermatorrhoea (Caius, 2003; Das & Agarwal, 1991; Karnic, 1994; Khare, 2007; Rastogi & Mehrotra, 1991; Sharma & Rana, 2005).

Azadirachta indica A. Juss. (Pl. 32A).

Syn.:M. *parviflora* Moon.,M. *indica* Brandis, *Melia azadirachta* L.

Family: Meliaceae.

Vern.: Neem.

Ayurvedic, English, Sanskrit and Regional Names:

Ayur.: Arishta, Arishtaphala, Nimba, Nimbaka, Paaribhadra, Pichumanda, Pichumandaka, Pichumarda, Sutiktak, Tiktaka;

Eng.: Indian Lilac, Margosa Tree, Neem Tree;

Sans.: Arishta, Avkapadapa, Chhardana, Dzaad-darakhte;

B.: Nim, Nimgachh;

G.: Nim;

H.: Nim, Nimgachh;

Kan.: Ariya-veppa, Vippa;

Mal.: Bawtamaka, Kamaku, Tamdbin, Thamaka;

Mar.: Danujhada, Kohumba, Libado, Limba, Limbado;

Tam.: Arulundi, Kinji, Nimbamu, Taruka, Vepa, Vepu, Vembu, Yapa, Yeppa;

Tel.: Bevina-marakadbevina-mara, Heb-bavu, Nimbamu, Vemu, Vepa, Yapa, Yeppa.

Distribution: Cultivated throughout greater part of India.

Description: Trees with short erect trunk and broad crown. Leaves alternate, imparipinnate. Flowers white, in axillary panicles, honey scented. Drupe and endocarp woody.

Flowering & Fruiting: June-September.

Habitat Ecology: Edges of cultivated fields; NIT, Anu, 900m.

Material Examined: HPU-EL-SKS-ABS-15045; 03.07.2008.

Parts Used: Leaves. Twigs. Fruits.

Folk Uses: Twigs used as dentrifice*. Bath with decoction of leaves recommended for scabies*, measles*, herpes* and other skin ailments. Dried leaves used as preservative for packing woollen* clothes. ¼ tsp of powdered leaves given once a day for 30 days to purify blood. One glass decoction of leaves taken daily for diabetes*; its poultice applied for suppuration of boils. 2-3 leaves given twice daily in malaria fever. Ripe fruits also considered good against skin ailments. A stick made of its wood ('danda') utilized for grinding* to flavour salt. Leaves burnt to repel mosquitoes from cowshed.

Active Constituents: Found to contain nimocinol, 2'-3' dehydrosalano, meliacins, nimocinolide, isoazadirolide and nimbocinolide (*leaves*).

Biological Activity: Leaves anthelmintic.

Uses in Literature: Useful in India as anthelmintic, antidote, antiseptic, astringent, blood purifier, diuretic, insecticidal, insect repellent, purgative, stimulant, tonic; and for boils, bronchitis in childern, chicken pox, chest pain, constipation, dermatitis, diarrhoea, eczema, epilepsy, foot and mouth diseases, fever, itching, insect bites, malaria, mastitis, miscarriage, piles, pimples, pyorrhoea, skin disease, small pox, sores, suppuration (vet.), syphilis, wounds (Ambasta, 1986; Bhatt *et al.*, 2001; Bondya *et al.*, 2006; Gogoi & Borthakur, 2001; Jadhav, 2006; Kadel & Jain, 2006; Kalpesh *et al.*, 2004; Khanna, 2002; Khanna & Kumar 2000; Khare, 2007; Kumar & Chauhan, 2006; Kumar & Goel, 1998; Kumar & Jain, 1998; Kumar *et al.*, 2004; Mohanty *et al.*, 1998; Pandey & Rout, 2006; Retnam & Martin, 2006; Sharma, 2003; Sharma *et al.*, 2002; Silori & Rana, 2000).

Major Ayurvedic Formulations: Shankhapuspi-taila, Jvarasamhara-rasa, Mahamari-chyadi taila, Arsoghna vati, Pathyadi-kvatha, Maha-manjisthadyarista, Punarnavadi-kvatha, Rakta Sodhaka-vati.

Bacopa monnieri (L.) Penn. (Pl. 32B).

Syn.: *Bacopa monnieria* (L.) Wettst., *Herpestis monnera* L., *Moniera cuneifolia* Michx.

Family: Scrophulariaceae.

Vern.: Jal Brahmi.

Ayurvedic, English, Sanskrit and Regional Names:

Ayur.: Aindri, Braahmi, Nirbraahmi;

Eng.: Thyme Leaved Gratiola;

Sans.: Brahmi, Nira Brahmi, Saraswati;

Ass.: Brahmi;

B.: Brihim-sak;

G.: Jalneveri, Kadaviluni, Neer brahmi;

H.: Brahmi;

Mal.: Barna;

Mar.: Birami, Jalnam, Nirbrahmi;

Tam.: Nirbrahmi, Sambra ne elaye;

Tel.: Sambranicettu.

Distribution: Throughout India up to 1,300m.

Description: A glabrous creeping herb rooting at joints. Leaves fleshy with dotted lower surface. Flowers pale blue, purple veined, small white, solitary and borne in leaf axils. Capsules 5mm long, ovoid, acute, included in persistent calyx. Seeds small, oblong, striate pale.

Flowering & Fruiting: June-September.

Habitat Ecology: Marshy areas; Neri, 381m, Brahmni, 787m.

Material Examined: HPU-EL-SKS-ABS-15046; 08.07.2009.

Parts Used: Whole Plant. Leaves.

Folk Uses: 1-3g powder of the plant given with cow's milk on empty stomach once a day for epilepsy, insanity, nervous disorders and as brain tonic. Plant used as fodder*.

Active Constituents: Brahmine, herpestine, saponins, monnierin, hersaponin, bacosides A & B, apigenin-7-glucuronide and luteolin-7-glucuronide (*leaves*).

Uses in Literature: Reported in Indian literature as adaptogenic, antispasmodic, antiulcer, astringent, cardio-tonic, cerebral activator, diuretic, immuno-modulator, nervine tonic, sedative; and for alzheimer's disease, antianxiety, asthma, bronchitis, diarrhoea, epilepsy, insanity, intellectual activity, tranquiliser (Ambasta, 1986; Islam, 2002; Khare, 2007; Kirtikar & Basu, 1994; Parrotta, 2001; Prajapati *et al.*, 2003; Rastogi & Mehrotra, 1991; Thakur *et al.*, 1989).

Bambusa arundinacea (Retz.) Roxb. (Pl. 32C).

Syn.: *Bambusa bambos* (L.) Voss., *Arundo bambos* L., *B. pungen* Blanco, *B. spinosa* Roxb., *B. neesiana* Arn. *ex* Munro.

Family: Poaceae.

Vern.: Bainjh.

Ayurvedic, English, Sanskrit and Regional Names:

Ayur.: Kichaka, Shatparvaa, Trinadhwaj, Vansha, Vanshalochana, Venu, Yavphala;

Eng.: Spiny or Thorny Bamboo;

Sans.: Vans;

B.: Mahlbans;

H.: Bans;

Tam.: Moongli.

Distribution: Throughout the plains and low hills of India and Myanmar; wild and cultivated.

Description: Perennials with tall stem upto 33m long. Stem sheaths and blade with long horizontal shoots and armed with 2-3 recurved spines. Internodes upto 45cm. Leaves long; tip sharp, stiff. Panicle enormous, often occupying the whole stem. Rachis usually stiff, shining. Lodicules 3, small fimbriate.

Flowering & Fruiting: Flowering sporadically.

Habitat Ecology: Roadsides, forests, dry areas; Dungi, 906m.

Material Examined: HPU-EL-SKS-ABS-15047; 29.02.2010.

Parts Used: Stem. Leaves. Plants.

Folk Uses: Bamboos used locally for making ladders*, baskets, 'chicks'* (window curtains) mats and outdoor furniture. Plant considered sacred* and used for performing religious rituals at the time of marriage and to make bier ('arthi') for carrying dead body for last rites. Twigs used for fencing agricultural fields and as fuel. 10-12 pieces of stem sprouts given to enhance the chances of pregnancy in case of 'abhjana' (anoestrus)* in animals. Paste of 2-3 leaves given for smooth expulsion of placenta (vet.). Young stems ('manu') pickled*, leaves used as fodder*. Gum ('banslochan') used as a substitute for 'shilajeet'*. Also, it is used for making 'panjiri' (a dish made up of dry fruits and gum in 'desi' ghee) given to ladies after delivery for strength.

Active constituents: *Bamboo-manna* contains a silicious crystalline substance cyanogenic glucoside and taxiphyllin.

Uses in Literature: Reported to be used for bleeding gums, bone fracture, cooling, cure of dysmenorrhoea, delivery (vet.), fencing, haemoptysis, house building, leucoderma, painful joints, pulp, ulceration; and as antiinflammatory, antileprotic, aphrodisiac, bechic, blood purifier, carminative, emmenagogue, expectorant, febrifuge and tonic (Ambasta, 1986; Geetha *et al.*, 1996b; Jain, 1991; Khare, 2007; Malik *et al.*, 2009; Punjani, 1998).

Major Ayurvedic Formulations: Shitolashadi-churn, Banslochan.

Trade: Bans.

Bambusa nutans Wall. (Pl. 32D).

Syn.: *Bambusa falconeri* Munro.

Family: Bambusaceae.

Vern.: Bans.

Ayurvedic, English, Sanskrit and Regional Names:

Ayur.: Kichaka, Shatparvaa, Trinadhwaj, Vansha, Vanshalochana, Venu, Yavphala;

Eng.: Thorny Bamboo;

Sans.: Vansh;

B.: Mahlbans;

H.: Bans.

Distribution: H.P., Sikkim; Commonly cultivated.

Description: A tall solitary stem arising from creeping rhizome. Leaves linear-lanceolate, tessellate and without spiny branches. Culms not crowded, nodes not much raised. Spikelets in bracteate fascicles, Caryopsis with adherent pericarp, oblong, hairy at apex.

Flowering & Fruiting: Flowering sporadically.

Habitat Ecology: Roadsides, forests, dry areas; Tikker Khatrian, 908m.

Material Examined: HPU-EL-SKS-ABS-15048; 29.04.2010.

Parts Used: Stem. Leaves. Plant.

Folk Uses: Plant considered sacred* and used for performing religious rituals at the time of marriage and making bier '(arthi)' for carrying corpse to cremation ground for final rites. Young stems ('manu') pickled* and also used as vegetable*. Twigs used for fencing agricultural fields and as fuel. Leaves used as fodder*. Bamboos used for making ladder, baskets, 'chicks' (window curtains) 'chataei' (mat) and light furniture*. Gum 'banslochan' used as a substitute for 'shilajeet'; and preparing 'panjiri' (a dish made up of dry fruits and its gum roasted in 'desi' ghee) and given after delivery to women for strength. 10-12 pieces of stem sprouts given to enhance the chances of pregnancy in case of 'abhjana'* (anoestrus) in animals.

Active constituents: *Bamboo-'manna'* contains a silicious crystalline substance cyanogenic glucoside and taxiphyllin.

Uses in Literature: Reported so far for bleeding gums, bone fracture, cooling, cure of dysmenorrhoea, delivery (vet.), fencing, haemoptysis, house building, leucoderma, painful joints, pulp, ulceration; and as antiinflammatory, antileprotic, aphrodisiac, bechic, blood purifier, carminative, emmenagogue, expectorant, febrifuge and tonic (Ambasta, 1986; Jain, 1991; Punjani, 1998).

Major Ayurvedic Formulations: Shitolashadi-churn, Banslochan.

Trade: Bans.

Bauhinia vahlii Wight & Arn. (Pl. 33A).

Syn.: *B. racemosa* Vahl, *Phanera vahlii* Benth.

Family: Fabaceae.

Vern.: Tor.

English and Regional Names:

Eng.: Camel's foot climber;

B.: Chehur;

G.: Chambelli;

H.: Jallur, Malijhan, Malghan, Mahul, Malo, Malu;

Mal.: Motanvally;

Mar.: Chambal, Chambura;

Oriya: Shiali;

P.: Taur;

Tam.: Mandarai;

Tel.: Adattige, Madapur.

Distribution: Sub-Himalayan tracts from Ravi eastwards, ascending to 1,000m in U.P., foot of Himalayas, Kashmir-Sikkim upto 1,500m.

Description: Enormous climbers with wide spreading stem, densely pubescent branchlets and circinate tendrils, rough, dark reddish brown bark and rigidly subcoriaceous leaves which are glabrescent above at maturity. Flowers white changing into cream yellow. Pods sublignose, rusty brown with velvet Hair. Seeds suborbicular, flat, dark brown.

Flowering & Fruiting: April-June.

Habitat Ecology: Forests, rocky slopes, dry areas; Dhaneta, 596m.

Material Examined: HPU-EL-SKS-ABS-15049; 25.04.2009.

Parts Used: Stem. Leaves. Wood.

Folk Uses: Leaves used as fodder*; and for making meal plates ('pattal')* and bowl ('duna')*. Also, powdered leaves (2-3g) taken with water to check dysentery. Dried and powdered seeds prescribed for abortion* in ladies. Pods given to livestock as medicine for stomach trouble. Wood used for cordages, ropes and as fuel*.

Active Constituents: Therapeutic properties attributed to glycosides of quercitol, kaempferol, quercitroside, isoquercitroside, rutoside, taxifoline rhamnoside, agathisflavone and its mono and di-o-merthyl derivatives, quercetin, betulinic acid, 3 glucoside tannins and proteins (*plant*).

Biological Activity: Stem antiacetylcholine and hypotensive.

Uses in Literature: Known in literature as an adhesive for fixing hair, anthelmintic, antiinflammatory, aphrodisiac, astringent, cholagogue (seed pulp and stem eaten raw), stomachic, tonic, vegetable, vermifuge; and for amenorrhoea, antifertility, bone fracture (vet.), cough, diarrhoea, dysentery, facial paralysis, gonorrhoea, local cigarrettes, making meal plates and rough table clothes, malaria, mucilagenous remedies, piles, rain hats and caps, ropes and cordage, syphilis and thatching (Ambasta, 1986; Banerjee, 1996; Chhetri, 2005; Goud & Pullaiah, 1996a; Jain, 1991; Khare, 2007; Kirtikar & Basu, 1994; Pandey & Rout, 2006; Pawar & Patil, 2004; Rama Rao & Henry, 1996; Sahu, 1996; Retnam & Martin, 2006; Saini, 1996b; Singh & Prakash, 1994, 1996; Tosh, 1996; Tirkey, 2004; Watt, 1972).

Bauhinia variegata L. (Pl. 33B).

Syn.: *B. purpurea* Wall., *B. candida* Roxb., *Phanera variegata* Benth.

Family: Fabaceae.

Vern.: Kachnar, Karale, Karayalae.

Ayurvedic, English, Sanskrit and Regional Names:

Ayur.: Ashmantaka, Gandhaari, Kaanchana, Kaanchanaara, Kaanchanaaraka, Sonapushpaka;

Eng.: Buddhist bauhinia, Mountain ebony, Orchid tree, Variegated bauhinia;

Sans.: Ashmantaka, Chamrika, Gandari, Girija, Kanchanara, Karbadra, Kaddara, Mahapushpa, Pakari, Tamrapushpa, Uddalaka, Yugmaptra;

B.: Raktakanchan;

Bo.: Kanaraj, Kanchan, Segapa-Munthari;

G.: Kovindara;

H.: Barial, Gurial, Gwiar, Kachnar, Kaniar, Padrian;

Kan.: Kanchavala;

Mal.: Chuvannamandaram;

Mar.: Kanchan, Raktakanchan;

Oriya: Boroda;

Tam.: Mandarai, Sagapumanchori, Vellaippuvatt;

Tel.: Devakanchanamo; Urdu-Kachnal.

Distribution: Throughout India.

Description: A moderate-sized erect deciduous tree with stout glabrescent branchlets, simple, suborbicular, rigidly subcoriaceous, shallow cordate leaves. Flowers lateral, sessile or short peduncled, white variegated with red and blue strips. Pods hard, flat, glabrous.

Flowering & Fruiting: March-August.

Habitat Ecology: Forests, roadsides, planted in villages and agricultural field; Tauni Devi, 1,050m.

Material Examined: HPU-EL-SKS-ABS-15050; 27.03.2008.

Parts Used: Leaves. Buds. Flowers.

Folk Uses: Buds, flowers used as vegetable and for making pickles, 'pakoras', 'chutney'*, 'rayata' (culinary preparations); also 2-5g of its powder consumed with milk, twice a day for 3-5 days, for menstrual problems*, diarrhoea and dysentery. Leaves employed as fodder for increasing lactation*.

Active Constituents: Isoquercitin, astragalin, quercitin, β-p-glucopyranoside, hentriacontene, octacosanol, reducing sugar, stigmasterol, β-sitosterol (*flower*), have been characterized.

Biological Activity: Seed extr. exhibits marked agglutinating activity for human blood. Antibacterial and CNS depressant activity also found to be +ve.

Uses in Literature: Considered as an alternative, anthelmintic, antidote to snake poison, antiinflammatory, astringent, blood purifier, culminative, edible, fodder, gentle laxative, pot herb, purgative, sacred, symbolic offering to Lord Buddha, tonic, vermifuge (vet.); and for agricultural implements, beedi manufacturing (leaves), bleeding gum, cold, cough, curd preparation (petals), cuts (vet.), diarrhoea, dysentery, enhancing lactation, expelling placenta (vet.), fatness, foot & mouth (vet.), gastropathies (vet.), goitre, gripping pains from stomach and bowels, indigestion, mouth ulcers, paintings (Mithila), pickling, piles, skin diseases, tanning, toothache, tuberculosis, uterine bleeding, ulcers and wounds (vet.) (Ambasta, 1986; Banerjee, 1996; Bondya *et al.*, 2006; Chettri, 2005; Chauhan, 1999;

Dixit, 1987; Gill & Nyawuame, 1994; Hosagoudar & Henry, 1996a ; Jain, 1991; Jain *et al.*, 1991; Jha & Basak, 1994; Jha *et al.*, 1996; Khanna *et al.*, 1996a, b; Khare, 2007; Kiruba *et al.*, 2006; Kurian, 1995; Lal *et al.*, 1996; Lalramnghinglova, 1996; Masih, 2003; Mohanty *et al.*, 1998; Negi *et al.*, 1985; Pal & Jain, 1998; Pandey & Rout, 2006; Pandey & Verma, 2002; Patole & Jain, 2002; Rao, 1996; Retnam & Martin, 2006; Saini, 1996b; Sikarwar *et al.*, 1994; Singh, 1996, 1999b; Sinha, 1996; Verma, 1998; Verma & Chauhan, 2006; Watt, 1972).

Major Ayurvedic Formulations: Chandanasava, Kanchanaaradi-Kwath, Kanchanara-guggulu, Kanchanara-gutika, Kanchanar-kwath, Ushirasava, Vidangasava.

Trade: Stem bark sold as 'Kachnar Chal'.

Adulterant or Substitute: Bark used as an adulterant or substitute for that of 'Ashoka' (*Saraca asoca* Roxb.).

Berberis asiatica Roxb. *ex* DC. (Pl. 33C).

Syn.: *Berberis aristata* DC., var. *Floribunda.*

Family: Berberidaceae.

Vern.: Kashmal.

Ayurvedic, English, Sanskrit and Regional Names:

Ayur.: Daaru, Daaruharidraa, Daarunishaa, Daarurajani, Daarvi, Pushtaphala, Sthirphala, Vrahitaphala;

Eng.: Indian Barberry, Tree Turmeric;

Sans.: Daruharidra, Kantakati, Kateri;

B.: Daruhaldi;

G.: Daruharidra, Daruhuldur;

H.: Chitra, Darhald, Kashnal;

Kan.: Daruhaladi, Naradarishana, Maradarishina;

Mal.: Naradarisina, Maramanjal;

Mar.: Daruhald;

Oriya: Daruhalidi, Daruharidra;

Tam.: Mullukala, Usikkala;

Tel.: Manupusupu.

Distribution: H.P., Nepal up to 3,500m.

Description: Erect, stout bushy shrubs possessing rarely acute, thick coriaceous leaves which are strongly reticulate beneath. Racemes peduncled or subsessile. Flowers yellow. Young fruits white, blue when ripe.

Flowering & Fruiting: May-September.

Habitat Ecology: Shrubberies, common; Bhaleth, 506m.

Material Examined: HPU-EL-SKS-ABS-15051; 12.03.2008.

Parts Used: Root. Stem. Leaves.

Folk Uses: Leaves make an excellent fodder. Leaf extr. used to wash infected eyes and also helpful in case of dysentery in animals. Wood used as fuel. Smoke of burnt plant helps to repel mosquitoes* from courtyards and from vicinity of animals. 2-4g powdered roots prescribed with luke warm milk after dinner as laxative, its decoction given to sugar patients* (one glass daily) on empty stomach for curing diabetes*. 1-3g powdered bark of the root taken with honey as blood purifier and skin problems, boils, scabies* and pimples*. Ripe fruits 'kashmal' edible and considered as blood purifier.

Active Constituents: Berberine, karachine, taxilamine, oxycanthine, umbellatine, oxycanthine, jatrorrhizina and resins, tannins (*plant*).

Biological Activity: Antibacterial, antidepressent and antitubercular activities have been attributed to berberine.

Uses in Literature: Reported to be used as an alternative, antiperiodic, antiscorbutic, blood purifier, cooling deobstruent, diaphoretic, fuel, laxative to children, psychoactive plant, stomachic, tonic; and for affections of eye, ammenorrhoea, anorexia, apthous sores, bites of rat and snake, boils, cholera, diarrhoea, dyeing, dyspepsia, edible purposes, enlargement of spleens, eye complaints, infection in horses, intermittent and remittent fevers, intestinal catarrh, jaundice, liver (vet.), malarial affections, menorrhagia, painful micturition, piles, rheumatism, skin diseases, sun blindness, tannins, ulcers, and vomitings during pregnancy (Ambasta, 1986; Anonymous, 1948-1976; Chauhan, 1999; Jain, 1968; Jain *et al.*, 1994; Khare, 2007; Kirtikar & Basu, 1994; Parmar & Kaushal, 1982; Sensarma, 1991; Shome *et al.*, 1996; Singh, 1996; Watt, 1972).

Major Ayurvedic Formulations: Ashwagandhyarista, Bhrargaraja-taila, Chandanaadivati, Darbhyadi tel, Darbhyadi-leha, Darbyadikwath, Dardyadi-taila, Dashanglepa, Jatyadi-taila, Khadira-gutika, Khadirarishta, Laghuvishgarbha-taila, Punarnava-mandoor, Rasanjanadi-churna, Sudershan-churna, Triphala-ghrita.

Trade: Traded in Indian markets as 'Rasaunt'.

Boerhavia diffusa L. (Pl. 33D).

Syn.: *B. procumbens* Roxb., *B. erecta* Gaertn., *B. repens* L., *B. repens* var. *diffusa* (L) Hook., *B. repens* var. *procumbens* (Roxb.) Hook.

Family: Nyctaginaceae.

Vern.: Itsat.

Ayurvedic, English, Sanskrit and Regional Names:

Ayur.: Katthilla, Punarnavaa, Rakta-punarnavaa, Shophaghni, Shothaghni;

Eng.: Hogweed, Pigweed, Spreading Hogweed;

Sans.: Nilini, Punarnava, Rakthkanda, Raktapatrika, Raktapunarnava;

H.: Gadapurana, Sant, Sarikinjngi;

Mal.: Thumizhazhma, Thazhuthama;

Tam.: Mukkurattai;

Tel.: Atikamamidi.

Distribution: Weed in drier parts of India.

Description: A prostrate herb having reddish branches. Leaves ovate, margin undulate. Flowers crimson-red, minute and in small bracteate umbels. Fruits clavate, penta-ribbed.

Flowering & Fruiting: July-September.

Habitat Ecology: Drier Areas; Anu khurd, 858m; Bramni, 858m.

Material Examined: HPU-EL-SKS-ABS-15052; 13.04.2010.

Parts Used: Root. Leaves. Whole Plant

Folk Uses: 50-100ml decoction of the plant recommended for urinary disorders (once daily till cure). Fresh leaves cooked as pot herb*. Paste of plant prepared in 'til' (*Sesamum indicum* L.) oil good for joint pains and insect bite*, also given as medicine for cudding* and stomach swelling* in livestock. Mixture of root and leaves in equal proportion considered diuretic and also given for curing jaundice. Boiled concentrate of its roots mixed with green peelings of stem of *Albizia lebbeck* Willd. and *Tinospora cordifolia* Miers *ex* Hook. f. & Thoms. taken as blood purifier*.

Active Constituents: Hentriacontane, β-sitosterol, ursolic acid, a phytoecdysome-β-ecdysterone and a new (methyal flavone, 5, 7-dihydroxy-3', 4'-dimethoxy-6-8-dimethyl flavone, triacontanol, a purine nucleoside-hypoxanthine-9-arabinofuranoside, 2 rotenoside, boeravinones A, B and C, punarvine, C-methyl flavone, punarnavoside (*root*) and crystaline amino acid (*leaves*) are the active principles.

Biological Activity: Antispasmolytic, antiarrhythmic, antiviral, anti-inflammatory activities confirmed.

Uses in Literature: Reported to be used for asthma, blood dysentery, cold, cough, colitis, cuts, deworming, diarrhoea, dislocated joints, epilepsy, elephantiasis, eye diseases, fever, gall bladder stones, jaundice, piles, pimples, leprosy, leucorrhoea, jaundice, joint diseases, odema, paralysis, spermatorrhoea, urinary troubles, wounds; and as abortifacient, analgesic, anthelmintic, antiarrhythmic, antiarthritic, anticonvulsant, antiinflammatory antiviral, CNS depressant, diuretic, expectorant, febrifuge, purgative and spasmolytic (Ambasta, 1986; Bondya *et al.*, 2006; Borthakur *et al.*, 2004; Dwivedi *et al.*, 2005; Gogoi & Borthakur, 2001; Khanna & Kumar, 2000; Khare, 2007; Kshirsagar *et al.*, 2003; Kumar & Chauhan, 2005; Mohanty *et al.*, 1998; Maya *et al.*, 2003; Retnam & Martin, 2006; Sen & Behera, 2003).

Major Ayurvedic Formulations: Punarnava, Punarnavasana, Punarnavadi, Mandoor, Guggulu, Abana Tablets.

Trade: Roots sold as 'punarnava' or 'itsat', 'utchata'.

Adulterant or Substitute: Leaves and shoots used as an adulterant or substitute for 'sweta punarnava' or 'varshabhu' (*Trianthema portul-castrum* L.).

Bombax ceiba L. (Pl. 34A).

*Syn.: B. heptaphylla Cav., B. heptaphyllum Rox*b., *B. malabaricum* DC., *Salmalia malabarica* (DC.) Schott & Endl.,

Gossampinus malabarica (DC.) Merr., *Gossampinus rubra* Ham.

Family: Bombacaceae.

Vern.: Simal, Simbal.

Ayurvedic, English, Sanskrit and Regional Names:

Ayur.: Kankataadhya, Mochaa, Mochaahva, Pichhila, Raktapushpa, Shaalmali, Sthiraayu, Tuulini;

Eng.: Red silk cotton tree, Semal tree, SilkCotton Tree;

Sans.: Apurari, Bahuvirya, Nirgandhapuspi, Mocha, Rakta-pushpa, Salmali;

B.: Roktosimul, Simul, Tula;

Bo.: Saer, Somr, Shembal;

G.: Ratoshemalo, Shimlo, Shimul;

H.: Kantisembal, Pagun, Semal, Simal, Somr;

Mal.: Illavu, Mocha, Mullilavu, Unnamurrika;

Mar.: Kanteri samar, Savara, Shevari, Simlo, Tamari;

Oriya: Buro, Mochorso;

Tam.: Agigi, Ilavam, Muli-ilava-maram, Pougar, Selavagu;

Tel.: Kondaburaga, Mundla-buraga-chettu, Pinnaburaga.

Distribution: Throughout hotter Parts of India up to 1,350m.

Description: A large deciduous tree with prickly stem, wide spreading branches, glabrous, lanceolate leaves and red flowers appearing before the leaves. Capsule 5-valved. Seeds enveloped in dense silky hair.

Flowering & Fruiting: February-May.

Habitat Ecology: Drier Areas, wastelands, roadsides; Jhanikker, 1,050m.

Material Examined: HPU-EL-SKS-ABS-15053; 27.02.2009.

Parts Used: Bark. Seeds. Spine. Wood.

Folk Uses: Wood employed for shuttering purposes and for making match sticks, 'toran dawar', altar ('bedi') in marriages* and musical instruments* like 'dholak' and 'drums'. Wooden sticks made from its stem used for grinding* paddy. One tsp of bark decoction recommended twice a day for 15 days to cure sex diseases. Crushed seeds mixed with wheat flour fed to livestock for stomach disorders. Silky hair used for stuffing cushions and pillows. Poultice made from its spine good for boils. Leaves given as fodder for cattle; also sweet infusion of its leaves taken as substitute for 'chawanprash'*. Poultice prepared by keeping leaves of 'kali basuti' (*Pogostemon benghalensis* Burm.) in its stem for a month, considered good for joint pains* and arthritis*. Flower buds cooked as vegetable.

Active Constituents: Lupeol, β-sitosterol (*stem*), hexaacosanol, tocopherol, terpenes (*seeds*), arbinose, galactose, galacturonic acid, catechu-tannic acid and traces of rhamnose (*gums*) have been isolated.

Biological Activity: Antiviral, hypoglycaemic (*flowers*) and oxytocic (seeds) activities found +ve.

Uses in Literature: Reportedly used earlier as an alternative, astringent, aphrodisiac, coolant, cordage, contraceptive, cushion, demulcent, diuretic, expectorant, fibre, firewood, fodder, laxative, styptic, tonic and vegetable; and for abdominal pains, acne, anaemia, antifertility, asthma, atrophy, blood dysentery, bleeding piles, boils, bone fracture, burns, carbuncle, chicken pox, cholera, cough, debility, diarrhoea, fever, filling material for jackets, gonorrhoea, gum trouble, hematuria, inflammation, leucorrhoea, liver enlargement, indigestion (vet.), mattresses and pillows, monorrhagia, pain in legs, pimples, pneumonia, prolapses of uterus, rheumatism, scabies, scorpion-bites, sex weakness, skin troubles, spermatorrhoea, toothache, ulcers, urinary complaints, vitality, ulceration of bladders, vitality, wounds, water troughs and window shutters (Ambasta, 1986; Aminuddin & Girach, 1991; Asolkar *et al.*, 1992; Aswal, 1996; Bajpayee & Dixit, 1996; Banerjee, 1996; Chauhan, 1999; Girach, 1992; Girach & Aminuddin, 1995; Gogoi & Borthakur, 1991, 2001; Kadel & Jain, 2006; Katewa & Arora, 1997; Khare, 2007; Lalramnghinglova, 1996; Maliya, 2004; Manandhar, 1989 a, b; Mishra *et al.*, 1996b; Mukherjee & Namhata, 1990; Patil, 2004; Rao, 1996; Retnam & Martin, 2006; Sarangi & Sahu, 2004; Sen & Behera, 2003; Shome *et al.*, 1996; Singh, 2003; Singh & Maheshwari, 1992; Singh & Zaheer, 1996; Watt, 1972).

Major Ayurvedic Formulations: Abhaya-dirista, Salmali-ghrita Laghugangadhar-churna, and several aphrodisiac, restorative and expectorant preparations.

Trade: Mochavasa, Suparikas Phool.

Adulterant or Substitute: Seed oil used as an adulterant or substitute for cotton seed oil.

Bougainvillea spectabilis Willd. (Pl. 34B).

Family: Nyctaginaceae.

Vern.: Garden-glory.

English and Hindi Names:

Eng.: Garden glory;

H.: Bogenvillea.

Distribution: Often cultivated.

Description: A woody straggling thorny shrub. Leaves orbicular-ovate. Flowers small, yellow in cymose heads having red coloured showy bracts in clusters with number exceeding the flowers. Anthocarp (achene surrounded by persistent perianth lobes) one-seeded with unequal cotyledons and non-endospermic seeds.

Flowering & Fruiting: January-June.

Habitat Ecology: Often cultivated; Bhira; 780m.

Material Examined: HPU-EL-SKS-ABS-15054; 13.04.2010.

Parts Used: Whole Plant. Flowers. Leaves. Wood.

Folk Uses: Grown for ornamental and fencing* purposes. Flowers offered to deities for worshiping*. Wood used as fuel. Leaves employed as fodder* for goats and for making manure*.

Active Constituents: Glycosides of isorhamnetin and quercetin (*flowers*), pinitol (*leaves*) have been isolated.

Biological Activity: Pinitol (0.01g/kgp.o.) exhibits significant hypoglycaemia in normal fasted mice.

Uses in Literature: Reportedly described earlier as an ornamental, and for paintings (Mithila), ulcers and diabetes (Gogoi & Borthakur, 1991; Jha & Basak, 1994; Khare, 2007; Kshirsagar *et al.*, 2003; Lalramnghinglova, 1996).

Brassica campestris L. (Pl. 34C).

Syn.: *Brassica rapa* L.

Family: Brassicaceae.

Vern.: Sarson.

Ayurvedic, English, Sanskrit and Regional Names:

Ayur.: Sarshapa, Siddhaartha;

Eng.: Yellow sarson;

Sans.: Bhutaghana, Bimbata, Duradharsha, Grahagna, Kadamba, Kadambaka, Rajika, Sarahapa, Sarishapa, Sidhartha, Tantuka, Ugragandha;

B.: Kalisarson, Sadrai, Sanshi, Shurshi, Sursi;

H.: Bangasarson, Baralai, Dain, Jariya, Kalisarson, Lahota, Pilasarson, Sarson, Sursi, Tori, Toriya;

Mal.: Karupakatuka;

Mar.: Kalamohare, Sherasa;

Tam.: Karuppukkadugu;

Tel.: Nallaavalu.

Distribution: Throughout India.

Description: Erect annual herbs with yellow flowers in corymbose heads. Pods glabrous, suberect; valves with midrib and flexuous veils. Seeds small, smooth, pale or dark.

Flowering & Fruiting: March-September.

Habitat Ecology: Cultivated as an escape; Kuthera, 900m.

Material Examined: HPU-EL-SKS-ABS-15055; 07.12.2010.

Parts Used: Leaves. Tender Shoots. Seeds.

Folk Uses: Young leaves and tender shoots cooked as vegetable, 'sarson-ka-saag'. Oil cake used as cattle feed. Mustard seeds heated in oil used for garnishing pulses. Seed oil utilized in cooking by the natives; oil combined with camphor applied externally for muscular rheumatism*, whereas in combination with garlic used for earache*. Hot massage of seed oil mixed with black pepper good against injuries* resulted during agricultural operations (vet.). Also, given with wheat flour to check abortion* in animals.

Active Constituents: Active constituents are glycerides of erucic acid, sinigrin, sinalbin, p-hydroxy benzyl isothiocynate and sinapine (*seeds*).

Uses in Literature: Known to be used as counter-irritant, diuretic, emetic, rubefacient, stimulant; and for cold, cough, epilepsy, fever, gastroentric inflammation, influenza fever, itching, leucoderma, piles, skin eruptions, toothache and ulcers (Chopra *et al.*, 1956; Chunekar & Hota, 1998; Chunekar & Pondel, 1999; Deeba *et al.*, 2009; Husain *et al.*, 1992; Kadal & Jain, 2006; Khare, 2007; Kirtikar & Basu, 1994; Kumar *et al.*, 2004; Mishra *et al.*, 1996b; Tiwari & Pande, 2006).

Brassica juncea (L.) Czern. & Coss. (Pl. 34D).

Syn.: *B. willdenovii* Boiss., *Sinapsis juncea* L., *S. integrifolia* Willd., *S. ramosa* Roxb., *S. rugosa* Roxb., *S. cuneifolia* Roxb.

Family: Brassicaceae.

Vern.: Rai.

Ayurvedic, English and Regional Names:

Ayur.: Aasuri rai, Raai, Raajika Tikshnagandhaa;

Eng.: Brown Mustard, Chinese musturd;

Tam.: Kadugu;

Folk.: Raai.

Distribution: Cultivated in Panjab, Uttar Pradesh and other parts of India.

Description: Tall, branching annuals which are more or less hispid. Lower leaves petioled, sometimes pinnatifid; upper toothed, subsessile. Flowers yellow in corymbose heads. Pods linear-lanceolate, valves with prominent midrib. Seeds small, dark, rugose.

Flowering & Fruiting: March-September.

Habitat Ecology: Cultivated as an escape; Ladraur, 978m.

Material Examined: HPU-EL-SKS-ABS-15056; 11.02.2010.

Parts Used: Leaves. Tender Shoots. Seeds.

Folk Uses: Young plants cooked as pot herb* and added to various edible preparations like pulses, 'chutney', 'raita'* (curd) and pickles* as rich source of vitamin B complex. 100-200g of its powdered seeds mixed with wheat flour given to check dehydration* whereas soaked seeds given with curd for expelling intestinal worms in livestock. Paste of grinded seeds in buttermilk given to children for deworming. Oil cakes used as fodder*. Seed oil massaged on nasal passage to check congestion and on soles as coolant*.

Active Constituents: Active constituents are glucosinolate, sinigrin, allyl isothiocyanate (*seeds*). Oil cake contains N, P_2O_5, K_2O, and mucilage: sinalbin, sinapine.

Uses in Literature: Known so far as anodyne, anthelmintic, antidysenteric, aperient, apertif, cattle feed, counter irritant, diaphoretic, diuretic, edible, emetic, galactagogue, rubefacient, stimulant, stomachic, substitute for mustard; and for anoestrus (vet.), arthritis, cough, complaints of nervous systems, dry and hot pharyngitis, foot & mouth, footache, headache, indigestion, loose motion (vet.), nostril blockage, pancreatic secretions, rabies (vet.), rheumatism, sneezing, strength (vet.), treating coryza, warmth (vet.) (Ambasta, 1986; Anonymous, 1948-1976, 1998b; Bhatt *et al.*, 2001; Burkill, 1966; Chauhan, 1999; Chinthu *et al.*, 1998; Duke & Wain, 1981; Jain, 1968; Jain *et al.*, 1994; Khare, 2007; Kirtikar & Basu, 1994; Mishra *et al.*, 1996b; Rekha *et al.*, 1998).

Brassica nigra (L.) Koch. (Pl. 35A).

Family: Brassicaceae.

Vern.: Rai.

Ayurvedic, English and Regional Names:

Ayur.: Banarasi Raai, Raajika;

Eng.: Black Mustard;

Tam.: Kadugu;

Folk.: Raai.

Distribution: Cultivated in Panjab, Uttar Pradesh and other parts of India.

Description: Erect, branched, rigid annual herbs which are more or less hispid with yellow flowers in corymbose heads. Leaves petioled, lower lyrate and upper ones entire. Pods appressed to the stem.

Flowering & Fruiting: March-September.

Habitat Ecology: Cultivated as an escape; Hamirpur, 787m.

Material Examined: HPU-EL-SKS-ABS-15057; 09.01.2010.

Parts Used: Leaves. Tender Shoots. Seeds.

Folk Uses: Young plants relished as pot herb* and added to various edible preparations like pulses, 'chutney', 'raita'* (curd) and pickles as rich source of vitamin B complex. 100-200g of its powdered seeds mixed with wheat flour given to check dehydration* whereas soaked seeds given with curd for expelling intestinal worms in livestock. Oil cakes used as fodder. Seed oil massaged on nasal passage to check congestion and on soles as coolant. Grinded seeds added to buttermilk and given to children for deworming. Seed oil considered as blood purifier*.

Active Constituents: Active constituents are thioglycoside, glucosinolate, sinigrin, allyl isothiocyanate, crotonyl isothiocyanate (*seeds*). Oil cakes contain N-4.9%, P_2O_5-2.5%, K_2O -1.5%.

Uses in Literature: Known so far as anthelmintic, antidysenteric, cattle feed, counter irritant, diaphoretic, edible, stomachic, substitute for mustard; and for anoestrus, cough, complaints of nervous systems, dry and hot pharyngitis, foot & mouth, indigestion, loose motion, nostril blockage, pancreatic secretions, rabies, sneezing, strength, treating coryza, warmth (vet.) (Ambasta, 1986; Anonymous, 1948-1976; Chauhan, 1999; Jain, 1968; Jain *et al.*, 1994; Khare, 2007; Kirtikar & Basu, 1994).

Bryophyllum calycinum Salisb. (Pl. 35B).

Syn.: *B. pinnatum* (Lam.) Kurz, *Kalanchoe pinnata* Pers., *Cotyledon rhizophylla* Roxb.

Family: Crassulaceae.

Vern.: Gillar pattar, Patharchat.

Ayurvedic, English and Regional Names:

Ayur.: Airaavati, Parnabija;

Eng.: Ahiravana-mahiravana, Ghayamari;

B.: Koppata;

G.: Ghayamari;

H.: Zakhm-Haiyat-Ka-Patta;

Kan.: Yo-kiya-pin-ba;

Mal.: Lonnuhadakana-gidd;

Tam.: Simajamunda, Runakalli;

Tel.: Ela-marunna, Elamarunga, Murikuti, Simaja-muda.

Distribution: Throughout tropical parts of India.

Description: Tall erect glabrous herbs with stems upto 1m. Flowers-reddish green or purple. Corolla globose octagonal. Buds develop in the crenatures of the leaves after droping produce new plants. Fruits enclosed in persistent pappery corolla and calyx.

Flowering & Fruiting: March-May.

Habitat Ecology: Drier areas; Krishna Nagar, 858m.

Material Examined: HPU-EL-SKS-ABS-15058; 08.07.2008.

Parts Used: Leaves. Whole Plant.

Folk Uses: Plant grown as an ornamental in pots. 2-3 leaves mixed with fruits of 'bil' *Aegle marmelos* (L.) Correa, (once a day for five days) given in dysentery. Leaves eaten raw and considered antilithic*. Toasted leaves applied to cure wound, boils, pimples and acne*.

Active Constituents: Active properties are due to bryophyllol, bryophollone, bryophollenone, bryophynol and two homogenous phenanthren derivatives-2 (9-decenyl) phenanthrene (I) and 2 (undecenyl), malic, isocitric and citric acid (*leaves*).

Biological Activity: Bersaldegenin-1, 3, 5, orthoacetate inhibited growth of several cancer cell lines.

Uses in Literature: Reported in literature as antibacterial, antifungal, antiseptic, astringent, carminative, disinfectant, emollient; and for boils, bite of venomous insects, bruises, discolouration of affected part, dysentery, healing wounds, kidney and liver troubles leprosy, sloughing ulcers and swelling (Ambasta, 1986; Jain & Patole, 2001; Khare, 2007; Retnam & Martin, 2006; Sharma *et al.*, 2001).

Butea monosperma (Lam.) Kuntze (Pl. 35C).

Syn.: *B. frondosa* Koen *ex* Roxb., *Erythrina monosperma* Lamk., *Plaso monosperma* Kuntze

Family: Fabaceae.

Vern.: Palah, Palash, Plah.

Ayurvedic, English, Sanskrit and Regional Names:

Ayur.: Brahmavriksha, Kimshuka, Kshaarshreshtha, Paalasha, Raktapushpaka, Samidbar;

Eng.: Bastard Teak, Bengal Kino, Butea gum, Flame of the forest;

Sans.: Bijasneha, Kinsuka, Ksharashreshtha, Palasa, Palasham, Vatapotha;

B.: Palas, Palasa;

Bo.: Khakara, Palasi;

G.: Khakara, Palasso;

H.: Chichra, Kankrej, Palas;

Kan.: Muttuga, Muttuga-mara, Thoras;

Mal.: Brahmavrikshan, Palasinsamatha;

Mar.: Kakradha, Phalas, Paras, Palas;

Oriya: Kinjuko, Polaso, Porasu;

Tam.: Kutumurukku, Murukkam, Parasa, Pora San.;

Tel.: Modugachettu, Mohtu, Palas, Tella-modugu; Tellamoduga, Vatapodhamu;

U.P.: Dhak.

Distribution: Throughout India.

Description: A medium-sized erect deciduous tree upto 13m and possessing pinnately trifoliate leaves. Shoots clothed with grey or brown silky pubescence. Leaflets coriaceous, silky tomentose beneath. Flowers bright orange red in terminal racemes and appearing before leaves. Pods 1-seeded, grey downy.

Flowering & Fruiting: March-April.

Habitat Ecology: Drier parts, borders of agricultural fields; Jhaniyari, Anu, 900m.

Material Examined: HPU-EL-SKS-ABS-15059; 23.03.2008.

Parts Used: Leaves. Flowers. Wood.

Folk Uses: One tsp decoction of leaves prescribed twice daily for seven days to cure amenorrhoea, also used in making 'beedi' (cigarette). Flowers tied as an amulet on stomach for relief from pain. Decoction of flowers (thrice daily for fifteen days) good against jaundice. Wood considered sacred and used for performing 'hawan'. Wood also used for shuttering, frames, making light structures*. Leaves given as fodder.

Active Constituents: Principle constituents are butrin, coreopsin, isobutrin-5, monospermoside isocoreopsin and three iso-derivatives, sulphurein, chalcones *(flowers)*, butin, butrin, isobutrin, palasonin, palastrin (*plant*).

Biological Activity: Antifungal, antibacterial, antiimplantation, antiovulatory, spasmolytic and anthelmintic (due to palasonin) activities +ve.

Uses in Literature: Used so far as an abortifacient, antiseptic, anthelmintic, aphrodisiac, astringent, beedi wrapper, colouring material, contraceptive, coolant, diuretic, febrifuge, fish poison, fibre, fodder, fuel, haemorrhagic, stomachic, tonic, vegetable, lac host, laxative, ornamental, sacred, steriliser, styptic; and for bad breath, boil, bone fracture, bowls, cuts, delirium, diabetes, diarrhoea, dog bite, dysentery, dummy corpse, eye complaints, expelling worms, fever, foot and mouth diseases, gastric problems, heart burns, inflammation, irregular menstruation, jaundice, leucorrhoea, ophthalmia, pains, piles, pimples, plates, ringworms, skin diseases, spermatorrhoea, sun stroke, tooth and gum disorders, traditional dwelling, tuberculosis, urinary complaints, water scoops, wounds, well curb and worshipping (Ambasta, 1986; Bajpayee & Dixit, 1996; Banerjee, 1996; Bhandary *et al.*, 1996; Bhatt *et al.*, 2001, 2002; Chopra *et al.*, 1956; Gill & Nyawuame, 1994; Goel *et al.*, 1984; Jain, 1968, 1991, 1996; Jain *et al.*, 1994; Jadhav, 2006; Jha & Varma, 1996; Kadel & Jain, 2006; Katewa *et al.*, 2001; Khanna & Kumar, 2000; Khare, 2007; Kothari & Moorthy, 1996; Kurian, 1995; Lalramnghinglova, 1996; Masih, 2003; Mohanty *et al.*, 1998; Pandey & Rout, 2006; Pandey & Verma, 2002; Patel *et al.*, 2003; Pawar & Patil, 2004; Rama Rao & Henry, 1996; Ravishankar & Henry, 1992; Reddy *et al.*, 1996; Retnam & Martin, 2006; Saini, 1996b; Sarangi & Sahu, 2004; Satapathy & Brahmam, 1996; Silori & Rana, 2000; Singh, 2003; Singh & Prakash, 1994, 1996; Singh & Zaheer, 1996; Varma, 1997; Venkataratnam & Venkata Raju, 2004; Watt, 1972).

Major Ayurvedic Formulations: Ayaskrti, Janmagautti, Krimighna-churn, Krimi-kuthara-rasa, Krimimudgar-ras, Mahanarayan-taila, Nyagrodhadi -Kwatha -churna, Palasabhijadi-churn, Palsa-Ksharaghirta, Plasa-puspasava, Plash-ghrit.

Trade: Traded in Indian market as 'Plashapapra' (*seeds*), 'Dhaak phool' or 'Tesh' flowers and 'Bengal kino' (*gum*).

Adulterant or Substitute: Dried trunk juice used as an adulterant or substitute to East Indian Kino (*Pterocarpus marsupium* Roxb.). Seeds used as substitute for Sanotin, Kirmala (*Artemisia maritime* L. var. *thomsoniana* Clarke).

Callistemon lanceolatus DC. (Pl. 35D).

Syn.: *C. citrinus* Skeels

Family: Myrtaceae.

Vern.: Bottle Brush.

English Name:

Eng.: Bottle Brush Tree.

Distribution: Cultivated in Himalayas upto 1,500m.

Description: A small evergreen tree. Leaves glossy, lanceolate. Flowers crimson in terminal spikes about 15cm long, branches growing through the spikes, filaments red long, giving appearance of brush. Sepals and petals inconspicuous.

Flowering & Fruiting: March-April.

Habitat Ecology: Planted. Bhota; 765m.

Material Examined: HPU-EL-SKS-ABS-15060; 22.05.2011.

Parts Used: Spike. Whole Plant.

Folk Uses* : Planted as an ornamental*. Spike used for decorative purposes*.

Uses in Literature: There is no mention of its use in literature.

Calotropis procera (Ait.) R. Br. (Pl. 36A).

Syn.: *Asclepias procera* Willd., *Calotropis hamiltonii* Wall., *C. heterophylla* Wall., *C. wallichii* Wight

Family: Asclepiadaceae.

Vern.: Ak, Madar.

Ayurvedic, English, Sanskrit and Regional Names:

Ayur.: Aasphota, Alarka, Arkaparna, Kshirparna, Surya, Suuryaahvya, Tapana, Tuulaphala, Vasuka,Vikirna;

Eng.: Milkweed, King's crown, Sodom Apple;

Sans.: Alarka, Arka, Binbora, Mandara, Vasuka; B.- : Akada; G.- : Akadoakdo, Ankado, Nanirui, Nanoakado;

H.: Akada, Ak, Mamiri;

Kan.: Bili yekkada gida, Yakkigida;

Mal..: Erukku;

Mar.: Mandara;

Oriya: Orko;

Tam.: Vellerukku;

Tel.: Erajiledu.

Distribution: Drier arid zones of western and central India.

Description: Evergreen shrubs upto 5m. Umbels long peduncled, cottony. Flowers pink coloured, purple spotted. Latex milky. Follicles covered with woolly pubescence.

Flowering & Fruiting: February-May.

Habitat Ecology: Wastelands, riverbeds, roadsides, forest areas; Salooni, 1,514m.

Material Examined: HPU-EL-SKS-ABS-15061; 23.04.2009.

Parts Used: Flower. Leaves. Roots. Seeds. Latex.

Folk Uses: Plant considered as sacred and worshipped in 'Navgarah vatika' and 'Shani puja', its wood used for hawan*. Milky latex used as a cure for skin problems, toothache (latex added to oil in 1:1 ratio and burnt then applied on dental caries for relief), ring worm and piles. Seed fibres used for stuffing* purposes. Seed oil massaged for curing rheumatoid arthritis.

Active Constituents: Active principles due to uzarigenin, syriogenin, proceroside, calotropin, calotoxin, uscharin, uscharidin, α and β-amyrin, β-sitosterol, kalotropagenin, uzarigenin (*latex*) and a new triterpene calotropenil acetate, cyanidin-3-rhamnoglucoside (*flowers*).

Biological Activity: Calotropin exhibits digitalis like action on the heart.

Uses in Literature: Reported to be used as an abortifacient, alterative, antidote in scorpion sting, antiseptic, bow-strings, camel fodder, cardiac, diaphoretic, emetic, gun powder charcoal, infanticide, laxative, magico-therapeutic (put on forehead of women in labour to achieve easy delivery), masticatory, psychoactive plant, symbolic (abode of God and taboo to offer to Godesses), tonic, tooth brush; and for asthma, boils, bronchitis, bronchitis (vet.), cholera, chronic rheumatism, cold, constitutional syphilis, corn, cuts (vet.), dental caries, diarrhoea, dropsy, dysentery, eczema, elephantiasis, epilepsy, expelling worms, fever, fibre, gastric trouble of children, guinea worm, idiopathic ulceration, insect bite, leprosy, malarial fever, menstrual disorders, mercurial cachexia, mouth diseases (vet.), neck injury (vet.), pain, piles, pyrexia (vet.), rheumatoid arthritis, ring worm, scorpion bite, skin diseases (vet.), small pox, sores, spleen complaints, stomach disorders, swellings, whooping cough and wounds (vet.) (Ambasta, 1986; Bhattacharya, 1996; Datt, 1996; Huidrom, 1996; Jain, 1991, 1996; Jain *et al.*, 1994; Kapur *et al.*, 1996; Khanna *et al.*, 1996a; Maheshwari & Singh, 1984; Masih, 2003; Mishra *et al.*, 1997; Pandey & Verma, 2003; Saini, 1996 a,b; Sharma & Borthakur, 2003; Sharma & Sood, 1997; Shome *et al.*, 1996; Sikarwar, 1996; Singh & Maheswari, 1983; Singh & Prakash, 1996; Srivastava *et al.*, 2003; Watt, 1972).

Major Ayurvedic Formulations: Arkadi-churna, Arka-lavana, Arkaolakanda, Arka-taila, Arkeswar, Dhanvantara-ghrita, Ekangavira, Mahavata-vidhavamba, Mahavisagarbha-taila, Pravala-pancamrta, Sutikabharana.

Canna indica L. (Pl. 36B).

Syn.: *C. chinensis* Willd., *C. indica* var. *orientalis* Rosc., *C. orientalis* Rosc.

Family: Cannaceae.

Vern.: Aeran, Hardarshan, Sudarshan.

English, Sanskrit and Regional Names:

Eng.: Indian Bead, Indian Reed, Indian Shot;

Sans.: Devkili, Kamkshi, Krishnamantara, Sarvajaya, Shilarambha, Vanakadali, Kamakshi, Lalsarbojaya, Sarbajaya; B.- : Akalabera; G.- : Sabbajaya, Sarvaliaya;

H.: Hudigana, Kalahu;

Kan.: Kattuvala;

Mal.: Devakali;

Mar.: Sorobojoya;

Oriya: Hakik;

Tam.: Kalvalai, Kalvalaimani, Kundimani, Puvalai, Siramalai;

Tel.: Guruginja.

Distribution: Mostly throughout India.

Description: Erect undershrubs with a creeping, subterranian, perennial, tuberous rhizome and fleshy thick roots. Leaves sub-sessile, oblong, large and acuminate. Corolla bright red. Stem simple, leafy. Capsule red.

Flowering & Fruiting: January-February.

Habitat Ecology: Oftenly cultivated in gardens as an ornamental; wild in wastelands; Baru, 800m.

Material Examined: HPU-EL-SKS-ABS-15062; 25.08.2010.

Parts Used: Flower. Leaves.

Folk Uses: Planted on the wetland areas for reclaimation of soil*. Leaves used as fodder* and for making thatched roofs. Poultice of leaves heated in oil used against stomachache and internal swelling*. Flowers used in worshipping.

Uses in Literature: Used in India as an antidote to snake bite, demulcent, diaphoretic, diuretic, febrifugal, fodder (vet.); and for burns, chest pain, cough and cold, cloth-like jute, dropsy, dyeing, edible, headache, hounding dogs (vet.), meal plates, necklace, offerings and adornments, ornaments, pin-worms, ropes, sacks, stomachache, swelling of abdomen in cattle, thatching houses, ulcers, worshipping

(Agarwal, 2003; Ambasta, 1986; Anonymous, 1948-1976; Bajpayee & Dixit, 1996; Dagar, 2003; Datt, 1996; Gogoi & Borthakur, 1991; Jain, 1991; Kapur & Nanda, 1996; Rama Rao & Henry, 1996; Watt, 1972).

Major Ayurvedic Formulations: Aragvahadi-kvatha-churna.

Cannabis sativa L. (Pl. 36C).

Syn.: *C. indica* Lamk.

Family: Cannabinaceae.

Vern.: Bhang.

Ayurvedic, English, Sanskrit and Regional Names:

Ayur.: Bhangaa, Indraasana, Maadani, Maatulaani, Trailokya-vijayaa, Tribhuvanavijayaa, Vijayaa;

Eng.: Hemp, Indian hemp;

Sans.: Bhanga, Chapola, Ganja, Ganjika, Hursini, Indrasana, Jaya, Ujaya, Viyaya, Vrijpatta;

B.: Bhang, Ganja, Sidhi;

H.: Bhang, Charas, Ghanja, Gur, Kinnab, Siddhi;

Kan.: Bhangi-gida;

Mal.: Ginjilachilachi, Kanchavachetto;

Mar.: Bhangachajhada;

Tam.: Ganja-chedi, Kalpam, Korkkarmuli;

Tel.: Ganjari-chettu.

Distribution: Throughout India.

Description: Erect aromatic annuals with palmately divided leaves and stem upto 2.5m. Leaves usually alternate; the lower ones sometimes opposite. Flowers yellowish-green; male in drooping panicles, female axillary. Achenes small, enclosed in perianth, glandular hairy.

Flowering & Fruiting: June-September.

Habitat Ecology: Wastelands, edges of fields; Krishna Nagar, 858m.

Material Examined: HPU-EL-SKS-ABS-15063; 12.03.2008.

Parts Used: Whole Plant. Leaves. Seed.

Folk Uses: Plant used in black magic*. Stem used for making fibres, ropes*, mats*, sacs, cordage* and weaving*. Poultice of leaves with mustard oil applied on joints for relieving inflammation* and rheumatic pains*; and its juice applied on insect bites as an antidote* both in humans and livestock. Smoke of burnt leaves given to livestock to repel insect pests*, leaves also used for making 'pakora'* and 'ghota' (hallucinagic drink) and offered* to appease Lord Shiva. Roasted seeds fed to cattle in diarrhoea.

Active Constituents: These are glucocapparins (*seeds*), pthalic acid, 1-stachydrine, cannabidiolic acid, canabidiol, cannabinol, tetrahydrocannabinol, eugenol, guaiacol, cannabigerol, delta-9-tetrahydrocannabinol (THC) (*flower, fruit husk*).

Biological Activity: Antibacterial, hypotensive and spasmolytic activity and effect on CNS found +ve.

Uses in Literature: Known in Indian literature as an anthelmintic, anthelmintic (vet.), appetiser, diuretic, edible (seeds), hallucinogenic, hypnotic, laxative, narcotic, nervine, sedative, stimulant, sleeping pills; and for asthma, blood dysentery (vet.), bowel complaints, bronchitis, cold, cough, convulsion, cramps, cuts, delirium, diarrhoea, dyspepsia, ear complaints, epilepsy, eye diseases, eczema (vet.), flatulence (vet.), gonorrhoea, hydrocele, indigestion (vet.), labour pains, paralysis of tongue, piles, rheumatism (vet.), skin diseases, sores, tetanus and making rough clothes (Agarwal, 2003; Bhattarai, 1992; Chopra *et al.*, 1956; Dastur, 1964; Drury, 1978; Jain, 1991; Kaul, 1997; Khanna & Kumar, 2000; Khare, 2007; Manandhar, 1988; Mishra *et al.*, 1996b; Mitre, 1997; Nath & Bordoloi, 1989; Retnam & Martin, 2006; Samwatsar & Diwanji, 1996b; Shah, 1997; Sharma, 2003; Tiwari & Pande, 2006).

Major Ayurvedic Formulations: Jatiphaladi-churn, Madnanandmodaka, Kameshwar-modak, Vedantak-rosa, Vijaya-vatika.

Adulterant or Substitute: Seed oil used as substitute for linseed oil in paints, varnish and making soft soap.

Canscora diffusa (Vahl) R. Br. (Pl. 36D).

Syn.: *Canscora lawii* Wt.

Family: Gentianaceae.

Vern.: Shankhpushpi.

Ayurvedic Name:

Ayur.: Daakuni.

Distribution: Throughout India extending to 1,000m.

Description: Annuals growing up to 50cm high with many cylindrical branches and stalkless leaves which are oppositely arranged on the branches. Flowers in lax cymes at the end of branches, rosy pink.

Flowering & Fruiting: October-November.

Habitat Ecology: Wild in fellow land; planted as medicinal herb; Khaggal, 920m.

Material Examined: HPU-EL-SKS-ABS-15064; 03.07.2008.

Parts Used: Whole Plant. Leaves.

Folk Uses: Decoction of the plant (15-25ml, twice a day) given in epilepsy. Whole plant used as a nerve tonic for mental disorders, in dysentery* and stomach troubles*.

Active Constituents: These are xanthones, xanthone glucosides (-) loliolide, methoxy and hydroxyl xanthones (*aerial parts*), beta-amyrin, friedelin, gentianina, xanthones and 16 xanthones including magniferin (*roots*), iron, potassium, calcium, nitrates and carbonates (*plant ash*).

Biological Activity: Xanthones have activity against *Mycobacterium tuberculosis* whereas magniferin is protective against liver injury.

Uses in Literature: Reported in literature as anticonvulsant, anti-inflammatory,

CNS depressant, hepatoprotective, substitute of *C. decussata* Schult. (Khare, 2007; Sethiya *et al.*, 2010).

Capsicum annuum L. (Pl. 37A).

Family: Solanaceae.

Vern.: Mirach.

Ayurvedic, English and Regional Names:

Ayur.: Katuviraa, Lankaa, Raktamaricha;

Eng.: Red chillies, Red pepper;

H.: Mirch;

Tam.: Milagay.

Distribution: Cultivated.

Description: Annual or perennial herbs with lanceolate-ovate, entire or repand leaves and white flowers in axillary cymes. Berries elongated which are red on maturity.

Flowering & Fruiting: Greater part of the year.

Habitat Ecology: Cultivated; Bhareri, 920m.

Material Examined: HPU-EL-SKS-ABS-15065; 04.09.2010.

Parts Used: Plant. Fruit.

Folk Uses: Plant considered good for magico-religious* purposes. Grounded fruits used for preparing pepper sauce* and various pickles*. Dried powder of fruits along with burnt 'reetha' (fruits of *Sapindus mukorossi* Gaertn.) applied on parts bitten by rabied dog. 200 ml decoction of fruits along with *Trigonella foenum-graecum* L., *Cuminum cyminum* L., *Allium cepa* L. and jaggery given twice daily for treating skin eruptions* in cattle.

Active Constituents: Lutein, capsorubin, cryptoxanthin, capasnthin, rubixanthin phytofluene, capasaicin, capsicin, capsaicin and solanin (*fruits*) are the main constituents.

Biological Activity: Capsaicin in conc. of 0.1 to 1.0 µm induces relaxation in precontracted longitudinal muscle strips of gastric fundus potentiated by hexamethonium whereas it activates sensory afferent neurons at conc. of (0.1 to 1.0µm).

Uses in Literature: Used as analgesic, carminative, expectorant, fruit, rubefacient, spice, stimulant; and for common cold, dog bite, diarrhoea, dysentery, dyspepsia, foot diseases, piles, rheumatism, scarlatina, sore throat, snake bite and yellow fever (Bhatt *et al.*, 2001; Gogoi & Borthakur, 2001; Khare, 2007; Kirtikar & Basu, 1994; Kumar *et al.*, 2004; Retnam & Martin, 2006; Samati, 2004; Sen *et al.*, 2002; Sharma, 2003).

Major Ayurvedic Formulations: Visham-jwaraghinivati, various liniments and embrocations.

Cardiospermum halicacabum L. (Pl. 37B).

Syn.: *C. microcarpum* (kunth) Blume

Family: Sapindaceae.

Vern.: Tikku-malu. Kanphuti.

Ayurvedic, English, Sanskrit and Regional Names:

Ayur.: Kaakatiktaa, Kaakaadani, Karnsphotaa, Shatakratulataa;.

Eng.: Heart pea, Heart seed;

Sans.: Bunu-uchchhe, Indravalli;

H.: Kanphuti, Kapalphuta;

Tam.: Mudukottan.

Distribution: Tropical and subtropical countries upto 1,212m.

Description: An annual climbing nearly glabrous herb with long, slender grooved branches, alternate, pinnate leaves and a few, small, white, irregular flowers in axillary long stalked corymbs having a pair of simple coiled tendrils near the base. Sepals 4 free, petals 4 in unequal pairs. Stamens 8 unequal. Capsule globose, 3-celled, membranous with 3 seeds which are heart-shaped, usually arillate, spherical and black. Testa crustaceous.

Flowering & Fruiting: August-September.

Habitat Ecology: Wild in fellow land, wasteplaces; Badhani, 950m.

Material Examined: HPU-EL-SKS-ABS-15066; 03.09.2009.

Parts Used: Whole Plant. Leaves.

Folk Uses: 2-3 drops of leaf juice prescribed as drops for earaches*. Poultice of leaves applied on lower abdomen of ladies after delivery for expelling placental waste*. Leaf extr. also used for washing hair for checking dandruff. Paste of its seeds given to livestock to check dysentery resulted due to consumption of 'kodri' grass (*Paspalum scrobiculatum* L.) or as an antidote against poisonous 'lakondru' plant (*Bryophyllum* sp.). Also, the mixture comprising of its seeds and leaves of 'gandhela' (*Murraya koenigii* Spreng.) considered good against 'patrara' (acute tympany)* and 'ban' (intestinal obstruction)* diseases of animals. Crushed seeds of this plant given with wheat bread ('chapatti') to horses for sexual problems*.

Active Constituents: Active constituents of aerial parts are beta-sitosterol and its D-glucoside, an alkaloid, oxalic acid, and amino acids, quebrachitol, saponin.

Biological Activity: Alkaloid fraction from seeds exhibited hypotensive and cardiac inhibition in dogs, blocked spasmogenic effects of acetyl-choline, histamine and 5-HT on guinea pig ileum, biphasic effort on frog rectus abdominis muscle.

Uses in Literature: Reported to be used as analgesic, antiarthritis, antibacterial, anti-inflammatory, antisickling, diaphoretic, diuretic, emmenagogue, laxative, refrigerant, rubefacient, stomachic; and for alopecia, amenorrhoea, anorexia (vet.), common cold, dandruff, darkening hair, diarrhoea (vet.), ephemeral fever (vet.), erysipelas, emetic, filaria (vet.), haemorrhoids, increasing body weight, lumbago, nervous diseases, repelling lice and parasite (vet.), rheumatism, rheumatism (vet.), skeletal fractures (Geetha *et al.*, 1996b; Goud & Pullaiah, 1996b; Khare, 2007; Reddy *et al.*, 1998; Sikarwar, 1996).

Carduus nutans L. (Pl. 37C).

Syn.: *Carduus lucidus* Wall.
Family: Asteraceae.
Vern.: Lhee.

English and Hindi Names:

Eng.: Musk Thistle;
H.: Kanchari.

Distribution: Western Himalayas (Kashmir-Shimla), Panjab.

Description: Tall, stout, erect thistles upto 1m with simple interruptedly spinulous winged stem. Leaves entire, 1-2 pinnatifid waved spinuous. Heads discoid, ovoid or globose, solitary or clustered, inclined or drooping. Flowers crimson. Achenes glabrous, obtusely 4-angled.

Flowering & Fruiting: Greater part of the year.

Habitat Ecology: Wild; Dhaneta, 596m-1,070m.

Material Examined: HPU-EL-SKS-ABS-15067; 05.10.2010.

Part Used: Plant.

Folk Uses: Planted for soil erosion* and as a hedge*. Plant given as a fodder to increase milk capacity* by 1.5 times in domestic animals. Also, cooked as pot herb* after removing spines.

Use in Literature: Used as a good fodder grass for camel (Agarwal, 2003; Khare, 2007).

Carica papaya L. (Pl. 37D).

Family: Caricaceae.
Vern.: Papita.

Ayurvedic, English, Sanskrit and Regional Names:

Ayur.: Erand-karkati, Papitaa;
Eng.: Melon tree, Papaw, Papaya, Paw-Paw;
Sans.: Chirbita, Eranda, Naikadala;
B.: Papai;
Bo.: Chibhado, Katha, Paputa;
G.: Chibda, Eranda kakdi, Thadchibadi;
H.: Anda Kharbuja, Papita;
Mal.: Kappalam, Nalikadala;
Mar.: Papaya;
Oriya: Omrytobhondo, Popoya;
P.: Popai;
Tam.: Pappli, Parangiyananakki;
Tel.: Madananaba, Madhurnakamu.

Distribution: Cultivated throughout India.

Description: Rapidly growing sub-herbaceous, branchless, small, soft wooded trees having succulent trunk with milk sap and large glabrous, palmatifid and palmi-nerved leaves. Flowers pale-yellow, fragrant. Fruits succulent, indehiscent, 1-celled. Seeds numerous, black, enclosed in sweet mucus pulp.

Flowering & Fruiting: Greater part of the year.

Habitat Ecology: Planted; Kalinjari, 900m.

Material Examined: HPU-EL-SKS-ABS-15068; 02.01.2010.

Parts Used: Fruit. Leaves. Seeds.

Folk Uses: Unripe fruits cooked as vegetable and its pulp stuffed in 'prantha' (fried 'chapatti'). Ripe fruits edible and considered good for liver ailments, diabetes and as cardiac tonic. Fruits also taken with milk to cure constipation*; and its latex applied externally for treating eczema. Latex of the plant also consumed to expel worms. Leaves used as fodder* for animals. 3-5 g powdered seeds prescribed with milk once daily to cure diabetes and heart troubles. Seeds used as an adulterant for black pepper.

Active Constituents: Hentricontane, β-sitosterol, benzoyl thiourea, carpaine (*seeds*), myristoleic, palmitoleic, arachidic acid, pectin, free sugar, papain, vit.c, nicotinic acid (*fruits*) have been characterized.

Biological Activity: Carpaine exhibits antitumour activity *in vitro* against mouse lymphoid leukaemia (1:210), lymphocytic leukaemia (P:338) and ehlrich ascites tumour cells, and also as antimicrobial against gram +ve and -ve bacteria.

Uses in Literature: Used earlier as an abortifacient, adulterant, anthelmintic, antidote for cancer, blood anticoagulant, carminative, diuretic, emmenagogue, face creams, febrifuge, galactagogue, hair shampoo, purgative, soft drinks, stomachic, source of vit. and papain, substitute for soap, tonic, vegetable; and for amoebic dysentery, asthma, beri-beri, bleeding piles, boils, bone fracture, candies, cardiac diseases, chest problems, cooling, cough, cuts, dental caries, diabetes, diptheria, dyspepsia, eczema, excessive bleeding of the kidney, flatulence, flavouring meat, gastric, gingivitis, gonorrhoea, haemptysis, heart depression, itching, jaundice, kidney stone, liver disorder, malaria, marmalades, pain in gums, pickling, piles, promoting menstrual flow, reducing elephantoid growth, rheumatis, ringworm, skin diseases, spleenic and hepatic enlargement, tonsils, tooth and gum ache, tuberculosis, ulcers, urinary trouble, warts and wounds (Ambasta, 1986; Bhardwaj & Gakhar, 2003; Borthakur, 1993; Borthakur *et al.*, 2004; Dagar & Dagar, 1999; Ghosh *et al.*, 1996; Gill *et al.*, 1993, 1997; Gogoi *et al.*, 2003; Hosagoudar & Henry, 1996a; Jain, 1991; Jain *et al.*, 1991; Khare, 2007; Kulkarni & Kumbhojkar, 1993; Kurian, 1995; Lalramnghinglova, 1996; Maliya & Singh, 2003; Pandey & Verma, 2002; Rama Rao & Henry, 1996; Retnam & Martin, 2006; Sen & Behera, 2003; Sharma *et al.*, 2002; Singh & Zaheer, 1996; Siwakoti & Verma, 1996; Watt, 1972).

Carissa carandus L. var. *congesta* (Wt.) Bedd. (Pl. 38A).

Syn.: *C. macrophylla* Wall.
Family: Apocynaceae.
Vern.: Badda Garnae.
Ayurvedic, English, Sanskrit and Regional Names:
Ayur.: Karinkaara, Karanarda, Krishnapaakphal, Kshirphena;
Eng.: Christ's Thorn, Bengal Currant;
Sans.: Bokkarambuka, Jalipushpa, Krishna Pakaphala, Vanalaya;

B.: Karamcha, Kurumia;
Bo.: Hartundi, Karinda;
G.: Karamadan, Karamarda;
H.: Karaunda, Karrona;
Kan.: Karekayi;
Mal.: Kalavu, Karakka;
Mar.: Boronda, Haradundi, Karavandi;
Oriya: Khyirokoli, Korondo;
Tam.: Kalakkay, Perungala;
Tel.: Paddavaka, Peddalive, Vaka.

Distribution: Throughout the drier sandy or rocky soils of India; cultivated from Panjab to Sri Lanka.

Description: Erect spinous shrubs or small gnarled trees with many dichotomous, rigid, spreading branches having two simple or forked thorns at its axils and nodes. Leaves subsessile, thinly coriaceous with rounded or retuse base. Cymes terminal. Peduncle stout. Flowers white, sweet-scented and in corymbose cymes. Thorns long and sharp. Berries ovoid, red -dark purple.

Flowering & Fruiting: April-July.

Habitat Ecology: Open slopes, wastelands in drier areas; Garne-ke-Gallu, 900m.

Material Examined: HPU-EL-SKS-ABS-15069; 15.05.2011.

Parts Used: Leaves. Fruit. Wood.

Folk Uses: Wood used for making combs, and its ash applied to cure pain in joints*. Leaves used as fodder* for goats and sheep. Fruits edible and consumed raw as well as pickled, and considered good for anaemic* problems and also used as blood purifier*.

Active Constituents: Carrissonecarissol glucose, galactose, serine, glutamine, alanine, valine, phenylalanine, glycine and carissic acid (*fruits*) are the chief active constituents.

Biological Activity: Hypotensive effect and cardiotonic activity confirmed.

Uses in Literature: So far used as an anthelmintic, antidote for snake bite, antiscorbutic, astringent, edible, expectorant, laxative, stomachic; and for ansarca, bite of rabied dog, blood dysentery (vet.), checking bleeding of internal organs, chutney, combs, convulsions, cooling in bilious complaints, curry, diarrhoea (vet.), dropsy, dyeing and tanning, epilepy, fencing, fever, indigestion, itching, jellies, pickling, madness, postnatal care, pudding, remittent fevers, rheumatism, scurvy, skin infection, sore (vet.), spoons, stimulating appetite, strengthening and giving tone to the stomach, tarts and pudding, tussar silk worms and wounds (vet.) (Ambasta, 1986; Bhandary *et al.*, 1996; Khanna *et al.*, 1996a; Khare, 2007; Kurian, 1995; Sahu, 1996; Upadhye & Kumbhojkar, 1996; Watt, 1972).

Carissa spinarum auct. non L. (Pl. 38B).

Syn.: *C. opaca* Stapf *ex* Haines.
Family: Apocynaceae.
Vern.: Garnae.

Ayurvedic, English, Sanskrit and Regional Names:

Ayur.: Karanarda, Karinkaara, Krishnapaakphal, Shirphena;
Eng.: Bengal currents;
Sans.: Bokkarambuka, Jalipushpa, Krishna, Pakaphala, Vanalaya;
B.: Karamcha, Kurumira;
Bo.: Hartundi, Karinda;
G.: Karamadan, Karamarda;
H.: Karaunda, Karrona;
Kan.: Karekayi;
Mal.: Kalavu, Karakka;
Mar.: Boronda, Haradundi, Karavandi;
Oriya: Khyirokoli, Korondo;
Tam.: Kalakkay, Perungala;
Tel.: Paddavaka, Peddakalive, Vaka.

Distribution: Throughout India.

Description: An erect, spinous shrub possessing leathery elliptic, ovate or rounded acute leaves and white, sweet-scented flowers in corymbose cymes. Thorns long and sharp. Berries ovoid, red to dark purple.

Flowering & Fruiting: May-June.

Habitat Ecology: Open slopes, wastelands in drier areas; Garne-ke-Galu, 900m.

Material Examined: HPU-EL-SKS-ABS-15070; 09.05.2010.

Parts Used: Stem. Leaves. Fruit.

Folk Uses: Gargles of stem decoction cure mouth ulcer*. Leaves used as fodder* for cattle and stem for fencing*. Fruits edible and good for anaemia*. Wood, twigs and dry leaves used as fuel*.

Active Constituents: A new triterpine carissol (*fruits*) along with glucose, galactose, serine, glutamine, alanine, valine, phenylalanine, glycine and carissic acid have been isolated.

Biological Activity: CNS depressant and hypotensive activity found (+)ve.

Uses in Literature: Known so far for combs, constipation, cough, curry, dyeing, dysentery, edible fruits, fever, indigestion, veterinary ailments, wounds; and as laxative (Jain, 1991; Khare, 2007; Pal, 1981; Retnam & Martin, 2006).

Major Ayurvedic Formulation: Kusthara.

Carthamus oxycantha Bieb. (Pl. 38C).

Family: Asteraceae.
Vern.: Pohli.

Ayurvedic, English and Regional Names:

Ayur.: Kusumbha, Vahinshikhaa, Vastraranjaka;
Eng.: Safflower;
Tam.: Chendurakam.

Distribution: Present wild in throughout India.

Description: Thistle-like herbs possessing white stem and branches, leaves alternate, stiff, sessile, lanceolate, lower shortly spinulose-toothed whereas upper half amplexicaul very spinous. Flowers orange yellow in discoid heads. Corolla-tube slender, deeply 5-lobed. Achenes glabrous, 4-angled.

Flowering & Fruiting: December-April.

Habitat Ecology: Forests, open fields, roadsides; Tauni Devi, 1,050m.

Material Examined: HPU-EL-SKS-ABS-15071; 18.02.2010.

Parts Used: Plant. Flowers. Seed.

Folk Uses: Plant yields a dye*. Oil obtained from seeds massaged for rheumatism, sores and open wounds*. Poultice of flowers applied on skin eruptions.

Active Constituents: Plant contains sesquiterpene glycoside, hinesol-beta-D-fucopyranoside, luteolin-7-glucoside.

Biological Activity: Oil aids prevention of arteriosclerosis and coronary heart diseases.

Uses in Literature: Reported earlier as diuretic, emmenagogue, sedative, stimulant; and for arteriosclerosis, coronary heart diseases, fevers, healing sores, kidney disorders, measles, rheumatism, skin eruptions (Ambasta, 1986; Chopra *et al.*, 1956; Chunekar & Pondel, 1999; Husain *et al.*, 1992; Khare, 2007; Kirtikar & Basu, 1994).

Carum copticum Hiern (Pl. 38D).

Syn.: *Ammi copticum* Boiss., *Athamantha ajowan* Wall., *Bunium aromaticum* L., *Ligusticum ajwain* Fleming, *L. ajouan* Roxb., *Ptychotis coptica* DC., *P. ajowan* DC., *Seseli ammoides* Jacq., *Sison ammi* Jacq., *Trachyspermum ammi* (L.) Sprague.

Family: Apiaceae.

Vern.: Jawain, Ajwain.

English, Sanskrit and Hindi Names:

Eng.: Ajowan, Carum, King's cumin;

Sans.: Ajmoda;

H.: Ajwain, Ajowan.

Distribution: Tropical and subtropical regions, cultivated from Panjab and Bengal S. Deccan.

Description: A perennial or annual herb with stem up to 90cm. Leaves 2-3 pinnate, ultimate segments linear. Flowers white, in umbels. Carpels subpentagonous, dorsally compressed, nearly plane on the inner face. Ridges usually distinct, vittae solitary small. Fruit muricate, subhispid and ovoid.

Flowering & Fruiting: April-May.

Habitat Ecology: Cultivated in fields; Bassi, 987m.

Material Examined: HPU-EL-SKS-ABS-15072; 15.08.2009.

Parts Used: Whole Plant. Leaves. Seeds.

Folk Uses: Seeds mixed with jaggery or sugar given for abdominal pain both in man and animals. About 50g powdered seeds mixed with 'triphala' (*Emblica officinalis* Gaertn., *Terminalia bellirica* (Gaertn.) Roxb., and *Terminalia chebula* Retz.) given for dysentery and gastric ailments. Also, the seed powder mixed with 'ginger'

(*Zingiber officinale* Rosc.), 'methi' (*Trigonella foenum graecum* L.), 'kaudi saunf' (*Anethum sowa* Roxb.) and 'hing' (*Ferula asafoetida* L.) considered good against bloating and abdominal pain in animals. Seeds soaked in cow's urine for 12-13 days given for cough* and sore throat* (vet.).

Active Constituents: Therapeutic properties attributed to carvone, limoline, flavonoids, quercetin, polysacharides and fixed oil.

Biological Activity: Essential oil shows moderate antibacterial and antifungal activity.

Uses in Literature: Known earlier for curing bloating and fullness, dyspeptic problems, dysentery, flatulence, oedema, spastic conditions of the gastrointestinal tract, tympany, wound (vet.); and as anthelmintic, antibacterial, antifungal (Ambasta, 1986; Anonymous, 1998b; Bhatt *et al.*, 2001; Choondal, 1988; Iqbal *et al.*, 2005; Khare, 2007; Reddy & Raju, 2000).

Cassia fistula L. (Pl. 39A).

Syn.: *C. rhombifolia* Roxb., *Cathartocarpus fistula* Pers., *Cassia fistula* Willd.

Family: Fabaceae.

Vern.: Aahali, Allia, Kanyar.

Ayurvedic, English, Sanskrit and Regional Names:

Ayur.: Aaragvadha, Aarevata, Chaturaangula, Karnikaara, Kritamaala, Kritmaalaka, Nripapaadapa, Praagraha, Raajadruma, Raajvrksha, Shampaaka, Vyaadhighaataka;

Eng.: Drum stick, Golden shower, Indian Laburnum, Purging Cassia;

Sans.: Aragvadha, Chatuangula, Kritamalluka;

Ass.: Sonaru;

B.: Amultas, Sonali;

G.: Balla, Garmala;

H.: Amaltas, Bandarlauri;

Kan.: Kakke;

Mal.: Saturangulam, Svuarnnakam;

Mar.: Bhawabaya, Chimkani;

Oriya: Soturongulo, Sunari;

P.: Alash, Karangal;

Tam.: Arogaram, Irayavinnadagam;

Tel.: Aragvadhamu, Sampakamu.

Distribution: Throughout India, common in Himalayas upto 1,400m.

Description: A medium-sized erect tree having straight trunk and long, pinnately compound leaves. Leaflets long, ovate, distinctly stalked, usually acute, margin entire, rigidly subcoriaceous, pale green and strongly veined. Main rachis pubescent. Flowers yellow and in hanging raceme. Pods 30-60cm long, cylindric, brownish black, indehiscent with numerous broadly-ovate seeds immersed in dark coloured sweet pulp.

Flowering & Fruiting: March-July.

Habitat Ecology: Forest, open fields, roadsides; Nadaun, 483m.

Material Examined: HPU-EL-SKS-ABS-15073; 18.05.2011.

Parts Used: Leaf. Seed. Fruit. Wood.

Folk Uses: 2-4g dried grinded pods mixed with *Terminalia chebula* Retz., *Terminalia bellirica* (Gaertn.) Roxb. and *Emblica officinalis* Gaertn., used for stomach disorders. One tsp decoction of fruit pulp taken twice daily for a week to relieve gastric ailments and constipation both in man and livestock. Decoction of 100g powdered fruits considered good against constipation and bloating (vet.). Poultice of tender twig applied for toe infection*. Seeds prescribed as laxative in animals. Wood used as fuel and for construction work for making doors, roofs and agricultural implements. Tree worshipped locally as Narsingh God.

Active Constituents: Fistucacidin identified as racemic or meso 3, 4, 4, 7, 8 pentahydroxyflavan (*bark*), leucoanthocyanidin-5-4-dihydroxy flavon-3-4-diol (I) (sap *wood*), rhein, glucose, sucrose and fructose (*fruit pulp*), galactomannan (*seed*), seven new biflavinoides β-sitosterol (I-VII) and two triflavinoids (VIII-IX), (-) epicatechin, procyanidin B-2,(-) epiafzelechin and its 3-O-glucosides (*leaves*), steroids, glycosides, flavonoids, tannins, quinones, disaccharides, phlobaphenes, oxyanthroquinone have been characterised.

Biological Activity: Aq. extr. of fruit pulp exhibits antibacterial activity.

Uses in Literature: Used earlier as antibacterial (vet.), antinarcotics, antiseptic, bloat (vet.), blood purifier, cooling agent, edible (vet.), emetic, febrifuge, insecticidal (vet.), laxative (vet.), mild cathartic, ornamental, purgative (vet.), psychoactive, symbolic, tonic (roots), vegetable, vermicides; and for abdominal pains, adjust bile, agricultural implements, anthrax (vet.), black water fever, blindness, body swelling, bloated stomach, body heat, boils, burns, cancer, carts, cataract, chest infection, chest pain, cold (vet.), diabetes, diarrhoea (vet.), dog bite, dyeing, dysentery, eczema, epilepsy, expelling hook worms, gout, house construction, indigestion, indigestion (vet.), intestinal worms, jaundice, liver complaints, malarial fever, pimples, ploughs, scorpion bite, swelling in throat (vet.), tool handles and wheels, toothache, tympany (vet.), whooping cough and wounds (vet.) (Ambasta, 1986; Anonymous, 1998b; Balasubramaniam & Prasad, 1996; Bhandary *et al.*, 1996; Bondya & Sharma, 2003; Chauhan, 1999; Dixit & Pal, 1996; Gill & Nayawuame, 1994; Girach & Aminuddin, 1995; Gogoi & Borthakur, 1991; Goud & Pullaiah, 1996b; Jain, 1968, 1991, 1996; Jain & Patole, 2001; Jain *et al.*, 1994; Khare, 2007; Kothari & Moorthy, 1996; Kumar, 1996; Kumar & Jain, 1998; Kurian, 1995; Lalramnghinglova, 1996; Malik *et al.*, 2009; Manandhar, 1989a; Masih, 2003; Molla & Roy, 1996; Pal, 1981; Pal & Jain, 1998; Patil & Merat, 2003; Patole & Jain, 2002; Pullaiah & Kumar, 1996; Rama Rao & Henry, 1996; Reddy & Raju, 2000; Retnam & Martin, 2006; Sahoo & Mudgal, 1995; Saini, 1996a; Sharma, 1996; Sharma & Sood, 1997; Sikarwar, 1996; Sikarwar *et al.*, 1994; Singh, 1999b; Siwakoti & Verma, 1996; Srivastava *et al.*, 2003; Tiwari & Tiwari, 1996; Tosh, 1996; Venkataratnam & Venkata Raju, 2004; Watt, 1972).

Major Ayurvedic Formulations: Aargavadhaadi-arishta, Agnikumar-churn, Aragvadhadi-leha, Aragvadharistaz, Argavadhaadi-taila, Arishta, Kankayan-gutika, Leha-taila, Mahamanjisth-yadyarista, Rag-vadhadi-kwath, Rasnadi-kwatha-yoga.

Trade: Traded for its dry fruits and pulp.

Cassia occidentalis L. (Pl. 39B).

Syn.: *Senna occidentalis* Roxb., *C. foetida* Pers., *C. sophera* Wall.

Family: Fabaceae.

Vern.: Alliuna, Bada-rellu.

Ayurvedic, English, Sanskrit and Regional Names:

Ayur.: Kaasaari , Kaasamarda;

Eng.: Fetid Cassia, Nergo coffee, Rubbish Cassia, Stinking weed;

Sans.: Arimarda, Kasari, Kasamara;

B.: KalKashunda;

Bo.: Hikal;

H.: Barikasondi, Chakunda, Kasunda;

Mal.: Karintakara;

Oriya: Kasundri;

Tam.: Nattandagarai, Peyavirai;

Tel.: Kasindha;

Urdu: Kasonji.

Distribution: Common throughout India; naturalised.

Description: Annual undershrubs possessing subglabrous, purplish branches, fetid leaves, yellow flowers in short peduncled racemes and recurved, glabrous, tortulose pods. Seeds hard, dark olive green, ovoid-compressed at one end and rounded at other.

Flowering & Fruiting: May-September.

Habitat Ecology: Wastelands, edges of cultivated areas; Bari Mandir, 1,060m.

Material Examined: HPU-EL-SKS-ABS-15074; 25.08.2010.

Parts Used: Leaves. Root. Pods. Seeds.

Folk Uses: Tender pods used for making pickles*. 'Chutney'* prepared by grinding its seeds with 'methi' (*Trigonella foenum-graecum* L.), 'ajwain' (*Trachyspermum ammi* L.) and 'gur' (jaggery) considered good for gastric disorders and diabetic patients*. Leaf paste applied as poultice to check swelling due to clotting*; and its juice with honey good against mouth ulcers*. 1-3g powdered roots with luke warm water taken on empty stomach against skin diseases and deworming*.

Active Constituents: These are aloe-emodin, a bianthraquinone, chrysophenol, dianthronic heterosides, jaceidin-7-rhamnoside, matteucino 1-7-rhamnoside rhein and potassium chelidonate (*leaves*), chrysolein, linoleic , linoleic and oleic acid, campesterol, 8-carbomethoxy-1-7-dihydroxy -3 methylxanthane, 1-8-dihydroxy-2 methyl anthraquinone, 1, 4, 5, trihydroxy anthroquinone derivatives, tannic acid and toxalbumin (*seeds*), steroid, glycosides, flavones, dianthronic, heteroside, sennoside A & B anthraquinone.

Biological Activity: Seed, root and leaf exhibit antibacterial properties.

Uses in Literature: Reported earlier as antibacterial, anticonvulsion, antifungal, antiperiodic, diuretic, edible, purgative, substitute for coffee, stimulant drink, stomachic, pot herb, tonic, tooth brush; and for bone fracture, dog bite, dropsy, diarrhoea, dysentery, eczema, fever, gastric complaints, lactation, rheumatism, ring worm, skin diseases, snake bites, throat infection, veterinary ailments, whooping cough, wounds (Ambasta, 1986; Chopra *et al.*, 1956; Jain, 1991; Khanna, 2002; Khare, 2007; Kshirsagar *et al.*, 2003; Mohanty *et al.*, 1998; Patole & Jain, 2002; Retnam & Martin, 2006; Sharma & Sood, 1997; Uphof, 1968).

Major Ayurvedic Formulation: Liv.-52.

Catharanthus roseus G. Don (Pl. 39C).

Syn.: *Vinca rosea* L., *Lochnera rosea* (L.) Rchb.

Family: Apocynaceae.

Vern.: Sadabahar.

English and Regional Names:

Eng.: The Periwinkle;

B.: Nayantara;

H.: Sadabahar;

Mal.: Ushamalari;

Mar.: Sadphul;

Oriya: Ainskati;

Tel.: Billaganneru.

Distribution: Cultivated throughout India, Western Himalayas: upto 700m, Upper Gangetic plains.

Description: Erect pale green annuals having acute stem branches, opposite, obovate, glossy, accuminate, membranous leaves with minutely scabrid margins. Flowers white or pink and in cymose axillary clusters. Corolla mouth narrow, hairy. Follicles cylindrical and many-seeded.

Flowering & Fruiting: Throughout Year.

Habitat Ecology: Wastelands, Krishna Nagar, 858m.

Material Examined: HPU-EL-SKS-ABS-15075; 18.07.2008.

Parts Used: Whole Plant. Flower. Leaves. Roots.

Folk Uses: Plant grown as an ornamental*. Flowers offered to appease deities. Leaf juice applied as antidote against wasp stings. Roots stomachic and good for diabetes (25g powder thrice daily till cure). Juice of leaves along with *Azadirachta indica* A. Juss., *Luffa aegyptica* Mill., *Lycopersicum esculentum* Mill. taken on empty stomach one glass daily for blood purification*, diabetes and gastric disorders*.

Active Constituents: Flowers rich in vinblastine, coronaridine, vindorosine, vindoline, catharanthine, mitraphlline, ajmalicine, tetrahydroal stonnie, 11-methoxytabersonineb and reserpine.

Biological Activity: Antitumorous (catharanthamine), anticancerous (vinblastine-i.p. or i.v.), (ã ohimbine-1mg/kg of rats), cytotoxic hypotensive and hypoglycaemic activities (aq. extr. of fruits and seeds) +ve.

Uses in Literature: Reportedly described for anthrax (vet.), behavioural disorders, boils, chronic dyspepsia, cold (vet.), cut (vet.), diabetes, diarrhoea, dysentery, fever, flatulence, gastric catarrh, high blood pressure, leucorrhoea, madness (vet.), mastitis (vet.), menorrhagia, scabies, skin infections, swelling of body (vet.), toothache, wasp sting, wounds; and as anticancerous, antihaemorrhagic, antileukaemic, astringent, haemostatic (vet.), hypotensive, mosquito repellent (vet.), purgative, sedative, stomachic (Ambasta, 1986; Bondya & Sharma, 2003; Hembrom & Goel, 2005; Hosagoudar & Henry, 1996b; Jain, 1991; Khare, 2007; Kurian, 1995; Lalramnghinglova, 1996; Prasad *et al.*, 1996; Retnam & Martin, 2006; Sharma & Sood, 1997; Siwakoti & Verma, 1996; Watt, 1972).

Cedrela toona Roxb. (Pl. 39D).

Syn.: *C. febrifuga* Blume, *C. hexandra* Wall., *C.longifolia* Blume, *C. serrata* Royle, *C. australis* F. Muell., *Toona ciliata* Roem.

Family: Meliaceae.

Vern.: Tooni.

Ayurvedic, English, Sanskrit and Regional Names:

Ayur.: Nandi-Vrksha, Tuunikaa;

Eng.: Happy tree, Indian Mahagany Tree, Moulmein Cedar, Red Cedar, Sandal, Toon;

Sans.: Apina, Apita, Kachla, Kanta, Kubheraka, Nanda, Nandi vriksha, Tooni,

Ass.: Henduripoma, Jia;

B.: Lud-Tunna, Tuni;

Bo.: Deodari, Kudaka, Limb, Mahanim, Tundu, Tuni, Tunna, Tupa;

H.: Kudaka, Limb, Mahanim, Tun, Tuna, Tupa, Tuni, Tunna;

Mal.: Akil, Arana, Devabaram, Malarveppu, Kacham;

Mar.: Deodari, Kuruk;

Oriya: Mona Limba;

Tam.: Madagirvempu, San.-davanembu;

Tel.: Gali, Nandi.

Distribution: Common throughout the greater parts of India, ascending to 1,000m.

Description: Deciduous tall trees with coloured wood, abruptly pinnate leaves, numerous, opposite or subopposite entire or serrated leaflets and drooping, terminal panicles. Flowers small, white, fragrant. Capsule 5-celled, ovoid, dehiscent.

Flowering & Fruiting: May-June.

Habitat Ecology: Roadsides, open slopes; Bassi, 987m.

Material Examined: HPU-EL-SKS-ABS-15076; 04.06.2010.

Parts Used: Leaves. Flower. Bark. Wood.

Folk Uses: 5-10 ml decoction of bark given to check fever, boils and for regulating the menstruation* (once a day for five days). Wood used for making doors, windows, furniture and agricultural implements, construction works. Leaves used as fodder, 'pattri' bed* for cattle and added to storage grains as insect repellent. Flowers used for making dye.

Active Constituents: Isolated constituents are cedrelone, β-sitosterol, (-)-alloaromadendrene, cadalene, (+) α calacorene, (-)-calamenene, (+) Ä cadinene, (-) Ä-cadinol, (-)-copaene, cubenol, farnesene, geranylgeraniol, (-)-ledol, leucocyanidin, T-murolol and siderin (*wood*).

Biological Activity: Stem bark anticancerous, spasmolytic whereas leaf exhibits antiameobic, CNS depressant, hypoglycaemic and spasmolytic activities.

Uses in Literature: Described in literature as an antiperiodic, antipyretic, antiseptic, astringent, cattle feed, dye (yellow), insect repellent, preservant, tonic; and for bridges, doors, floors, furniture, dysentery, fever, gastric troubles, hoops for sieves, mouth ulcers, plywood, tonsilitis, window, wood carving, wounds (Ambasta 1986; Chopra *et al.*, 1956; Jain, 1991; Kapur & Singh, 1996; Khare, 2007; Pandey & Verma, 2002; Siwakoti & Verma, 1996; Watt, 1972).

Celastrus paniculatus Willd. (Pl. 40A).

Syn.: *Celastrus alnifolia* Don, *C. dependens* Wall., *C. nutans* Roxb.

Family: Celastraceae.

Vern.: Sankheere, Sankhiryan.

Ayurvedic, English, Sanskrit and Regional Names:

Ayur.: Jyotishmati, Kaakaandaki, Kangunikaa, Katuveekaa, Maakaanguni, Paaraavatpadi, Svarnalatikaa, Vegaa;

Eng.: Black Oil, The Oleum Nigrum Plant;

Sans.: Myinkhoung-na-young;

B.: Kujari, Kujri;

Bo.: Malkangana;

G.: Kangani, Malkangoni, Mlkanganitela, Pigavi;

H.: San.-khu, San.-khii;

Kan.: Jiotishmati;

Mal.: Kariganne;

Mar.: Ruglim;

P.: Malkakni;

Tam.: Ruglim;

Tel.: Valushuva.

Distribution: Tropical and subtropical Himalayas, ascending to tropical Himalayas.

Description: A large woody scandent climber with terete branches, oval-oblong, serrate, membranous and coriaceous glabrous leaves. Flowers yellowish-green and in pendulous panicles. Fruits globose, small, green-yellow when ripe. Seeds 1-2 in each cell, enveloped in red fleshy coat.

Flowering & Fruiting: April-June.

Habitat Ecology: Forests, roadsides; Bhota, 765m; NIT, 900m.

Material Examined: HPU-EL-SKS-ABS-15077; 28.06.2008.

Parts Used: Leaves. Seeds.

Folk Uses: One tsp of seed oil in a glass of milk given once a day to cure hidden injuries* whereas one tsp of its powdered seed given twice a day till cure for cough and bronchitis*. Seed oil mixed with that of mustard in equal ratio applied externally for skin affections. Leaves added to horse feed as fodder*.

Active Constituents: Main ingredients are celastrin, β-dihydrogaroturan sesquiterpenoids (VI, VII, VIII, IX), sesquiterpenoids, 1 β, 6-diacetoxy-9 -β-benzoyloxy-8 β-connamoyloxy-β-dihydroagarofuran (V), malkanguinol, sesquiterpene, ester-malkangunin, celapanine and celapanigine, paniculatadiol along with β-amyrin; β-sitosterol and fatty acids (seed *oil*).

Biological Activity: Antiviral, hypotensive and spasmolytic activities found to be positive.

Uses in Literature: Reported in India as abortifacient, aphrodisiac, brain tonic, diaphoretic, febrifugal, laxative, mosquito repellent, stimulant; and for amenorrhoea, arthritis, blood clotting, body pains, cold, cough, eczema, fencing, fever, headache, hiccups, hysteria, impotency, kidney troubles, leprosy, leucoderma, leucorrhoea, mental depression, paralysis, piles, pneumonia, psoriasis, rheumatism, scabies, skin diseases, spermatorrhoea, stomach pain, syphilis, tuberculosis and urinary inflammation (Ambasta, 1986; Jain & Patole, 2001; Khanna & Kumar, 2000; Khare, 2007; Kirtikar & Basu, 1994; Pandey & Rout, 2006; Punjani, 1998; Rao & Pullaiah, 2001; Sarangi & Sahu, 2004; Silori & Rana, 2000; Tirkey, 2004).

Major Ayurvedic Formulations: Geriforte tablets, Himcolin cream, Joytismati tel, Remem tablets and syrups and Seotex.

Trade: Seeds sold in the drug markets under the name of 'Malkangani' or 'Jyotismati beej'.

Celosia argentea L. (Pl. 40B).

Syn.: *Celosia cristata* L.

Family: Amaranthaceae.

Vern.: Kukkar phul.

Ayurvedic, English, Sanskrit and Regional Names:

Ayur.: Shitivaaraka, Vitunna;

Eng.: Cox comb plant, Quail Grass;

Sans.: Pothura;

B.: Swetmurga;

G.: Lapadi;

H.: Safed murgha;

P.: Sarwali sil;

Tel.: Gurugu.

Distribution: Throughout India.

Description: Annual erect herbs. Leaves alternate, elliptic, ovate below and linear lanceolate above. Flowers crowded in terminal pink shining heads, spikes long peduncled, cylindric or oblong or ovoid. Nutlets with many seeds having rough or polished testa and opening irregularly or by lid.

Flowering & Fruiting: August-December.

Habitat Ecology: Cultivated, as an escape; Rangas, 765m.

Material Examined: HPU-EL-SKS-ABS-15078; 17.06.2010.

Parts Used: Whole Plant. Flower. Seeds.

Folk Uses: Mainly planted for ornamental* purposes. Flowers used for making bouquet*. One tsp of powdered seed prescribed twice daily for kidney stones*.

Active Constituents: Celosianin and isocelasianin.

Uses in Literature: Used earlier as antipyretic, antiscorbutic, aphrodisiac, coolant, galactagogue, source of potassium, vit. B1& B6; and for blood dysentery, diarrhoea, eye diseases, inflammation, menorrhagia, stomatitis, uterine bleeding, wasp and scorpion stings (Chopra *et al.*, 1956; Chatterjee & Pakrashi, 1997; Khare, 2007; Kirtikar & Basu, 1994).

Centella asiatica (L.) Urb. (Pl. 40C).

Syn.: *Hydrocotyle asiatica* L.

Family: Apiaceae.

Vern.: Brahmi.

Ayurvedic, English, Sanskrit and Regional Names:

Ayur.: Brahma-manduuki, Manduukaparni, Manduukaparnikaa, Maanduuki, Saraswati;

Eng.: Indian Pennywort;

Sans.: Bhekaparni, Divya, Manduki;

Ass.: Manimini;

B.: Brahmamanduki;

Bo.: Karinga, Karivana;

H.: Brahmamanduki, Khulakhudi;

Kan.: Von-de-laga;

Mal.: Kodagam, Muthal;

Mar.: Brahmi;

Tam.: Babassa,Vallari;

Tel.: Bokkudu.

Distribution: Common throughout India. Tropical and subtropical regions, in marshy places up to 700m.

Description: A prostrate mat forming herb rooting at nodes. Leaves reniform, margin toothed. Flowers purplish-white, and in axillary umbels which are 3-6 flowered. Fruits minute, orbicular, reticulated, laterally compressed.

Flowering & Fruiting: April-May.

Habitat Ecology: Along water channels in fields; Bhira, 780m; Bassi, 987m.

Material Examined: HPU-EL-SKS-ABS-15079; 15.07.2008.

Parts Used: Whole Plant. Leaves.

Folk Uses: One tsp of powdered leaves with cow's milk taken in the morning to improve memory and as astringent*. 100 ml leaf juice with sugar candy given thrice a day for cooling. 50-80ml decoction of the plant considered good for stress, fatigue* and as brain tonic; also supplemented with juice of 'neem' (*Azadirachta indica* A. Juss.) and 'black pepper' (*Piper nigrum* L.) (once daily 20-30 ml for 30 days.) for blood disorders.

Active Constituents: Therapeutic properties attributed to centoic, centellic, centic acid, brahmoside, brahminoside, two triterpenic acid-brahmic and iso-brahmic acid, betulic acid, stigmasterol, 2 saponins-brahmoside and brahminoside (whole *plant*).

Biological Activity: EtOH (50%) extr. of plant antiprotozoal and spasmolytic.

Uses in Literature: Reportedly used for blood disorders, bodyache, boils, cholera, cough, diarrhoea, dysentery, eczema, eye complaints, fever, gastric disorders, headache, improving speech and imagination, insanity, jaundice, leprosy, liver complaints, memory, mouth sores, nervine disorder, kidney stone, respiratory disorders, scabies, skin diseases, spermatorrhoea, syphilis, tubercles, tumour, urinary complaints, wounds; and as anthelmintic, blood purifier, brain tonic, coolant, diuretic, emmenagogue, insectcidal, laxative, local drink, postnatal tonic, stomachic, vegetable (Ambasta, 1986; Augustine & Sivadasan, 2004; Bondya *et al.*, 2006; Borthakur *et al.*, 2004; Chettri, 2005; Chopra *et al.*, 1956; Gogoi & Borthakur, 2001; Jain, 1968, 1991; Jain & Patole, 2001; Khanna & Kumar 2000; Khare, 2007; Kshirsagar & Singh, 2000; Kumar & Pullaiah, 1998; Rao & Pullaiah, 2001; Sharma & Sood, 1997; Sharma *et al.*, 2002; Singh, 2003; Retnam & Martin, 2006; Upadhyay & Singh, 2005).

Major Ayurvedic Formulations: Brahmi-vati, Brahmi-ghrit, Brahmi-tel, Brahmi-amala-taila, Brahmi-rasayana, Geriforte tablets and Saraswatarista.

Trade: Marketed as 'Brahmi'.

Adulterant or Substitute: Plant used as an adulterant or substitute to Herpestris, 'Brahmi' (*Bacopa monnieri* Wettst.)

Centipeda minima (L.) A. Br. & Asch. (Pl. 40D).

Syn.: *C. orbicularis* Lour., *Artemisia minima* L., *Artemisia sternutatoria* Roxb. *Centipeda minuta* Benth., *Cotula sternutatoria* Wall., *Dichrocephala schmidii* Wight, *Myrigyne minuta* Less., *Sphaeromorphaea centipede* DC.

Family: Asteraceae.

Vern.: Nakchikdu.

Ayurvedic, English and Sanskrit Names:

Ayur.: Chhikkikaa, Chhikkini, Kshavaka;

Eng.: Sneezewort;

Sans.: Chhikika, Chikkani.

Distribution: Damp places throughout the plains and low hills in India.

Description: Annual or perennial prostrate, woolly, herbs having obovate-oblong, alternate, entire or toothed leaves and globose, solitary axillary, subsessile heads. Achenes 4-angled, tip-obtuse, angles hairy.

Flowering & Fruiting: September-November.

Habitat Ecology: Along water channels in damp places, moist slopes; Bassi, 987m.

Material Examined: HPU-EL-SKS-ABS-15080; 15.07.2008.

Parts Used: Whole Plant. Leaves.

Folk Uses: Extr. of the plant prescribed for cough and congestion (15-20ml, twice a day till cure). Decoction of leaves given in cold and asthma. Leaves alongwith 'Jaiphal' (*Myristica fragrans* Houtt.) considered good for 'dandu'* (prolapsed uterus) and placental evagination (vet.)*.

Active Constituents: Therapeutic properties attributed to flavonoids, sesquiterpenes and amide.

Uses in Literature: Known to be used for curing asthma, cold, hemicrania, nasopharyngeal tumours, obstructions, rhinitis, sinusitis, white diarrhoea (vet.); and as antitussive and expectorant, cutaneous anaphylaxis (Ambasta, 1986; Gogoi & Borthakur, 2001; Jain, 1968, 1991; Khare, 2007).

Cestrum nocturnum L. (Pl. 41A).

Family: Solanaceae.

Vern.: Rat-Ki-Rani.

English and Regional Names:

Eng.: Lady of Night, Night Blooming Jasmine;

B.: Hasna;

H.: Rat ki Rani.

Distribution: Commonly cultivated throughout India upto 1,500m.

Description: Evergreen shrubs. Leaves long, pointed, ovate-lanceolate, shining. Flowers yellowish-green, in axillary branched cluster. Fruit a berry.

Flowering & Fruiting: August-October.

Habitat Ecology: Agricultural fields, wastelands; Hira Nagar, 900m; Sujanpur, 567m.

Material Examined: HPU-EL-SKS-ABS-15081; 01.07.2011.

Parts Used: Whole Plant. Leaves. Flowers.

Folk Uses: Planted by locals for ornamental* purposes and fragrance. Flowers considered sacred* for offering to deities and dried ones burnt as incense. Decoction of leaves used as a good wash for mouth sores*; its powder (2-3 g) given with cow's milk on empty stomach for sinusitis*. A pinch of snuff of powdered leaves taken to check headaches*.

Active Constituents: Active properties attributed to benzaldehyde, benzyl alcohol, benzyl acetate, methyl benzoate, phenyl acetaldehyde, phenyl ethyl alcohol, linalool and α-farnesene (*flowers*), tigogenin, yuccagenin, nornicotine, cotinine and myosmine (*leaves*).

Biological Activity: Aerial parts spasmolytic, diuretic and hypotensive.

Uses in Literature: Recorded in India for epilepsy; and as perfumes, cosmetics (Sharma, 2003; Uphof, 1968).

Chenopodium album L. (Pl. 41B).

Syn.: *C. giganteum* Don, *C. nepalense* Hort., *C. viride* L., *C. laciniatium* L., *C. purpurascens* Ham.

Family: Chenopodiaceae.

Vern.: Bathu, Kunnoha.

Ayurvedic, English, Sanskrit and Regional Names:

Ayur.: Vaastuuka;

Eng.: Bacon Weed, Dirty Dick, White Goose foot, Wild Spinach;

Sans.: Agralohika, Chielika, Kankella, Mridapatri, Shakarata, Vastuki;

B.: Bathu sag, Chandan betu;

Bo.: Chakwit;

H.: Bathua, Khartua Sag;

P.: Bathu;

Tam.: Parupukire;

Tel.: Pappukuras.

Distribution: Throughout India, tropical and temperate Himalayas ascending to 4,000m.

Description: Erect glabrous herbs having striped stems, irregularly lobulate leaves and flowers tinged with purple and in cluster forming lux paniculate. Sepals keeled. Utricles orbicular, smooth shining.

Flowering & Fruiting: January-March.

Habitat Ecology: Cultivated fields, wasteplaces; Kanjiyan, 910m.

Material Examined: HPU-EL-SKS-ABS-15082; 25.03.2010.

Parts Used: Whole Plant. Stem. Leaves. Seeds.

Folk Uses: Plant extr. (25-50ml) taken with sugar twice a day for stones (antilithic)*. Plant also given as fodder*. Young shoot and leaves used as vegetable and considered good for rheumatic pains* and constipation. Stem along with leaves fed to animals for facilitating expulsion of placenta. Paste of its leaves along with 'koru' grass (*Roylea cinerea* Baillon) applied to repel exoparasites* in animals. One tsp of seed powder with honey taken on empty stomach to expel worms.

Active Constituents: Constituents identified are ascorbic acid, β-carotene, catechins, gallocatechin, caffeic acid, p-coumaric acid, ferulic acid, β-sitosterol, campesterol, xanthotoxin, stigmasterol, n-tri-acontanol, scopoletin, imperatorin, spinasterol and 7-stigmasterol (*plant*).

Biological Activity: Plant EtOH (50%) extr. CNS depressant.

Uses in Literature: Recorded so far as anthelmintic, antiscorbutic, antiseptic, aphrodisiac, blood purifier, diuretic, laxative, local drinks, tonic and vegetable; and for anaemia, cardiac disorders, cut and delivery (vet.), debility, eye troubles, galactagogue (vet.), hepatic disorders, intestinal ulcers, piles, reviving taste, skin diseases, skin problems (vet.), sore, spleen and bilious disorders, throat, treatment of burns, urinary complaints, worms, and wounds (Ambasta, 1986; Asolkar *et al.*, 1992; Bhatt, 1990; Bondya *et al.*, 2006; Chhetri, 2005; Chopra *et al.*, 1956; Dey, 1973; Gaur *et al.*, 1992; Jain, 1991; Khare, 2007; Sen & Batra, 1997; Sharma, 1996).

Chenopodium ambrosioides L. (Pl. 41C).

Syn.: *Anobrina ambrosioides* (L.) Spach, *Blitum ambrosioides* Crantz.

Family: Chenopodiaceae.

Vern.: Kah-ajwain.

Ayurvedic, English and Regional Names:

Ayur.: Kshetra-vaastuuka, Sugandh-Vaastuuka;

Eng.: Indian Wormseed, Mexican Tea., Sweet Pigweed;

H.: Kah-Ajwain;

Tam.: Kattasambadam.

Distribution: Widely distributed in wasteplaces in south India; In Himachal below 1,500m elevation.

Description: Much branched, erect aromatic herbs upto 100cm. Flowers minute and clustered in pale green leaf spikes. Fruits somewhat globular, slightly compressed with a thin pericarp having glandular hair surrounding the seed.

Flowering & Fruiting: July-November.

Habitat Ecology: Wasteplaces; Kanjiyan, 910m.

Material Examined: HPU-EL-SKS-ABS-15083; 25.10.2009.

Parts Used: Whole Plant. Aerial Parts. Seeds.

Folk Uses: 20-30ml plant extr. given with jaggery twice a day for intestinal worms (vet.) and stones (antilithic)*.Young shoot and leaves used as vegetable and considered good for rheumatic pains* and constipation*. Plant also given as fodder*. One tsp of seed powder with honey given on empty stomach to expel worms. Powdered seeds mixed with wheat flour given to check constipation in animals; also its decoction given in abdominal pain and 'basaihar' (mastitis)* (vet.) and dysentery*.

Active Constituents: Constituents identified are ascaridole (*seed*), kaempferol-7-rhamnoside, ambroside (*leaves*), quercetin, calcium, phosphorous, oxalates and essential oil (*plant*).

Biological Activity: Oil effective with tetrachloro-ethylenein (1:3) against *Ascaris* infection.

Uses in Literature: Recorded as an anthelmintic, antifungal, antimicrobial, antispasmodic, diuretic, emmenagogue, haemostatic, Mexican tea, nervine, sudorific, vegetable; and for amoebic dysentery, cholera, haemorrhoids, influenza, intestinal ulceration, pectoral, pneumonia, round and hookworms, toothache, typhoid, veterinary medicines and wounds (Ambasta, 1986; Asolkar *et al.*, 1992; Bondya *et al.*, 2006; Chhetri, 2005; Chopra *et al.*, 1956; Dey *et al.*, 1973; Hajra & Baishya, 1981; Jain, 1991; Joshi, 1993; Khare, 2007; Sen & Batra, 1997; Uphof, 1968).

Cicer arietinum L. (Pl. 41D).

Family: Fabaceae.

Vern.: Chollae.

Ayurvedic, English, Sanskrit and Regional Names:

Ayur.: Chanaka, Chanakaa, Harinantha, Jivan, Sakal-priya, Vajinantha;

Eng.: Bengal Gram, Chick Pea, The Common gram;

Sans.: Kalapai;

Bo.: Chenna;

G.: Harbara;

H.: But, Channa, Chola, Santali;

Kan.: Kadalai, Kadale;

Mar.: Harbara;

P.: Chana, Chola;

Tam.: Harimandhakam; Sannagala;

Tel.: Kadale, Karikadale, Kempu, Kudoly.

Distribution: Extensively cultivated throughout in northern India and Nilghiris.

Description: A viscose, much branched annual herb with pinnate leaves and solitary bluish flowers in the axils of the leaves. Pods small and green.

Flowering & Fruiting: March-April.

Habitat Ecology: Cultivated; Bijari, 823m, Nadaun, 483m.

Material Examined: HPU-EL-SKS-ABS-15084; 15.03.2010.

Part Used: Seeds.

Folk Uses: Grains eaten raw, roasted or boiled, and fed to horses; husk used as cattle feed. A handful of overnight soaked grams given in 'dona' (leaf plate) to each woman attending ladies 'sangeet' (song and dance programme held for the marriage function) or as 'prasad' in 'kirtan'(programme in which religious songs are sung in praise of deities) and considered auspicious*. Also, seeds consumed as 'dal'(curry). Powdered seeds ('basin') employed for various culinary preparations,* and for preparing facial pack* ('butna') in combination with sandal powder and 'sarson' (mustard) oil for both groom and bride at the time of marriage. A tsp of fried 'basen' boiled in cow's milk given for cure of cough and cold. Seeds considered important in religious ceremonies to appease lord 'Saturn'. Soaked seeds and leaves given to animals to overcome weakness* and to check frequent abortions*.

Active Constituents: Principle constituents are oxalic, acetic, malic acid, isoaliquiritigenin, isoaliquiritigenin-4′-glucoside, 3, 4, 7,-trihydroxyflavone, daidzein, pratensein, p-coumaric acid, biochanin-7-glucoside, pyrimidine nucleoside-uracil-5-B-D fructofuranosyl 1-monophosphate and uracil-5-B-D ribofuranosyl, 2′-3-cyclic monophosphate and β-sitosterol (*seeds*), lecithin, phytin, saponin (*grain*), biochanin A, biochanin B, biochanin C (*sprouting gram*).

Uses in Literature: Seeds eaten as antibilious, antifungal, antihyperlipidemic, antistress, atherosclerosis, dal or after roasting and in form of basin, hypocholesteremic, prophylactic feed to horses; and for acidity, blood dysentery (vet.), indigestion, diarrhoea, constipation, dyspepsia, jaundice and snake bite (Ambasta, 1986; Khare, 2007; Kirtikar & Basu 1994; Mishra *et al.*, 1996b; Rastogi & Mehrotra 1990, 1991, 1993, 1995a, b).

Cinnamomum camphora (L.) Nees & Eberm. (Pl. 42A).

Syn.: *Laurus camphora* L.

Family: Lauraceae.

Vern.: Kapoor, Kapur.

Ayurvedic, English, Sanskrit and Regional Names:

Ayur.: Chandra, Chandra Prabhaa, Ghanasaara, Gandhadravya, Himakara, Hima-valuka, Himopala, Indu, Karpura, Shashi, Shitalraja, Sitaabhra, Tushaara;

Eng.: Camphor, Camphor Cinnamomum, Official camphor;

Sans.: Karpura;

Mal.: Ghanarasam, Himavaluka;

Oriya: Chondrongyon, Romokorpura;

Tam.: Indu, Kirasemiran, Pugalvisurandiram;

Tel.: Chandramu, Ghanasaramu, Karpuramu.

Distribution: Cultivated in Himalayas upto 1,500m.

Description: Aromatic evergreen trees possessing elliptic, long pointed, 3 nerved leaves. Buds enclosed in silky, caducous, imbricating scales. Flowers tiny hairy within, greenish-yellow. Fruits globular with thick club-shaped base.

Flowering & Fruiting: April-August.

Habitat Ecology: Planted; NIT, 900m, Neri, 381m.

Material Examined: HPU-EL-SKS-ABS-15085; 05.06.2010.

Parts Used: Fruit. Bark. Wood.

Folk Uses: 100g mixture of camphor and equal quantity of mustard oil applied for checking skin irritation* and for suppuration of boils*. Wood used for performing 'hawan'*. 10-15ml of its fruit juice given in diarrhoea. 10ml decoction of its bark mixed with 50g black pepper (*Piper nigrum* L.) and 25g dried ginger powder (*Zingiber officinale* Rosc.) given once or twice daily for a week to check high fever.

Active Constituents: Ylangene, b-santalene, calacorone, calamenene, d-cadinene (*camphor oil*) are the active ingredients.

Biological Activity: Plant sedative, anodyne, antispasmodic, and diaphoretic.

Uses in Literature: Known in published literature as acro-necrotic poison, analeptic, anthelmintic, antileprotic, antiseptic, antispasmodic, aphrodiasic, carminative, antidote to strychnia, natural camphor, psychoactive plant, irritant in muscular strains, sedative, stimulant; and for affections of genitourinary system, angina pectoris, bed sores, cancer, curing wounds, diarrhoea, delirium, dysmenorrhoea, perfumery, nervousness, inhoofs, fever, gout, hysteria, impotency, inflammation, malaria, nervous headache, nymphomania, painful affections of uterus, perfumery, sleeplessness, smallpox, spermatorrhoea, whooping cough and wounds (vet.) (Ambasta, 1986; Jain *et al.*, 1984; Khare, 2007; Mishra *et al.*, 1996b; Sharma, 2003; Sharma & Sood, 1997; Vihari, 1995; Watt, 1972).

Cinnamomum tamala Nees & Eberm (Pl. 42B)

Syn.: *C. albiflorum* Nees, *C. cassia* Don, *C. pauciflorum* var. a Tazia., *Persea tamala* Spreng., *Laurus cassia* Roxb., *L. albiflora* Wall., *L. tamala* Ham., *L. tazia* Ham., *L. somcaurium* Ham., *L. sailyana* Ham.

Family: Lauraceae.

Vern.: Tajpatta.

Ayurvedic, English, Sanskrit and Regional Names:

Ayur.: Naalikaa, Naalukaa, Patra, Patraka, Tamaalpatra, Tejapatra, Utkat;

Eng.: Cassia cinnamomum, Indian Cassia Lignea;

Sans.: Dalavhaya, Patrakhya, Vasanavhaya;

Ass.: Dopati;

B.: Tejpat, Chota sinkoli;

Bo.: Darchini, Tamala;

G.: Tamalpatra;

H.: Dalchini, Tezpata;

Kan.: Thit-kya-bo;

Mal.: Karuvapatta patran;

Mr. : Dalchinitiki, Sambharpana;

Oriya: Tejapatra;

P.: Tajpater;

Tam.: Talishappaalttiri;

Tel.: Aakupatri, Tallishapatri.

Distribution: Tropical and subtropical Himalayas: 450-2,100m.

Description: Trees with leathery leaves, white flowers in pubescent panicles. New leaves pinkish tinged. Drupes ovoid, black when ripe.

Flowering & Fruiting: June-August.

Habitat Ecology: Forests; Nadaun, 483m, Neri, 381m.

Material Examined: HPU-EL-SKS-ABS-15086; 08.07.2008.

Parts Used: Leaves. Bark.

Folk Uses: Leaves used for flavouring rice, pulses and other dishes; its decoction considered good for mouth wash*. 10-20ml of its decoction in equal combination with 'mulathi' (*Glycyrrhiza glabra* L.), black pepper (*Piper nigrum* L.), and 'soa' (*Anethum sowa* Roxb.) given once daily for five days for asthma, bronchitis, cold, cough, flatulence, menstrual disorders*, tuberculosis* and as carminative.

Active Constituents: Essential oil from *leaves* contains cinnamic aldehyde, phellandrene, eugenol, linalool, cinnamaldehyde, 3, 4´, 5, 7-tetrahydroxy-flavone, 3, 3´, 4´, 5, 7-pentahydroxy flavone, kaempferol-3-O glucopyranoside, kaempferol-3-O-sophoroside, quercetin-3-O-rutinoside.

Uses in Literature: Known earlier as adulterant of true cinnamon, carminative, colic, condiment (leaves), expectorant, ingredient of chywanprash, narcotic, pan stimulant, stomachic, sudorific; and for affection of nerves or heart, asthma, bites of serpents, bronchitis, calico printing, cold, cough, debility of stomach, diarrhoea, dropsy, dyeing, enlargement of spleen, essential oil, fever, flatulence, gonorrhoea, headache, heart and throat complaints, indigestion, manufacturing of soap, oil in perfumery, pain in womb, piles, poisoning by opium, retention of urine and catanenia, rheumatism, scorpion sting and suppression of lochia after child birth (Ambasta, 1986; Chauhan, 1999; Ghosh *et al.*, 1996; Jain, 1991; Khare, 2007; Lalramnghinglova, 1996; Mehrotra *et al.*, 1995; Negi & Pant, 1990; Rana *et al.*, 1996; Sharma & Sood, 1997; Singh, 1996, 1999b; Sinha, 1996; Watt, 1972).

Major Ayurvedic Formulations: Amrtabindu, Amrtadhara, Ashwagandharista, Chandrapradhavati.

Trade: Leaves collected and sold in the market by the name of 'tejpat'.

Adulterant or Substitute: Used as substitute of true Cinnamon.

Cissampelos pareira L. (Pl. 42C)

Syn.: *Cissampelos pareira* L. var. *hirsuta* (Buch.-Ham. *ex* DC.) Forman, *C. caapeba* L., *C. convolvulacea* Willd., *C. hirsuta* Buch.-Ham. *ex* DC., *C. orbiculata* Buch., *C. heradiflora* Wall., *C. obectata* Wall., *C. discolor* Buch., *C. diversa* Miers, *C. elata* Miers, *C. grallatoria* Miers, *C. eriantha* Miers, *C. subpeltata* Thwaites, *C. delicatula* Miers, *C. nepalensis* Rhodes, *Cocculus orbiculatus* DC., *C. villous* Wall., *Menispermum orbiculatum* L.

Family: Menispermaceae.

Vern.: Bhatindu, Pathindu.

Ayurvedic, English, Sanskrit and Regional Names:

Ayur.: Ambashthaa, Aviddhakarni, Paathaa, Piluphalaa, Shreyashi , Varatiktaa, Vriki;

Eng.: False Pareira Brava, Ice Vine, Periera Browa, Velvet leaf;

Sans.: Ambashtha, Ambashthika, Brihattika, Ruchaishya, Shishira, Vidhakarai;

A.: Saruthubukilata;

B.: Akanadi,Tejomalla;

Bo.: Pahadmul, Pahadvel;

H.: Akauadi, Dekhnirbissi, Pari;

Kan.: Padavali;

M.P.: Karepar;

Mal.: Kattuvalli, Pattuvalli;

P.: Bat bel, Batindu path, Haiyat, Pataki, Pilajur;

Tam.: Pomuskite, Punmushtie, Vatatirupi;

Tel.: Pata.

Distribution: Throughout tropical and subtropical parts of India upto 1,500m.

Description: Climbing shrubs possessing orbicular-reniform leaves. Flowers small, greenish-yellow. Male cymes long-peduncled whereas female ones are in racemes. Drupes ovoid, compressed, scarlet-red.

Flowering & Fruiting: May-November.

Habitat Ecology: Rocky slopes, wastelands, moist hills; Patta, 1,020m.

Material Examined: HPU-EL-SKS-ABS-15087; 14.08.2010.

Parts Used: Root. Leaves. Seeds.

Folk Uses: Roots of the plant used in making local wine. 250mg powdered roots alongwith black pepper (*Piper nigrum* L.) given for seven days to cure leucorrhoea*. Leaf juice (15-20ml) mixed with jaggery and salt given twice daily as cure of dysentery. 5-10ml decoction of leaves twice a day considered good for constipation and fever. Leaves along with wheat dough given to livestock for dysentery. 20-40ml decoction of seeds prescribed on empty stomach for kidney stones*.

Active Constituents: Hayatine, 1-curine, iso-chondrodendrine, thymol, 11 quaternary, pareirine, hyatinin, hyatin and d-iso-chondrodendrine cycleamine, d1-beberine, hayatidine, hayatine-5-tertiary alkaloids and d-quercitol (*leaves*), hyatin, furnished hydrochloride, picerate, chloroplatirate, methiocide and methochloride (*roots, vines*), isoquinoline, pelosine, berberine, reserpine, cissampetive, quercitel and sterol (*roots*) are the active constituents.

Biological Activity: CNS depressant and muscle relaxant properties exhibited.

Uses in Literature: Known in literature for abscess, antifertility, bodyache, boils, bronchitis, burns, carbuncle, catarrhal, child birth, cold, cough, cystitis, diarrhoea, dysentery, epilepsy, eye complaints, headache, heart burn, gastric complaints, indigestion, infection of bladder, inflammation, jaundice, liver disorders, pimples, piles, scabies, spermatorrhoea, snake bites, stabilising soil, stomach ulcer, urinary troubles, wounds; and as carminative, colic, diuretic, febrifuge, laxative, local drinks, sedative, stomachic, symbolic, tonic (Ambasta, 1986; Borthakur, 1993; Borthakur *et al.*, 2004; Brahmam & Saxena, 1990; Brahmam *et al.*, 1996; Chauhan, 1999; Datt, 1996; Dey, 1973; Dey *et al.*, 1973; Drury, 1978; Jain, 1991; Katewa *et al.*, 2001; Manandhar, 1980, 1996; Masih, 2003; Maya *et al.*, 2003; Saxena *et al.*, 1981; Sharma, 2003; Sharma & Sood, 1997; Singh, 2003; Uphof, 1968; Varma, 1997; Verma & Chauhan, 2006; Watt, 1972).

Major Ayurvedic Formulations: Agnimukh-churn, Brhat-gangadhar-churna, Ganga-dhar-churna, Kutajastak-kwath, Mahayograj-guggulu, Pathadi-kwath, Pippalaya-sav, Prada-rantaka-lauha, Pushayug-churn, Sarasvata-ghrita, Stanyasodhana-kasaya-churna.

Trade: Roots sold in the drug market as 'patha'.

Adulterant or Substitute: Root used as substitute for true 'Pareira Brava' of commerce (*Chondodendron tomentosum* Ruiz. & Pav.).

Citrus aurantium L. (Pl. 42D).

Syn.: *C. aurantium* var. *bigardia* Hook. f.

Family: Rutaceae.

Vern.: Sangtara, Santra.

English, Sanskrit and Regional Names:

Eng.: Bigarde orange, China orange, Orange tree, Portugal orange, Seville Sour, Sweet orange;

Sans.: Airavata, Chakradhivasi, Kirmira, Nagar, Tragagandha;

B.: Kamalanembu, Narungi, Naranga, Varishtha;

Bo.: Cantara, Naranghi;

G.: Narangi;

H.: Amritphal, Narunge, Sangatra;

Kan.: Hearlay;

Mal.: Madhuranaranna;

Mar.: Nargungasala, Sakulimba;

Oriya: Nagoronga, Narongi;

P.: Naranj, Santara;

Tam.: Narangam, Puttansini;

Tel.: Battavinarinja, Narinja.

Distribution: Hot valleys along the foot hills of Himalayas.

Description: Small slender trees. Leaves foliolate. Flowers pure white, in racemes. Fruits globose, oblate, orange coloured. Rind loose with sweet, yellow pulp.

Flowering & Fruiting: February-May.

Habitat Ecology: Planted; Bijari (Cha kmoh), 823m.

Material Examined: HPU-EL-SKS-ABS-15088; 03.12.2010.

Parts Used: Fruit. Rind.

Folk Uses: Fruit edible and considered excellent source of vit. C; also employed for preparing various confections. Rind powder given with honey to check cough, and constitutes one of important components of face pack* employed by the locals.

Active Constituents: These are 5-hydroxyauranetin, 5-o-demethylnobiletin, 6-7-dimethoxy-coumarin, aurantin, hesperidin (*peel*), hesperidin, naringin and poncirin, neo-hesperidin aglycone and limonene (*fruits*).

Biological Activity: α, β-pinenes, limonene, octanol, citronellal inhibit seed germination and seedling growth in *Amaranthus retroflexus.* Naingen and naringenin selectively inhibit *in vitro* growth of cancer cells in human (breast, colon, liver).

Uses in Literature: Used earlier as allxiphasmic, antispasmodic, blood purifier, carminative, contraceptive, disinfectant, edible, flavouring agent, juice, psychoactive plant, refreshing drink, stimulating, stomachic, tonic; and for acne, allaying thirst in fever, bilious affections, cold, cough, eczema, gout, improving appetite, indigestion, infusion, intestinal worms, jaundice, liquors, marmalades, perfumery, preparation of confections, psoriasis, lime marmalades, oil, other drinks, perfumery, confectionary items, reddish eye, skin cracks and vomiting (Ambasta, 1986; Jain *et al.*, 1994; Khare, 2007; Samati, 2004; Sen & Batra, 1997; Sharma *et al.* 2002; Sikarwar, 2002; Srivastava *et al.*, 2003; Watt, 1972).

Citrus grandis (L.) Osbeck. (Pl. 43A).

Syn.: *C. maxima* (Burm.) Merr., *C. decumana* L., *Aurantium maximum* Burm.

Family: Rutaceae.

Vern.: Chakotra.

Ayurvedic, English, Sanskrit and Regional Names:

Ayur.: Madhukarkatikaa;

Eng.: Forbidden fruit, Paradise-Apple, Pommel-Moose

Shadock, Pummelo;

Sans.: Madhukarkati;

B.: Batavinebu, Botornebu;

Bo.: Papanass;

G.: Obakotru;

H.: Batavinebu, Chakotra, Sadaphal;

Kan.: Chaoktre, Sakkota;

Mal.: Bambitinarakam;

Mar.: Panis;

Oriya: Batabilembu;

P.: Chakotra;

Tam.: Pambalimash;

Tel.: Pampanpanasa.

Distribution: Cultivated in India.

Description: Trees upto 13m tall. Leaflets ovate-oblong. Petiole broadly winged. Flowers white. Fruits globose, large-sized, pale yellow. Rind thick, pulp varying in colour from crimson to pale pink, sweet or acrid.

Flowering & Fruiting: November-March.

Habitat Ecology: Fields, near dwellings; Kangoo, 649m, Anu Khurd, 900m.

Material Examined: HPU-EL-SKS-ABS-15089; 12.05.2011.

Parts Used: Leaves. Fruits.

Folk Uses: Leaves used for scouring teeth*. Fruits edible, pickled and also taken with black pepper (*Piper nigrum* L.) for stomach ailments*, its juice considered good against vomiting. Fruit juice also given to livestock (500ml-1ltr mixed with equal quantity of water) to remove indigestible* matter from the stomach.

Active Constituents: Principle constituents are 5-geranyloxy-7-methoxy-coumarin, isoimperatorin, bergapten and 5-geranyloxypsoralen (bergamottin) in (*fruits, leaves*).

Uses in Literature: Used as edible nutritive, refrigerant; and for cholera, cough, constipation, diarrhoea (vet.), epilepsy, indigestion (vet.), inflamed nerves, insanity, jams and marmalades, morbid cancer, nervous disorders, offering, painful swellings, rheumatism, scurvy, ulcers, vomiting, waist and back pains (Ambasta, 1986; Borthakur, 1996 a,b; Gogoi & Borthakur, 1991; Khare, 2007; Kurian, 1995).

Citrus jambhiri Lush. (Pl. 43B).

Family: Rutaceae.

Vern.: Jambirdi, Jhamirdi, Kimb.

Ayurvedic and English Names:

Ayur.: Jambira, Jambh,Jambhir, Jaamphal;

Eng.: Rough lemon.

Distribution: Hot Valleys of Himalayas.

Description: A small tree with stout spines. Leaf pale green. Flowers white, fragrant arranged in cymes. Fruits yellow with thick rind.

Flowering & Fruiting: October-March.

Habitat Ecology: Cultivated, as well as wild; Dimmi, 1,006m.

Material Examined: HPU-EL-SKS-ABS-15090; 05.03.2009.

Parts Used: Leaves. Fruit.

Folk Uses: Plant used as stock for grafting* *Citrus aurantium* L. Leaves used for scouring teeth*. Fruit edible* and also pickled; its juice given with a pinch of salt and black pepper for liver problems* and also used as 'chukh' (substitute for lemon) for imparting acidic* taste to various preparations in marriage. Sacred vermillion ('kungu')* made by mixing its juice with turmeric (*Curcuma longa* L.) and 'suhaga' (borax).

Active Constituents: A new flavone-tangeretin along with β-sitosterol, hesperidin characterized from (*peel oil*).

Uses in Literature: So far used as anthelmintic, antibacterial, antihistaminic, antiscorbutic, carminative, stomachic; and for cold, cough, influenza, treatment of white spots, and pickles (Chopra *et al.*, 1956; Husain *et al.*, 1992; Khare, 2007; Kirtikar & Basu, 1994).

Citrus limon (L.) Burm. f. (Pl. 43C).

Syn.: *C. medica* var. *leomonum* L. (Risso) Hook. f., *C. limonum* Risso., *C. medica* L. var. *limon* L.

Family: Rutaceae.

Vern.: Khatta.

Ayurvedic, English, Sanskrit and Regional Names:

Ayur.: Airaavata, Dantashatha, Limpaka, Naaranga, Nimbu, Nimbuka;

Eng.: Lemon;

Sans.: Bara Nebu, Goranebu, Jambirah;

G.: Motalimbu;

H.: Bara Nibu, Jambhira, Paharinembu, Paharikagzi;

Kan.: Bijapura, Bijuri;

Mar.: Idalimbu, Thoralimbu;

Oriya: Periya Yelumichai;

Tel.: Bijapuram.

Distribution: Hot Valleys of Himalayas.

Description: An evergreen tree with white, scented flowers in cymes. Leaflet elliptic-oblong. Fruits yellow, globose, mammilate with abundant acidic juice. Rind thin, pulp abundant.

Flowering & Fruiting: November-February.

Habitat Ecology: Cultivated; Nalti Galore, 787m.

Material Examined: HPU-EL-SKS-ABS-15091; 11.11.2009.

Parts Used: Leaves. Fruit.

Folk Uses: Fruits edible* and also used for refreshing drinks,* culinary preparations, pickling and magico-religious* purposes. Heated half fruit bandaged to heal twisted ankle*. Cut pieces of fruit mixed with washing powder and rock salt rubbed over body of animals to expel ectoparasites. Leaves used for scouring teeth* and flavouring 'desi' ghee.

Active Constituents: These are apigenin, luteolin, chrysoseriol, quercitin, isorhamnetin, new flavonols limcitrin and limocitrol-sinapic acid, p-coumaric acid, scopoletin, umbelliferone, hesperidin, citrusin A, B, C and D, coniferin, syringin and dehydro-deconeferyl alcohol-4-β-glucoside (*fruit*), eriocitrin (*peel*).

Biological Activity: Coniferin and syringin (1.0mg / 100g i.v.) found to raise blood pressure of SHR-sp rats whereas dehydrodiconiferyl alcohol -4 -β glucoside (I.V.) lowers it in spontaneously hypertensive rats.

Uses in Literature: Reported in literature as an anthelmitic (vet.), carminative, citric acid, disinfectant, febrifuge, juice (vet.), nervine tonic, peat; and for cold, corns, culinary purposes, dysentery, flavouring liquors, indigestion, lemon oil (obtained as by product), lowering body heat, pickles, preparation of beverages, scabies, skin diseases, stomachache, throat diseases, vitality to body and warts (Ambasta, 1986; Dagar & Dagar, 1999; Kadel & Jain, 2006; Khare, 2007; Kurian, 1995; Mishra *et al.*, 1996a,b; Retnam & Martin, 2006; Shome *et al.*, 1996; Tirkey *et al.*, 2003; Vishwanathan & Singh, 1996).

Citrus medica L. (Pl. 43D).

Syn.: *C. medica* var. *medica.*

Family: Rutaceae.

Vern.: Nimbu.

Ayurvedic, English, Sanskrit and Regional Names:

Ayur.: Bijaahva, Bijpuura, Lunga, Maatulaka, Maatulunga, Nahaalunga;

Eng.: Adam's Apple, Cedrat , Citron, Melon Lime;

Sans.: Amlakshera, Bijaphalaka, Rochanaphala;

B.: Baranimbu, Nimbu;

Bo.: Bijapura, Limbu;

H.: Baranimbu, Nimbu;

Kan.: Madala, Mahaphala, Rusaka;

Mal.: Gilam, Matalanarakam;

Mar.: Mahalunga;

P.: Bajauri;

Tam.: Kogilachan, Kommattimadalai;

Tel.: Lungamu.

Distribution: Valley along the foot hills of Himalayas.

Description: Evergreen small shrubs or trees with elliptic leaves having round apex. Flowers scented, in axillary cymes, white tinged pink outside. Fruits obovoid, yellow with a leathery rind.

Flowering & Fruiting: April-September.

Habitat Ecology: Cultivated; Barsar, 815m.

Material Examined: HPU-EL-SKS-ABS-15092; 15.05.2010.

Parts Used: Leaves. Fruit.

Folk Uses: Lemon squash ('shakanjabi') prepared by adding salt, sugar, black pepper (*Piper nigrum* L.) and lemon juice to water considered a refreshing drink against heat strokes. Fruits pickled* and its juice used as preservative*; given ½ tsp with sugar pellets, thrice daily to check diarrhoea, flatulence, indigestion, jaundice* and vomiting. Fruit juice mixed with salt rubbed over scorpion bite in animals. Leaves used for scouring teeth*. Powdered dried rind mixed with sandal powder and 'basin' (*Cicer arietinum* L.) used as face packs*.

Active Constituents: New triterpenoid-citrusin along with palmitic and succinic acids, citropten, α-sitosterol and daulosterol (*fruits*), coumarins, limettin, scopoletin, limonin, diosmin, β-sitosterol and β-D-glucoside (*peels*) have been isolated.

Biological Activity: Anthelmintic activity found to be +ve.

Uses in Literature: Reported earlier as anthelmintic, animal poison, antidote to acronacrotic, antiscorbutic, carminative, refreshing drink, stomachic, vermifuge (vet.); and for abdominal pains, acnes, acute rheumatism, agricultural implements, biliousness, bleeding gums, cardiac diseases, checking constipation, congestion of the liver, diarrhoea, dysentery, dyspepsia, essential oil, freckles, fatigue, foul breath, indigestion, inflammatory affections, improving general weakness, haemorrhage, hypertrophy of spleen, kidney disease, lemon syrup, marmalades, scorpion bites, trembling of limbs, urinary disorders and vomiting (Ambasta, 1986; Borthakur & Sarma, 1996; Bhatt *et al.*, 2001; Chowdhery, 1996; Jain, 1991; Khare, 2007; Kurian,

1995; Lalramnghinglova, 1996; Mohanty & Padhy, 1996; Mohanty *et al.*, 1998; Negi & Gaur, 1991; Samati, 2004; Sen *et al.*, 2002; Sharma & Singh, 2001; Singh *et al.*, 1997; Tiwari & Tiwari, 1996; Watt, 1972).

Major Ayurvedic Formulations: Gandhak-vati, Hingol-shodhan, Nimbuka-phala-rasa, Sankh-vati.

Clematis grata Wall. (Pl. 44A).

Family: Ranunculaceae.

Vern.: Dhagli bel.

Hindi Names:

H.: Biliri, Ghantiali.

Distribution: Subtropical and temperate W. Himalayas.

Description: Climbing shrubs with pinnate, opposite leaves having three hairy leaflets and greenish-white or cream coloured, fragrant flowers in loose panicles. Achenes densely pubescent, styles hairy.

Flowering & Fruiting: July-September.

Habitat Ecology: Shrubberies, cultivated areas; Ladraur, 978m.

Material Examined: HPU-EL-SKS-ABS-15093; 30.09.2010.

Parts Used: Plant. Leaves. Stem. Flowers.

Folk Uses: Paste of leaves and flowers applied to cure skin eruptions* and itching*. Powdered leaves and flowers mixed with mustard oil prescribed for skin ailments. Pieces of stem kept in dental caries for relief from toothache*. Plant used for making ropes*.

Active Constituents: These are a new saponin-clematoside S isolated (*roots*), quaternary apophine alkaloids, magnofluoine (*aerial parts*) and protoanemonin a fungitoxic compound (*leaves*).

Uses in Literature: Used in India for cancerous and foul ulcers, gout, rheumatic pains, spleen affections and varicose veins (Chopra *et al.*, 1956; Chunekar & Hota, 1998; Chunekar & Pondal, 1999; Kapur, 1996; Khare, 2007; Kirtikar & Basu, 1994).

Colebrookia oppositifolia Sm. (Pl. 44B).

Syn.: *C. ternifolia* Roxb., *Elsholtzia oppositifolia* Poir.

Family: Lamiaceae.

Vern.: Dashane.

Regional Names:

Bo.: Bhahmini, Dasai, Dasari;

H.: Binda, Bindu, Pansra;

Kum. : Dulshat;

Oriya: Bosika, Darigopi;

P.: Bamera, Basuti, Dashane, Dhisbekkar, Sampru, Suali.

Distribution: Subtropical Himalayas.

Description: Hoary shrubs upto 3m high with oblong-elliptic, acuminate, opposite leaves, stout trunk and terete branches, often whorled in trees. Flowers white, in dense clusters. Spikes numerous, 10cm long. Fruits schizocarpic carcerulus with small, non-endospermic seeds.

Flowering & Fruiting: April-November.

Habitat Ecology: Common on dry slopes, wastelands, roadsides; Bhoranj, 987m.

Material Examined: HPU-EL-SKS-ABS-15094; 12.03.2008.

Parts Used: Whole Plant. Leaves. Roots. Wood.

Folk Uses: Plant juice mixed with mustard oil used as drops to relieve earache*. Decoction of leaves used for gargles to check mouth ulcers*. Leaf paste applied as poultice to heal wounds. Roots used in making local wine 'sur'*. 250 ml extr. of leaves used for washing wounds (vet.). Also, its leaf and stem paste applied for healing burn injuries (vet.). Roasted leaves mixed with rock salt and leaves of 'banna' (*Vitex negundo* L.) added to water in earthern pot considered good appetizer* for domestic animals. Wood used as fuel*.

Active Constituents: Active constituents are flavones-5-hydroxy-4', 6, 7, 8-tetramethoxy flavone and hydroxy-4', 6, 7-tetramethoxy flavone (*aerial parts*).

Uses in Literature: Used for asthma, bruises, burns, catarrah (vet.), conjuctivitis (vet.), cough, cuts, digestion, eye complaints, eye diseases (vet.), epilepsy, extracting worms from sores on the leg, giddiness, gun powder and charcoal, hysteria, sores, toothache, wounds, wounds (vet.) and as anthelmintic, anthelmintic (vet.) (Ambasta, 1986; Gaur *et al.*, 1992; Jain, 1984, 1991; Jha & Varma, 1996; Kulkarni & Kumbhojkar, 1996; Manandhar, 1988; Ramadas *et al.*, 2000; Rama Rao & Henry, 1996; Shrestha, 1985; Siwakoti & Verma, 1996; Varma, 1997; Watt, 1972).

Colocasia esculenta (L.) Schott (Pl. 44C).

Syn.: *C. antiquorum* Schott

Family: Araceae.

Vern.: Kachalu.

Ayurvedic, English, Sanskrit and Regional Names:

Ayur.: Pindaaluka, Aaluki;

Eng.: Cocoyam, Dasheen, Eddo, Taro;

Sans.: Kachwi, Kachu;

B.: Kachu;

H.: Arvi, Ghuiya, Kachalu;

Kan.: Kachhi, Shamagadde;

Mal.: Shembu;

Mar.: Alu;

Oriya: Sam, Saru;

Tam.: Seppan , Kizhangu;

Tel.: Chamadumpa, Chemagadda.

Distribution: Cultivated upto 2,500m.

Description: Tall, coarse tuberous herbs with a stout short caudex, heart-shaped leaves and petiole sheathing at the base. Flowers in spadix; flowering and leafing occur together. Spadix shorter than yellow spathe. Berries ovoid.

Flowering & Fruiting: July-October.

Habitat Ecology: Cultivated; Sameerpur, 1,196m.

Material Examined: HPU-EL-SKS-ABS-15095; 27.08.2009.

Parts Used: Whole Plant. Leaves. Rhizome.

Folk Uses: Leaves cooked as pot herb* and for making 'patora'* (a regional delicacy prepared by applying paste of powdered Bengal grams ('basin') along with salt and condiments on each of the leaf which is then rolled, steamed and served hot). Heated leaves bandaged and considered good for joint pains*. Leaves given as fodder for 'dandu' (prolapsed uterus) in animals. Leaves also used for carrying water* by village children while playing. Decoction prepared by boiling its leaves alongwith 'ajwain' (*Carum copticum* Benth.), 'ban tobacco' (*Verbascum thapsus* L.) and 'kah-ajwain' (*Chenopodium ambrosioides* L.) considered good for bloating* in animals. Rhizome also consumed as vegetable.

Active Constituents: Phytoalexin-9, 12, 13-trihydroxy-10 (E-octadecenoic acid, dihydrosterol, aliphatic compound, nonacosane, cyaniding-3-gluciside, β-sitosterol and stigmasterol (*rhizome*) have been isolated.

Uses in Literature: Reported earlier as anodyne, astringent, demulcent, edible, galactagogue, laxative, rubefacient, stimulant, styptic, vegetable; and for abscess, alopaecia, bees sting, boils, chronic dysentery, congestion, fertility (vet.) gravies, haemorrhoids, insect bite, kidney stone, lactation (vet.), rheumatism and soups (Ambasta, 1986; Augustine & Sivadasan, 2004; Girach *et al.*, 1998; Gogoi & Borthakur, 2001; Khare, 2007; Khanna, 2002; Kumar & Goel, 1998; Samati, 2004; Tiwari & Pande, 2006).

Cordia dichotoma Forst. f. (Pl. 44D).

Syn.: *C. obliqua* Willd., *C. latifolia* Roxb., *C. myxa* Thwaites, *C. lowriana* Brandis, *C. indica* Lamk.

Family: Cordiaceae.

Vern.: Lasora, Lasura.

Ayurvedic, English, Sanskrit and Regional Names:

Ayur.: Bahuvaara, Bahuvaaraka, Bhutvrkshak, Lisodaa, Picchila, Shelu, Shleshmaataka, , Uddaalaka shita;

Eng.: Sebastan Plum;

Sans.: Bahuvarah, Kshudrashleshmataka, Madhubhutadruma;

B.: Bahubara, Chottabohnaari;

Bo.: Bargund;

G.: Rayagundo, Vadagundo;

H.: Chhotalasura, Lasura;

Kan.: Chikka Challe;

Mal.: Madaviriyasam, Vidi;

Mar.: Chokri, Goden;

Oriya: Domolo, Kophokoturi;

P.: Laswara;

Tam.: Naruvilli, Viriyam;

Tel.: Chinnanakkera, Nekkara.

Distribution: Throughout India.

Description: A medium-sized tree. Leaves alternate, stalked, coriaceous, broad ovate, toothed at the apex, entire, upper surface glabrous, lower pubescent. Flowers whitish (greenish), 0.62cm in diameter, in forked spiked or racemed coiled cymes. Berries green, yellow when ripe.

Flowering & Fruiting: April-July.

Habitat Ecology: Drier areas, wastelands, open slopes, moist forest; Tal, 861m.

Material Examined: HPU-EL-SKS-ABS-15096; 29.05.2011.

Parts Used: Inflorescence. Fruit. Gum.

Folk Uses: Inflorescence and fruits cooked as vegetable and considered good for dysentery. Fruits pickled and considered refrigerant; its mucilagenous fluid used as gum and applied for curing mouth ulcers*, also given to livestock mixed with feed for expelling worms.

Active Constituents: β-sitosterol (*roots*), protein and oil, mixture of palmitic, oleic, stearic, behenic, linoleic acids and other fatty acids (*kernel*), allantoin and beta-sitosterol (*stem bark*), taxifolin-3-rhamnoside (*seeds*) are the active principles.

Uses in Literature: Known so far as anthelmintic, antidysentric, astringent, blood purifier, colic, diuretic, expectorant, fibre yielding, mucilaginous, purgative, vegetable; and for agricultural implements, boats, bone fracture (vet.), burning of throat, chest diseases, cholera, cooling, constipation, diarrhoea, dropsy, dysentery, edible purposes (fruits), expelling worms (vet.), fever, flatulence, gonorrhoea, headache, increasing sperm count, lung disease, pickling, ulcer, urination, urinary diseases urticaria and well curb (Ambasta, 1986; Dagar & Dagar, 1996; Girach *et al.*, 1997; Hosagoudar & Henry, 1996a; Jain, 1991; Katewa & Sharma, 1998; Khare, 2007; Kumar & Pullaiah, 2003; Lalramnghinglova, 1996; Mokat & Deokule, 2004; Negi & Gaur, 1991; Parmar & Kaushal, 1982; Patole & Jain, 2002; Pawar & Patil, 2004; Rama Rao & Henry, 1996; Rao, 1996; Saini, 1996b; Singh, 1996; Tirkey, 2004; Watt, 1972).

Coriandrum sativum L. (Pl. 45A).

Family: Apiaceae.

Vern.: Dhania.

Ayurvedic, English, Sanskrit and Regional Names:

Ayur.: Chahatraa, Dhaanyaka, Kustumburu, Dhaanyeyaka, Dhanika, Dhanikaa, Dhaanaa, Dhaanya, Dhaniyaa, Kunati, Vitunnaka;

Eng.: Coriander;

Sans.: Dhanyaka, Dhanyakam;

H.: Dhaniya;

Mal.: Kothambala;

Tam.: Kotamalli;

Tel.: Danyalu, Kotimiri.

Distribution: Throughout India.

Description: Erect, annual, branched, glabrous herbs. Leaves ternetly lobed, ovate; apex obtuse. Flowers white, in compound umbels. Petals white or puplish. Outer flowers zygomorphic, inner ones actinomorphic. Fruits sub-globose.

Flowering & Fruiting: October-December.

Habitat Ecology: Cultivated; Sujanpur, 567m.

Material Examined: HPU-EL-SKS-ABS-15097; 02.12.2010.

Parts Used: Whole Plant. Leaves. Seeds.

Folk Uses: Plant considered sacred* and often hung at the entrance of newly constructed house. Decoction of seeds and plant (5ml) given thrice a day as coolant and for flatulence*. Leaves used for flavouring* and its paste used for suppuration of boils*.

Active Constituents: Active constituents are umbelliferone, scopoletin, rutin, coumarins, quercitin, kaempferol, β-sitosterol, D-mannitol and flavonoid glycosides (*seeds*).

Biological Activity: Antimicrobial activities found to be positive.

Uses in Literature: Used so far as anthelmintic, antimicrobial, antipyretic, aphrodisiac, aromatic, astringent, carminative, diuretic, laxative, stimulant, stomachic, tonic; and for bronchitis, burning sensation, diarrhoea, dysentery, inflammation, jaundice, pain in eyes, piles, sore throat, stomatitis, toothache, thirst (Khare, 2007; Kirtikar & Basu, 1994; Sharma, 2003; Retnam & Martin, 2006).

Cucumis sativus L. (Pl. 45B).

Syn.: *C. muricatus* Wall., *C. hardwickii* Royle.

Family: Cucurbitaceae.

Vern.: Kakri, Kheera.

Ayurvedic, English, Sanskrit and Regional Names:

Ayur.: Tiktakarkatikaa, Traapusha, Trapusha, Trapushi;

Eng.: Cucumber, Musk Melon, Sweet Melon;

Sans.: Bahpuspha, Kandalu, Pitapuslpa, Sudhavasa, Trapusha, Tundilaphala;

B.: Khira, Sasa;

Bo.: Kakri, Kankari;

G.: Khira;

H.: Khira;

Mar.: Kakdi, Khira, Tavase;

Oriya: Kavnai;

P.: Khira, Khiyar;

Tam.: Piping-kay;

Tel.: Dozakaya.

Distribution: Cultivated throughout India.

Description: A climbing herb. Tendrils simple. Leaves petaloid, palmately lobed, hispid on both sides. Flowers yellow. Fruits oblong, cylindric, covered with harsh hair of soft spines. Seeds many, oblong, compressed.

Flowering & Fruiting: May-September.

Habitat Ecology: Grown locally; Dhaneta, 596m.

Material Examined: HPU-EL-SKS-ABS-15098; 05.09.2010.

Parts Used: Fruit. Seeds.

Folk Uses: 15-25ml of fruit juice prescribed thrice daily for 30 days to expel kidney stones*. Fruits edible; its slice used in facial cosmetics to remove dark circle around eyes and as astringent* and cleansing agent*. Fruit paste applied for healing burn injuries* in cattle. Dried seeds eaten and its extr. considered a refreshing drink*. Powdered decorticated seeds added to various culinary preparations as 'magaj' and considered nutritive*.

Active Constituents: These are palmitic, stearic and oleic acids, their glycerides, sterols, squelene (*fruits*).

Uses in Literature: Used as diuretic, fruit, refrigerant, salad, tonic; and for confectionary, dysentery, facial cosmetics, fever, gastric (vet.), pneumonia, sunstroke, sweets, throat affection, urinary infections and whooping cough (Anonymous, 1998a, b; Ambasta, 1986; Jain, 1991; Khare, 2007; Kulkarni & Kumbhojker, 1993, 1996; Negi & Gaur, 1991; Samati, 2004; Satapathy & Brahmam, 1996; Vihari, 1995; Watt, 1972).

Cucurbita hispida Wall. (Pl. 45C).

Syn.: *Benincasa cerifera* Savi, *B. hispida* (Thunb.) Cogn.

Family: Cucurbitaceae.

Vern.: Dhuda Petha, Petha.

Ayurvedic, English, Sanskrit and Regional Names:

Ayur.: Kunshmaanda;

Eng.: Ash pumpkin, Melon, Tallow gourd, White Gourd;

Sans.: Gramya, Karkati, Kushmanda, Nagapuspha, Phala;

B.: Chalkumra;

Bo.: Kohala;

H.: Golkaddu, Petha;

Kan.: Budagumbala Kayi;

Mal.: Kumpalanna;

Mar.: Kohala;

P.: Chalkumra, Golkaddu;

Tam.: Kalytanppushinikkay;

Tel.: Burdagumuda, Pendigummadikaya.

Distribution: Cultivated.

Description: Large climbers with rigid petiole having pungent hair and tendrils upto 60cm long. Leaves reniform, more or less five lobed. Flowers large and yellow. Fruits fleshy, white, cylindric, without ribs, hairy ultimately covered with waxy bloom.

Flowering & Fruiting: June-October.

Habitat Ecology: Grown locally; Nadaun, 483m.

Material Examined: HPU-EL-SKS-ABS-15099; 12.10.2008.

Part Used: Fruit.

Folk Uses: Pieces of fruit fed to check frequent pregnancy loss* (vet.). Vegetable of fruits considered good for expelling worms. Fruits used in confectionary* preparations and its pieces added in the preparation of 'badian'*(small round dried paste of grinded pulse along with pieces of fruit of this plant) added to curries to make it nutritious. Also, fruits employed as symbolic* for animal sacrifices* and magico-religious puja* at the time of house warming ceremonies and dying* black hair into grey.

Active Constituents: β-sitosterol, lupeol, n-tricontanol, mannitol, arginine, aspartic acid, glutamic acid, asparagine, glutamin, proline, hydroxyl proline, isoleucine, cysteine, L-leucine, glucose and rhamnose have been characterized from *fruits*.

Uses in Literature: Reported earlier as anthelmintic, aphrodisiac, diabetic, diuretic, general tonic, laxative, vegetable (flowers); and for asthma, dysentery, epilepsy, intestinal worms, piles, preserving corpse, stomach problems, and tuberculosis (Ambasta, 1986; Bhardwaj & Gakhar, 2003; Ghosh *et al.*, 1996; Jha *et al.*, 1996; Khare, 2007; Lalramnghinglova, 1996; Mahato & Mahato, 1996; Rama Rao & Henry, 1996; Retnam & Martin, 2006; Sharma & Borthakur, 2003; Siwakoti & Verma, 1996).

Cucurbita pepo L. (Pl. 45D).

Syn.: *C. pepo* Roxb.

Family: Cucurbitaceae.

Vern.: Kaddu.

Ayurvedic, English, Sanskrit and Regional Names:

Ayur.: Karkaaruka, Kuushmaandaka, Kuusmaandi, Peeta kuushmanda, Seetaaphal;

Eng.: Pumpkin, Pompion, Squash red gourd vegetable Marrow;

Sans.: Karkaru, Kurkara;

B.: Kadimah, Kumra;

Bo.: Kaula;

G.: Dudia;

H.: Kadimah, Kaddu, Kadimah, Lal kumra, Safed, Kaddu, Sitaphal;

Kan.: Kumbalakai;

Mal.: Mathan;

Mar.: Kohala;

Oriya: Panikakhara;

Tam.: Parangikayi, Suraikayi;

Tel.: Budadegummedi, Gummadi.

Distribution: Cultivated.

Description: Climbers having stems and leaves with harsh prickly armour, a broad triangular-pointed outline and large, solitary, yellow, monoecious flowers. Fruits large-sized; peduncle woody, strongly grooved.

Flowering & Fruiting: June-September.

Habitat Ecology: Cultivated; Ladraur, 978m.

Material Examined: HPU-EL-SKS-ABS-150100; 16.09.2008.

Parts Used: Whole Plant, Fruits, Seeds and Rind.

Folk Uses: Dried plant used as fuel. Roasted seeds eaten as 'magaj' and added to culinary and confectionary* preparations. Fruits cooked as vegetable and also offered* to appease Lord Shiva. Dried rind used as utensil to keep water for birds.

Active Constituents: Cholesterol, ergosterol, campesterol, β-sitosterol (*seeds*), p-hydroxybenzaldehyde, anisyl alcohol, p-hydroxybenzyl alcohol, veratryl alcohol have been isolated.

Biological Activity: Seeds exhibit anthelmintic activity *in vitro* and *in vivo* against *Hymenolepis nana* and *Dicrocrelium denddriticum.*

Uses in Literature: Useful as anthelmintic, diuretic, galactagogue, poison (vet.), psychoactive, taenicide, tonic, vegetable; and for boils, burns, diarrhoea, edible purposes, inflammation, scorpion sting, utensils, worshipping (Ambasta, 1986; Jain, 1991; Jain *et al.*, 1994; Jha *et al.*, 1996; Khare, 2007; Samati, 2004; Watt, 1972).

Curcuma domestica (Medik.) Valeton (Pl. 46A).

Syn.: *Curcuma longa* L.

Family: Zingiberaceae.

Vern.: Haldi.

Ayurvedic, English, Sanskrit and Regional Names:

Ayur.: Gauri, Hattavilaasini, Haridraa, Haridruma, Kaanchani, Krimighna, Kshanda, Priyaka, Varavarnini, Yoshitapriyaa;

Eng.: Turmeric;

Sans.: Aneshta, Aurukesafur, Bahula, Gauri, Haridra, Kurkum, Nisa, Rajani, Ralri, Sarsud;

B.: Halder, Halja;

H.: Haladu, Haldi, Halud, Hardi;

Kan.: Halede;

Mal.: Arishina;

Mar.: Halada;

P.: Haridru;

Tam.: Pasupu;

Tel.: Mannal, Marinal.

Distribution: Throughout India.

Description: A tall herb possessing long stiputate tubers which are orange coloured inside. Leaves large-sized, oblong-lanceolate, tapering to the base. Flowers in dense compound spikes, concealed by sheathing petiole. Capsule with many, arillate seeds.

Flowering & Fruiting: September-October.

Habitat Ecology: Cultivated; Cha kmoh, 736m.

Material Examined: HPU-EL-SKS-ABS-150101; 25.08.2010.

Parts Used: Leaves. Rhizome.

Folk Uses: One tsp of powdered rhizome recommended with milk twice daily to cure internal injuries and after mixing with fairly hot mustard oil massaged* for quick healing of wounds both in man and animals. Paste of rhizomes applied for healing burns*, and its paste in combination with mustard oil considered good for strengthening gums*. Paste prepared by taking equal proportion of turmeric, sandal wood and leaves of *Melia azedarach* L. valued as facial pack and for skin affections (pimples, acne, skin boils). Sweet 'ladoos' prepared with 'haldi' considered good for rheumatism. Turmeric considered auspicious and used in all religious ceremonies*. Leaves employed for wrapping 'patoras'* (a delicacy prepared from leaves of *Colocasia esculenta* L.) to impart its characteristic flavour during steaming. 'Batana'* (facial pack) made from 'haldi' (turmeric) and 'basin' (powdered seeds of *Cicer arietinum* L.) usually employed to improve skin complexion of both bride and bridegroom at the time of marriage.

Active Constituents: Three antioxidants, curcumin-4-hydroxycinnamoy (feruloyl) methane and bis (4-hydroxycinnamoy) methane, α and β-turmerones, zingiberene, β pinenes, camphene, limonene, terpinene, caryophyllene, curcumene, linalool, turmenone, artermenone, borneol, isoborneol, camphor, eugenol, curdione, curzerenone, curcumin, monodemethoxy-curcumin and didemethoxy curcumin, turmerone (*rhizome*) have been characterized.

Biological Activity: Curcumin on tropical application markedly inhibited TPA and arachidonic acid induced epidermal inflammation (ear oedema in mice). Turmerin exhibited efficient antioxidant and DNA protectant activities whereas curcumin-4-hydroxycinnamoy (feruloyl) methane and bis-(4-hydroxycinnamoy) methane showed anticoagulant activity.

Uses in Literature: Known to be useful as anthelmintic (vet.), antiarthritic antibacterial, antiparasitic, antiperiodic, appetiser (vet.), antiprotozoal, antiseptic, antiseptic (vet.), antispasmodic, carminative, CNS active, condiments, cooling agent, spasmolytic, stomachic, tonic, vermifuge (vet.); and for anthrax (vet.), bad taste of mouth, bad taste of mouth cavity (vet.), bloat, bone fracture, bone fracture (vet.), chest pain, cold, cough, conjunctivitis, conjunctivitis (vet.), constipation (vet.), crack palate (vet.), cuts, deworming, diarrhoea, dislocated bone (vet.), dysentery (vet.), exoparasite (vet.), gastric disorder, heat strokes (vet.), horn cancer, hump sore (vet.), immunity (vet.), indigestion (vet.), injuries (vet.), insect sting, joint diseases, liver disorder, mastitis, opacity of cornea (vet.), rejuvenating milk veins, rheumatism (vet.), skin infection, sprain and bruises, swelling (vet.), ulcer (vet.), urinary problems, vomiting, whooping cough, wounds (Ambasta, 1986; Bhatt *et al.*, 2001; Bhattarai, 1992; Borthakur *et al.*, 2004; Chinthu *et al.*, 1998; Choondal, 1988; Deeba *et al.*, 2009; Dwivedi *et al.*, 2005; Gogoi & Borthakur, 2001; Jha *et al.*, 1991; Kadel & Jain, 2006; Khare, 2007; Khanna, 2002; Mishra *et al.*, 1996b; Mohanty *et al.*, 1998; Mokat & Deokule, 2004; Pal, 1981; Ramadas *et al.*, 2000; Rama Rao & Henry, 1996; Ranade, 1998; Reddy *et al.*, 1998; Reddy & Raju, 2000; Rekha *et al.*, 1998; Retnam & Martin, 2006; Samati, 2004; Sen & Batra, 1997; Sen *et al.*, 2002; Sharma, 2003; Sharma & Singh, 2001; Sharma *et al.* 2002; Sikarwar, 1996; Tiwari & Pande, 2006; Varshney, 1998).

Cuscuta reflexa Roxb. (Pl. 46B).

Syn.: *C. grandiflora* Wall., *C. verrucosa* Sweet, *C. hookeri* Sweet, *C. macrantha* G. Don, *C. elatior* Chois.

Family: Cuscutaceae.

Vern.: Akashbel, Amarvela, Halwan.

Ayurvedic, English, Sanskrit and Regional Names:

Ayur.: Anarvalli;

Eng.: Doddar;

Sans.: Akashab'hawan'a, Akashavalli, Akshspavana, Amaravallika, Nilatar;

Ass.: Amarlati;

B.: Algusi, Haldi-algusi-lutta;

G.: Amarbel, Akaswel;

H.: Akashabela, Amarabela, Jaunsar, Sarabali;

Mar.: Kashavela, Nirmulia;

Oriya: Kolanirmuli;

P.: Amil, Nilathari, Niradhar, Zarbutti;

Tam.: Kodiya gundal;

Tel.: Lanjasavaramu, Savarap, Sitamapurgonallu;

U.P.: Banwar, Kasbanwar;

Urdu: Akashabel, Imalbel.

Distribution: Throughout India.

Description: Leafless parasitic twiners with fleshy yellowish stems, sweetly-scented, white, crowded flowers in small clusters. Corolla ovoid or campanulate lobed, short, imbricate in bud. Fruits fleshy, globose, 4-seeded, black.

Flowering & Fruiting: September-October.

Habitat Ecology: Parasitic climber on *Duranta*, *Citrus*; Thana Drogan, 950m.

Material Examined: HPU-EL-SKS-ABS-150102; 09.05.2010.

Parts Used: Aerial Plant Parts.

Folk Uses: A bath in warm decoction of stem considered good to relieve swelling and rheumatic pains whereas massage of the mustard oil boiled with powdered climber good for internal injuries and rheumatic pains. 5-10ml decoction of the aerial plant parts (twice daily for 3 days) checks dysentery. Poultice of the plant paste applied to cure red spots* due to blood clotting. The hot water fomentation of the plant boiled along with *Vitex negundo* L., *Cedrela toona* Roxb., *Dodonaea viscosa* L. considered good for, for curing gout, joint pains and internal injuries*.

Active Constituents: Cerotic, linolenic, linolic, oleic, stearic and palimitic acid, phytosterol (*seeds*), abscisic acid (*leaves*), dulcitol, luteolin, and its glycosides, quercetin, cuscutin (*stem*), amino acid, cusculatin, cuscutin (*plant*) have been isolated.

Biological Activity: Antiviral (aerial parts), relaxant and spasmolytic (aq. extr. of the plant) activities +ve. Spasmolytic and acetylcholine-like action in atropinised rabbit auricles and frog's heart also exhibited.

Uses in Literature: Earlier known to be used as an anodyne, anthelmintic, carminative, diaphoretic, diuretic, fodder, haematuria, haemostatic, purgative, stomachic; and for biliousness, bodyache, bone fracture, burns, cuts, cold, cough, diarrhoea, diptheria, displaced knee cap, eye disorders, falling hair, fever, headache, increase blood, indigestion, itching, jaundice, killing lice (vet.), liver complaints, nervine weakness, odema, rheumatism, skin diseases, sprain, swelling of legs, testicles, ulcers, urticaria, veterinary diseases and wounds (Ambasta, 1986; Chauhan, 1999; Datt, 1996; Dixit & Pal, 1996; Girach, 1992; Girach & Aminuddin, 1995; Girach *et al.*, 1998; Hembrom, 1991; Jain, 1991; Jain *et al.*, 1991; Katewa & Arora, 1997; Khare, 2007; Kumar & Chauhan, 2005; Kumar & Jain, 1998; Mahato & Mahato, 1996; Manandhar, 1985, 1987a; Maya *et al.*, 2003; Sahoo & Mudgal, 1995; Saini, 1996 a, b; Satapathy & Brahmam, 1996; Sen & Behera, 2003; Sharma, 2003; Singh, 1996; Singh & Prakash, 1996; Siwakoti & Verma, 1996; Srivastava *et al.*, 2003; Tirkey *et al.*, 2003b; Tiwari & Tiwari, 1996; Watt, 1972).

Major Ayurvedic Formulations: Itriafal Ustakhuddus (unani drug) beside being used in traditional drugs.

Cycas revoluta L. (Pl. 46C).

Family: Cycadaceae.

Vern.: Sago Palm.

Sanskrit and Regional Names:

Sans.: Varaguna;

H.: Jangali-madan-mast-ka-phul;

Kan.: Mundichalu;

Oriya: Maddu;

Tam.: Canningay, Madanagama;

Tel.: Intalappama.

Distribution: Planted.

Description: Evergreen palm-like trees with heavy trunk and upwardly rarely forked, an elegant rossette of stiff shining fronds divided to the midrib into 80-100 long narrow segments. Leaflets reflexed. Seeds orange or red, naked and directly attached with megasporophylls.

Flowering & Fruiting: February-May.

Habitat Ecology: Cultivated; Bhota, 765m.

Material Examined: HPU-EL-SKS-ABS-150103; 07.05.2010.

Parts Used: Whole Plant. Leaves. Seeds.

Folk Uses: Planted as an ornamental*. Fruits (seeds) edible. Juice of leaves given to check indigestion* whereas young circinately coiled leaves cooked as vegetable.

Active Constituents: These are cyasin (*roots*), amentoflavone, podocarpus flavone A, 2, 3-dihydroamentoflavone, 2, 3-dihydro-hinokiflavone, isoginkgetin and bilobetin (*leaves*).

Uses in Literature: Recorded earlier as food in time of scarcity (*seeds*), psychoactive plant, vegetable; and for edible seeds, eradicating rats, flatulence, rheumatic pains, sago starch (Ambasta, 1986; Jain, 1991; Jain *et al.*, 1994; Nazarudeen *et al.*, 1996; Rama Rao & Henry, 1996; Watt, 1972).

Cymbopogon citratus (DC.) Stapf (Pl. 46D).

Syn.: *Andropogon citratus* DC.

Family: Poaceae.

Vern.: Khatti ghas.

Ayurvedic and English Names:

Ayur.: Bhuutika, Bhuutikaa;

Eng.: Lemon grass.

Distribution: Cultivated in Panjab, H.P., Maharashtra, Gujarat, Karnataka.

Description: Grasses having narrow leaves with fragrance, of lemon due to presence of volatile oil, and rounded or cordate base. Inflorescence fascicled or panicled spikes. Fruits a achenial caryopsis. Seeds endospermic with single cotyledon.

Flowering & Fruiting: September-January.

Habitat Ecology: Cultivated; Khaggal, 920m.

Material Examined: HPU-EL-SKS-ABS-150104; 03.07.2008.

Parts Used: Whole Plant. Leaves. Stem.

Folk Uses: Planted as an ornamental*. 3-5g powdered leaves with honey taken thrice a day for digestive problems* and as antispasmodic. Poultice of fresh leaves checks headaches and body aches. 15-20ml decoction of its leaves along with benafsha given for cough and cold (twice daily, three days). Also, used for flavouring tea* and other refreshing drinks*. Pieces of its stem and leaves used by healer while chanting mantras to cure 'patrara'* (acute tympany) disease (vet.).

Active Constituents: Volatile oil, citral (70%), citronellal, geraniol, myrcene, d-limonene isolated from *leaves*.

Biological Activity: d-limonene in the oil induces activity of glutathione s-transferase, a detoxifying enzyme for chemical carcinogen detoxification.

Uses in Literature: So far reported to be known as analgesic, antibacterial, anticatarrhal, anticholerin, antifungal, antiperiodic, antipyretic, carminative, depressant, detoxifying, spasmodic, stimulant and sudorific substance (Khare, 2007).

Cynodon dactylon (L.) Pers. (Pl. 47A).

Syn.: *C. stellatus* Willd., *Panicum dactylon* L., *Paspalum dactylon* DC., *Digitaria dactylon* Scop.

Family: Poaceae.

Vern.: Dhrub, Doob, Droova.

Ayurvedic, English, Sanskrit and Regional Names:

Ayur.: Anantaa, Bhaargavi, Duurvaa, Golomi, Sahastravirya, Shatparvaa, Shatvalli, Shatviryaa, Shitaa,Tiktaparvaa;

Eng.: Bermuda Grass, Conch Grass, Dhub Grass, Doob grass, Dubh, Dubla;

Sans.: Durva, Haritali;

B.: Dhobi-ghas, Durba, Dubh, Dubla;

H.: Dhub, Hariali;

Kan.: Kudi garikai, Garikaihallu;

Mar.: Arugam-pilla, Hariali;

P.: Barawa, Dhub, Khabbal, Tallu;

Tam.: Gherricha, Haryali;

Tel.: Gerichagaddi, Harvali.

Distribution: Throughout India upto 3,000m.

Description: A prostrate, perennial grass; rooting at the nodes with stem forming matted tufts and short ascending branches. Leaves linear-lanceolate, flat, narrow, soft and smooth. Flowers in digitate spikes. Fruit a caryopsis. Seeds with one cotyledon.

Flowering & Fruiting: August-November.

Habitat Ecology: Edges of fields, open slopes; Kot, 950m.

Material Examined: HPU-EL-SKS-ABS-150105; 14.05.2011.

Parts Used: Whole Plant. Leaves.

Folk Uses: Plant considered sacred and employed in all religious ceremonies, and considered as good fodder and soil binder. Its paste applied for suppuration of boils. 3-4 drops of its leaf juice check nose bleeding*. Sucking a few fresh leaves checks thirst. Bare footed walk on its field early in the morning improves eyesight*. Paste of stem and leaves used by healer while chanting mantras to cure 'patrara' (acute tympany) disease (vet.)*.

Active Constituents: Principle constituents are 2-(3-methoxy-4-hydroxy phenol propionic acid, 3-methoxy-4-hydroxy-benzoic acid, 4-hydroxy-benzoic acid, 2-(4-hydroxy phenyl), phytol, β-sitosterol, stigmasterol acetate (*leaves*), alkaloids, steroid, glycoside, flavone, terpenes, organic acid, alcohol, wax, sugars. β carotene, vit. C, palimitic acid, triterpenoids, alkaloids, ergonovine, p-coumaric acid and glucose (*plant*).

Biological Activity: Plant exhibits antiviral properties.

Uses in Literature: Important as anasarca, anticatarrhal, astringent, cooling, diuretic, expectorant, galactagogue (vet.), pasture and lawn grass, soil binder, stomatitis, vulnerary; and for acrid, bad taste in mouth, biliousness, bruises, burning sensation, cuts, cystitis, diarrhoea (vet.), dropsy, dysentery, dyspepsia (vet.), fatigue, fever, genito-urinary disease, haematuria, hysteria, increasing lactation in animals, inflamed tumours, internal injuries (vet.), leprosy, leucorrhoea, micturation, menorrhagia, metrorrhagia, nephritis, odema, ophthalmia, piles, prickly heat, scabies, skin diseases, sweet thirst, trypanosomiasis (vet.), ulcers and urticaria, vomiting, wounds, wounds (vet.) (Ambasta, 1986; Borthakur & Sarma, 1996; Girach *et al.*, 1998; Gogoi & Borthakur, 2001; Kadel & Jain, 2006; Khare, 2007; Ramadas *et al.*, 2000; Ravi, 1994; Reddy & Raju, 2000; Retnam & Martin, 2006; Sen & Behera, 2003; Sharma, 2003; Sharma *et al.* 2002; Tiwari & Pande, 2006; Upadhyay & Singh, 2005).

Major Ayurvedic Formulations: Dooravadi-kwath, Doorvadi-ghrita, Doorvadi-taila, Chandrakala-rasa.

Adulterant or Substitutes : Rhizome used as an adulterant or substitute to Couch grass (*Agropyron repens* Beauv.).

Dalbergia sissoo Roxb. (Pl. 47B).

Syn.: *D. pendula* Ten.

Family: Fabaceae.

Vern.: Shisham, Tahli.

Ayurvedic, English, Sanskrit and Regional Names:

Ayur.: Krishnashimshapaa, Picchilaa, Shimshapaa;

Eng.: South Indian Red wood, The Sissoo;

Sans.: Aguru, Dhira, Kapila, Mandalapatri, Sasam, Sasim, Shisham, Sissu, Sissai, Sisam;

Ass.: Sissu;

B.: Sissu;

G.: Nukku-Kattai, Sissom, Tanach, Yette;

H.: Shisham, Shisu, Sisam, Sissai;

Oriya: Sisu;

Kan.: Sisapa, Singsupa;

Mal.: Iruvil;

P.: Shewa, Shin, Sisham, Tali, Tahli;

Tam.: Gette, Itti, Karra, Nukku, Sissu, Tali;

Tel.: Biridi, Cishmbage, Ettasissu, Sinsupa.

Distribution: Throughtout India, ascending to 1,650m.

Description: Trees with numerous, finely grey downy branches and alternate pinnate leaves having zig-zag rachis. Flowers yellowish-white and in axillary panicles which are shorter than leaves. Pods strap-shaped, narrow at base into long stalk, pale brown, glabrous. Stamens 9 monoadelphous. Flat indehiscent pods with non-endospermic seeds.

Flowering & Fruiting: May-July.

Habitat Ecology: Open grassy slopes; Nalti, 851m.

Material Examined: HPU-EL-SKS-ABS-150106; 28.03.2008.

Parts Used: Twigs. Leaves. Roots. Wood.

Folk Uses: Dried twigs used as fuel. Leaves used as fodder, and also good against weakness and abortion* in animals. Its poultice good against joint pains* and cracked bones*. Wood employed for making furniture, agricultural implements, building works, tool handles, doors and windows. Powdered roots given for flatulence*. Its paste applied to heal burn injuries* of cattle and decoction given twice or thrice daily (about 1-2 litre per dose) for dysentery (vet.).

Active Constituents: Biochanin A-7-glucosido (sissortin), isoflavone-o-glycoside (*leaves*), a new allylphenol of latifolin type-dalbergiphenol along with dalbergenone, OMe-dalbergin, hydroxyl and OMe-dalbergenone, methyl dalbergin (*heartwood*), biochanin A and tectorigenin (*flowers*) have been characterized.

Biological Activity: Spasmolytic (plant) activity confirmed.

Uses in Literature: Described earlier as an alternative, antiemetic, charcoal, fodder, fuel wood, ornamental (ear pendant), sacred; and for agricultural implements, building works, boils, cabinet wood, carved work, charpai legs, cuts, diarrhoea (vet.), dysentery, gonorrhoea, headache, high class furniture, leucorrhoea, musical instruments, railway sleepers, spasm (vet.), skin diseases, spermatorrhoea (Ambasta, 1986; Bhattarai, 1992; Borthakur & Sarma, 1996; Jain, 1991; Katewa *et al.*, 2001; Khanna & Kumar 2000; Khare, 2007; Kumar & Chauhan, 2005; Mandal & Basu, 1996; Ranade, 1998; Siwakoti & Verma, 1996).

Datura metel auct. non L. (Pl. 47C).

Syn. : *D. innoxia* Mill.

Family: Solanaceae.

Vern.: Datura, Dhatura.

Ayurvedic, English, Sanskrit and Regional Names:

Ayur.: Dhattuura;

Eng.: Thornapple;

Sans.: Dhustura;

B.: Kaladhutura;

Bo.: Kaladhutura;

G. : Kalodhaturo;

H.: Kaladhatura;

Mal.: Kariyummatta, Madanam;

Mar.: Kaladhotra;

Oriya.: Jambunodo;

Kan.: Ummatta;

Tam.: Vellumattai;

Tel.: Datturamu.

Distribution: N.W. Himalayas.

Description: Coarse, scented, glabrous, undershrubs with broadly-ovate, entire or deeply toothed, pubescent, pale green leaves and creamish flowers in axillary panicles. Capsule 4-chambered, armed with spines, with poisonous seeds which contain alkaloid.

Flowering & Fruiting: June-August.

Habitat Ecology: Wastelands, drier areas, Krishana nagar; 900m.

Material Examined: HPU-EL-SKS-ABS-150107; 23.08.2008.

Parts Used: Twigs. Leaves. Flowers. Seeds.

Folk Uses: Flowers offered to appease Lord Shiva; and its smoke considered a good remedy to ward off the effect of magic and evil eyes* in livestock. 3-5 ml extr. of flower and fruits taken twice daily for 5 days against phlegm*. 1-3g ash of roots mixed with honey considered good for pneumonia* (twice daily, 3-7days). Smoke of burnt dried leaves and twigs inhaled to cure asthma, cough and cold*. Application of fruit juice on scalp helps in checking dandruff*. Poultice of its seeds mixed with mustard oil applied externally for rheumatic pains.

Active Constituents: Important constituents are hyoscine (scopolamine) and hyoscymine (*plant*) and scopolamine (*seeds*).

Biological Activity: Seeds antimicrobial; whole plant spasmolytic.

Uses in Literature: Known in literature as anodyne, antidote for narcotics, antihydrophobic, antiseptic, antispasmodic, aphrodiasic, cerebral depressant, country liquor, emollient, sedative, septic, suppurative boils; and for amenorrhoea, asthma, bleeding wounds and burns (vet.), cholera, chronic bronchitis, chronic rheumatism, cold (vet.), cough, cough (vet.), diarrhoea (vet.), dog bite, eczema (vet.), excessive secretion of milk, glandular enlargement, gonorrhoea, gout, high fever, inducing permanent sterility in females, inflammation of breast, insanity, insect bites, insect bite (vet.), intermittent fevers, mumps, offering to Lord Shiva, rabies, relieving pain of patella and swelling in horses, rheumatism, scabies, skin (vet.), skin diseases, sprains, swelling, tightening flabby breast, toothache, waist pain and wounds (Ambasta, 1986; Agrawal, 1981; Aminuddin & Girach, 1991; Banerjee, 1996; Bhatt *et al.*, 2001, 2002; Borthakur, 1993; Chauhan, 1999; Dagar & Dagar, 1996; Dwivedi *et al.*, 2005; Hajra, 1987; Jain, 1968; Jain *et al.*, 1991; Kadel & Jain, 2006; Katewa *et al.*, 2001; Khanna *et al.*, 1996a; Khare, 2007; Kshirsagar & Singh, 2000; Kshirsagar *et al.*, 2003; Kumar & Pullaiah, 1998; Mahato & Mahato, 1996; Maliya & Singh, 2003; Mandal & Basu, 1996; Masih, 2003; Mishra *et al.*, 1997; Mohanty & Padhy, 1996; Mohanty *et al.*, 1998; Pandey & Verma, 2002; Rama Rao & Henry, 1996; Sahoo & Mudgal, 1995; Saini, 1996a; Saxena & Vyas, 1981; Shah & Joshi,

1971; Sharma, 2003; Shekhawat & Anand, 1984; Singh & Prakash, 1994, 1996; Singh *et al.*, 1987; Siwakoti & Verma, 1996; Srivastava *et al.*, 2003; Sundaravelu *et al.*, 2003; Venkataratnam & Venkata Raju, 2004; Watt, 1972).

Datura stramonium L. (Pl. 47D).

Syn. : *D. stramonium* var. *tatula* (L.) Cl., *C. tatula* L., *D. ferox* Nees, *D. wallichii* Dunal, *D. inermis* Jacq., *Stramonium vulgatum* Gaertn.

Family: Solanaceae.

Vern.: Datura, Dhatura.

Ayurvedic, English, Sanskrit and Regional Names:

Ayur.: Dhuurta, Krishnadhattuura;

Eng.: Apple of Peru, Devil's Apple, Devil's Trumphet, Mad Apple, Thorn Apple;

Sans.: Devika, Dhattura, Dhurta, Dhurtakrit, Dhustura, Dhultura;

B.: Sadadhutura;

P.: Dattura, Tattura;

G.: Dholol Dhaturo;

H.: Dhatura;

Mal.: Mtulam, Ummam, Ummata;

Oriya. : Dhutura, Sukladhutura;

Kan.: Ummatta;

Tam.: Emanamam, Simalyumattai, Turutturam, Umattai, Vellumattai;

Tel.: Datturamu, Tellavummetta, Ummetta.

Distribution: N.W. Himalayas: 900-2,200m.

Description: Undershrubs possessing stalked, irregularly lobed toothed leaves and white flowers in axillary panicles. Capsule ovoid, covered with rigid, sharp prickles.

Flowering & Fruiting: June-October.

Habitat Ecology: Wastelands, drier areas; Rangas, 765m.

Material Examined: HPU-EL-SKS-ABS-150108; 10.07.2009.

Parts Used: Leaves. Twigs. Flowers. Seeds.

Folk Uses: Fruits intoxicating, and its juice applied on scalp to check dandruff*. Flowers considered sacred for offering to appease Lord Shiva. Inhalation of smoke of its burnt leaves believed to cure asthma. An amulet of its roots tied on the body to cure fever. 1g powdered seeds prescribed with cow's milk for a few days to check fever. Poultice of its seeds mixed with mustard oil applied externally for rheumatic pains whereas its smoke considered a good remedy to ward off the effect of magic* and evil eyes* to livestock.

Active Constituents: Important constituents are hyoscine (scopolamine) and hyoscymine (*plant*) and scopolamine, fluorodaturative and homofluoro-daturatine, a new vitanolide-vitastra-monolide along with daturalactone and its keto derivative (*seeds*).

Uses in Literature: Well known in India for asthma, bleeding wounds and burns (vet.), boils, bronchitis, controlling salivation and exoparasite, cold(vet.), cough, cough (vet.), dental caries, diarrhoea, dislocation of joints, dog bite, dysemenorrhoea, high fever, inducing permanent sterility in females, inflammation of breast, insanity, insect bite (vet.), intermittent fevers, lactation (vet.), milching (vet)., mydriatic, nervous disorders (vet.), nymphomania, painful diseases of rectum, paralysis, post encephalitic parkinsonianism, purpurea, rheumatism, scabies, sciatica, skin diseases, sprains, swelling, tonsillitis, toothache, travel sickness, ulcers, vicious indulgence, waist pain, wounds and offering to Lord Shiva; and as an anodyne, antidote for narcotics, antihydrophobic, antiseptic, antispasmodic, aphrodisiac, cerebral depressant, country liquor, psychoactive plant, septic, sedative (Ambasta, 1986; Chopra *et al.*, 1956; Jain, 1968, 1991; Jain *et al.*, 1994; Jha *et al.*, 1991; Khare, 2007; Manandhar, 1996; Singh, 1999b; Upadhye *et al.*, 1994; Watt, 1972).

Daucus carota L. var. *sativa* DC. (Pl. 48A).

Family: Apiaceae.

Vern.: Gajjar.

Ayurvedic and English Names:

Ayur.: Gaajaara, Garjara, Granjana;

Eng.: Carrot, Cultivated Carrot.

Distribution: Cultivated in India, extensively in Punjab, Haryana, Uttar Pradesh.

Description: An annual or biennial, hispid herb with 2-4 pinnate leaves having ultimate segments narrow or small. Umbels compound. Fruits elliptic, terete or dorsally compressed, ridges all prominent.

Flowering & Fruiting: November-March.

Habitat Ecology: Cultivated; Jhaniari, 900m.

Material Examined: HPU-EL-SKS-ABS-15109; 23.03.2008.

Parts Used: Leaves. Roots.

Folk Uses: Leaves consumed as potherb* along with *Brassica campestris* L. and *Spinacia oleracea* L. and pulses. Leaves and roots also given to livestock as fodder*. Roots largely cooked as vegetable*; grated carrots mixed with milk cooked as 'halwa'* and also employed for making 'murabba'* (preserve), 'pickle'*, 'soup' and refreshing drink* for anaemic* and hepatic patients*.

Active Constituents: Beta-carotenes, interferon flavones (*roots*), the furanocoumarins, 8-metho-xypsoralen, and 5-methooxypsoralen are present (*plant*), tertinen-4-ol (*oil/seeds*).

Uses in Literature: Reported earlier for bronchitis, burning micturation, cardiac muscle relaxant, cough, gout, hot flushes, lowering blood sugar, menopause, palpitation, reducing uric acid, resistance to virus, urinary troubles; and as antilithic, cystitis, diuretic, edible (rich in carotene), emmenagogue, hypotensive, spasmolytic (Ambasta, 1986; Khare, 2007).

Dendrocalamus hamiltonii Nees & Arn. _ex_ Munro (Pl. 48B).

Syn.: *Bambusa monogyna* Griff., *B. maxima* Ham., *B. falconeri* Munro.

Family: Bambusaceae.

Vern.: Baingh, Maggar.

Regional Names:

Ass.: Kokua;

H.: Kokua; Pecha;

Tam.: Kagzi-Bans.

Distribution: N. W. Himalayas: 1,500-2,400m.

Description: Sturdy shrubs having glabrous stem sheath, 3-nerved, long pointed leaves with lower surface covered with white tomentose. Spikelets ovoid. Flowers yellow, unisexual. Anthers black, hairy, twisted. Drupes yellow, aggregated in small heads.

Flowering & Fruiting: May-August.

Habitat Ecology: Forests, shrubberies (common); Tikker Khatrian, 908m.

Material Examined: HPU-EL-SKS-ABS-150110; 12.08.2009.

Parts Used: Whole Plant. Leaves. Wood. Gum.

Folk Uses: Plant grown as hedge. Stems extensively employed for making baskets, houses and bier for carrying corpse* to funeral ground. Pieces of stem also used for setting fractured bones* (vet.). Bamboo sticks employed as drenching bottle to give liquid jaggery in 'ban' (intestinal obstruction)* and other diseases of animals. Leaves given as fodder* to check dysentery* (vet.). 10-20 ml decoction of leaves prescribed once daily for 15-20 days against hernia*. 2-3 leaves fed to facilitate explusion of placenta in animals. Wood used as fuel*. 'Banslochan' a gum obtained from it used for making 'sund' (a preparation made up of dry fruits and gum) given to ladies after delivery*.

Uses in Literature: Used in India for basket, bow, building material, burning sensation, edible purposes, expelling placenta (vet.), fencing, fishing lines, hunting, sore throat, spermatorrhoea, stomach pain, thread, traditional dwelling, ulcers and killing maggots in wounds of cattle and goats; and as cordage (bark), symbolic in religious ceremonies, vegetables (young culms), water conduits (Ambasta, 1986; Borthakur, 1996a; Chhetri, 2005; Chowdhery, 1996; Jadhav, 2006; Jain, 1991; Kapur, 1996; Kumar *et al.*, 2004; Lalramnghinglova, 1996; Patel *et al.*, 2003; Patole & Jain, 2002; Punjani, 1998; Rama Rao & Henry, 1996; Samati, 2004; Sen & Batra, 1997; Singh, 2003; Watt, 1972).

Dioscorea bulbifera L. (Pl. 48C).

Syn.: *D. sativa* auct. non Thunb., *D. pulchella* Roxb., *D. pulchella* (Roxb.) Prain, *D. tamnifolia* Salisb., *Helmia bulbifera* Kunth, *D. decemangularis* Herb., *D. thunga*, Herb., *Helmia bulbifera* Kunth.

Family: Dioscoreaceae.

Vern.: Tardi.

Ayurvedic, English, Sanskrit and Regional Names:

Ayur.: Banaalu, Grshti, Raktaalu, Suraalu, Vaaraahi, Vaaraahikanda;

Eng.: Black bearing Yam;

Sans.: Badar Kachna, Vishvaksenapriya;

Ass.: Kathalu;

B.: Banalu;

G.: Shalvinavelya;

H.: Bhirvolikanda;

Oriya: Sisu;

Kan.: Onthalaigasu;

Mal.: Kattukachil;

P.: Zamikand;

Tel.: Pannukelangu.

Distribution: Subtropical Himalayas.

Description: Climbers with opposite, ovate-cordate leaves and large-sized tubers. Male racemes pubescent or woolly, female spikes solitary. Capsules with truncately rounded top. Seeds 3-winged.

Flowering & Fruiting: June-August.

Habitat Ecology: Scrub Jungles, hedges, moist rocky slopes; Mair, 840m.

Material Examined: HPU-EL-SKS-ABS-150111; 05.06.2011.

Parts Used: Plant, Tuber.

Folk Uses: Plant used in religious pooja and its poultice applied for curing arthritis*. Tubers considered as poor man's vegetable and believed to repel ghosts*.

Active Constituents: Active properties due diosgenin, diosbulbins A, diosbulbin B and diosbulbin C (*tuber*).

Biological Activity: Antiinflammatory and antifertility activities found +ve.

Uses in Literature: Known for its edible tubers, diabetes, intestinal worms, skin diseases; and as contraceptives, psychoactive (Ambasta, 1986; Gangwar & Ramkrishnan, 1990; Jain, 1991; Khare, 2007; Lal *et al.*, 1996; Sharma, 2003, Singh, 1996; Singh & Prakash, 1994; Singh & Zaheer, 1996; Sinha, 1996; Siwakoti & Verma, 1996; Watt, 1972).

Major Ayurvedic Formulations: 'Chyavanprash', Garbhadharini-vati, Mahamash-tel, Narsingh-tel.

Dodonaea viscosa L. Jacq. (Pl. 48D).

Syn. : *D. angustifolia* L., *D. dioca* Roxb., *D. burmanniana* DC., *D. pallida* Miq., *D. micropoda* DC., *D. wightiana* Blume, *D. pentandra* Griff., *D. heterophylla* Hortul, *D. spathulata* Sm., *D. arabica* Hochst. & Steud., *Ptelea viscosa* L.

Family: Sapindaceae.

Vern.: Mehandru.

Ayurvedic, English, Sanskrit and Regional Names:

Ayur.: Raasnaa;
Eng.: Jamaca Switch Sorrel;
Sans.: Rasna;
B.: Bandari, Zakhlumi;
Bo.: Jakhmi, Lutchmi;
H.: Alliar, San-atta, Sonalta, Walaytinahndi;
Kan.: Bandara;
Mal.: Unnataruvi, Virali;
Mar.: Lutchmi;
Oriya: Mohra;
P.: Banmehndu, Mendhar, Sonatta;
Tam.: Velari, Virali;
Tel.: Bandaru, Pullenu.

Distribution: Afghanistan-H.P. upto 1,500m.

Description: Evergreen shrubs. Leaves alternate, viscid-shining with yellow resinous dots. Flowers in short cymes, greenish-yellow, without petals and sepals in two series. Capsule yellowish-brown, three-celled, compressed and 3-winged. Seeds lenticular or subglobose and compressed.

Flowering & Fruiting: February-April.

Habitat Ecology: Open slopes, roadsides, often gregarious; Tikker Khatrian, 908m.

Material Examined: HPU-EL-SKS-ABS-150112; 13.03.2008.

Parts Used: Twigs. Leaves. Wood.

Folk Uses: Decoction of the plant (15-25ml) given to regularize menstruation* (twice daily, 5-7days). Twigs employed for separating* maize grains from cobs. Leaves used as cattle feed. Poultice of boiled leaves applied for healing fractured bones and sprains, also its mixture with *Vitex negundo* L., *Cuscuta reflexa* Roxb. considered good for hot water fomentation to cure aching joints and rheumatism. Poultice of leaves applied to heal scald* injuries and for dandruff. Hot poultice of same alongwith stem bark of 'rahyan' (*Soymida febrifuga* A. Juss.), corm of 'haldi' (*Curcuma domestica* Valeton.) and leaves of 'bhangaru' (*Vanda roxburghii* R.Br.) applied with bamboo sticks to heal fractured bones both in man and animals. Wood used as fuel.

Active Constituents: These are β-sitosterol, stigmasterol and isorhamnetin, a leucocyanidin (*bark*), two prosapogenins-RI barrigenol-21, 22-diangelate and jegosapogenol-21-(2, 3, dihydroxy-2-methyl-butyroyl) and 22-angelates (*stem bark*).

Biological Activity: Plant extr. exhibits cardio-inhibitory and spasmolytic activity on smooth muscle and intestine. It also has sedative effect on virgin Guinea pig uterus and hypotensive effect unaffected by atropine.

Uses in Literature: Useful earlier as aromatic, astringent, carminative, febrifuge, fire wood, fodder, manure, purgative, substitute for hops in preparation of wine and beer, sudorifice in gout and rheumatism; and for backache, bone fracture, bone fractures (vet.), broken legs of livestock, burns, cold, cuts, dandruff, dental diseases, epilepsy, fracture, intestinal worms, muscular pains, paralysis, sprains, ticks (vet.), waist pain, worms (vet.) and wounds (Ambasta, 1986; Geetha *et al.*, 1996b; Goud & Pullaiah, 1996a,b; Jain 1991; Kemparaja *et al.*, 1998; Khare, 2007; Kumar & Pullaiah, 1998; Rama Rao & Henry, 1996; Reddy *et al.*, 1998; Retnam & Martin, 2006; Sensarma, 1989; Sharma, 2003; Sharma & Sood, 1997; Sudarsanam *et al.*, 1995; Vishwanathan, 1997; Watt, 1972).

Adulterant or Substitute: Fruits used as substitute for true Hops (*Humulus lupulus* L.).

Duranta repens L. (Pl. 49A).

Syn.: *D. plumieri* Jacq.

Family: Verbenaceae.

Vern.: Neel Kanthi (Neel Kanta).

English, Sanskrit and Regional Names:

Eng.: Creeping skin flower, Golden Dew Drop, Pigeon Berry;

Sans.: Vriti Kanta;

H.: Daranta.

Distribution: W. Himalayas: 200-1,600m.

Description: A spiny shrub. Leaves opposite, broadly elliptic, entire. Spines stout. Flowers bright, lavender-blue with pale centers, in cymes or spikes, often compound or paniculate. Fruits golden yellow, globular, enclosed by persistent 4-calyx. Seeds exalbuminous.

Flowering & Fruiting: April-August.

Habitat Ecology: Common as a hedge plant; Saur, 840m.

Material Examined: HPU-EL-SKS-ABS-15113; 15.05.2011.

Parts Used: Aerial Parts. Leaves.

Folk Uses: Plant grown as hedges for fencing, repelling mosquitoes, and as pollution absorbent*. Gargles of decoction of its leaves cure toothache*, 10-20ml decoction of the leaves taken with a pinch of salt and black pepper once daily during mid-day for joint pains*.

Active Constituents: β-sitosterol, α-amyrin, scuttellariein, 4'-methoxy-scutellarein, scutellarein-7-O-rhamnoside, pectolinarigenin and 7-O-rutinoside are the active constituents (*leaves*).

Uses in Literature: Recorded earlier as an antifungal, antirheumatic, hedge plant, insecticidal, mosquito repellent and turnery (Ambasta, 1986; Khare, 2007).

Eclipta prostrata Roxb. (Pl. 49B).

Syn.: *Eclipta alba* (L.) Hassk.

Family: Asteraceae.

Vern.: Bhringraja.

Ayurvedic, English and Regional Names:

Ayur.: Bhringa, Bhringaja, Bhringaraaja, Bhrngaara, Bhrngaaraka, Kesharaaja, Keshranjana, Maarkava;

Eng.: Trailing Eclipta plant;

Tam.: Karisalaankanni.

Distribution: Throughout India up to 2,000m.

Description: Erect, slender roughly pubescent herbs having opposite, sessile, narrowly lanceolate leaves. Peduncles 1-2-axillary, short or long and slender. Heads radiate, flowers white. Achenes narrowly oblong, ribbed and tipped with a pappus of 2-5 minute teeth.

Flowering & Fruiting: August-November.

Habitat Ecology: Present along water channels, 'kuhls', ditches upto 1,300m; Khaggal, 920m.

Material Examined: HPU-EL-SKS-ABS-15114; 03.07.2008.

Parts Used: Whole Plant. Leaves. Root.

Folk Uses: Juice of plant recommended for jaundice* and as tonic* whereas bath with decoction of the plant good against skin diseases. 2-5g of powdered plant consumed with honey daily in morning as hair tonic* and for improving eye sight*. Extr. of the plant mixed with luke warm mustard oil massaged* on head to cure insomnia*.

Active Constituents: Plant contains alpha-terthienyl-methanol, stigmasterol, beta-amyrin, mixture of desmethyl wedelolactone.

Uses in Literature: Known to be antihepatotoxic, antipyretic, antiseptic, coolant, digestive, dyeing, emetic, hypotensive, immunoactive, mycocardial depressant, purgative; and for arthritis, gallbladder, inflammatory diseases, nematodes (roots), rheumatoid scorpion bites, skin infections and tactic ulcer (Ambasta, 1986; Chauhan, 1988a, b; Khare, 2007).

Major Ayurvedic Formulations: Bhringarajasava, Maha-bhringraja-taila, Gandhaka-rasayana, Shadabindu-taila, Ashvakanchuki-rasa, Sutashekhara-rasa, Ananda-bhairva-rasa.

Elaeocarpus ganitrus Roxb. *ex* G. Don (Pl. 49C).

Syn.: *E. sphaericus* K. Schum., *Ganitrus sphaericus* Gaertn.

Family: Elaeocarpaceae.

Vern.: Rudraksha.

Ayurvedic, English and Regional Names:

Ayur.: Panchmukhi, Rudraaksha;

Eng.: Utrasum Bead tree;

Tam.: Rudraaksham.

Distribution: Native of West Bengal,M.P., Maharashtra, Orissa, A.P., and Western Ghats.

Description: A tree with glabrescent, elliptic acute leaves. Flower buds ovoid-conical, elongate, white. Flowers usually hermaphrodite and in axillary racemes. Stamens many, anthers bearded. Drupes purple, stone tubercled, grooved, five-celled, five-seeded.

Flowering & Fruiting: December-April.

Habitat Ecology: Cultivated; Neri, 381m.

Material Examined: HPU-EL-SKS-ABS-15115; 13.03.2008.

Parts Used: Whole Plant. Leaves. Fruits. Seeds.

Folk Uses: Plant considered sacred and worshipped. Powdered fruits 0.5g mixed with warm water given twice or thrice daily for asthma, dysentery*, epileptic fits and headaches. An amulet of seeds around the neck believed to keep evils* away. Poultice of leaves good for rheumatic pains*. Its ornamental* seeds strung into beads by 'faqirs' and 'pujaris' as rosary* for religious worshipping and appeasing various deities.

Active Constituents: Constituents isolated are palmitic, isopalmitic, linoleic and myristic acids (*plant*). rudrakine (+)-elacocarpine and (+)-iso-elacocarpine, phenolics-quercetin, gallic acid and ellagic acid (*leaves*).

Uses in Literature: So far used for headaches, epileptic fits, worshipping; and as antiasthmatic, anticonvulsant, broncho-dilatory, cardiostimulant, choleretic, hypotensive, sedative, spasmolytic (Ambasta. 1986; Khare, 2007; Kirtikar & Basu, 1994).

Elettaria cardamomum (L.) Maton (Pl. 49D).

Syn.: *Amomum cardamomum* L., *Alpinia cardamomum* (L.) Roxb.

Family: Zingiberaceae.

Vern.: Elaichi.

Ayurvedic, English, Sanskrit and Regional Names:

Ayur.: Bhrngaparnikaa, Draavidi, Elaa, Kshudrailaa, Prihtvikaa, Sukshmailaa, Triputaa, Truti, Tutthaa, Upkunchikaa;

Eng.: Lesser cardamom, Cardamom;

Sans.: Bahoola, Chundruvala, Ela, Elo;

B.: Ilachi;

G.: Elachi;

H.: Chottielachi.

Mal.: Malabari elachi;

Mar.: Veldode, Velloda;

P.: Illachi;

Tam.: Aila-cheddi, Ela-kaya, Ellakay;

Tel.: Ellakay, Elaki chettu, Ela-kaya.

Distribution: Throughout India.

Description: Perennial herbs. Leaves distichous. Flowers in panicle, white or pale green in colour. Fruits trilocular, pale green to yellow in colour. Seeds arillate, many, endospermic.

Flowering & Fruiting: April-August.

Habitat Ecology: Cultivated, different parts of Hamirpur; 787-944m.

Material Examined: HPU-EL-SKS-ABS-15116; 29.05.2011.

Parts Used: Fruits. Seeds.

Folk Uses: Powdered seeds used as spice* and for flavouring tea and coffee; its paste prepared with glycerine considered good against mouth ulcers. Fruits chewed raw for flatulence,* indigestion, bronchitis* and upper respiratory tract infection. Its decoction also regarded good for cough, cold and throat infection*.

Active Constituents: Active constituents are α, β-pinene, sabinene, myrcene, limonene, cineol, cymene, linalool, borneol, nerylacetate, geraniol, nerol, neolidol (*seeds*).

Uses in Literature: Described in Indian literature as antiemetic, antigripe, antiseptic, antispasmodic, aphrodisiac, aromatic, bechic, beverages and liquors, carminative, colic, coolant, diuretic, emmenagogue, orexigenic, purgative, stimulant, stomachic; and for asthma, bronchitis, confectionary, flavouring, flatulence, husk for rheumatism, indigestion, liver disorders, loss of appetite, paste as balm for headache, piles, scabies, toothache and vomiting (Ambasta. 1986; Khare, 2007; Sharma, 2003; Retnam & Martin, 2006).

Major Ayurvedic Formulations: Chyawanprash, Elachi Bati.

Emblica officinalis Gaertn. (Pl. 50A).

Syn.: *Phyllanthus emblica* L., *Cicca emblica* Kurz, *Dichelastina nodicaulis* Hance

Family: Euphorbiaceae.

Vern.: Ambla, Amla.

Ayurvedic, English, Sanskrit and Regional Names:

Ayur.: Aamalaa, Aamalaka, Aamalaki, Amla, Amoghaa, Amritphala, Dhaatri, Dhaatriphala, Hattha, Kaayasthaa, Shiva, Vayasyaa, Vrshya;

Eng.: Emblic Tree, Indian Goose Berry, Myrobalan Tree;

Sans.: Adiphala, Akara, Amalaka, Amalaki, Amaraphala, Amlika, Amritaphala, Bahuphali, Dhatri, Dhatrika, Jatiphala, Kayastha, Pancharasa, Rochani, Shirphala, Shiva, Triphala,Vrishya, Vrittaphala;

A. : Amlaki, Sohmyrlain;

B.: Amla, Amlaki;

H.: Amalaki, Amla, Amlika, Anola, Anuli, Aungra, Bhoza, Daula;

Kan.: Amalaka, Nelli, Nilika, Kayi;

Mal.: Amalakam;

Mar.: Anvala, Anoli, Avala, Arola, Bhuiawali;

Oriya: Amlaki, Gondhona, Ohalu;

Tam.: Amalagam, Nelli, Sirottam;

Tel.: Amalakamu, Amalaki, Usiriki;

Urdu: Anwala.

Distribution: Throughout India; often cultivated 1,400m.

Description: Deciduous trees with crooked trunk, conchoidally flaking bark, light green, glabrous, obtuse, subsessile leaves and greenish-yellow flowers in axillary fascicles on the leaf bearing branchlets. Fruits succulent, globose, pale yellow, acidic, 6-lobed.

Flowering & Fruiting: April-September.

Habitat Ecology: Common along borders of agricultural fields, dry soil; Gallu Amb Tarkawari, 980m.

Material Examined: HPU-EL-SKS-ABS-15117; 02.05.2008.

Parts Used: Plant. Branches. Fruit. Wood.

Folk Uses: Plant considered sacred and worshipped on 'Panch Bhi kma'* (specific religious days of Hindus). One tsp of powdered fruit taken with honey (twice a day) for seven days to cure cough, piles* and as appetite enhancer. Fresh fruits or its pickle or jam considered good as rich source of vit. C, memory enhancer, hepatoprotective,* refrigerant* and for checking constipation, cardiac troubles. Paste of mixture of its powdered dry fruits and leaves of 'henna' (*Lawsonia alba* L.) applied as cure for dandruff, hair dye and conditioner. 'Triphala churna', comprising its dried fruit as one of the important constituents along with 'harad' (*Terminalia chebula* Retz.) and 'bhera' (*Terminalia bellirica* Roxb.) considered good against dysentery, gastritis, intestinal obstruction, indigestion and constipation in both man and animals. Hot decoction (1-2 liters) prepared from 'triphala' powder also given to animals to check cough and cold ('dhaknu'). Small pieces of its wood along with sandal wood offered in the last rites by one and all to the funeral pyre,* of the deceased and in pooja* ceremonies.

Active Constituents: Active constituents are tannins, polyphenolic compounds 1, 3, 6-trigalloyl glucose, terchebin, corialgin, ellagic acid, phyllantidine and phyllantine, terchebin (*fruits, leaves*), vit. C, I.A.A., four other auxins al, a3, a4, a5, two growth inhibitors R1 and R2 (*fruits*).

Biological Activity: Leaf extr. antibacterial. 50% EtOH extr. of fruits anticancerous, antiviral, carminative and stomachic. Water soluble fraction of methanol extr. of leaves exhibits antiinflammatory activity in carrageenin and dextran induced rat hond pan odema.

Uses in Literature: Recorded to be anticancerous, aphrodisiac, appetiser (vet.), astringent, blood purifier, detergent, diuretic, edible (eaten raw or cooked), expectorant, hair dye, hair shampoo, hair oil, important ingredient of 'chayawanprash' and 'triphala', increasing appetite, inferior quality furniture, laxative, magico-therapeutic to treat toothache, rich source of vit. C, symbolic abode of God (sacred), tonic; and for agricultural implements, anthrax, bronchial problems (vet.), burn (vet.), cold, cough, diarrhoea, dysentery (vet.), dyspepsia (vet.), garments, gastropathies (vet.), gonorrhoea, headache, impaction (vet.), indigestion (vet.), inflammation, inflammation (vet.), liver infection (vet.), rheumatism (vet.), rinder pest (vet.), spermatorrhoea, stomachache (vet.), tanning, ulcer, ulcer (vet.), vaginal complaints, vomiting, washing head, well rings, wounds (vet.) and wound maggots (Ambasta, 1986; Aminuddin & Girach, 1991, 1993; Bhardwaj & Gakhar, 2003; Banerjee & Pal, 1996; Borthakur, 1996a; Borthakur & Gogoi, 1994; Chauhan, 1999;

Dixit, 1987; Ghosh *et al.*, 1996; Girach, 1992; Gogoi & Borthakur, 2001; Hosagoudar & Henry, 1996a; Jain, 1968, 1991; Jain *et al.*, 1994; Khanna *et al.*, 1996b; Khare, 2007; Kothari & Moorthy, 1996; Kulkarni & Kumbhjokar, 1996; Kumar & Pullaiah, 2003;Mandal & Basu, 1996; Mehrotra *et al.*, 1995; Mohanty *et al.*, 1998; Negi *et al.*, 1993; Pal, 1981; Pal & Jain, 1998; Pandey & Verma, 2003; Patole & Jain, 2002; Pawar & Patil, 2004; Ramadas *et al.*, 2000; Ranade, 1998; Retnam & Martin, 2006; Reddy *et al.*, 1998; Sahoo & Mudgal, 1995; Sharma, 2003; Sharma & Singh, 2001; Sharma & Sood, 1997; Singh, 1998, 2003; Sudarsanam *et al.*, 1995; Tarafder, 1983; Tirkey *et al.*, 2003; Upadhyay & Singh, 2005; Varshney, 1998; Vihari, 1995; Vishwanathan & Singh, 1996; Watt, 1972).

Major Ayurvedic Formulations: Agnitundi-vati, Ambalyadi-churn, Amlaki-churn, Amla-kyadi-churn, Amlaki-rasayan, Arogya-vardhini, Brahma-rasayan, Brahmi-amal tail, Brahmyarasayana; fresh fruit pulp is the main ingredient of 'chayavanprash', 'sanjivani-vati', 'triphala-churn', 'triphala-ghrit', 'triphala-karanjaadi yog', 'kabz-care'.

Trade: Amla, Emblic Myrobalan.

Equisetum debile Roxb. *ex* Vauch. (Pl. 50B).

Syn.: *E. ramossisimum* Desf.

Family: Equisetaceae.

Vern.: Ghornali.

Ayurvedic and English Names:

Ayur.: Ashwa-puchha;

Eng.: Field horsetail.

Distribution: Himalayas: high altitutes along sandy streams.

Description: Perennial shrubs with lax scramling stem upto 3m, long slender branches in a whorl, 5cm long internodes, leaf teeth black with scarious margin and sessile spike in the funnel-shaped tip of the branch. Sporophylls oblong, pale with black centre. Sporangia oblong, yellow.

Habitat Ecology: Along borders of well, agricultural fields, roadsides; Dho, 1,034m.

Material Examined: HPU-EL-SKS-ABS-15118; 08.02.2010.

Parts Used: Whole Plant. Stem. Branches.

Folk Uses: Poultice of plant applied for healing bones*. Decoction of plant (10-20ml) considered good for inflammation of urinary tract (twice daily, for 5-7days). Stem and branches given as fodder to check evagination of uterus.

Active Constituents: These are silicic acid, silicates, alkaloids, nicotine, palustrine and palustrinine, flavonoid, saponin, and minerals (*whole plant*).

Uses in Literature: Reported as antihaemorrhagic, astringent, diuretic, haemopoietic, haemostatic; and for genitourinary affections, cystitis, enuresis, leucocytosis, prostatitis, styptic, urethritis, vulnerary (Ambasta, 1986; Chopra *et al.*, 1956; Khare, 2007; Islam, 2002; Watt, 1972).

Eriobotrya japonica Lind l. (Pl. 50C).

Syn.: *Mesilpus japonica* Banks

Family: Rosaceae.

Vern.: Laukhat, Lukat.

Ayurvedic, English and Regional Names:

Ayur.: Lottaala;

Eng.: Chinese Medlar, Japan Medlar, Loquat;

H.: Logat;

Kan.: Lokkte;

Tam.: Ilakota;

Urdu: Lakhota.

Distribution: Throughout India; often cultivated.

Description: Evergreen, small, robust, woolly cultivated trees. Leaves obovate, subsessile, coriaceous, strongly ribbed and toothed. Flowers white, crowded, fragrant in terminal panicles. Fruits pyriform, succulent orange yellow. Seeds non-endospermic, embryo with fleshy cotyledons.

Flowering & Fruiting: July-September.

Habitat Ecology: Along borders of agricultural fields, roadsides; Bhota, 765m.

Material Examined: HPU-EL-SKS-ABS-15119; 08.05.2011.

Parts Used: Leaves. Fruit. Wood.

Folk Uses: Fruits edible and considered as good appetizer*. Leaves used as fodder*. Wood used for construction*, low grade furniture and agricultural implements*.

Active Constituents: These are loquatoside, glucose, fructose and sucrose (*fruit pulp*), sorbitol-6-phosphate dehydrogenase, ascorbic acid oxidase and vit., sesquiterpeneglycoside, polyhydroxy triterpenoids. (*leaves*).

Biological Activity: Ursolic acid exhibits potent anticarcinogenic activity against sarcoma 180 cells and significantly suppresses myconcogene 0 expression. Leaf extr. exhibits marked antifungal activity against *Ustilago tritici.*

Uses in Literature: Described earlier as an antiscorbutic, aphrodisiac, diuretic, expectorant, ruler and drawing material, sedative; and for allaying vomiting, desert, jams, jellies, piles, skin eruptions, thirst, and toothache (Ambasta, 1986; Chopra *et al.*, 1956; Khare, 2007; Watt, 1972).

Eruca sativa Lam. (Pl. 50D).

Syn.: *Brassica eruca* L., *B. erucoides* Roxb.

Family: Brassicaceae.

Vern.: Taramira.

Ayurvedic, English, Sanskrit and Regional Names:

Ayur.: Bhutaaghana, Daradharsha, Shveta-Sursaa, Siddaartha, Tuvari, Tuvarikaa;

Eng.: Rocket;

Sans.: Tambeh;

B.: Duan, Sahwan, Taramira, Tira, Tura;

H.: Shwet-surska, Suffed shorshi, Taramir;

P.: Mandao.

Distribution: N.W. Himalayas.

Description: Erect, branching annual herbs. Leaves lyrate, pinnatifid. Flowers lilac or yellow, veined with violet. Petals veined. Pods erect and appressed to stem. Seeds globose.

Flowering & Fruiting: February-April.

Habitat Ecology: Wastelands, roadsides (cultivated); Barsar, 815m.

Material Examined: HPU-EL-SKS-ABS-15120; 10.02.2010.

Parts Used: Whole Plant. Seed.

Folk Uses: Plant used as fodder*. ½ teaspoon of seed powder given for 5 days, twice a day, to check muscular pains*. Seed oil applied externally in animals for expelling lice. Powdered seeds and oil cakes mixed with those of 'binnaula' (*Gossypium arboreum* L.) in ratio of 1:5 given for enhancing yield of milk*. About 500g powdered seeds mixed with wheat flour considered good to check foul smell from milk, bloating* and 'basaihar' (mastitis)* diseases (vet.).

Active Constituents: Therapeutic properties attributed to isorhamnetin and its 3-glucosides (*leaves*), (4-Me-thio)-Bu-glucosinolate (glucoerucin) (*seeds, whole plant*).

Biological Activity: Seeds antibacterial and also agglutinate human red cells without specificity to any particular group.

Uses in Literature: Useful as antibacterial, antiscorbutic, aphrodisiac, diuretic, rubefacient, stimulant, stomachic, vesicant; and for dropsy (vet.), ectoparasite attachment (vet.), epilepsy, inflammation, itching, leucorrhoea, piles, skin eruptions, toothache, vomiting and ulcers (Chopra *et al.*, 1956; Deeba *et al.*, 2009; Khare, 2007; Kirtikar & Basu, 1994).

Eucalyptus umbellata Domin (Pl. 51A).

Syn.: *E. globulus* L.; *E. tereticornis* Sm.

Family: Myrtaceae.

Vern.: Safeda.

Ayurvedic, English, Sanskrit and Regional Names:

Ayur.: Haritaparna, Neelaniryaasa, Nishotraa, Rechani, Saralaa, Sugandhapatra, Suvahaa, Tailaparna, Tilaparna, Tribhandi, Triputaa;

Eng.: Blue Gum, Eucalyptus;

Sans.: Tailaparnah;

H.: Safeda, Yikeliptas;

Tam.: Karupuramaram.

Distribution: Introduced in India upto 1,200m.

Description: A tree having bark peeling in long thin layers. Leaves alternate. Flowers white and in umbellate cluster of three. Capsule with locullicidal dehiscence.

Flowering & Fruiting: March-April.

Habitat Ecology: Wastelands, slopes, commonly planted along roadsides; Pucca Bharo, 840m.

Material Examined: HPU-EL-SKS-ABS-15121; 03.04.2008.

Parts Used: Twigs. Whole Plant. Leaves. Wood.

Folk Uses: Trees planted for reclaiming* wet-land areas. Leaves used in storage houses for killing pests*. Leaf bed used for cattle to remove skin ectoparasites. Powdered leaves with honey used for cure of cough and cold. Poultice of boiled leaves checks headache. Smoke of burnt leaves good for repelling mosquitoes*. Dried twigs used as fuel. Wood used for construction, poles and furniture*.

Active Constituents: Isolated constituents are cis-and trans-pinacarveol, a-pinene, an unsaturated a-ketone, oleanoic, gallic, ferulic and maslinic acids (*leaves*), gallic acid, 7-B-D-glucosyloxy-5-hydroxy-2 methyl-chromone, β-sitosterol and (+) catechin (*wood*).

Biological Activity: Oil astringent, thermogenic, antiseptic, stimulant, anodyne, cardiotonic and diuretic.

Uses in Literature: Described earlier for asthma, burns, cough, cold, dysentery, dyspepsia, flatulence, intermittent fever, headache, pneumonia, pyorrhoea, sleepers, engineering construction, ship building and flooring; and as antiseptic, cardiotonic, deodorant , diuretic, dyspepsia (vet.) (Ambasta, 1986; Khare, 2007; Khanna, 2002; Kumar & Jain, 1998; Pal, 1980; Sharma, 2003).

Major Ayurvedic Formulations: Panchgun-taila, Nilgiri-taila.

Eugenia cerasoides Roxb. (Pl. 51B).

Syn.: *Syzygium cerasoideum* (Roxb.) Raizada; *S. cerasoides* var. *obovatum* (Kurz) Santapau & Jar.; *E. operculata* Duthie.

Family: Myrtaceae.

Vern.: Kathman.

English, Sanskrit and Regional Names:

Eng.: Black plum, Java plum;

Sans.: Jambaram, Rajaphala;

Ass.: Godhajam, Mokra-Jam;

H.: Dugdugia, Piaman;

B.: Chotajam, Kalajam;

Bo.: Jhambul;

G.: Jambudi, Ravanam;

Oriya: Pitajam, Monisiagamo.

Distribution: Sub-Himalayasn tract.

Description: Medium-sized evergreen trees possessing elliptic leaves, greenish-white, sessile, regular, pale green, small and crowded flowers borne on 5-10cm long branches of lateral panicles. Berries globose or ovoid. Seed single, endospermic.

Flowering & Fruiting: April-July.

Habitat Ecology: Border of agricultural fields; Anu Khurd, Ballooni, 958m.

Material Examined: HPU-EL-SKS-ABS-15123; 30.05.2010.

Parts Used: Twigs. Leaves. Fruit. Wood.

Folk Uses: Leaves used as fodder*. Twigs employed as 'pattri'* (bedding for animals). Fruits taken for cure of diabetes*, mouth sores* and anaemia*. Wood employed for making agricultural implements*, construction work and low grade furniture*.

Uses in Literature: Useful for boards, bronchitis, building works, dysentery, joint pains; and as acrid, astringent (Ambasta, 1986).

Eugenia jambolana Lam. (Pl. 51C).

Syn.: *Syzygium cumini* (L.) Skeels, *S. jambolanum* DC., *Calyptranthes jambolana* Willd., *C. capitellata* Ham., *Myrtus cumini* L., *E. frondosa* Wall.

Family: Myrtaceae.

Vern.: Jaamun, Jaman.

English, Sanskrit and Regional Names:

Eng.: Black Plum, Java Plum;

Sans.: Jambu, Jambula, Mahaskandha;

Ass.: Jamu;

B.: Kala-jam;

Bo.: Jambul;

G.: Jambura, Jamli, Ravanam;

H.: Jam, Jaman, Jamun, Paiman, Phalinda;

Mal.: Naga, Nava, Perinnaral;

Oriya: Jamkuli, Jamo;

Tam.: Naval, Narvel, Naga;

Tel.: Nairuri, Naredu, Nasudo.

Distribution: Throughout India.

Description: A large evergreen glabrous tree with close grained white and durable wood, elliptic-oblong, firm, opposite stalked, shining, acuminate, coriaceous leaves which are smooth and shining above. Cymes short and compact or laxly paniculate with greenish-white, fragrant, sessile, numerous flowers. Berries succulent, obliquely-oblong, black with pink juicy pulp and single, endospermic seed.

Flowering & Fruiting: May-July.

Habitat Ecology: Roadsides, wastelands, borders of agricultural fields; Brahmni, 858m.

Material Examined: HPU-EL-SKS-ABS-15124; 15.06.2010.

Parts Used: Leaves. Fruit. Seed. Wood.

Folk Uses: Leaves used as fodder, also given to animals for expelling intestinal worms. 3-5g mixture of its powdered leaves and those of *Aegle marmelos* Correa. and 'neem' (*Melia azadirachta* L.) taken with luke warm water on empty stomach for treating arthritis*. Fruits edible and good for mouth ulcers and anaemia. One tsp decoction of its fruit with onion (*Allium cepa* L.) considered antiemetic* (twice a day). 10-12g of its powdered fruit given as single dose in snake bite. Also, tsp of powdered fruit and seed taken twice a day with water to control diabetes. Wood used for construction works, agricultural implements and furniture.

Active Constituents: Active principles associated to jambolin, ellagic acid, tannin (19%), gallic acid, chlorophyll, fatty oil, starch, resin, sugar , traces of oil (*seed*), anthocyanin, delphinidin-3-gentiobioside, malvidin-3-laminaribioside-petunidin-3-gentiobioside, glucose, fructose, gallic acid, malic acid, eugenia triterpenoids A & B, oleanic acid (*fruit*), isoquercitrin, quercetin, kaempferol, myricetin (*flowers*), myristic, linoleic, stearic, palmitic, vernolic, lauric, sterculic, malvalic acids (*seed oil*), betulinic acid, friedelin, friedelinol, kaempferol, quercetin and sitosterol (*stem-bark*).

Biological Activity: Seeds hypoglycaemic whereas aerial parts hypotensive.

Uses in Literature: Described in published literature as antidote to Nux vomica poisoning, astringent, colic, colouring material in Mithila paintings, constuction wood, fodder, foliage for silkworms, green manure, mouthwash, props for shafts and galleries in mines, sacred (temple sculptures and symbolic abode of gods), tonic, vegetable; and for agricultural implements, anaemia, asthma, blisters in mouth, boat building, branches (festivals and marriages), bronchitis, calculous affections, cancer, cart-wheels, cholera (vet.), colouring fish nets, diabetes, diarrhoea, diarrhoea (vet.), digestive complaints, dyeing and tanning, dysentery, edible purposes (fruits), enlargement of spleen, expelling placenta after delivery, fermentation for rice beer, furniture, giddiness, gums, indigestion, intestinal worms, jams, jellies, leucorrhoea, liver disorders, natural shiny black hair, piles, pimples, preserves, purifying blood, redness of eyes, sleepers, sore (vet.), squashes, stomach pains, stored grain pest, swelling (vet), thatching, tool handles, troughs, ulcers, vinegar (ripe fruits), well-curbs, wound maggots and wounds (vet.) (Ambasta, 1986; Aminuddin & Girach, 1993; Anonymous, 1998a, b; Bhattarai, 1992; Borthakur & Sarma, 1996; Chowdhery, 1996; Chowdhury *et al.*, 1996; Deeba *et al.*, 2009; Dixit & Pal, 1996; Gogoi *et al.*, 2003; Henry *et al.*, 1996; Hosagoudar & Henry, 1996a; Jain, 1968, 1991; Jha & Basak, 1994; Kadel & Jain, 2006; Khanna *et al.*, 1996b; Kulkarni & Kumhojkar, 1996; Lalramnghinglova, 1996; Malik *et al.*, 2009; MandaI & Basu, 1996; Masih, 2003; Mishra *et al.*, 1996b; Pal & Jain, 1998; Radhakrishnan *et al.*, 1996; Rama Rao & Henry, 1996; Shome *et al.*, 1996; Sikarwar & Kaushik, 1992; Singh, 1996, Sinha, 1996; Siwakoti & Varma, 1996; Sudarsanam *et al.*, 1995; Tirkey *et al.*, 2003a; Watt, 1972).

Major Ayurvedic Formulations: Jambarista, Jambuadi-taila, Panchapallara-yoga,

Panchapallava-yoga, Pusyanuga-churna, Usirasava.

Trade: Jaamun-guthli, Jambul-stone.

Eupatorium adenophorum Spreng. (Pl. 51D).

Syn.: *Ageratina adenophora* (Spreng.) King & Rob.
Family: Asteraceae.
Vern.: Kalipadadi.
Ayurvedic, English and Regional Names:
Ayur.: Ayaapaana, Vishalyakarani;
Eng.: Croften weed, Eupatory;
H.: Kalibasuti;
Tam.: Ayapanai.
Distribution: Native of Mexico; found in Himalayass upto 1,800m.
Description: Branched herbs having homogamous heads. Peduncles glandular hairy. Achenes 5-ribbed. Pappus 1-seriate.
Flowering & Fruiting: February-May.
Habitat Ecology: Moist slopes; Lambloo, 850m.
Material Examined: HPU-EL-SKS-ABS-15122; 08.10.2009.
Parts Used: Leaves.
Folk Uses: Leaves used externally for eye diseases*. Decoction of leaves (500-700ml) given in case of constipation* in animals (thrice daily till cure).
Active Constituents: Isolated constituents are ayapanin, ayapin, carotene, free vit. C (plant).
Biological Activity: Plant spasmolytic.
Uses in Literature: Used so far for body swellings, cancerous growth, cuts, injuries, snake bite, stomachache, toothache, tumours, wounds; and as shampoo (Abraham, 1981; Ambasta, 1986; Gangwar & Ramkrishnan, 1990; Hajra & Baishya, 1981; Joseph & Kharkongor, 1981; Khare, 2007; Rao & Jamir, 1982).

Euphorbia pulcherrima Willd. (Pl. 52A).

Syn.: *Poinsettia pulcherrima* (Willd. *ex* Klotzsch) Graham

Family: Euphorbiaceae.

Vern.: Lalpatta.

Ayurvedic and Regional Names:

Ayur.: Gudaa, Sehunda, Snuhi, Snuk;

H.: Shakar-pitan, Sohund, Surai;

Kum. : Sohund, Suli;

P.: Chu, Chula, Suli, Thar, Thor, Tordanda.

Distribution: Himalayas, upto 2,000m.

Description: Erect, cactus-like fleshy shrubs having angled branches, alternate, sessile, spathulate leaves with thorny, persistent stipules. Involucres yellow. Capsule pale grey, pedunculate.

Flowering & Fruiting: June-July.

Habitat Ecology: Wastelands, drier areas, rocky slopes; Lohardi, 858m.

Material Examined: HPU-EL-SKS-ABS-15125; 22.11.2010.

Parts Used: Whole Plant. Latex.

Folk Uses: Plant considered good for fencing* purposes and as an ornamental*. Milky latex applied externally for suppuration of boils, skin tumours*; also given to animals for throat problems*. A few of its drops applied for removing thorns* from the body.

Active Constituents: These are euphrol, cycloeucalenol, taraxerol, sitosterol, coumarins, aromatic carboxylic acid (*latex*), 2, 3-di-O Me ellgic and succinic acid (*plant*), taraxeryl-OAc, glut-5-en-3β-yl-OAc, β-sitosterol, friedelan-3-α-ol, lupeol, β-amyrin and epitaraxerol (*stem*).

Biological Activity: 50% EtOH extr. of stem exhibits antiviral and spasmogenic properties.

Uses in Literature: Known to be useful as abortifacient, anthelmintic, cathartic (latex), emetic, fish poison, fuel wood, vegetable; and for boils, burns, congestion, cough, cuts, diarrhoea, dysentery, ear complaints, increasing milk yield, leprosy, leucorrhoea, skin diseases, sprains, stomach problems, temporary blindness, warts and wounds (Ambasta, 1986; Binoj & Balakrishnan, 1996; Gogoi & Borthakur, 2001; Jadhav, 2006; Jain, 1984, 1991; Jain *et al.*, 1991; Kadel & Jain, 2006; Kalpesh *et al.*, 2004; Kapur & Nanda, 1996; Krenzelok *et al*, 1996; Manandhar, 1987a; Patil, 2004; Shah & Joshi, 1971; Sharma, 2003; Singh, 1996b; Upadhyay & Singh, 2005; Watt, 1972).

Euphorbia royleana Boiss. (Pl. 52B).

Syn.: *E. pentagona* Royle

Family: Euphorbiaceae.

Vern.: Chparchoo, Chui, Chu.

Ayurvedic and Regional Names:

Ayur.: Gudaa, Sehunda, Snuhi, Snuk;

H.: Shakar-pitan, Sohund, Surai;

Kum. : Sohund, Suli;

P.: Chu, Chula, Suli, Thar, Thor, Tordanda.

Distribution: Himalayas, upto 2,000m.

Description: Erect, cactus-like fleshy shrubs having angled branches. Leaves alternate, sessile, spathulate with thorny, persistent stipules. Involucres yellow, cyathia large and green with 4 lobes and rounded glands. Capsule pale grey, pedunculate. Schizocarpic regma opening into three cocci. Seed endospermic.

Flowering & Fruiting: June-July.

Habitat Ecology: Wastelands, drier areas, rocky slopes; Tikker Khatrian, 908m.

Material Examined: HPU-EL-SKS-ABS-15126; 15.04.2010.

Parts Used: Whole Plant. Latex. Pith.

Folk Uses: Plant considered good for fencing* purposes. Milky latex used by natives for blinding fish; also applied for suppuration of boils, skin tumours, and considered good for throat problems* in animals. Also, a few of its drops applied on the proximal end of infected tail* to check infection ('nanain') in livestock. Poultice of crushed stems considered antiseptic*. Pith slices mixed with wheat dough fed to cattle for stimulating ovulation*; and also to avoid abortion* after conception.

Active Constituents: Properties associated to triterpenoids, euphol, cycloeucalenol, four tetra- and four tri-esters of macrocyclic diterpene ingenol, octacosanol, tetra-cosanol, beta-sitosterol, stigmasterol, alpha-amyrin and campesterol (*latex*), ingenol, 2, 3-di-O Me ellgic and succinic acid (*plant*), taraxeryl-OAc, glut-5-en-3β-yl-OAc, β-sitosterol, friedelan-3-α-ol, lupeol, β-amyrin and epitaraxerol (*stem*).

Biological Activity: 50% Et OH extr. of stem exhibits antiviral and spasmogenic properties.

Uses in Literature: Known to be used as abortifacient, anthelmintic, cathartic (latex), fish poison, fuel wood, vegetable; and for boils, burns, congestion, cough, cuts, diarrhoea, dysentery, ear complaints, increasing milk yield, leprosy, leucorrhoea, skin diseases, sprains, warts and wounds (Ambasta, 1986; Binoj & Balakrishnan, 1996; Gogoi & Borthakur, 2001; Jadhav, 2006; Jain, 1984, 1991; Jain *et al.*, 1991; Kadel & Jain, 2006; Kalpesh *et al.*, 2004; Kapur & Nanda, 1996; Manandhar, 1987a; Patil, 2004; Shah & Joshi, 1971; Sharma, 2003; Singh, 1996; Upadhyay & Singh, 2005; Watt, 1972).

Ficus benghalensis L. (Pl. 52C).

Syn.: *F. elastica* Roxb., *F. indica* L., *Urostigma benghalense* Gasp. *Urostigma elasticum* Miq., *Visiana elastica* Gasp.

Family: Moraceae.

Vern.: Bad, Barh, Bare.

Ayurvedic, English, Sanskrit and Regional Names:

Ayur.: Bahupaada, Dhruv, Nyagrodha, Vata;

Eng.: Banyan Tree;

Sans.: Avaroha, Bahupada, Bhandira, Bhringi, Chira, Dhruva, Jatala, Karmaja, Mana, Nila, Nandi, Rohina, Sungi, Vata, Vanaspati, Yamapriya;

A. : Bot;

B.: Bar, Bot, But;

G.: Vad, Vadlo, Vor;

H.: Bar, Bargad, Bargat, Ber, Bor;

Mal.: Ala, Nyagrodham, Peral, Vatam;

Mar.: Vada, Wad, War;

P.: Bargad, Bera, Bohar, Bohir, Bor;

Tam.: Al, Ala, Eguvasam, Kadavam, Kavasugam, Pali, Parul, Vadam;

Tel.: Bhandiramu, Mari, Matti, Pedaimatti, Rohi, Vati, Vitapi.

Distribution: Throughout India.

Description: A large ambrageous tree with numerous aerial prop roots and drooping branches. Leaves leathery, ovate, obtuse, entire. Figs pubescent, sessile and axillary; red on ripening.

Flowering & Fruiting: February-June.

Habitat Ecology: Usually planted along with *Ficus religiosa* as a shade tree; Tauni Devi, 1,050m.

Material Examined: HPU-EL-SKS-ABS-15127; 18.02.2011.

Parts Used: Plant. Leaves. Latex. Wood.

Folk Uses: Tree considered sacred and worshipped; natives tie holy thread around its trunk for getting their wish fulfilled. Dried latex chewed as a chewing gum* by children. Mustard oil smeared heated leaves of the plant, applied externally for suppuration of under arm boils* ('kachhali'). Garland of its leaves hung on the main entrance of house at the time of house warming* ceremony. Wood used for performing 'hawan'*.

Active Constituents: Known to contain leuco-anthocyanin (*stem bark*), quercitin-3-galactoside, rutin, b-sitosterol, taraxasterol, lupeol, β-amyrin along with psoralen and bergapten (*leaves*) and glycoside (*fig*).

Biological Activity: Bark exhibits hypoglycaemic activity in both normal and alloxan diabetic rats.

Uses in Literature: So far described as an antidote, aphrodisiac, astringent, edible, fodder, tonic, tooth brush, vegetable (leaf bud), well curbs; and for atrophy, bangy poles, bleeding and swelling of gums, boils, cart shafts, cart yokes, diabetes, diarrhoea, dysentery, fever, furniture, gastric (vet.), gonorrhoea, leprosy, menorrhagia, nausea, piles, plates, scabies, sexual diseases, sores in mouth, snake bites, spermatorrhoea, toothache, urinary complaints, vomiting, wounds, wood in tent poles and religious ceremonies (Ambasta, 1986; Bondya & Sharma, 2003; Datt, 1996; Girach, 1992; Hembrom & Goel, 2005; Huidrom, 1996; Jain, 1991; Jain & Sharma, 1996; Khanna & Kumar 2000; Khare, 2007; Kulkarni & Kumbhojkar, 1996; Kumar & Chauhan, 2005; Maliya & Singh, 2003; Manandhar, 1996; Masih, 2003; Mohanty & Padhy, 1996; Mohanty *et al.*, 1998; Mukherjee & Namhata, 1990; Pandey & Verma, 2002; Rao & Pullaiah, 2001; Sahoo & Mudgal, 1995; Saini, 1996a; Sen & Behera, 2003; Singh, 2003; Singh & Pandey, 1996; Sinha, 1996; Tirkey *et al.*, 2003; Upadhye *et al.*, 1994; Vihari, 1995; Watt, 1972).

Major Ayurvedic Formulations: Nyagrodhadi-kvetha-churna, Nyagrodhadi-churna, Vatajatadi-taila.

Adulterant or Substitute: Fruit powder used as an adulterant for 'kamala' (*Mallotus phillippensis* (Lam.) Muell-Arg.)

Ficus carica L. (Pl. 52D).

Syn.: *F. palmata* Forssk., *F. caricoides* Roxb., *F. pseudo-sycamorus* Decne., *F. virgata* Roxb.

Family: Moraceae.

Vern.: Dura, Anjir.

Ayurvedic, English, Sanskrit and Regional Names:

Ayur.: Bhadrodumbara, Manjul, Phaglu, Raajodumbara;

Eng.: Common Fig;

Sans.: Anjira, Ten;

B.: Fagari, Fagu, Faguri;

Bo.: Anjir;

G.: Anjura, Anjuri;

H.: Anjir;

Kan.: Anjura, Tie-thie;

Mal.: Simayatti;

P.: Anjira;

Tam.: Simaiyatti, Tenatti;

Tel.: Anjuru, Manjimedi.

Distribution: Cultivated in many parts of India, Punjab, W. Himalayas.

Description: Small deciduous trees having cordate, deeply lobed leaves. Lobes obtuse, rough on upper side. Flowers inconspicuous, incomplete, unisexual, in hypanthodium. Figs in pair on long leafless scaly shoot, globose, reddish brown when ripe.

Flowering & Fruiting: March-August.

Habitat Ecology: Wastelands, edges of fields, drier areas; Sujanpur, 567m.

Material Examined: HPU-EL-SKS-ABS-15128; 30.05.2011.

Parts Used: Leaves. Latex. Fruits.

Folk Uses: Fruits edible, unripe cooked as vegetable*, dried fruits also added to nonvegetarian preparations for taste. Leaves used as fodder. Latex of the plant used for removing prickles* from the skin.

Active Constituents: Calotro phenyl acetate, lupeil acetate, oleanolic acid, β-sitosterol and bergapten (*leaves*) are the active constituents.

Uses in Literature: Reported earlier for building purposes, curing boils, constipation, delivery, diabetes, dysentery, indigestion, leucorrhoea, liver disorders, lung and bladder diseases; and as anthelmintic demulcent, edible, emollient, fodder, laxative, nutritive, wine and brandy (Ambasta, 1986; Khanna & Kumar, 2000; Khare, 2007; Kirtikar & Basu, 1994; Maliya & Singh, 2003; Negi *et al.*, 1985; Shah & Joshi, 1971; Sikarwar, 2002).

Ficus glomerata Roxb. (Pl. 53A).

Syn.: *F. chittagonga* Miq., *F. racemosa* Wall., *F.mollis* Miq., *F. goolereca* Roxb., *Covellia glomerata* Miq., *C. mollis* Miq.

Family: Moraceae.

Vern.: Umara.

English, Sanskrit and Regional Names:

Eng.: Cluster Fig;

Sans.: Jantuphala, Phalasambendha, Suprastish, Thida;

Ass.: Jangedumuru, Yagyadimru;

B.: Jagyadumar, Loa, Lowa;

Bo.: Umbar;

G.: Umbro;

H.: Gular, PaRao, Umar, Umarai;

Kan.: Athi;

Kash. : Rumbal;

Mal.: Rumadi, Umbaram;

Oriya: Dimri, Odumboro;

P.: Batbar, Kakallam, Kathagular;

Tam.: Adavu, Anjiram, Malaiyinmmunivam, Tumbaram;

Tel.: Atti, Medi.

Distribution: Throughout India.

Description: An evergreen tree. Leaves ovate-oblong, membranous, glabrous above and beneath, tapering to a bluish point at apex with entire margin. Receptacle reddish with depressed umbilicous.

Flowering & Fruiting: April-July.

Habitat Ecology: Mostly found growing in moist localities and bordering streams; Bhoranj, 950m.

Material Examined: HPU-EL-SKS-ABS-15129; 17.05.2011.

Parts Used: Leaves. Stem.

Folk Uses: Planted as shadow tree on roadsides; also considered sacred and worshipped. Leaves used as fodder; its paste applied for healing open wounds. Fruits edible, unripe cooked as vegetable. Wood used as shuttering for house construction*, and for performing 'hawan'* during religious ceremonies.

Active Constituents: Lupeol, friedelin, β-sitosterol, stigmasterol, cerylbehenate, lupeol, its acetate, α-amyrin acetate (*trunk bark*), a tetracyclic triterpene, glucanol acetate (*leaves*) have been identified.

Biological Activity: A glycoside rich fraction from leaves reported to have hypotensive, vasodilator and cardiac depressive effect. Also, found to be antiprotozoal, hypoglycaemic and anticancerous.

Uses in Literature: Described earlier as an astringent, febrifugal, fodder, edible (fruits), powerful tonic, sacred, stomachic, tooth brush, vegetable, wood for shopping wells; and for asthma, blood dysentery (vet.), boils, bronchitis, piles, diarrhoea, disease of kidney and spleen, dislocation of joints, doors, dry cough, fusebox, gonorrhoea, gum, healthy foetus (vet.), intestinal worms, leucorrhoea, match box, menorrhagia, mouth sores, mumps, rinderpest (vet.), shutters, smallpox, urinary and venereal complaints, well frames and wounds (Ambasta, 1986; Hosagoudar & Henry, 1996a; Jain, 1991; Jain *et al.*, 1991; Jha *et al.*, 1996; Khanna &

Kumar 2000; Khanna *et al.*, 1996b; Kumar, 1996; Kumar & Chauhan, 2005; Kumar & Jain, 1998; Kumar & Pullaiah, 1998; Kurian, 1995; Mishra *et al.*, 1996a; Mohanty *et al.*, 1998; Mokat & Deokule, 2004; Painuli & Maheshwari, 1996; Rama Rao & Henry, 1996; Saini, 1996a; Sharma, 1996; Singh & Pandey, 1996; Sinha, 1996; Watt, 1972).

Major Ayurvedic Formulations: Mutrasamgrahaniya -kasaya-churn, Nyagrodhi-kvatha-churn, Udumbara-kavaleha, Udumbaramata, Udumbarasara.

Ficus hispida L. f. (Pl. 53B).

Syn.: *F. oppositifolia* Willd., *F. daemonum* Koem., *F. prominens* Wall., *F. molis* Willd., *F. scabra* Jacq., *Covellia daemonum* Miq., *C. oppositifolia* Gasp. Rich., *C. setulosa* Miq., *C. courtallensis* Miq., *C. wightiana* Miq., *C.assamica* Miq., *C. dasycarpa* Miq., *Sycomorpha roxburghii* Miq.

Family: Moraceae.

Vern.: Daebreyan.

Ayurvedic, Sanskrit and Regional Names:

Ayur.: Kaakodumbara, Kaashtodumbara, Malapu, Malayu, Phalgu;

Sans.: Bhadrodumbarika, Kushthaghni, Phalguni;

A. : Khoskadumar;

B.: Dumoor, Kakdumar;

Bo.: Dhedu, Mira;

G.: Dhedaumura, Janglianjir;

H.: Gobla, Katugularia, Titmila;

Kan.: Adaviatti, Kadatti;

Mal.: Erumanakku, Peyalti;

Mar.: Bhokada, Dherumbar, Kharoti;

Oriya: Bhaidimiri, Korotosamo;

Tam.: Ottanalam, Sonatti;

Tel.: Bottamatti, Kukkabodda.

Distribution: Throughout India; 450-1,100m.

Description: A small tree with broadly ovate opposite leaves and shortly-stalked, globose, pale yellow figs.

Flowering & Fruiting: June-August.

Habitat Ecology: Rocky slopes usually along borders of agricultural fields; Kanjian, 910m.

Material Examined: HPU-EL-SKS-ABS-15130; 17.05.2011.

Parts Used: Leaves. Fruits. Latex. Wood.

Folk Uses: Ripe fruits eaten as such and pickled; unripe ones and tender leaves cooked as vegetable for checking flatulence*. 3-4g powdered bark in combination with 'soa' (*Anethum sowa* Roxb.) given with cow's milk for spermatorrhoea*. Latex applied in dental caries* for relief from toothache. Wood used for making agricultural implements* and a roller in 'ghrat'* (countryside flour grinder), construction* purposes and as fuel*.

Active Constituents: Therapeutic properties of *bark* due to bergapten, psoralen, beta-amyrin, beta-sitosterol, hispidine phenanthrain-dolizidines, n-triacontanyl, and gluanol acetate.

Biological Activity: Stem bark exhibits antiprotozoal, antibacterial, anthelmintic and antiviral whereas root bark found to be CNS stimulant and hypoglycaemic.

Uses in Literature: Described earlier as anthelmintic, antiperiodic, cattle fodder, edible, emetic, fibre, galactagogue (vet), purgative, tonic, vegetable (leaves sometimes cooked with pork); and for blood dysentery, bubo (vet), curries, cuts, delivery (vet), earache, easy sheding of placenta (vet), giddiness, gonorrhoea, headaches, jams, milching cattles to dry up their milk, preserving foetus in the womb, removal of placenta (vet), stomach ulcers, warts and wounds (Ambasta, 1986; Geetha *et al.*, 1996a; Girach *et al.*, 1996, 1997; Jain, 1991; Khare, 2007; Mohanty & Padhy, 1996; Rama Rao & Henry, 1996; Saini, 1996a; Singh *et al.* 1996b; Sinha, 1986, 1987; Siwakoti & Verma, 1996; Watt, 1972).

Ficus palmata Forsk. (Pl. 53C).

Syn.: *F. caricoides* Roxb., *F. pseudo-sycamorus* Decne., *F. virgata* Roxb.

Family: Moraceae.

Vern.: Duda, Fegura, Phegura.

Ayurvedic, English and Regional Names:

Ayur.: Anjiri, Phalgu;

Eng.: Indian fig;

P.: Daholia, Jamir, Phaguro;

G.: Pepri;

H.: Anjiri, Khabara;

Tel.: Manjimedi;

U.P.: Ban-Gular.

Distribution: Throughout India; Himalayas upto 2,300m.

Description: A bush or small tree with broadly ovate, opposite, three-nerved leaves. Male flowers on hairy pedicels. Gall flowers sessile or pedicelled. Figs axillary, stalked, solitary, pear-shaped and purple coloured.

Flowering & Fruiting: April-September.

Habitat Ecology: Rocky slopes along agricultural fields; Kalinjhari, 900m.

Material Examined: HPU-EL-SKS-ABS-15131; 14.04.2011.

Parts Used: Twigs. Leaves. Fruits. Latex.

Folk Uses: Brush prepared from its twig used for appling hot 'gachal' (jaggery) for treating 'padjibhi' (glossitis)*. Leaves fed to facilitate evagination of placenta*. Unripe fruits and tender leaves cooked as vegetable* and considered as laxative and appetizer* whereas ripe ones good for curing diabetes* and lung affections.

Powdered fruits prescribed twice daily for 3-5 days with cow's milk to aid in gastro-intestinal disorders*. Latex applied as an emollient for ringworm* and to draw out thorns from skin.

Active Constituents: *Leaves* yield beta-sitosterol and bergapten.

Uses in Literature: Useful in India as cattle fodder, demulcent, edible *(fruits)*, laxative; and for bladder, building purposes, diseases of lungs, dysentery, high blood pressure, indigestion, jam, lactation (vet.), pimples and drawing thorns, squash making (Ambasta, 1986; Ansari, 1991; Aswal, 1996; Jain, 1991; Kapur & Nanda, 1996; Khare, 2007; Lal *et al.*, 1996; Negi *et al.*, 1993; Parmar & Kaushal, 1982; Rana *et al.*, 1996; Saini, 1996a; Sinha, 1996; Tiwari & Pande, 2006; Watt, 1972).

Ficus religiosa L. (Pl. 53D).

Syn.: *F. affinior* Griff., *Urostigma religiosum* Gasp., *U. affine* Miq.

Family: Moraceae.

Vern.: Pipal

Ayurvedic, English, Sanskrit and Regional Names:

Ayur.: Ashvattha, Bodhidru, Bodhivriksha, Chalapatra, Gajabhaksha, Kshiradruma, Peeppal, Sebya;

Eng.: Pepal Tree, Peepul Tree, Pipul Tree;

Sans.: Asvatthah, Chaityavriksha, Pavitraka, Vriksharaja, Yojnika;

A. : Ahot;

B.: Ashathwo, Hesar, Hesak, Pipar;

Bo.: Arli, Jari, Pimpala;

G.: Arasa,Aswartham;

H.: Ashathwa, Asud, Asvattha, Aswat, Pipal;

Kan.: Arali, Nyaungbawdi, Nyanugleaudi, Nyoungbaude, Nyoungbaudi;

Mal.: Arayal, Chlodalam;

Mar.: Ashvatha, Pimpala, Pipul;

Oriya: Bor, Burr;

Tam.: Achuvattam, Podoroganam, Rai, Raiga, Ragi, Ravi, Vangardi;

Tel.: Arali, Arle, Asvalta, Basri, Bogaha, Haspath, Rangi.

Distribution: Throughout India; widely cultivated in Himalayas upto 1,400m.

Description: A large glabrous tree with grey bark having entire, long petioled leaves narrowed into linear-lanceolate tail. Figs axillary, in globular pair, dark purple when ripe.

Flowering & Fruiting: March-October.

Habitat Ecology: Usually planted along with *Ficus benghalensis*; Didwin Tikker, 800m.

Material Examined: HPU-EL-SKS-ABS-15132; 15.05.2011.

Parts Used: Stem. Leaves. Wood.

Folk Uses: Tree considered sacred and worshipped. Leaves used as fodder for increasing lactation* in animals. Garland of its leaves hung at the main door of house during 'grih-parvesh' (house warming ceremony) Wood used for performing 'hawan' ceremonies.

Active Constituents: Tannins, β-sitosterol and D-glucose, laurosterol (*bark*) have been isolated.

Biological Activity: Stem bark exhibits antiprotozoal activity against *E. histolytica*; anthelmintic activity against *Ascardia galli* and antiviral against Ranikhet disease virus.

Uses in Literature: Well known in literature as abortifacient, absorbent in inflammatory swelling, anthelmintic, antidote to snake bite, astringent, blood purifier, colouring material, contraceptive, edible (fruits and tender leaf buds eaten in times of scarcity), fixative, fodder, gum, laxative, psychoactive plant, sacred tree (cutting tree more sinful than killing a saint), spoons, tonic, tree of fertility, vermifuge (vet.); and for after pains, asthma, boils, conception, constipation (vet.), dysentery, expelling foetal membrane, falling of hair, foot and mouth diseases (vet), gonorrhoea, hiccups, jaundice, leucorrhoea, menstrual disorders, painting (Mithila), snake bite, sores, tanning, throat infection, toothache, urinary complaints, whooping cough and wounds (Ambasta, 1986; Bhatt *et al.*, 2001; Borthakur, 1996a; Borthakur *et al.*, 2004; Gogoi & Borthakur, 1991; Gogoi *et al.*, 2003; Hosagoudar & Henry 1996a; Jadhav, 2006; Jain, 1991; Jain *et al.*, 1991; Jha & Basak, 1994; Kalpesh *et al.*, 2004; Khanna, 2002; Khanna & Kumar 2000; Khanna *et al.*, 1996 a, b; Khare, 2007; Kumar & Chauhan, 2006; Kumar & Jain, 1998; Masih, 2003; Mishra *et al.*, 1996a; Pawar & Patil, 2004; Rama Rao & Henry, 1996; Satapathy & Brahmam, 1996; Sikarwar & Kaushik, 1992; Singh & Rao, 2003; Srivastava *et al.*, 2003; Virjee *et al.*, 1984; Watt, 1972).

Major Ayurvedic Formulations: Nyagrodhadi-churn, Nyagrodhadi-kvatha-churn.

Adulterant or Substitute: Stem bark and roots used as substitute for Soma (*Ephedra gerardiana).*

Ficus roxburghii Wall. *ex* Miq. (Pl. 54A).

Syn.: *F. auriculata* Lour., *F. macrophylla* Griff., *F. regia* Miq.,*F. scleroptera* Griff., *Covellia macrophylla* Miq.

Family: Moraceae.

Vern.: Rumal, Trayambalu, Triamble.

Regional Names:

B.: Demur, Doomoor, Sapai;

H.: Tirmal, Trimmal;

P.: Ber, Burh, Dadari, Timbal, Tirnal, Tirmal, Trimbal.

Distribution: Sub-Himalayan tract, outer Himalayas ascending upto 1,600m.

Description: Trees having brown bark and 5-7 nerved leaves with cordate base. Hypanthodium with unisexual, minute and inconspicuous flowers. Figs stalked, ridged.

Flowering & Fruiting: April-August.

Habitat Ecology: Commonly found growing in moist localities and bordering streams; Brada, 1,120m.

Material Examined: HPU-EL-SKS-ABS-15133; 17.05.2011.

Parts Used: Leaves. Fruits. Wood.

Folk Uses: Leaves used as fodder and for making 'pattals' (meal plates). Ripe fruits edible whereas unripe ones cooked as vegetable ('bhruni'). Wood used as fuel*, for construction* and agricultural implements*.

Active Constituents: Tetracyclic triterpene, glucanol acetate (*leaves*), β-sitosterol, friedelin and epifriedelanol (*plants*), lupeol, friedelin, β-sitosterol and stigmasterol, cerylbehenate, lupeol, its acetae, amyrin acetate (*trunk bark*) have been isolated.

Biological Activity: Known as galactagogue (vet.) and for leucoderma and skin diseases.

Uses in Literature: Mentioned in literature as edible, fodder, galactagogue (vet.), vegetable; and for breast ulcers, delivery (vet.), leucoderma, meal plates for feasts in village, ring worm, ropes (bark), skin diseases (Jain, 1991; Parmar & Kaushal, 1982; Rama Rao & Henry, 1996; Watt, 1972).

Flacourtia ramontchi L. Herit (Pl. 54B).

Syn.: *Flacourtia indica* (Burm.f.) Merr., *F. latifolia* Cooke, *F. occidentalis* Blatt., *F. sapida* Roxb., *F. sepiara* Roxb., *Gmelina indica* Burm. f.

Family: Flacourtiaceae.

Vern.: Kangu.

Ayurvedic, English, Sanskrit and Regional Names:

Ayur.: Gopakantaa, Sruva-vriksha.Vikankata, Yajnyavrksha;

Eng.: Governor's Plum, Madagascar Plum;

Sans.: Swadukantaka;

B.: Binja, Katail, Tambat;

Bo.: Kakad, Swadu;

H.: Bhanber, Bincha, Iambat, Kakein, Katai, Katar;

Kan.: Hattarimullu, Humunki;

M.P.: Kankar;

Mar.: Kaikun, Kaker, Paker;

Oriya: Bonisha, Potnabonisho;

P.: Bhutankas, Bavachi, Kakao;

Tam.: Mullikkarai, Sottaikala;

Tel.: Nakkaneredu;

U.P.: Katai.

Distribution: Dry hills throughout India upto 1,000m.

Description: Small deciduous trees armed with axillary thorns. Leaves broadly ovate-crenate or toothed, apex acute, pubescent. Flowers greenish-yellow, dioecious. Fruits globose, red or dark brown, endocarp hard.

Flowering & Fruiting: March-June.

Habitat Ecology: Wasteslopes, rocky slopes; Sujanpur, 567m.

Material Examined: HPU-EL-SKS-ABS-15134; 05.06.2011.

Parts Used: Leaves. Fruits. Bark. Wood.

Folk Uses: Leaves used as fodder. Fruits edible. Gargles done with decoction of bark useful for mouth sores*. Wood used as fuel*, hedge* and for making agricultural implements and fencing*.

Active Constituents: Important constituents are tannin, flacourtin, phenolic glucoside ester (*bark*), 4-o-(β-D-glucosyl), (1-4) -β-2-3-di-o-methyl-D-xyloside (*heartwood*).

Biological Activity: Aerial parts diuretic and spasmogenic. Hypotensive activity and effect on CNS found +ve.

Uses in Literature: Known so far as an antidote to snake bite, appetiser, astringent, bite for mad dog, cattle fodder, diuretic, edible; and for agricultural implements, burning sensation in chest, carbuncle, cholera, diarrhoea, digestion, dysentery, eczema, excess body heat, facilitating child birth, fever, gout, indigestion, jaundice, liver disorder, reshaping abdomen after delivery, rheumatic pain, skin diseases, sores, stab wounds, tanning, tonsilitis, turnery, foot and mouth diseases, wounds (Ambasta, 1986; Aminuddin & Girach, 1996; Brahmam & Saxena, 1990; Jain, 1991; Jain *et al.*, 1991; Katewa & Arora, 1997; Khare, 2007; Maheshwari *et al.*, 1986; Mohanty *et al.*, 1998; Nazarudeen *et al.*, 1996; Parmar & Kaushal, 1982; Patole & Jain, 2002; Rama Rao & Henry, 1996; Ramachandran & Nair, 1981; Rao & Pullaiah, 2001; Saini, 1996a; Shah & Singh, 1990; Sharma & Singh, 2001; Sikarwar & Kaushik, 1992; Singh, 1996).

Foeniculum vulgare Mill. (Pl. 54C).

Syn.: *Foeniculum panmorium* DC., *F. officinale* All., *Anethum foeniculum* L., *A. panmorium* Roxb., *F. capillaceum* Gilib., *Ozodia foeniculum* W. & A.

Family: Apiaceae.

Vern.: Meethi-Saunf.

Ayurvedic, English, Sanskrit and Regional Names:

Ayur.: Madhuraa, Madhurikaa, Mishi, Mishreyaa, Shataahvaa, Shatapushpaa;

Eng.: Fennel;

Sans.: Madhurika;

B.: Bari-shopha, Panmohuri;

Bo.: Bari-Shopa, Variari, Variyali, Wariaree;

G.: Badishep;

H.: Badhesoppu, Bari-saunf, Mauri, Pan-muhori;

Kan.: Aspa, Badi-Sopu, Badyan;

Mar.: Badisapu, Badishep;

Tam.: Pedda-jila-kura, Sohikire;

Tel.: Smadhurika, Pedda-jiľa-kurra.

Distribution: Throughout India upto 2,000m.

Description: Tall, glabrous, biennial or perennial herbs. Leaves tripinnate, ultimate segments linear. Umbel having 15 or more rays. Petals notched inflexed. Vittae conspicuous. Fruits a schizocarpic cremocarp, splitting into two indehiscent mericarps, an albuminous seed present inside each mericarp.

Flowering & Fruiting: August-September.

Habitat Ecology: Cultivated, escape in wasteplaces; Gasota, 850m.

Material Examined: HPU-EL-SKS-ABS-15135; 10.08.2009.

Parts Used: Leaves. Fruits. Seeds.

Folk Uses: Leaves consumed as pot-herb*, also added to sauce and pulses*. Fruits used as carminative for stomach troubles. Seeds used as condiments and its infusion as refrigerant, 25-40ml of its hot decoction given to ladies after delivery; also the same mixed with 'benafsha' (*Viola odorata* L.), 'mulathi' (*Glycyrrhiza glabra* L.) and 'tulsi' (*Ocimum sanctum* L.), considered good for cough, cold and fever. One tsp roasted seeds given (twice a day, till cure) with water for diarrhoea. 25g seed powder alongwith 10g seed powder of 'jeera' (*Cuminum cyminum* L.) given to check 'mastitis'* in animals.

Active Constituents: Columbianetin, 7 hydroxy and 6, 7-dihydroxy coumarins, marmesin, osthenol, psoralen, scoparone, seselin, xanthotoxin, α and β-amyrins, oleanolic acid, Δ5 avenasterol, campesterol, Δ7 stigmasterol, stigmastadienol, linoleic, oleic, palmitic and petroselinic acid (*fruit, seeds*).

Biological Activity: Antimicrobial and haematinic activities found to be confirmed.

Uses in Literature: Considered useful as an abortifacient, antibacterial, anticarcinogenic, antifungal, anti-inflammatory, antioxidant, antispasmodic, aromatic, carminative, chromoprotective, colic, diuretic, emollient, emmenogouge, endoparasite (vet.), febrifugal, galactagogue, ophthalmic, stimulant, stomachic; and for amenorrhoea, anorexia (vet.), bloat (vet.), burning sensation in body, chest diseases, cough, diarrhoea, diarrhoea (vet.), dysentery, dyspepsia, emollient, fever, flatulence, flavouring, haemorrhoids, headache, kidney problems, menstruation, nausea, spleen complaints, retention of urine, thirst, toothache, uterine prolapse (vet.), wounds (Ambasta, 1986; Bhalla *et al.*, 1982; Deeba *et al.*, 2009; Khare, 2007; Kumar & Chauhan, 2005; Retnam & Martin, 2006; Sharma, 2003; Uniyal, 1968).

Adulterant or Substitute: Oil from fruit used as an adulterant for Anise oil (*Pimpinella anisum* L.).

Fragaria chiloensis (L.) Duchesne (Pl. 54D).

Syn.: *F. vesca* auct.

Family: Rosaceae.

Vern.: Strawberry.

English Names:

Eng.: Alpine strawberry, Perpetual strawberry.

Distribution: Throughout India.

Description: A creeping soft and silky herb with trifoliate long stalked leaves, erect stem, perennial rootstock, producing long runners which are rooting at the joints. Flowers white, large and showy. Fruits red with globose, elongated succulent receptacle, dotted with numerous seed like glabrous achenes, 1-seeded, indehiscent.

Flowering & Fruiting: May-June.

Habitat Ecology: Cultivated; Bhota, 765m

Material Examined: HPU-EL-SKS-ABS-15136; 15.04.2009.

Parts Used: Leaves. Fruit.

Folk Uses: Fruits edible and considered good in anaemia*, also added for preparing desert. Leaves used as fodder.

Active Constituents: *Fruit* contains isofucosthol, sitosterol, compesterol, cholesterol and stigmasterol.

Uses in Literature: Noted so far as edible; and for desert and ice-creams (Ambasta, 1986; Watt, 1972).

Fumaria indica (Haussk.) Pugsley. (Pl. 55A).

Syn.: *F. parviflora* Lam., *F. vaillantic* Loisel.

Family: Fumariaceae.

Vern.: Papra, Pitpapda, Pitpapra.

Ayurvedic, English, Sanskrit and Regional Names:

Ayur.: Kavacha, Parpata, Parpataka, Renu, Sukshmapatra, Varatikta;

Eng.: Fine-leaved-Fumitory;

Sans.: Araka, Charaka, Kalapanga, Krishnashakha, Nakra, Pansha, Panshu;

B.: Bansulpha;

Bom.: Pitpara;

Guj.: Khasudlio, Pitpapda;

H.: Pitpapada, Pitpapra, Pittapapara, Shatra;

Mar.: Pitpapda;

Tam.: Chata-rashi, Tura, Tusa;

Tel.: Bagaltulmulk, Chatarashi.

Distribution: Throughout India.

Description: Annual diffuse herbs having much-divided glaucous leaves with linear segments, sessile, pale pink, irregular flowers in lux raceme. Fruits globose, indehiscent.

Flowering & Fruiting: February-March.

Habitat Ecology: Moist slopes, weed in cultivated fields; Jhaniker, 1,050m.

Material Examined: HPU-EL-SKS-ABS-15137; 16.02.2010.

Part Used: Whole Plant.

Folk Uses: 50ml decoction of plant taken once every morning for 2-5 days to cure fever. Plant also used as fodder*.

Active Constituents: Protopine, nonacosanol, sitosterol, n-alkanes and sterols (*leaves, stem*); protopine and 1-tetrahydrocopitsine, d-8-OMe-dihydroxy sanguinarine, fumarilinc and oxy-sanguinarine (*seeds*), sanguinarine, protopine, cryptopine, fumaridine, fumaramine, perfumidine, perfumine, d and dl-bicuculline, dl-tetrahydrocopyisine, fumariline, fumarilicine, narceimine, dehydro-cheilanthifoline, copticine and narlumidine (*plant*) have been isolated.

Biological Activity: Herb extr. antiarrhythmic in rabbit.

Uses in Literature: Described in literature for bodyache, diarrhoea, dyspepsia, fever, flu, indigestion, liver complaints, mouth ulcers and skin diseases; and as an anthelmintic, aperient, astringent, blood purifier, diaphoretic, diuretic, laxative, cool drink and vegetable (Ambasta, 1986; Chopra *et al.*, 1956; Dhiman, 1976; Drury, 1978; Hembrom & Goel, 2005; Sharma, 2003; Uphof, 1968; Watt, 1972).

Major Ayurvedic Formulations: Arvindasava, Chandrakala Ras, Chandana Sava, Mahasudarshan Kwath, Livomyn drops, Parpatadi-Kwath, Parpatadyarista, Sadanga-paniya.

Trade: Plant sold as 'pitpapara' in drug markets of India.

Glycine max Merr. (Pl. 55B).

Syn.: *G. soja* Sieb. & Zucc., *G. hispida* Maxim.

Family: Fabaceae.

Vern.: Soyabean, Soya.

English Names:

Eng.: Soyabean, Soya.

Distribution: Cultivated as pulse crop mainly in Punjab, H.P., Maharashtra and other parts of India.

Description: Twining or sub erect herbs with 3-7 stipulate leaflets, flowers in axillary racemes. Short peduncled, dense or lax. Calyx densely silky. Corolla little exerted. Pods linear, straight, 4-5 seeded, clothed with adpressed Hair.

Flowering & Fruiting: September-December.

Habitat Ecology: Cultivated, planted near houses and gardens; Dhaneta, 596m.

Material Examined: HPU-EL-SKS-ABS-15138; 30.09.2010.

Parts Used: Whole Plant. Leaves. Seeds.

Folk Uses: Seeds cooked for its rich protein supplement. Paste of soaked, grinded seeds used for making 'badian'*, 'pakoras'* and 'soya milk'. Soya 'paneer'* considered good for sick people for its rich protein contents and low uric acid formation. Flour of its seeds mixed with that of wheat flour cooked as 'chapatis'*. Also, eaten as 'snacks'*. Leaves, soaked seeds and cake after extraction of soya milk given to animals as an excellent fodder*.

Active Constituents: These are soyasapogenol A, B and E.

Phospholipids, lecithins, proteins and minerals, glysine, beta-conglycinin (*seeds*).

Biological Activity: In studies pertaining to experimental carcinogenesis in animals, soybean isoflavones exhibited protective effect in 65% animals.

Uses in Literature: Noted earlier as anticarcinogenic, antifatigue, antiobesity, hepatoprotective, hypoglycaemic, protein supplement; and for disturbances of fat metabolism and gallstones (Ambasta, 1986; Khare, 2007; Watt, 1972).

Glycyrrhiza glabra L. (Pl. 55C).

Family: Fabaceae.

Vern.: Mulathi, Malathi.

Ayurvedic, English, Sanskrit and Regional Names:

Ayur.: Atirasaa; Madhuli, Madhuyashtikaa, Madhuyashtyaahvaa, Yashtimadhu;

Eng.: Licorice, Liquorice;

Sans.: Madhuyashti;

B.: Jashtimadhu;

H.: Mulathi;

G.: Jethimadh;

Kan.: Athi Madhura, Jeshtamadhu;

Mal.: Iratimadhuram;

Tam.: Atimadhuram;

Tel.: Yashtimadhukam.

Distribution: Occurs wild on the hills of H.P. and J. & K.; native of Mediterranean region.

Description: Perennial herbs having alternate, pinnate, yellow green leaves, lilac or light violet flowers in slender axillary spikes which are equal to or slightly longer than the leaves. Fruits 1-3cm long, flat, densely covered all over with small spinous outgrowths.

Flowering & Fruiting: May-September.

Habitat Ecology: Hill slopes, moist lands, also cultivated for medicinal purposes; Chabutra, 797m.

Material Examined: HPU-EL-SKS-ABS-15139; 07.06.2011.

Parts Used: Leaves. Roots.

Folk Uses: Leaves act as good fodder*; its poultice applied on head against sun burns* with fresh curd; and also consumed for gastric troubles and ulcers. Roots of the plant used as medicine for sore throat, cough, chest congestion and fever. 15-25ml decoction of its roots mixed with 'benafsha' (*Viola canescens* Wall.), 'elachhi' (*Elettaria cardamomum* (L.) Maton.) and 'long' (*Syzygium aromaticum* Merr. & Perry.) given once or twice a day, for cold, bronchitis, sore throat, chest congestion and fever. Roots sucked for quick relief from tonsillitis*.

Active Constituents: Glycyrrhizin, glycyrrhetinic, isoflavonoids, chalcones, coumarins, triterpenoids and sterols, lignans, amino acids, amines, gums and volatile oils (*roots*).

Biological Activity: Glycyrrhetinic acid shows activity against viral hepatitis, herpes, eczema, psoriasis.

Uses in Literature: Recorded earlier as antiallergic, antidiabetic, antidepressive, antiinflammatory, antistress, antiulcer, demulcent, emmenagogue, estrogenic, expectorant, mild laxative, spasmolytic; and for abdominal pain, adrenocorticoid inefficiency, arthritis, bronchitis, catarrh, cough, duodenal ulcers, eczema, genitourinary diseases, herpes, inflamed stomach, lever protective, mouth ulcers, psoriasis, respiratory tract infections, rheumatism, tuberculosis, urinary tract infection, viral hepatitis (Ambasta, 1986; Khare, 2007).

Gossypium arboreum L. (Pl. 55D).

Syn.: *G. indicum* Tod., *G. neglectum* Tod., *G. sanguineum* Hassk., *G. intermedium* Tod., *G. obustifolium* Roxb., *G. nanking* Meyen

Family: Malvaceae.

Vern.: Kapah.

Ayurvedic, English, Sanskrit and Regional Names:

Ayur.: Kaarpaasi;

Eng.: Tree Cottan;

Sans.: Karpasamu;

Bo.: Deokapas;

H.: Deokapas, Nurma;

Kan.: Hathi;

Mal.: Chemparutti;

Mar.: Devakapusa;

Oriya: Karposa, Kopa, Kapas;

P.: Kapas;

Tam.: Sembaruti;

Tel.: Patti, Karpasmu.

Distribution: Cultivated all over India upto 1,800m.

Description: A shrub possessing 5-7 palmatly lobed leaves and purple or yellow flowers with a crimson centre. Capsule ovoid, acute.

Flowering & Fruiting: July-September.

Habitat Ecology: Cultivated, planted near houses and gardens; Cha kmoh, 736m.

Material Examined: HPU-EL-SKS-ABS-15140; 07.07.2010.

Part Used: Seed.

Folk Uses: Cotton fibres used for stuffing* pillows, quilts, making sacred thread for religious purposes, wick for oil lamps, and clothes* for wearing. Seed oil used for cooking*. Soaked seeds mixed with 'desi' ghee given to check abortion* or seeds and oil cakes mixed with taramira (*Eruca sativa* Mill.) and wheat husk considered good for enhancing milk* yield in cattle.

Active Constituents: These are 1-gossypol (*seed*), steroids, organic acid, pigment and vit.B.

Biological Activity: Gossypium-8-o-rhamnoside along with quercitin, 3-7-o-glucoside of quercitin exhibit antibacterial activity against *Pseudomonas maltophila* and *Enterbacter cloacae*. Seed extr. spasmolytic and hypotensive.

Uses in Literature: Described in published literature as emollient, galactagogue (vet.), soothing agent and symbolic for making sacred thread; and for asthma, arthritis, conjuctivitis of infants, earache, easy delivery, easy delivery (vet.), fever, fibre, gonorrhoea, insanity, labour complaints, leucorrhoea, regulating fertility, regulating menses, skin eruptions, stomach pains, stranguary and patellar dislocations (Bhatt *et al.*, 2001; Jadhav, 2006; Khanna *et al.*, 1996b; Khare, 2007; Pawar & Patil, 2004; Sharma, 2003; Sharma & Borthakur, 2003; Sikarwar, 2002; Tirkey, 2004; Watt, 1972).

Grevillea robusta A. Cunn. *ex* R. Br. (Pl. 56A).

Family: Proteaceae.

Vern.: Silver oak.

English and Regional Names:

Eng.: Silky oak, Silver Oak;

Tam.: Savukkumaram.

Distribution: Planted in gardens.

Description: Evergreen trees. Leaves spinnate. Flowers orange in cluster. Fruits brown, obliquely-oblong, compressed.

Flowering & Fruiting: March-July.

Habitat Ecology: Often planted along roadsides; Bhota, 765m.

Material Examined: HPU-EL-SKS-ABS-15141; 26.05.2011.

Parts Used: Plant. Wood.

Folk Uses: Planted for ornamental purposes for its large yellow flowers. Wood considered good for ornamental work.

Active Constituents: These are grevillo, sitosterol, epilyeniresinol, its 2-o-glucoside, rhamnocitrin-3-o-rutinosid and kaemferol-3-o-rutinoside (*wood*).

Uses in Literature: Known so far as edible, green manure, ornamental; and for furniture, gum, panelling, paraquet, plywood, tannin, toys and veneering (Ambasta, 1986; Sinha, 1996).

Grewia oppositifolia Roxb. (Pl. 56B).

Syn.: *G. optiva* Drumm.

Family: Tiliaceae.

Vern.: Beul.

Ayurvedic and Regional Names:

Ayur.: Dhanvana, Parushaka;

H.: Bahul, Bemal, Bewal, Bhengul, Bhimal, Biul, Biung;

Kan.: Thidsal;

Kum.: Bimal, Dhamnu;

P.: Bahul, Biul, Behal, Dhaman, Pharwa.

Distribution: N.W. Himalayas, ascending to 2,000m.

Description: A moderate-sized tree upto 10m. Leaves distichous; stipules longer than the petioles. Flowers large, yellowish and in leaf-opposed umbellate cymes. Drupes smooth, olive black, succulent when ripe.

Flowering & Fruiting: May-June.

Habitat Ecology: Edges of agricultural fields, roadsides, wasteplaces; Tikker Khatrian, 1,000m.

Material Examined: HPU-EL-SKS-ABS-15142; 25.06.2009.

Parts Used: Leaves. Stem. Bark.

Folk Uses: Leaves lopped for fodder for increasing fat content* of milk. Stem retting yields fibre for making ropes and cordages, Crushed bark kept in buccal cavity and mouth for inducing chewing of cud* in livestock. Wood used for cot frames and tool handles, spears, bows, poles, construction* of houses, furniture, agricultural implements and fuel.

Active Constituents: Properties attributed to crude proteins (10.1%), fat (6.8%), crude fibre (14.1%), N-free extr. (54.8%), total carbohydrates (68.9%), ash (14.1%), calcium (4.18%), phosphorus (0.25%) (dry mature *leaves*).

Uses in Literature: Used so far as cordage, detergent, dye, fodder, fuel, galactagogue (vet.) and torch in religious ceremonies; and for agricultural implements, axe handles, bone fracture (vet.), clothing, constipation, eruptions, facilitating child birth, gastric problem (vet.), increasing lactation, indigestion (vet.), oarshaft and ropes (Ambasta, 1986; Bhatt, 1990; Gaur *et al.*, 1992; Hosagoudar & Henry, 1996a; Jain, 1991; Kapur, 1996; Khare, 2007; Lal & Dube, 1984; Lata, 1996; Rajwar, 1983; Rana *et al.*, 1996; Shah, 1987; Shah & Joshi, 1971; Sharma, 2003; Singh, 1996; Tiwari & Pande, 2006; Watt, 1972).

Habenaria pectinata Don (Pl. 56C).

Family: Orchidaceae.

Vern.: Orchid.

Ayurvedic and English Names:

Ayur.: Riddhi, Vriddhi;

Eng.: Orchid.

Distribution: Himalayassn region, up to 2,000m.

Description: Glabrous, robust, leafy weeds up to 70cm flowering stem. Leaves narrowly sword-shaped. Flowers pale yellow, green or white, crowded in spike. Petals glabrous; labellum 3-lobed, middle lobe forms the spur which is green and swollen at the tip. Fruit a capsule, opening by 3-6 longitudinal slits. Seeds numerous, minute, testa drawn out or winged.

Flowering & Fruiting: July-September.

Habitat Ecology: Common pasture weed of rainy season; Mattani, 920m.

Material Examined: HPU-EL-SKS-ABS-15143; 17.08.2010.

Parts Used: Whole Plant. Leaves. Flowers.

Folk Uses: Plant grown for ornamental* purposes. Leaves used as fodder*. Decoction of leaves regarded as a good heart and nervine tonic (10-20ml, once in the morning, 25-30 days). 2-4g of powdered tuberous roots with honey given against cardiac ailments.

Uses in Literature: Used so far as cardiac and nervine tonic (Khare, 2007).

Helianthus annuus L. (Pl. 56D).

Family: Asteraceae.

Vern.: Sooraj mukhi.

Ayurvedic, English, Sanskrit and Hindi Names:

Ayur.: Suurajmukhi, Suuryaavarta;

Eng.: Sunflower;

Sans.: Adityabhakta;

H.: Surajmukhi.

Distribution: Throughout India.

Description: An annual herb with rough hairy stems and large ovate leaves. Flowers yellow, in heads or capitulum of ray and disc florets. Achenes cylindrical and compressed.

Flowering & Fruiting: April-July.

Habitat Ecology: Cultivated; Rangas, 765m.

Material Examined: HPU-EL-SKS-ABS-15144; 20.07.2010.

Parts Used: Flower. Seed.

Folk Uses: Mainly planted in gardens for ornamental purposes. Poultice of its flower prepared in mustard oil considered good for external application in skin diseases*. Seed oil applied externally to bruises.

Active Constituents: Active principles associated due to kaurenic, trachylobanic, ent-trachylobanic acids, grandiflorolic acid, ψ-taraxene derivatives-heliantriol C and heliantriol F (*flowers*), palimitic acid (10.0), stearic (5.7) and linoleic acid (40.0) (*seed oil*).

Uses in Literature: Employed earlier as cooking oil, diuretic, edible, febrifuge, moisturizer; and for bronchitis, bruises, chronic renal diseases, cold, cough, diabetes, obesity, pulmonary and laryngeal afflictions, reducing serum and hepatic cholesterol, whooping cough (Khare, 2007; Kirtikar & Basu, 1994; Watt, 1972).

Helicteres isora L. (Pl. 57A).

Syn.: *Helicteres chrvsocalyx* Miq., *H. roxburghii* G. Don

Family: Sterculiaceae.

Vern.: Marorphali.

Ayurvedic, English, Sanskrit and Regional Names:

Ayur.: Aavartaki, Aavartani, Aavartphalaa;

Eng.: Indian Screw Tree;

Sans.: Avartani, Mriga-shinga;

B.: Atmora;

Bo.: Kevana, Kewan, Khiran;

G.: Murdasing;

H.: Jonkaphal, Morarphali;

Mal.: Isvarmuri, Kavum;

Mar.: Kevani, Kewan, Varkati;

Oriya: Murmuria, Murmuri;

P.: Jonkaphal, Maror phalli;

Tam.: Kaiva, Valampiri;

Tel.: Kavanchi, Nuliti.

Distribution: Dry forests throughout India.

Description: Sub-deciduous shrub or a small tree with stem up to 1.5-4.5m tall, grey bark and covered with stellate hair in young plants. Leaves simple, palmately nerved or obliquely cordate with serrated margin, scabrous above and pubescent beneath. Flowers solitary or in clusters, red petals turning pale blue when old. Fruits 5cm long.

Flowering & Fruiting: May-August.

Habitat Ecology: Plant grows gregariously in dry forests throughout India, wastelands; Bhota, 765m.

Material Examined: HPU-EL-SKS-ABS-15145; 29.06.2008.

Parts Used: Roots. Root Bark. Stem Bark. Leaves. Pod. Seed.

Folk Uses: Leaves lopped for fodder*; its poultice applied for skin diseases. Pods given for fever and cold. Decoction of seeds recommended for dysentery (2-5g, twice daily, 3-5 days). Stem bark used as astringent whereas 1-3g of root bark mixed with honey prescribed for cough and asthma. Root bark fibres used for sacks*, canvas* whereas stem fibres valued for making strings*, cattle harness* and container bags*.

Active Constituents: Active properties attributed to pro-vit. A, crude proteins 13-25 and CaO 3.15, tetratriacontanoic acid, tetratriaeontanol and sitosterol. (*leaves*). 4-quinolone alkaloid, malatyamine, diosgenin, cucurbitacin B, (*fruits, seeds*).

Uses in Literature: Used earlier as an anthelmintic, antibilious, antidiarrhoeal, antigalactic, antipyretic, astringent, demulcent, expectorant (used in cough and asthma); and for dysentery and skin diseases (Agarwal, 2003; Ambasta, 1986; Anonymous, 1998b; Bhattacharjee, 1998; Dash & Mishra, 2000; Jain, 1991; Khare, 2007; Kothari & Rao 2000; Mittre, 1997; Oommachan, 1977; Parker, 1993; Prasad *et al.*, 2002; Raju, 1998; Rastogi & Mehrotra, 1995b; Singh & Pandey, 1998; Sur & Halder, 2002; Varghese, 1996; Watt, 1972).

Major Ayurvedic Formulations: Gandharva Churna, Sidha Praneswarrasa.

Hibiscus rosa-sinensis L. (Pl. 57B).

Family: Malvaceae.

Vern.: China Rose, Shoe Flower.

Ayurvedic, English, Sanskrit and Regional Names:

Ayur.: Arunaa, Japaa, Javaa, Odrapushpa, Rudrapushpa;

Eng.: Chinese Hibiscus, Shoe Flower;

Sans.: Hariballabha, Odhrapushpa;

Ass.: Jobra;

B.: Jiwa, Oru;

Bo.: Jasavanda;

G.: Jasuva;

H.: Gulhar, Jasum, Jasut;

Kan.: Dasavala;

Mal.: Ayamparutti, Jumpa;

Mar.: Dasindachapula, Jasavanda;

Oriya: Mondaro, Ongribollika;

P.: Jasum;

Tam.: Irattaichegappuchembrat.

Distribution: Cultivated throughout India.

Description: Perennial shrubs. Leaves coarsely toothed at apex and entire at base. Flowers solitary, terminal or axillary with red corolla. Capsule many-seeded. Seeds numerous, non endospermic with naked seed coat.

Flowering & Fruiting: February-October.

Habitat Ecology: Cultivated in gardens; Salooni, 1,514m.

Material Examined: HPU-EL-SKS-ABS-15146; 15.04.2010.

Parts Used: Plant. Flower.

Folk Uses: Cultivated as an ornamental*. Paste of its flower good ointment for suppuration of boils.

Active Constituents: Active properties due to quercetin-3-diglucoside 3, 7-diglucoside, cyanidin-3, 5-diglucoside and cyanidin-3-sophoroside-5-glucoside (deep yellow *flowers*), and kaempferol-3-xylosylglucoside (ivory white *flowers*).

Biological Activity: Hypotensive and antispasmodic activities found (+) ve.

Uses in Literature: Known to be used for abdominal pains, boils, bronchitis, cough, diabetes, leucorrhoea, menstrual irregularity, menorrhagia, piles, whooping cough; and as emollient, refrigerant (Gogoi & Borthakur, 2001; Katewa & Sharma, 1998; Khare, 2007; Maliya, 2004; Pawar & Patil, 2004; Retnam & Martin, 2006; Sen & Behera, 2003; Sharma, 2003; Sharma *et al.*, 2001).

Major Ayurvedic Formulations: Japa-taila.

Holarrhena pubescens Wall. _ex_ Don (Pl. 57C).

Syn.: *H. codaga* G. Don, *H. antidysenterica* (Buch. -Ham.) Wall., *Echites antidysentrica* Roxb., *E. pubescens* Buch.-Ham., *Wrightia antidysentrica* Grah.

Family: Apocynaceae.

Vern.: Inderjaun, Keur.

Ayurvedic, English, Sanskrit and Regional Names:

Ayur.: Girimallikaa, Indravriksha, Kaalinga, Kalingaka, Kutaja, Shakra, Vatsa, Vatsaka;

Eng.: Antidysentric Holarrhena, Conessi Bark, Kurchi;

Sans.: Girimallika, Indra, Indrayavaphala, Kalinga, Kutaja, Mallikapushpa;

Ass.: Dhutkhuri, Dudcory;

B.: Kurchi, Titaindaraju;

Bo.: Dolokura, Dowla, Kura, Pandhrakura;

G.: Dhowda, Hath, Indrajwannujad;

H.: Dhudi, Karchi, Karra, Kora, Kureya;

Kash. : Keor, Kewar;

Mal.: Kaippakkotakappala, Venpala;

Mar.: Bedaki, Pandharakuda;

Oriya: Kherwa, Krya, Potrokrya;

P.: Kewar, Koeva, Kogur;

Tam.: Erukkalaipalai, Kudasam, Vattagam;

Tel.: Amkudu, Kakakodise, Peddankuda.

Distribution: Throughout India.

Description: A small deciduous tree. Leaves opposite, subcoriaceous. Flowers creamy-white, sweet-scented. Bark pale brown. Follicles white dotted.

Seeds glabrous, tipped with brown hair.

Flowering & Fruiting: June-February.

Habitat Ecology: Field borders, open slopes, drier areas; Anu, 858m.

Material Examined: HPU-EL-SKS-ABS-15147; 03.07.2008.

Parts Used: Bark. Leaves. Fruit. Seeds.

Folk Uses: Planted near temples for religious purposes*. 'Halwa' (sweet porridge) of its fruits given for regulating menstrual disorders. Paste of bark mixed with wheat dough fed to buffaloes for treating uterine problems* whereas its poultice with 'taramira' (*Eruca sativa* Lam.) applied for eczema and suppuration of boils*. Leaves used for performing 'sharaadh'* rituals (ceremony for remembering departed souls). Watery extr. of seeds (30-50ml, twice a day) taken for relieving body heat*.

Active Constituents: Active properties attributed to arginine, aspartic acid, concuressine, 3-epihetero-conessine, 11-α-20-dihydroxy -18, 20-epoxypregna, 1-4-dien-3-one kurchamine, kurchiminine, kurcholessine, protein, holarrifine, holadienine, lupeol, β-sitosterol, tannin (*plant*), kurchiphylline, kuchiphyllamine, kurchaline, holadysine and holadysamine (*leaves*), conessine, kurchicine, kurchine, conamine, conkurchine, holarrhenine, holarrhinine, holarrhin, lettocine (*bark*), linoceric, linolenic, oleic, palmitic and stearic acid, holarricine and 9-D-hydroxy -cis-12-octadecenoic acid (*seeds*).

Biological Activity: Hypotensive (bark), antiprotozoal (fruit), anticancerous, hypoglycaemic and antifungal (seed oil) activities +ve.

Uses in Literature: Described in Indian literature as an anthelmintic, antidote to snake bite, appetiser, CNS active, colic, edible, expectorant, haematic, lactagogue, ornamental; and for anaemia, asthma, bronchial fever, cold, constipation, cough, curdling milk, dental caries, diarrhoea, digestive complaints, dysentery, eczema, epilepsy, fever, gastric disorders, gastroenteritis, gout, jaundice, leprosy, leucoderma, making small articles, malaria, menstrual complaints, nose bleeding, piles, reviving taste, skin diseases, stomachache and veterinary problems (Ambasta, 1986; Chhetri, 2005; Gogoi & Borthakur, 2001; Jain, 1991; Khare, 2007; Pandey & Rout, 2006; Retnam & Martin, 2006; Sharma, 2003; Sharma *et al.*, 2002; Venkataratnam & Venkata Raju, 2004; Verma & Chauhan, 2006).

Major Ayurvedic Formulations: Kutajarista, Kutaghan vati, Kutawleha, Krimikuthar Ras, Karpur Ras, Punarnava-mandur, Sarvajawarhava-laghuganga-dhar-churna, Sudarshan-churna, Yograja-gugglu.

Trade: Bark and seeds traded in Indian markets as 'koora chhal' and 'karwa inderjau'.

Hordeum vulgare L. (Pl. 57D).

Syn.: *H. aegiceras*, *H. coeleste*, *H. gymnodistichum* of Watt., *H. dificiens* Steud., *H. distichon* L., *H. hexastichon* L., *H. intermedium* Kornioke, *H. sativum* Jess., *H. zeocriton* L.

Family: Poaceae.

Vern.: Jau, Satu.

Ayurvedic, English, Sanskrit and Regional Names:

Ayur.: Hayapriya, Hayeshtha, Shuka-dhaanya, Tikshnashuka, Yava;

Eng.: Barley;

Sans.: Akshata, Divga, Kanchuki, Yaya;

B.: Jab, Jau;

Bo.: Jav, Satu;

G.: Jau, Jav, Ymvah;

H.: Jau, Jav;

Kan.: Jave godhi;

Mal.: Yavam;

Mar.: Java;

P.: Jaon;

Tam.: Barliya-risi;

Tel.: Yaya.

Distribution: Throughout dry parts of India up to 4,000m.

Description: Erect, annual grasses with flat leaves, one-flowered spikelets arranged in groups of three, which are again arranged in two rows. Inferior palea of flower of the central spikelet is longest of the three, some of the spikelets imperfect. Fruit a caryopsis, seed endospermic with single cotyledon.

Flowering & Fruiting: November -February.

Habitat Ecology: Cultivated; Barsar, 815m.

Material Examined: HPU-EL-SKS-ABS-15148; 12.02.2011.

Parts Used: Leaves. Stem. Whole Plant. Grains.

Folk Uses: Grains edible*, eaten as 'sattu'. 15-20ml juice of tender leaves given for liver problems* and jaundice*. Seeds and stems used as fodder*. Stem and pieces of grounded roasted seeds given to increase lactation in animals*. 50-100ml decoction of the plant given against kidney stones*, urinary problems and reducing obesity (thrice daily, 7-15days).

Active Constituents: Beta-glucan (*grains*), flavonoid, 2''-O-glucosyl-isovitexin, indole alkaloid gramine (*leaves*).

Biological Activity: Plant extr. exhibits cyclohexanase activity and effective against treatment of thrombosis and antherosclerosis.

Uses in Literature: Known earlier as antiallergic, antibacterial, anti-inflammatory, demulcent, immuno-protective, nutritive; and for antherosclerosis, asthma, bowel inflammation, chronic sinusitis, cough, diarrhoea, dysentery (vet.), foot and mouth, lipid disorder, loose motion and weakness (vet.), lowering of cholesterol, muscular rigidity, obesity, urinary disorders, thrombosis, vit. C (Agarwal, 2003; Ambasta, 1986; Anonymous, 1998b; Deeba *et al.*, 2009; Khare, 2007; Mishra *et al.*, 1996b; Prajapati *et al.*, 2003, Sensarma, 1989; Sood *et al.*, 2005b; Watt, 1972).

Hypericum cernuum Roxb. (Pl. 58A).

Syn: *H. oblogifolium* Choisy, *H. speciosum* Wall., *H. govanianum* Wall.

Family: Hypericaceae.

Vern.: Pinli.

Distribution: W. Himalayas: 800-2,100m.

Description: Glabrous shrubs having terete branches. Leaves ovate, each pair of leaves placed at right angles to the next pair, above or below it. Flowers yellow in terminal cymes. Capsule conical, opening by 3-5 valves. Seeds numerous, small.

Flowering & Fruiting: April-October.

Habitat Ecology: Shrubberies, open slopes; Kangoo, 649m.

Material Examined: HPU-EL-SKS-ABS-15149; 17.05.2010.

Parts Used: Twig. Seeds.

Folk Uses: Twig used as toothbrush* and seeds as condiments for flavouring curries.

Uses in Literature: Useful earlier as an aromatic, poison (vet), stimulant; and for old sores, eczema and toothache (Kapur, 1996; Khare, 2007).

Impatiens balsamina L. (Pl. 58B).

Family: Balsaminaceae.

Vern.: Teur.

Ayurvedic, English, Sanskrit and Regional Names:

Ayur.: Tarini;

Eng.: The Garden Balsam;

Sans.: Dushpatri-jati;

B.: Dupati;

Bom. : Terada;

Guj. : Gulmendi; Pantambol;

H.: Gulmendi;

Mal.: Mecchingoon;

Mar.: Terada;

Oriya: Mujethi, Haragaura;

P.: Bantil, Halu, Tatura, Tilphar;

Tam.: Kasittumbai.

Distribution: Throughout India.

Description: Annual pubescent herbs. Leaves alternate. Flowers axillary, spur long. Seeds reticulate.

Flowering & Fruiting: July-September.

Habitat Ecology: Wasteplaces, field borders; Hamirpur, 787m.

Material Examined: HPU-EL-SKS-ABS-15150; 17.05.2011.

Parts Used: Leaves. Flowers.

Folk Uses: Leaves used as a substitute for henna. Poultice of flowers heals burns and regarded good for hair growth*.

Active Constituents: These are 3-mono-glucosides of cynidin, delphindin, kaempferol, pelargonidin and quercitin (*stem*), caffeic, p-coumaric, ferulic, gentisic, p-syringic and vanillic acid, 4-2 (2-carboxyphenyl)-4-oxobutyric acid, 2-hydroxy-1, 4-napthaquinone and napthoglycosidase (*flower, leaves*).

Biological Activity: 50% Et OH extr. of flowers exhibit marked antibiotic activity against *Sclerotinia fructicola* and *Colletotrichum lindemuthianum*.

Uses in Literature: Reported so far as cathartic, cooling, diuretic, demulcent, edible, emetic, laxative, substitute for henna and tonic; and for burns, lumbago and intercostal neuralgia (Ambasta, 1986; Anonymous, 1998b; Chopra *et al.*, 1956; Islam, 2002; Khare, 2007; Kirtikar & Basu, 1994).

Ipomoea cairica (L.) Sweet (Pl. 58C).

Syn.: *I. palmata* Forsk., *I. senegalensis* Lamk., *I. stipulaceae* Jacq., *I. pendula* Br., *I. pulchella* Roth, *I. tuberculata* Roem. & Sch., *Convulvulus cairicus* L., *C. tuberculatus* Lamk., *C. bellus* Spreng., *C. digitalis* Roxb., *C. heptaphyllus* Rottl., *C. lupulifoilus* Griff.

Family: Convolvulaceae.

Vern.: Railway creeper.

English Name:

Eng.: Railway creeper.

Distribution: Deccan Peninsula, Nepal, H.P.

Description: A twinner. Leaves palmately lobed. Flowers whitish pink in small stalked clusters or solitary. Capsule 1.5cm in diameter. Seeds numerous, endospermic.

Flowering & Fruiting: June-November.

Habitat Ecology: Roadsides, forest edges, wasteplaces, drier areas; Awahdevi, 1,232m.

Material Examined: HPU-EL-SKS-ABS-15151; 26.06.2008.

Parts Used: Plant. Leaves.

Folk Uses: Plant used for fencing* agricultural field borders and known for possessing antifire* property. Leaves used as fodder*.

Active Constituents: Leaves possess ergosine, ergocornine and ergocistrine.

Uses in Literature: Known todate as purgative and for cough and pain (Ambasta, 1986; Jain, 1991).

Ipomoea carnea Jacq. (Pl. 58D).

Syn.: *I. crassicaulis* (Benth.) Robinson, *I. fistulosa* Mart. *ex* Choisy, *Balatas crassicaulis* Benth.

Family: Convolvulaceae.

Vern.: Basuti.

Distribution: W. Himalayas upto 1,400m.

Description: Shrubs with hollow stem, long-stalked, ovate triangular leaves, and light pink flowers which are solitary or in stalked clusters. Seeds brown.

Flowering & Fruiting: May-July.

Habitat Ecology: Roadsides, forest edges, wasteplaces, drier areas; Bhota, 765m.

Material Examined: HPU-EL-SKS-ABS-15152; 20.07.2010.

Parts Used: Whole Plant. Leaves.

Folk Uses: Plant used as hedge plant for fencing* agricultural fields and to check soil erosion*. Poultice of leaves good against headache*. Leaf heated in oil tied for suppressing pimples, acne* and boils; also used as packing* material for early ripening* of bananas.

Active Constituents: Rich in anthracene glucoside, β-sitosterol, gum, ipomose, jalapin, triacontane and saponin (*leaves*).

Biological Activity: Et OH extr. of plant CNS depressant.

Uses in Literature: Known to be used as poison (vet.), purgative; and for arthritis, boils, cuts, sprain, swelling, ulcer and wounds (Ambasta, 1986; Jain, 1991;

Khanna, 2002; Khare, 2007; Kumar & Chauhan, 2005; Kumar & Jain, 1998; Singh & Prakash, 1994).

Jacaranda acutifolia Humb. & Bonpl. (Pl. 59A).

Syn.: *J. mimosifolia* D. Don, *J. ovalifolia* R.Br.

Family: Bignoniaceae.

Vern.: Nili gulmohra (Jackranda).

Distribution: Outer foothills of W. Himalayas; also cultivated in Indian gardens.

Description: Trees having pinnate leaves, blue or purplish flowers in panicles. Capsule nearly circular, flat, woody dehiscent loculicidal with non-endospermic, flattened, winged seeds.

Flowering & Fruiting: April-June.

Habitat Ecology: Commonly planted along roadsides and railway tracks; Sujanpur, 567m.

Material Examined: HPU-EL-SKS-ABS-15153; 13.04.2009.

Parts Used: Plant. Leaves. Bark. Wood.

Folk Uses: Planted as an ornamental* avenue plant. Wood used for low grade furniture*, construction works* and tool handles. Powdered leaves dusted for quick healing of wounds. Poultice of leaves applied on head for dandruff* and hair fall*. Decoction of bark and leaves (1 tsp, thrice a day till cure) given for syphilis and leucorrhoea*.

Active Constituents: These are scutellarein, hydroxyquinone, jacaranose, jacaranone, verbascoside and phenylacetic-b-glucoside (*leaves*), lupenone, β-sitosterol and 7′, 2, 3′, 4′-tetrahydroxyflavone-3-0-neohesperidioside (I) (*bark*).

Uses in Literature: Used earlier for blennorrhagia, pectoral, syphilis, tool handles, venereal diseases; and as vulnerary (Ambasta, 1986; Jain & Sharma, 1996; Khare, 2007).

Jasminum humile L. (Pl. 59B).

Syn.: *J. chrysanthemum* Roxb., *J. revolutum* Sims, *J. bignoniaceum* Wall., *J. wallichainum* Lindl., *J. inodorum* Jacq., *J. pubigerum* D. Don var. *glabrum* DC.

Family: Oleaceae.

Vern.: Chamaeli.

Ayurvedic, English, Sanskrit and Regional Names:

Ayur.: Hemapushpikaa, Svarnajaati, Svarna-yuuthikaa, Vaasanti;

Eng.: Golden Jasmine, Italian Jasmine, Nepal Jasmine, Yellow Jasmine;

Sans.: Henapushpika, Svarnajuthica,

B.: Svarnajui;

Bo.: Hemapushpika;

H.: Peelajuli, Peeli Chameli, Malto;

Kan.: Hasarunallige;

Mal.: Ponmallika;

Tam.: Semmalligai;

Tel.: Pachche-Abavimalle.

Distribution: Subtropical Himalayas: 1,500-3,000m.

Description: Erect, stout shrubs. Leaves alternate. Leaflets 3-7, oblong or ovate. Flowers yellow, in terminal clusters. Fruit a juicy black berry. Seeds solitary in each cell.

Flowering & Fruiting: May-September.

Habitat Ecology: Common in forests and gardens; Cha kmoh, 736m.

Material Examined: HPU-EL-SKS-ABS-15154; 03.04.2008.

Parts Used: Plant. Leaves. Flowers. Root. Fruits.

Folk Uses: Plant grown as an ornamental*. Flowers used in perfumery, considered sacred* and offered to appease deities. Paste of leaves and roots massaged on gums for relief from toothache. Decoction of roots, 1 tsp prescribed twice daily for a week against skin ailments.

Active Constituents: Alpha-amyrin, betulin, friedelin, lupeol, betulinic, oleanolic and ursolic acids; beta-sitosterol, 10-cinnamoyloxyoleoside-7-methyl-ester and a secoiridoid glycoside (*plant*).

Uses in Literature: Described till date as antidote in poison cases, aromatic essential oil, astringent, tonic to heart; and for chronic sinuses, dyeing, earache, eye diseases, fistulas, headache, perfumery and cosmetics, ringworm affections, skin diseases, stomatitis, toothache, ulcerations and wounds (Ambasta, 1986; Chauhan, 1999; Chopra *et al.*, 1965; Jain, 1991; Khare, 2007; Watt, 1972).

Jasminum multiflorum Roth. (Pl. 59C).

Syn.: *J. hirsutum* Willd., *J. pubescens* (Retz.) Willd., *Morgorium pubescens* Lamk., *Nyctanthes mutiflora* Burm., *N. pubescens* Retz.

Family: Oleaceae.

Vern.: Chamaeli, Malti.

Ayurvedic, English, Sanskrit and Regional Names:

Ayur.: Kasturi, Kunda, Mogaraa;

Eng.: Downy Jasmine;

Sans.: Kunda;

B.: Kunda;

H.: Kundphul;

Kan.: Kaak-avenako, Katta-vanakka;

Mal.: Katutsijiregam-mulla;

Mar.: Mogra;

Tam.: Malligai;

Tel.: Gujari, Kundamu, Malle.

Distribution: India, Common.

Description: A scandent diffuse shrub. Leaves opposite, hairy. Flowers white, in dense cyme. Fruits a berry.

Flowering & Fruiting: February-April.

Habitat Ecology: Roadsides, open slopes; Ladraur, 978m.

Material Examined: HPU-EL-SKS-ABS-15155; 21.03.2009.

Parts Used: Twigs. Leaves. Flowers.

Folk Uses: Leaves and flowers used for offering* to deities on religious ceremonies. Twigs along with leaves of 'garnae' (*Carissa spinarum* Hook. f.), fruits of 'chilli' (*Capsicum annuum* L.), 'kali mirach' (*Piper nigrum* L.) and wheat dough given to livestock as febrifuge*.

Active Constituents: Properties attributed to 10-hydroxyligustroside, 10-hydroxy-oleoside -11-methyl ester mutifloroside and multrioside (*plant*), α-amyrin, betulin, friedelin, lupeol, betulinic, oleanolic and ursolic acid and β-sitosterol (*leaves*).

Biological Activity: 50% Et OH extr. of aerial plant diuretic.

Uses in Literature: Noted earlier as an antidote to cobra venom, emetic; and for boils, burns, skin diseases, ulcer and wounds (Ambasta, 1986; Chopra *et al.*, 1956; Khare, 2007; Maheshwari *et al.*, 1981; Sharma, 2003; Retnam & Martin, 2006).

Jasminum sambac (L.) Ait. (Pl. 59D).

Family: Oleaceae.

Vern: Motia.

Ayurvedic, English, Sanskrit and Regional Names:

Ayur.: Madayanti, Madyantikaa, Mallikaa, Nava-, Shita-bhiru, Vaarshiki;

Eng.: Arabian jasmine, Tuscan jasmine;

Sans.: Mallika;

H.: Banmallika;

Mal.: Cheru pitchakam;

Tam.: Adukkumalli, Gundumalli;

Tel.: Boddumalle, Gundumalli.

Distribution: Subtropical regions of India.

Description: A scandent shrub with broadly ovate-elliptic leaves. Flowers white in terminal clusters, fragrant. Fruits a juicy black berry. Seed solitary in each cell.

Flowering & Fruiting: February-April.

Habitat Ecology: Roadsides, open slopes; Bhota, 765m.

Material Examined: HPU-EL-SKS-ABS-15156; 15.04.2009.

Part Used: Flowers.

Folk Uses: Flowers used for offering* to deities on religious ceremonies; its paste applied as poultice on forehead to cure headache.

Uses in Literature: Known so far for headache, fever, skin infection, syphilis, wounds (Khare, 2007; Retnam & Martin, 2006).

Jatropha curcas L. (Pl. 60A).

Syn.: *J. moluccana* Herb. Russ.

Family: Euphorbiaceae.

Vern.: Rattanjot, Jamnotta.

Ayurvedic, English, Sanskrit and Regional Names:

Ayur.: Kaanan-eranda, Sthula-eranda, Vyaaghrairanda;

Eng.: Purging Nut, Phusic Nut;

Sans.: Dravanti, Kananaerand, Nepala;

B.: Maraharalu, Mogali eranda, Rana-yerandi;

G.: Erundi, Ehandrajot, Jangli-arundi, Jangli-Yarandi, Ratanjota;

H.: Bagbherenda, Ban-bheranda, Eranda-gachh, Pahari errand, Safed-arnad;

Kan.: Kaak-avenako, Katta-vanakka;

Mal.: Kaak-avenako, Kisu-gi, Thinbau-kyeksu, Thinbau-kyetsu;

Mar.: Mogalieranda;

Oriya: Kaddam;

P.: Jamalgota, Japhrota, Kurukarlu, Mogalieranda, Maraharalu, Rattanjot;

Tam.: Kaat-amunck;

Tel.: Bettadaharlu, Maranarulle, Maraharalu.

Distribution: Cultivated and naturalized in most parts of India.

Description: Small deciduous trees with alternate, 3 or 5-lobed leaves. Flowers in terminal cymes, yellowish-green. Fruits ovoid, black. Seeds ovoid-oblong, dull brownish black.

Flowering & Fruiting: February-May.

Habitat Ecology: Frequent in wasteplaces, roadsides, field borders; Jhanikker, 1,050m.

Material Examined: HPU-EL-SKS-ABS-15157; 03.08.2010.

Parts Used: Twigs. Fruit. Seeds. Latex.

Folk Uses: Twigs used as tooth brush. Paste of powdered seeds with 'ghee' considered a good emollient* for chapped skin and feet. ½ tsp of roasted seeds with water prescribed as purgative. Latex good for mouth sores, ulceration, decaying teeth and healing injuries*.

Active Constituents: Isolated ingredients are β-sitosterol, β-amyrin, taraxerol (*stem bark*), Ca, Mg, Na, K traces of P, arachilic, linoleic, myristic, oleic, palmitic and stearic acid, arabinose, galactose, glucose, rhamanose, xylose, galacturonic acid and β-sitosterol (*seeds*).

Biological Activity: 50% Et OH extr. of aerial parts CNS depressant and diuretic.

Uses in Literature: Reported as an abortifacient, anasarca, dye, fish poison, lactagogue, purgative, rubefacient, tooth brush, symbolic to ward off evil spirits; and for body swelling, bloat, burns, cancer, carbuncle, cramps, diarrhoea, dropsy, dysentery, earache, eczema, fencing, headache, herpes, inflammation, piles, pleurosy, pneumonia, rheumatism, ringworm, sciatica, scabies, skin troubles, sores, syphilis, tooth trouble, whitlow and wounds (Ambasta, 1986; Chopra *et al.*, 1956; Dey, 1973; Jain, 1991; Kadel & Jain, 2006; Khare, 2007; Mohanty *et al.*, 1998; Mokat & Deokule, 2004; Punjani, 1998; Retnam & Martin, 2006; Sharma, 2003).

Adulterant or Substitute: Seed oil used as an adulterant for ground nut, olive oil, linseed oil and diesel.

Juglans regia L. (Pl. 60B).

Syn.: *J. regia* L. var. *kumaonica* Cas. DC., *J. arguta* Wall.

Family: Juglandaceae.

Vern.: Akhrot.

Ayurvedic, English, Sanskrit and Regional Names:

Ayur.: Akshoda, Akshodaka, Akshota, Karparaal, Pilu, Shailbhava, Vrantphala;

Eng.: Common Walnut Tree, European Walnut, Himalayassn Walnut, Walnut Tree;

Sans.: Akhoda, Akschota;

Ass.: Kabsing;

B.: Akhrut;

G.: Akhrot;

H.: Akhor, Akhrot, Khor;

Kash. : Akhor, Krot;

Mar.: Akroda;

P.: Akhori, Charmagz, Kabotang;

Tam.: Akrottu;

Tel.: Akrotu.

Distribution: Kashmir-S.E. Tibet: 1,500-3,000m.

Description: A large deciduous tree upto 30m tall and 7m girth. Leaves imparipinnate, aromatic. Male and female flowers present on the same tree. Drupes ovoid with green, thick, fleshy rind enclosing a woody, wrinkled two-valved nut.

Flowering & Fruiting: April-October.

Habitat Ecology: Forests, widely cultivated in W. Himalayas; Anu Khurd, 858m.

Material Examined: HPU-EL-SKS-ABS-15158; 05.05.2011.

Parts Used: Leaves. Fruit. Bark. Wood.

Folk Uses: Fruits edible; one fruit eaten on empty stomach in the morning to check joint pains* during old age. Powdered burnt fruit added in the preparations like 'mahani' and nonvegetarian dishes for flavouring and served especially to guests attending marriage in the family. Wood used for ornamental works*, furniture and construction*. Leaves used for scouring teeth and bark for reddening lips as substitute for lip-stick.

Active Constituentss : Principle constituents are 5-hydroxy-1, 4-napthoquinone, berberine (0.08%) (*stem bark*), oxalic acid, resin (*fruit*).

Biological Activity: Juglons inhibit growth of liver ascites carcinoma and 5-180 solid tumour in mice at 0.1 mg/ml *in vitro.*

Uses in Literature: Noted so far as an alexipharmic, alternative in scrofula, anthelmintic, aphrodisiac, astringent, cattle feed, detergent, edible (fruits), illuminant, insect repellent, laxative, nerve tonic, oil paints, psychoactive plant, tooth brush; and for bed wetting, carving, confectionary (pastries), cough, culinary purposes, dysentery, decoloring charcoal and active carbon, diarrhoea, dispelling flies, drawing instruments, dyeing, eczema, frames, foot and mouth diseases (vet.), furniture, gunstocks, hoof (vet), ice creams, kidney stones, lacquer work, menorrhagia, musical instruments, ploughs, printing, reddening the lips, rheumatism, skin disease, sores of toes, syphilis and tanning (Ambasta, 1986 Chhetri, 2005; Khare, 2007; Kirtikar & Basu, 1994; Lalramnghinglova, 1996; Negi & Pant, 1990; Negi *et al.*, 1993; Rana *et al.*, 1996; Saini, 1996b; Sharma, 2003; Singh, 1999b; Verma & Chauhan, 2006; Vishwanathan & Singh, 1996; Watt, 1972).

Major Ayurvedic Formulations: Amrtipras-ghrit, Ashoka-taila, Dandasa (bark), Dantasodhak, Yarin.

Trade: Dandasa, Akhrot, Akhor, Akhor, Akrut, Krot.

Lagenaria siceraria *(Molina) Standl.* (Pl. 60C).

Syn.: *Cucurbita siceraria* Molina., *Lagenaria leucantha* (Duchesne) Rusby, *Lagenaria vulgaris* Ser.

Family: Cucurbitaceae.

Ayurvedic, English, Sanskrit and Regional Names:

Ayur.: Ikshavaaku, Katu-tumbi, Pindaphalaa, Tiktaalaavu, Tumbini;

Eng.: Bottle gourd, Calbash gourd;

Sans.: Alabu;

Ass.: Lau-bogalau;

B.: Lau;

G.: Dudhi, Tumada;

H.: Lauki, Tum

Kan.: Halagumbala, Sorekayi;

Mar.: Bhopala, Dudh;

P.: Ghiya;

Tam.: Shorakkai;

Tel.: Sorrakaya.

Distribution: Commonly cultivated throughout India.

Description: A softly pubescent herb with angled stem and 3-5 lobed, hairy leaves. Flowers large, white, solitary. Fruit a fleshy pepo with hard pericarp and compressed exalbuminous seeds.

Flowering & Fruiting: July-September.

Habitat Ecology: Cultivated; Kanjiyan, 910m.

Material Examined: HPU-EL-SKS-ABS-15159; 14.09.2008.

Part Used: Fruits.

Folk Uses: Fruits cooked as vegetable and employed for making 'pakoras'*. Its juice, twice a day, considered good for heart problems*, digestive disorders* and jaundice.

Active Constituents: Active properties due to cucurbitacin B, D, G, & H, mainly B, algycones (*fruits*), beta-glycosidase, rhamnose, fructose, glucose, galactose, sucrose, raffinose and stachyose (*seed*).

Uses in Literature: Used earlier as anthelmintic, diuretic, emetic, purgative, vegetable; and for bowls, dropsy, jaundice and water bottles (Ambasta, 1986; Khare, 2007).

Lagerstroemia indica L. (Pl. 60D).

Syn.: *L. elgans* Wall., *Velaga globosa* Gaertn.

Family: Lythraceae.

Vern.: Har-Shingar.

Ayurvedic, English and Regional Names:

Ayur.: Siddha, Siddhaka;

Eng.: China Privet, Crape, Myrllle;

B.: Phurush, Telinga Ghina;

Bo.: Chinai-Mendhi, Dhayti;

P.: Saoni, Dhaura;

Tam.: Pavalakurinji, Sinappu.

Distribution: Throughout India.

Description: Trees with glabrous leaves and bright pink flowers. Panicle not condensed, minutely pubescent or glabrous. Capsules nearly globose with calyx teeth erect on them.

Flowering & Fruiting: June-October.

Habitat Ecology: Planted; Naduan, 483m.

Material Examined: HPU-EL-SKS-ABS-15160; 07.06.2009.

Parts Used: Plant. Flowers. Wood.

Folk Uses: Grown as an ornamental. Flowers offered* especially to appease Lord Shiva. 15-20 ml of its decoction given thrice daily to cure back pain*. Wood used as fuel*.

Active Constituents: Delphinidin-3-arabinoride, petnnidin-3-arabinoside, malridin-3-arabinoside, gallic acid, methyl gallate and ellagic acid (*flowers*) have been characterised.

Uses in Literature: Reccorded so far as an astringent, febrifugal, hydragogue, narcotic, ornamental, purgative, stimulant; and for apathe of mouth, dysentery (Ambasta, 1986; Bhardwaj & Gakhar, 2003; Khare, 2007; Sharma, 2003; Watt, 1972).

Lannea coromandelica (Houtt.) Merr. (Pl. 61A).

Syn.: *L. grandis* (Dennst) Engel, *Odina wodier* Roxb.

Family: Anacardiaceae.

Vern.: Kehmble.

Ayurvedic, English, Sanskrit and Regional Names:

Ayur.: Gudamanjari, Jhingan, Jingini;

Eng.: Egyptian Privet, Henna;

Sans.: Hinna, Yoranna;

Ass.: Jia;

B.: Bhadi, Jeol, Jiyal;

G.: Mavedi;

H.: Jhingan, Kaimil, Mohin;

Kan.: Ajasringi, Kuratige, Gojal;

Mal.: Odiya, Marum;

Mar.: Moi, Shimti;

Oriya: Indramai, Moi;

Tam.: Wodier, Kalasan;

Tel.: Appriyada.

Distribution: Throughout hotter parts of India.

Description: A medium-sized deciduous tree. Leaves odd-pinnate. Flowers greenish-yellow. Drupes oblong, red.

Flowering & Fruiting: July-September.

Habitat Ecology: Edges of fields; Bhota, 765m, Nadaun, 483m.

Material Examined: HPU-EL-SKS-ABS-15161; 07.05.2011.

Parts Used: Leaves. Bark. Wood.

Folk Uses: Leaves used as fodder*; poultice of boiled leaves applied on sprain. Powdered bark applied to cuts and wounds as antiseptic*. Wood utilised in construction, agricultural implements* and as fuel*.

Active Constituents: These are Dl-epicatechin and leucocyanidin Cluytyl ferulate, lanosterol, ellagic acid, quercetin, quercetin-3-arabinoside, iso-quercetin and morin (*bark*), beta-sitosterol, leucocyanidin and leucodelphinidin (*leaves*).

Uses in Literature: Known in India as astringent, dye, flocculating agent, wheel spokes, wood for building; and for asthma, astringent, bruises, carving and turnery, cattle yolks, child birth, clarification of cane juice, combs, cuts, dyeing and textile printing, elephantiasis, furniture, gout, gum for lactation in ladies, gum trouble, oil processors, packing cases, plough and spear shaft, printing, railway sleepers, sprain, swelling, toothache, yielding paper pulp, wounds (Ambasta, 1986; Augustine & Sivadasan, 2004; Jain & Patole, 2001; Khare, 2007; Retnam & Martin, 2006; Sharma, 2003).

Trade: Jhingan, Wodier.

Lantana camara L. (Pl. 61B).

Syn.: *L. aculeata* L., *L. myxa* L., *L. camera* L. var. *aculeata* Moldenke.

Family: Verbenaceae.

Vern.: Chudail buti, Phul lakdi.

Ayurvedic, English and Regional Names:

Ayur.: Chaturraangi, Vanachchedi;

Eng.: Lantana, Wild Sage;

G.: Ghanidalia;

H.: Desi-Lantana, Raimunia;

Kan.: Hesike, Kakke, Nata-hu-gida;

M.P.: Raimuniya;

Mal.: Arippu;

Mar.: Chadurang;

Oriya: Naga airi;

P.: Desilantana;

Tam.: Arippu;

Tel.: Pulikampa.

Distribution: Common weed throughout India.

Description: A straggling aromatic shrub having prickly branches. Leaves simple, opposite, ovate and acute. Flowers yellowish or orange red and in dense umbel. Corolla hairy. Drupes small, black, fleshy, shining, or purple.

Flowering & Fruiting: June-October.

Habitat Ecology: Wasteplaces, rocky slopes; Bhaleth, 506m.

Material Examined: HPU-EL-SKS-ABS-15162; 17.09.2010.

Parts Used: Whole Plant. Flower. Leaves. Wood.

Folk Uses: Planted as hedge plant along farm boundaries. Plant decoction applied on fresh cuts to check bleeding. Poultice of leaves used in joint pains. Concentrated and boiled extr. of flowers used in killing bedbugs and lice* in animals. Wood used as fuel*.

Active Constituents: Isolated costituents are arachidic, linoleic, myristic and palimitic acids, 1-triacontanol, 1, 8 cineole, citral, eugenol, furfural, geraniol, linalool, α-phellandrene, phellandrone, α-amyrin, β-sitosterol, lantadene (*flowers, leaves, stem*), lantadene A, lantadene B, lantanolic acid, lantic acid, leucine, ursonic acid, valine, tyrosine, and glycine (*leaves*).

Biological Activity: Acid extr. of shoot exhibits antibacterial activity. Antitumour activity of verbascoside *in vitro* might partly be due to inhibition of protein.

Uses in Literature: Recorded to be useful as antiseptic, carminative, diaphoretic, fish bait, green manure, insect repellent, stimulant, tooth brush and vulnerary; and for basketry, bleeding gums, burns, cold, cough, curing itching, cut, decaying teeth, diabetes, dyspepsia (vet), fencing, fever, fits, gum care, jaundice, malaria, mumps, rheumatic pains, sprains swelling, tetanus, tumours, ulcers of mouth, veterinary problems, vulnerary, whooping cough and wounds (Ambasta, 1986; Aminuddin & Girach, 1991; Augustine & Sivadasan, 2004; Bhattacharya, 1996; Dwarakan & Ansari, 1996; Geetha *et al.*, 1996a; Girach & Aminuddin, 1995; Goud & Pullaiah, 1996a; Henry *et al.*, 1996; Hosagoudar & Henry, 1996b; Jain, 1991; Khare, 2007; Kshirsagar *et al.*, 2003; Kumar & Goel, 1998; Kurian, 1995; Lal *et al.*, 1996; Lalramnghinglova, 1996; Masih, 2003; Mishra *et al.*, 1996a,b; Pandey & Verma, 2002; Punjani, 1998; Retnam & Martin, 2006; Saini, 1996b; Upadhyay & Singh, 2005; Watt, 1972).

Adulterant or Substitute: Fruits used as an adulterant to black pepper (*Piper nigrum*).

Lavandula angustifolia Mill. (Pl. 61C).

Syn.: *L. officinalis* Chaix, *L. spica* L.

Family: Lamiaceae.

Vern.: Lavender.

English Name:

Eng.: Lavender.

Distribution: Cultivated in many parts of India.

Description: A small herbaceous bushy shrub having woody branches, opposite, long, narrow, lanceolate leaves which are greyish or green with downy appearance. Flowers violet with bluish tinge in verticillaster or condensed cymose head. Fruit a schizocarpic carcerulus with non endospermic seeds.

Flowering & Fruiting: February-June.

Habitat Ecology: Cultivated for its oil content; Khaggal, 920m.

Material Examined: HPU-EL-SKS-ABS-15163; 03.07.2008.

Parts Used: Leaves. Flowers.

Folk Uses: Plant considered as insect repellent* and its powder (2-4g, twice daily till cure) prescribed with goat's milk for abdominal complaints and breast tumour*, also its poultice regarded as antiseptic and for curing headaches. Flowers used in extraction of oil* for perfumery* and cosmetics*.

Active Constituents: Plant contains linalool, linalyl acetate, borntol, camphor, caryophyllene, lavandulyl, limonenene, betaocimene, terpene-4-ol-and alpha-terpineol, ursolic acid lactone, betulin, betulinic acid, 3 beta-formyl ursolic acid.

Uses in Literature: So far known as antibacterial, antiflatulent, antiseptic, antispasmodic, cholagogue, sedative; and for abdominal complaints, depressive states, digestive disfunction, insomnia, restlessness (Khare, 2007).

Lawsonia alba Lam. (Pl. 61D).

Syn.: *L. inermis* Roxb., *L. spinosa* L., *L. purpurea* Lamk.

Family: Lythraceae.

Vern.: Mehendi.

Ayurvedic, English, Sanskrit and Regional Names:

Ayur.: Madayanti, Madayantikaa, Mendika, Ranjaka;
Eng.: Egyptian Privet, Henna;
Sans.: Hinna, Yoranna;
B.: Manghati, Muljuyati;
G.: Aivanam, Marithondi, Marutonri;
H.: Mehedi, Mendi, Shudi;
Kan.: Daunlacca, Pontaletshi, Mayilanchi, Marutonni;
Mal.: Dan, Danbin;
Mar.: Midi, Mindi;
Oriya: Mihndi;
P.: Mehendi;
Tam.: Goranta, Iveni, Pachcha, Pedda;
Tel.: Gorantlu, Gorante.

Distribution: Cultivated throughout India; wild in W. regions.

Description: A shrub with sessile, elliptic-lanceolate leaves, creamish-white, fragrant flowers in corymbose. Capsule size of a pea. Seeds many, angular, brown.

Flowering & Fruiting: June-September.

Habitat Ecology: Edges of fields; Neri, 381m.

Material Examined: HPU-EL-SKS-ABS-15164; 08.07.2008.

Parts Used: Whole Plant. Leaves.

Folk Uses: Paste of leaves alone (50g powdered 'mehendi') or in combination with 25g 'amla' (*Emblica officinalis* Gaertn.) and 25-50ml of curd applied on hair twice a week as a dye and for healing cuts between toes. Also, henna paste applied on hands and feet of young bride, girls and married womenfolk at the time of marriage, 'karwachauth', 'sawan'* and other ceremonial occasions and also for checking burning sensation. Gargles of decoction of its leaves considered good for gum diseases*. 25-30ml of plant extr. given once daily in the morning as blood purifier*.

Active Constituents: Henna contains coumarins, naptha-quinones, flavonoids, sterol and tannin.

Biological Activity: Antibacterial, antiinflammatory and antihyaluronidase activities confirmed.

Uses in Literature: Literature reveals its use as antibacterial, antifertility, antihaemorrhagic, antispasmodic, astringent, hair dye, oxytoxic, prophylactic, tonic; and for boils, burning feet, coloring hair, conjunctivitis, contraceptive, cooling effect, cuts between toes, debility, diarrhoea with blood, fissures, foot and mouth diseases, fungal infection, gonorrhoea, gynaecological diseases, headache, jaundice, leucoderma, piles, loose motions, pimples, sickle cells anaemia, skin inflammation, spermatorrhoea, spleen enlargement, syphilis and tool handles (Ambasta, 1986; Bhatt *et al.*, 2001; Bennet, 1985; Chopra *et al.*, 1956, 1983; Gogoi & Borthakur, 2001; Khare, 2007; Kirtikar & Basu, 1994; Kumar & Chauhan, 2005; Maya *et al.*, 2003; Mohanty *et al.*, 1998; Pawar & Patil, 2004; Rao & Pullaiah, 2001; Rastogi & Mehrotra, 1993; Retnam & Martin, 2006; Sen & Behera, 2003; Sharma, 2003; Singh & Pandey, 1998; Vihari, 1995; Watt, 1972).

Lens culinaris Medic. (Pl. 62A)

Syn.: *Ervum lens* L., *Lens esculenta* Moench

Family: Fabaceae.

Vern.: Masar.

Ayurvedic, English, Sanskrit and Regional Names:

Ayur.: Masura, Masurikaa, Mangalyaa, Mangalyak, Adaasa;

Eng.: Lentil;

H.: Masser, Masur, Masuri;

B.: Masser, Masur, Masuri;

G.: Masser, Masuri, Mussur;

Sans.: Masurah;

Kan.: Chanangi, Masur;

Mal.: Athanapparippu, Canampayar, Vattupparippu;

Mar.: Masser, Masur, Masuri;

P.: Malka, Masur, Masri;

Tam.: Atanapparuppu, Misur, Parupu;

Tel.: Chirisanagalu, Misurpappu.

Distribution: Throughout sub-tropical regions of India (U.P.,M.P., Bihar, W. Bengal, smaller extent in Punjab, Rajasthan, Maharashtra and Gujarat).

Description: Annual or perennial herbs with flowers in sparse peduncled racemes, rarely solitary. Leaves pinnate, rachis ending in a bristle or tendrils. Pods compressed, 1-2 seeded.

Flowering & Fruiting: October-December.

Habitat Ecology: Commonly cultivated for pulses; Joh 980m.

Material Examined: HPU-EL-SKS-ABS-15166; 15.9.2009.

Parts Used: Leaves. Seeds.

Folk Uses: Leaves employed as fodder* for milching cattle. Seeds used for making curry; its soup considered good for recovery in sick people. Boiled seeds mixed with 'patase' (large sized sugar pellets) given to animals for bloating*.

Active Constituents: Main ingredients of seeds are proteins 30%, lecithin, minerals and amino acids.

Uses in Literature: Known in literature for constipation, cure of gastric troubles, edible purposes, foul and indolent ulcers, injuries, lactation, swelling (vet.) (Agarwal, 2003; Ambasta, 1986; Anonymous, 1998b; Bhattarai, 1992; Khare, 2007; Mishra *et al.*, 1996b; Pal, 1981; Prajapati *et al.*, 2003).

Lepidium sativum L. (Pl. 62B)

Syn.: *Cardaria draba* (L.) Desv., *L. drada* L.

Family: Brassicaceae.

Vern.: Hallon.

Ayurvedic, English, Sanskrit and Regional Names:

Ayur.: Aashaalin, Chandrashuura, Chandrikaa, Nandini, Pashumehankaarikaa, Vass-pushpaa;

Eng.: Gardencress, Water cress;

Sans.: Ashalika, Bhadra, Chandrashura, Nandini;

B.: Halin;

G.: Asalio, Halin;

H.: Halin, Huruf;

Mal.: Asail, Asali;

Mar.: Ahliva;

P.: Halin, Shargundei, Tezak;

Tam.: Aliverai;

Tel.: Adalawtulu.

Distribution: Cultivated throughout India as a salad plant.

Description: Annual glabrous herbs possessing erect, branched stem with radical leaves, twice pinnatisect, Flowers small, white in long racemes. Sepals erect, equal at the base. Fruit a siliqua or silicula. Seeds one in each cell.

Flowering & Fruiting: November-January.

Habitat Ecology: Cultivated in all temperate regions; Lambloo, 850m.

Material Examined: HPU-EL-SKS-ABS-15167; 25.12.2010.

Parts Used: Whole Plant. Seeds.

Folk Uses: Herb used as fodder* for horses and camals. About 500g seeds mixed with one litre of milk given as 'kheer' (porridge) to animals for curing 'basaihar' (mastitis)* disease. 15-20ml decoction of the plant prescribed twice a day for treating asthma, bronchial affections and bleeding piles. Water extr. of seeds (10-20ml) considered good against urinary troubles, fever* and dysentery (thrice daily till cure).

Active Constituents: These are proteins, fats, carbohydrates, mineral matters, vitamin A, thiamine, riboflabin, niacin ascorbic acid, pantothenic acid, pyridoxin, rutin (*plant*), alkaloid, glucotropaeolin, sinapin, sinpac acid, mucilaginous matter, and uric acid (*seeds*).

Biological Activity: Et OH extr. of the plant showed antiviral activity against rinderpest virus.

Uses in Literature: Known so far as antiscorbutic, antiviral, diuretic, emmenagogue, lactagogue, stimulant; and for amoebic dysentery, asthma, bleeding piles, bone fractures, bronchial affections, dropsy (vet.), fever, gout, secondary syphilis, skin disorders, tenesmus and uterine prolapse (vet.) (Agarwal, 2003; Ambasta, 1986; Anonymous, 1998b; Deeba *et al.*, 2009; Parrotta, 2001; Prajapati *et al.*, 2003; Sood & Thakur, 2004; Watt, 1972).

Leucaena leucocephala (Lam.) de Wit. (Pl. 62C).

Syn.: *Mimosa leucocephala* Lam., *L. glauca* DC., *Acacia frondosa* Willd., *A. caringa* Ham., *A. biceps* DC., *A. leucocephala* DC.

Family: Fabaceae.

Vern.: Alseenia.

English and Regional Names:

Eng.: White popinae, Lead Tree;

G.: Lasobaval, Pardesi Baval, Vilaytibaval;

Mal.: Takaranniram;

Oriya: Rajokasundiri;

Tam.: Tagarai-virai.

Distribution: Throughout India.

Description: Trees with bipinnate leaves and white flowers in short peduncled heads. Pods linear flat, shining, dehiscent.

Flowering & Fruiting: September-November.

Habitat Ecology: Common on dry slopes, wasteplaces, roadsides, edges of fields; NIT, 900m.

Material Examined: HPU-EL-SKS-ABS-15168; 15.10.2008.

Parts Used: Whole Plant. Leaves. Wood.

Folk Uses: Tree planted for soil conservation* and shade*. Leaves employed as fodder for milching cattle. Wood used for construction purposes*.

Active Constituents: Active principle due to kaempferol-3-arabinoside, kaempferol-3-xyloside, quercetin-3-arabinoside, quercitin-3-rhamnoside and quercitin-3-galactoside (*leaves*).

Uses in Literature: Literature reveals its use as edible, fibre, fodder, ornaments, source of protein, stimulant, tonic, vegetable; and for baskets, carotene, fancy bags, pods, poultry, tender leaves (Ambasta, 1986; Gill & Nyawuame, 1994; Khare, 2007; Singh *et al.*, 1987; Sinha, 1986; Watt, 1972).

Linum usitatissimum L. (Pl. 62D).

Syn.: *L. trinervium* Roth.

Family: Linaceae.

Vern.: Alsi.

Ayurvedic, English and Regional Names:

Ayur.: Atasi, Kshumaa, Masrnaa, Nilapushpi, Umaa;

Eng.: Flax, Linseed;

Tam.: Ali, Sirrali (Seed), Virai.

Distribution: Cultivated mainly for oil throughout India.

Description: Annual herbs with cylindric erect stem and leaves which are simple below and narrow upper or lanceolate ones, blue petals and stigmas free, with linear clavate styles and flowers in broad cymes. Capsule scarcely exceeding the narrowly white-margined, acuminate sepals.

Flowering & Fruiting: September-October.

Habitat Ecology: Cultivated for oil, fibers and seeds; Dhaneta, 596m.

Material Examined: HPU-EL-SKS-ABS-15169; 10.12.2010.

Parts Used: Whole Plant. Stem. Leaves. Seeds.

Folk Uses: Plant used as fodder*. Ropes made from stem utilized for moving stirrer for churning butter*. Roasted seeds eaten for overcoming internal weakness*. 'Halwa' of seeds prepared along with 'suji'(fine granules of wheat flour) given for back-aches*. Seed oil used for massaging aching joints*, hidden injuries* and rheumatism. Oil cake used as fodder. Seed oil (1-2 litre once a day) given orally for 'basaihar' (mastitis)* to animals.

Active Constituents: Chlorogenic acid, palmitic, stearic oleic, linoleic acids, amino acids, sugars (*plant*). mucilage, fatty oil, cyanogenic glycosides, mainly linustatin, neolinustatin and linamarin, lignans, phenylpropane, linusitamarin (*seeds*).

Uses in Literature: Recorded earlier as antipidamic, antitussive, cardiac tonic, demulcent, emollient, laxative, nervine, pectoral; and for atherosclerosis, burns, cancer, chronic dropsy (vet.), dysentery (vet.), hypercholesterolemia, lupus nethritis, renal diseases, skin injuries, sores and stomachis (vet.) (Ambasta, 1986; Deeba *et al.*, 2009; Khare, 2007; Kirtikar & Basu, 1994; Mishra *et al.*, 1996b).

Litchi chinensis Sonn. (Pl. 63A).

Syn.: *Nephelium litchi* Camb., *Dimocarpus longan* Lour.

Family: Sapindaceae.

Vern.: Litchi.

English and Regional Names:

Eng.: Litchi, Lychee;

Bo.: Lichi;

H.: Litchi;

Oriya: Lishi;

Tam.: Illichi.

Distribution: Cultivated.

Description: Evergreen small trees having alternate, exstipulate leaves, which are shining above and glaucous beneath. Flowers greenish and in terminal panicles. Fruits ovoid, red when ripe. Seeds large, ovoid brown enclosed in a fleshy succulent sweet edible aril.

Flowering & Fruiting: April-June.

Habitat Ecology: Cultivated in gardens; Salasi, 805m.

Material Examined: HPU-EL-SKS-ABS-15170; 27.05.2011.

Parts Used: Leaves. Fruit.

Folk Uses: Poultice of leaves applied against animal bites*. Fruits edible and considered good for heart*, neuralgic disorders and as refrigerent*.

Active Constituents: *Fruit* contains cyanidin-3-rutinoside, cyaniden-3 glucoside and malvidin-3 acetylglucoside.

Uses in Literature: Described earlier as anodyne in neuralgic disorders, edible (fruits), orchitis; and for offering (Ambasta, 1986; Gogoi & Borthakur, 1991; Khare, 2007).

Luffa aegyptica Mill. (Pl. 63B).

Syn.: *Luffa pentandra* Roxb., *L. racemosa* Roxb., *L. clavata* Roxb., *L. petola* Ser., *L. cattu-picinna* Ser., *L. parvala* Wall., *L. gosa* Wall., *L. hederaces* Wall., *L. satpatia* Wall., *L. acutangula* W. & A., *L. cylindrica* (L.) Roem., *Bryonia cheirophylla* Wall., *Momordica cylindrica* L.,M. *luffa* L.

Family: Cucurbitaceae.

Vern.: Ghiya Tori.

Ayurvedic, English, Sanskrit and Regional Names:

Ayur.: Dhaamaargava, Mahaajaalini, Mahaakoshtaki, Raajakoshataki;

Eng.: Gourd Towel, Sponge Gourd, Vegetable, Sponge, Washrag Sponge;

Sans.: Aibhi, Brihatkoshataki, Dhamargava, Dirghakoshataki, Luff, Mahaphala, Rajakoshataki, Sapitaka;

Ass.: Palo;

B.: Bhol, Bhatkerela, Bhatkarkel;

Bo.: Turia;

G.: Ghutti-bira, Neti-bira;

H.: Dun-dul, Dhundul, Ghiatarui, Nenua, Purula;

P.: Liasada, Turi;

Tam.: Pichukku, Pikku;

Tel.: Guttibira, Netibira, Nunebira,Thabwot, Thapwotkha.

Distribution: Throughout India.

Description: Extensively climbing hairy herbs with stout, twisted, glabrous stem, trifid tendrils, orbicular-reniform, palmately lobed leaves and yellow flowers, monoecious at the summit of peduncles (male), solitary (female). Fruit cylindric, blunt at the ends and marked with longitudinal lines. Seeds grey, compressed, narrowly winged.

Flowering & Fruiting: May-September.

Habitat Ecology: Cultivated, wasteplaces; Patta, 1,020m.

Material Examined: HPU-EL-SKS-ABS-15171; 04.10.2010.

Part Used: Fruit.

Folk Uses: Fruit cooked as vegetable and also used in culinary preparations*; its juice along with *Momordia charantia* L., *Azadirachta indica* A. Juss. and *Lycopersicum esculentum* Mill. considered good for heart ailments* and diabetes* (20-30ml, empty stomach). Dried fruits yield a sponge used as a scrubber for bathing. Grated fruits rubbed on hands and feet to relieve the body heat.

Active Constituents: Saponins-lucyosides A, E, F, I, J, K, L and M (*aerial parts*), pentacyclic triterpene, bryonolic acid (*roots*).

Uses in Literature: Found to be as anthelmintic (vet.), carminative, cathartic, diuretic, edible, emetic, fibre, galactagogue, purgative, scrubber, vegetable; and for abnormal menstruations, anaemia, burns, cold and cough (vet.), diarrhoea, dog bites, dyspepsia (vet.), expulsion of worms, insect bites, killing white ants, liver diseases, mastitis, oedema, pharyngitis, piles, premature ejaculation, prickly heat, rhinitis, scabies, snake bite, swelling, and tonsilitis (Ambasta, 1986; Jain, 1991; Jha *et al.*, 1996; Katewa *et al.*, 2001; Khare, 2007; Kulkarni & Kumbhojkar, 1993; Kumar & Chauhan, 2006; Kurian, 1995; Maya *et al.*, 2003; Samati, 2004; Sarangi & Sahu, 2004; Satapathy & Brahmam, 1996; Singh & Maheshwari, 1992; Upadhye & Kulkarni, 2004; Watt, 1972).

Adulterant or Substitute: Seed oil substituted for olive oil. The sponge is employed in place of cellulose sponges.

Lycopersicum esculentum Mill. (Pl. 63C).

Syn.: *L. lycopersicon* (L.) Brit & Brown, *L. lycopersicum* (L.) Farewell, *L. lycopersicum* (L.) Karsten, *Solanum lycopersicum* L., *S. humboldtii* Willd., *S. pseudolycopersicon* Jacq.

Family: Solanaceae.

Vern.: Tamatar.

English and Regional Names:

Eng.: Love Apple, Tomato;

Ass.: Belatibengana;

B.: Gur-begun, Tamati;

Bom. : Vilaytivengan;

G.: Vilaytivengun;

H.: Tamatar;

Kan.: Chapar-badne;

Mar.: Velvangi;

P.: Bhatte;

Tam.: Simie-takalie-palam.

Distribution: Cultivated or as an escape over the greater parts of India.

Description: Pubescent herbs. Leaves pinnate. Flowers yellow and in cymes. Berries red when ripe. Seeds compressed.

Flowering & Fruiting: May-June.

Habitat Ecology: Cultivated; Sakand Gram, 1,000m.

Material Examined: HPU-EL-SKS-ABS-15172; 08.05.2008.

Parts Used: Plant. Fruit.

Folk Uses: Fresh ripe fruits eaten raw with black pepper (*Piper nigrum* L.) and salt early in the morning as wormicides*. Juice of the plant or its soup given against anaemia* and jaundice*. Pieces of ripe fruit fried along with onion in ghee or oil as vegetable*. Fruits considered appetiser* and consumed raw in salads; also used in the preparation of juice, ketchup, sauce, soup*, 'chutney' purries, salads*.

Active Constituents: Founds rich in tomatine, traces of solanin, glucose, fructose and some sucrose and ketoheptose, all essential amino acids except tryptophan and citric acid (ripe) and narcotine (unripe).

Biological Activity: Leaves antifungal.

Uses in Literature: Considered useful as antiseptic, aperient, blood purifier, chelagogue, digestive, edible, fodder, juices, renal stimulant, salad, soup; and for asthma, bronchitis, child birth, dyspepsia and apthous stomatitis (Ambasta, 1986; Augustine & Sivadasan, 2004; Chatterjee & Pakrashi, 1997; Khare, 2007; Maji & Sikdar, 1982; Shah, 1984).

Adulterant or Substitute: Roots used as an adulterant for that of Periwinkle (*Catharanthus roseus* G. Don).

Madhuca indica J.F. Gmel. (Pl. 63D).

Syn.: *Bassia longifolia* Koeing., *B. Latifolia* Roxb., *Maduca latifolia* (Roxb.) Mac Bride.,M. *longifolia* (L.) Mac Bride.

Family: Sapotaceae.

Vern.: Mahua.

English, Sanskrit and Regional Names:

Eng.: Butter Tree, Mahua;

Sans.: Madhuka;

B.: Mahul, Mahwa;

G.: Mahuda;

H.: Mahua, Mohwa;

Kan.: Hippe;

Mal.: Ilupa, Poonam;

Mar.: Mahwa, Mohwra;

Oriya: Madgi, Mahula, Moha;

Tam.: Elupa, Ilupei;

Tel.: Ippa.

Distribution: Cultivated in tropical and subtropical parts up to 1,200m.

Description: A large deciduous tree with rather shorter bole, larger crown. 12-25cm long, thick leathery leaves which are having pointed tip. Nerves quite prominent. Flowers small fleshy, pale or dull white, in clusters near the end of branches, stocks of flowers bent downwards. Musk scented.

Flowering & Fruiting: November-January.

Habitat Ecology: Cultivated along the roadside; Naduan, 483m.

Material Examined: HPU-EL-SKS-ABS-15173; 15.07.2009.

Parts Used: Leaves. Stem. Flowers. Bark. Seeds.

Folk Uses: Leaves used as astringent, its ash mixed with butter or ghee applied on burn injuries. Flowers eaten* raw or cooked*; powdered flowers (1-3g) with honey recommended for asthma; also utilized in making 'sur' (wine)*, syrups* and jams*. Seed oil used in perfumery, oil cake given as fodder*. Poultice of bark applied for skin rashes, bleeding gums and ulcers whereas its decoction (15-20ml, on empty stomach) good against diabetes. Wood regarded good for construction*.

Active Constituents: Constituents identified as triterpenoids, alpha and beta amyrin acetate, n-hexacosanol, beta-D-glucoside and beta-sitosterol (*fruit* pulp), quercertin, and dihydroquercetin (*nut shell*), vitamins, phosphorus, calcium, iron, magnesium, copper, sugars; sucrose, glucose, fructose, arabinose and rhamnose (*flower*), s. aponins-2, 3-di-o-glucopyranoside of bassic acid (*seed*).

Uses in Literature: Valued in India as anthelmintic, anticethalgic, antirheumatic, astringent, bechic, demulcent, emetic, emollient, galatogenic, laxative, stimulant; and for asthma, cure of pneumonia, diabetes, gum troubles, piles, pthisis, skin diseases, tonsillitis and ulcers (Agarwal, 2003; Ambasta, 1986; Chatterjee & Pakrashi, 1997; Chopra *et al.*, 1956; Dwivedi, 1993; Jain, 1968; Khan & Khanum, 2005; Khare, 2007; Kirtikar & Basu, 1994; Parrotta, 2001; Prajapati *et al.*, 2003; Rastogi & Mehrotra 1991; Sahni, 1998; Singh, 1982).

Mallotus philippinensis (Lam.) Muell.-Arg. (Pl. 64A).

Syn.: *Rottlera tintorii* Roxb., *R. aurantiaca* Hook. & Arn., *R. affinis* Hassk., *R. montana* and *mollis* Wall., *Croton philippinensis* Lam., *C. punctatus* Retz., *C. coccineus* Vahl., *C. montanus* Willd., *C. cascarilloides* Rauesch. *ex* Steud., *C. distans* Wall.

Family: Euphorbiaceae.

Vern.: Kambhal.

Ayurvedic, English, Sanskrit and Regional Names:

Ayur.: Chandra, Kampilla, Kampillaka, Karkash, Raktaanga, Rechan;

Eng.: Monkey Face Tree, Kamala Tree;

Sans.: Bahupushpa, Kampilla, Kapila, Pikaksha, Rechanaka;

Ass.: Chinderpang, Gangai, Jaggaru, Machugan, Puddum;

B.: Bosonto-gundi, Kumala, Sundragundi, Tung;

Bom. : Kamala, Ruhin, Shendri;

G.: Kapli, Kapila, Kapilo, Kamella-mavu;

H.: Dhola Sindur (Birbhum), Kamala, Kambhal Kambila, Rohni, Rula, Rulu, Sinduri, Rori;

Kan.: Hullchelu, Kasari, Kurku, Rangamali;

Mal.: Chenkolli, Piponnakam, Poonagam;

Mar.: Kapilo, Shendri;

Oriya: Bosntogundi, Kumala, Rora, Sundragundi;

Tam.: Chendrasinduri, Kapilla, Kapli, Kunkuma, Vassuntagunda;

Tel.: Kunkuma, Kurku, Rangamale, Sarankasari.

Distribution: Throughout tropical India.

Description: A small evergreen tree with slender branches. Leaves alternate, ovate-lanceolate, acuminate, entire toothed, orbicular red glands beneath. Flowers red, in stiff spikes, small. Capsule trilobed, covered with a bright red powder. Seeds subglobose, black.

Flowering & Fruiting: November-April.

Habitat Ecology: Dry slopes, moist forests; Anu, 900m.

Material Examined: HPU-EL-SKS-ABS-15174; 13.04.2008.

Parts Used: Leaves. Fruit. Seeds. Bark.

Folk Uses: Poultice of young leaves applied for itching, ringworms and skin ailments. Decoction of leaves used as coolant for washing hands and feet. Leaves lopped for fodder and also form good bedding for cattle ('pattri')*. ½ tsp of powder fruit with 'ajwain' (*Trachyspermum ammi* L.) and black pepper (*Piper nigrum* L.) taken for 12-15 days as blood purifier. One tsp of 'kamala' powder (fruit) prescribed with cow's milk or curd twice a day for deworming and jaundice. Red glandular powdery coating on fruit surface used as dye ('kungu') for supporting a'tilak' and colouring animals; and given along with jaggery for stomach pains or with butter milk to check constipation* in animals. An amulet of pieces of its bark tied around neck, wrist and feet for treating skin rashes* ('badalbis'). Wood used as fuel and for making tool handles*. Seed pulp taken as such for gastric ailments. Seed cake used as manure.

Active Constituents: Properties attributed to rottlerin, isorottlerin, homorotteler and octacosanol, traces of wax, volatile oil, starch, tannins, sugar, oxalic and citric acid (*plant*), crotoxigenin and its rhamnoside coroglaucigenin, rottlerin, isorottlerin (*seeds*), betulin-3-acetate, lupeol, lupeol acetate, sitosterol, bergenin (*heartwood*), rottlerin, isorottlerin and red compound (*leaves*).

Biological Activity: Alc. extr. of fruits show antibacterial activity against *Micrococcus pyogenes* var. *aureus* whereas aq. extr. of red resinous powder (0.03g) exhibits lithotropic effect in adult albino rats.

Uses in Literature: Known so far as an alternative, anthelmintic, antibacterial, anticancer, antioxidant for ghee, aphrodisiac, blood purifier, contraceptive, cathartic, dye, fuel, fodder, manure, purgative, red dye from it (adorned by women as vermilion on forehead as a sign of being married), styptic, tonic for married woman; and for eczema, jaundice, rapid drying of varnish (*seeds*), rheumatism, scabies, skin diseases, snake bite, sores, stomachache, tannins, thoracic pains, ulcers, and wounds (vet.) (Ambasta, 1986; Banerjee & Pal, 1996; Chhetri, 1994; Hosagoudar & Henry, 1996b; Jain, 1968, 1991; Kapur & Singh, 1996; Khare, 2007; Kumar & Jain, 1998; Mukherjee & Namhata, 1990; Rama Rao & Henry, 1996; Rana *et al.*, 1996; Retnam & Martin, 2006; Sharma, 2003; Singh, 1996; Singh & Prakash, 1994; Tirkey, 2004; Tirkey *et al.*, 2003a,b; Tiwari & Tiwari, 1996; Vihari, 1995; Watt, 1972).

Major Ayurvedic Formulations: Dhanvantara-ghrita, herbenol cream, Kampilla-ka-churna, Kampilaka-taila, Kash-isadighrit, Krimighatini-vati, Krimikuthar-ras, Misraka-sneha, Tinctura kamla.

Trade: Kamila powder, Kainda, Kanikaa.

Adulterant or Substitute: Red powder from fruits adulterated with annato dye (*Bixa orellana* L.).

Malvaviscus conzatii Greenm. (Pl. 64B).

Family: Malvaceae.

Vern.: Gudhal.

Distribution: Throughout India.

Description: A shrub with serrated leaves, red orange flowers in solitary terminal or axillary positions and staminal column projected out of petals. Fruits mostly abortive, seeds many and non-endospermic.

Flowering & Fruiting: Major part of the year.

Habitat Ecology: Hedges, near houses; Thankari, 1,000m.

Material Examined: HPU-EL-SKS-ABS-15175; 12.08.2010.

Parts Used: Whole Plant. Flowers.

Folk Uses: Largly planted for its ornamental* flowers which are also considered sacred* and offered to various deities. 5-10ml of its juice prescribed in impotency, once daily for a few months.

Active Constituents : Found to contain betaine, β-sitosterol, and its glucoside, allantoin, α-kaempferol and their 3-β-glucoside and 3 β-sophoroside (*flowers*).

Biological Activity: Betaine exhibits sperm immobilising action *in vitro.*

Uses in Literature: Used for premature ejaculation (Khanna, 2002; Khare, 2007).

Mangifera indica L. (Pl. 64C).

Syn.:M. *domestica* Gaertn.

Family: Anacardiaceae.

Vern.: Amb, Aam.

Ayurvedic, English, Sanskrit and Regional Names:

Ayur.: Aamra, Amb, Mangottai, Paruppu, Pikavallabha, Rasaal, Sahakaar;

Eng.: Cuckoo's Joy, Cupid's Favourite, Mango Tree, Spring Tree;

Sans.: Adisairrabha, Amrah, Manmathalaya, Parapushtama hotsava;

Ass.: Am, Gharium;

B.: Ambra, Uli;

Bo.: Amba, Amb;

G.: Am, Amb, Ambo, Amri, Anb;

H.: Am, Ambra, Amb;

Kan.: Mampalam, Manna, Mava, Mavu;

Mal.: Amram, Gomanne, Nattumavu;

Mar.: Ambo;

Oriya: Amb, Boula, Chuto;

P.: Amb, Amu;

Tam.: Adishelara yam, Elamavi, Guggu-mamidi, Mamadi, Mamid, Mamidi, Makandamu;

Tel.: Amba, Gujjumamidi, Makandamu, Mavina, Mavu.

Distribution: Tropical Himalayas; cultivated in tropics generally.

Description: Large evergreen trees. Leaves linear or oblong, acute shining, entire, margin undulate. Panicals usually tomentose. Flowers yellow, odorous arranged in many flowered pubescent panicles. Drupes large, fleshy, obliquely pyriform, yellow. Stones compressed, hard.

Flowering & Fruiting: March-June.

Habitat Ecology: Field borders, roadsides (planted); Ukhli, 840m.

Material Examined: HPU-EL-SKS-ABS-15176; 27.03.2008.

Parts Used: Leaves. Fruit. Wood.

Folk Uses: Ripe fruits edible and its pulp used for making candies ('aampapar').Young and unripe fruits acidic and used for making various types of pickle* and 'chutney'*. Decoction of unripe fruits with red carrot, a pinch of salt, sugar and spices prescribed to check sun stroke*. 'Aamchur'* (powdered drupe except kernel) used in various culinary preparations for imparting acidic taste and also important component of one of local delicacies-'mahani' which is served during feasts and marriage ceremonies. A garland of leaves hung on the main door at the time of house warming ceremony and considered sacred. Wood used as 'samadas' (sacred wood) for performing 'hawan' and also for furniture* and construction of house. Leaves given for facilitating evagination of placenta* ('jer girana') (vet.).

Active Constituents: Isolated constituents from different plant parts are homomangiferin and characterised as 2-β-D-glucopyranosyl-3-methoxy-I, 6-7-trihydroxyxanthone, leucine, tyrosine and valine (*leaves*), alanine, glycine, aminobutyric acid, kinic and shikimic acids (*leaves, bark*), triterpenoids (mangiferin, taraxerone, b-sitosterol), flavonoids, phenolic compounds and tannins.

Biological Activity: Gallic acid (1mg/ml) and quercitin (4mg/ml) showed marked antiinfluenza virus activity. Antifungal and antidiabetic activity also confirmed.

Uses in Literature: So far used as abortifacient, anthelmintic, antisyphilitic, appetiser, astringent, cremation wood, diuretic, edible, febrifugal, kibe, laxative, magico-therapeutic, sacred (Lord Buddha reposed under this tree), stomachic, symbolic (abode of god), tonic, vegetable; and for abdominal pain, agricultural implements, ansarca, anthrax (vet), antifertility, arthritis, asthma, blood cancer, burns, cholera, cold, constipation (vet.), cough, cracking of foot, cut, decoration in marriage and festivals, diarrhoea, diarrhoea (vet.), diabetes, dysuria, dysentery, excessive urination, fever, hiccups, house construction, hepatitis, indigestion, indigestion (vet.), intestinal worms, jaundice, liver disorders, piles, pneumonia (vet.), pyorrhoea, musical instrument, rabies (vet.), rheumatism, sun stroke, swelling of throat (vet.), wounds and writing inks (Ambasta, 1986; Aminuddin & Girach, 1991; Bhatt *et al.*, 2001; Borthakur & Gogoi, 1994; Borthakur *et al.*, 2004; Chowdhery, 1996; Ghosh *et al.*, 1996; Gill *et al.*, 1993, 1997; Gogoi *et al.*, 2003; Goud & Pullaiah, 1996a; Henry *et al.*, 1996; Hosagoudar & Henry, 1996 a,b; Jain, 1991; Jain *et al.*, 1994; Jha *et al.*, 1996; Kadel & Jain, 2006; Khare, 2007; Kiruba *et al.*, 2006; Kumar & Goel, 1998; Kapur *et al.*, 1996; Lalramnghinglova, 1996; Maliya & Singh, 2003; Manandhar, 1996; Masih, 2003; Mishra *et al.*, 1996b; Mohanty & Padhy, 1996; Mohanty *et al.*, 1998; Pal & Jain, 1998; Pandey & Rout, 2006; Pandey & Verma, 2002, 2003; Rekha *et al.*, 1998; Retnam & Martin, 2006; Samati, 2004; Sen & Batra, 1997;

Sen & Behera, 2003; Shah & Singh, 1990; Sikarwar, 1996; Sudarsanam *et al.*, 1995; Tirkey *et al.*, 2003b; Verma & Chauhan, 2006; Viswanathan *et al.*, 2001; Watt, 1972).

Trade: Mango.

Martynia annua L. (Pl. 64D).

Syn.: **M.** *diandra* Glox.

Family: Martyniaceae.

Vern.: Kaun.

Ayurvedic, English and Regional Names :

Ayur.: Kaakanaasikaa, Kaakaangi, Shirobal;

Eng.: Devil's Plant, Snake Head, Tiger Claw, Unicorn Plant;

B.: Bagh-noki;

Bo.: Vinchu;

H.: Bichu, Hathajori;

Mar.: Vinchu;

P.: Bichu;

Tam.: Thelkodukk-ukai, Nagavidakkai;

Tel.: Garudamukku.

Distribution: Tropical weed; in Himalayas upto 1,200m.

Description: Tall viscid herbs. Leaves cordate, sticky, sinuate toothed with unpleasant smell. Flowers pink, blotched with purple. Fruits black with a pair of hooked spinous beak on maturity.

Flowering & Fruiting: June-October.

Habitat Ecology: Shrubberies, wastelands; Neri, 381m.

Material Examined: HPU-EL-SKS-ABS-15177; 08.07.2008.

Parts Used: Leaves. Fruit.

Folk Uses: Ash of fruits mixed with coconut oil applied to heal burns* and abscess*. Poultice of fruit paste good against scorpion-stings. Gargles of leaf juice prescribed twice daily for sore throat* till cure.

Active Constituents: Pelargonidin-3-5 diglucoside, apigenin, luteolin, apigenin-7-O-beta-D-glucuronide, luteolin-7-O-beta-D-gluconide, pellargoni-din-3-5-diglucoside and cyanidin-3-galactoside have been isolated (*flower*).

Uses in Literature: Known so far as antidote for blisters, boils, epilepsy, inflammation, snake and scorpion stings, tuberculosis, ulcers, wounds (Ambasta, 1986; Goud & Pullaiah, 1996a; Khare, 2007; Kirtikar & Basu, 1994; Patil, 2004; Reddy *et al.*, 1998; Retnam & Martin, 2006).

Melia azedarach L. (Pl. 65A).

Syn.: *Melia sempervirens* Sw.,M. *bukayun* Royle

Family: Meliaceae.

Vern.: Draek.

Ayurvedic, English, Sanskrit and Regional Names:

Ayur.: Dreka, Mahaanimba, Ramyaka;

Eng.: Barbados Lilac, Bead Tree, Indian Lilac, Persian Lilac, Pride of India;

Sans.: Akshadru, Arishta, Brihannimba, Dreka, Gairia, Hinadruma, Karmuka, Mahanimba, Nimbaka, Parvatta-nimba-vrikshha, Vikshaha;

Ass.: Bakainu, Thamage;

B.: Gara Nim, Ghornim, Mahanim;

Bo.: Bakayam, Drek, Vilayatinim;

G.: Bakan, Bakanlimbodo, limbodo, Malai, Malai: veppam, vembu;

H.: Bakain, Bakarja, Drek, Deikna, Ghoranim, Mahanim, Mahanimba;

Kan.: Arevebu, Hutchu Bevu;

Mal.: Ka-ma-ka, Malaveppu, Ta-ma-ka;

Mar.: Bakananimb, Bakan limbodo, Limbara;

P.: Bakain, Chein, Dek, Drek;

Tam.: Konda-vepa, Makanim, Malaivembu, Tarakavepa;

Tel.: Turakavepa, Vettiveppa;

Urdu: Bakayan.

Distribution: Common throughout India.

Description: Trees with short trunk, finely marked hard wood, broad crown having bipinnate, opposite leaves, ovate, obtusely serrate, deciduous leaflets. fragrant, lilac flowers in long axillary panicle. Drupes subglobose, wrinkled.

Flowering & Fruiting: March-May.

Habitat Ecology: Roadsides, edges of cultivated fields; Kohli, 851m.

Material Examined: HPU-EL-SKS-ABS-15178; 25.12.2008.

Parts Used: Leaves, Seeds, Wood.

Folk Uses: Leaves used as fodder, its poultice along with powdered seeds considered good against dandruff*. Seeds utilized for making rosaries*. Wood used for making furniture (doors, window) and agricultural implements* (plough, jugala).

Active Constituents: Chief constituents are kaemferol-3-l-rhamno-D-glucoside, rutin (*leaves*), kuline, tetracyclic triterpenoides kulactone, kulalactone, kulinone (*bark*), azardine, aresin, tannin, meliotannic acid, benzoic acid, bakayanin sterol, a bitter principle margosine and a fixed oil containing sulphur (*fruits*).

Biological Activity: Stem bark anticancerous, antiviral and spasmolytic. Azedarchin A and 12-o-acetylazedarchins A & B exhibit antifeedant activity on larvae of *Spodoptera exigua*.

Uses in Literature: Reported in Indian literature as an antilithic, antiseptic, astringent, blood purifier, brain tonic, diuretic, emmenagogue, fodder, fuel wood, laxative, ornamental plywood, psychoactive, stomachic, symbolic abode of god; and for cold, fever (vet.), furniture, gout, hair fall, hypertension, hysteria, malarial fever, making idols, musical instruments and boxes, spasm (vet.), spleen enlargement, sprain (vet.), skin disease, tumours, ulcer (vet.), wounds (Ambasta, 1986; Borthakur

& Sarma, 1996; Chhetri, 1994; Dey, 1973; Goud & Pullaiah, 1996b; Jain, 1991; Jain *et al.*, 1994; Jha & Basak, 1994; Jha *et al.*, 1991; Kapur & Singh, 1996; Khare, 2007; Kurian, 1995; Lalramnghinglova, 1996; Rana *et al.* 1996; Retnam & Martin, 2006; Sharma, 1996; Siwakoti & Verma, 1996; Srivastava *et al.*, 2003; Watt, 1972).

Trade: Persian lilac.

Adulterant or Substitute: Leaves and bark substituted for true 'Neem' (*Melia azedirachta* L.).

Melissa officinalis L. (Pl. 65B).

Family: Lamiaceae.

Vern.: Balm.

English Names:

Eng.: Mountain Balm, Sweet or Lemon Balm.

Distribution: Introduced in India.

Description: Erect, branched, evergreen aromatic herbs having opposite, crenate leaves; tall, erect, hirsute stem and leaves with acute rounded or cordate base. Whorls numerous, few or many fid. Flowers pedicelled, blossoms small, white or light rose in colour. Fruits schizocarpic, carcerulus. Seeds small, non-endospermic.

Flowering & Fruiting: August-October.

Habitat Ecology: Cultivated; Khaggal, 920m.

Material Examined: HPU-EL-SKS-ABS-15179; 03.07.2008.

Parts Used: Whole Plant. Leaves.

Folk Uses: Leaves consumed as such for goiter. Oil obtained from the plant used in perfumery*, headaches, cosmetics* and skin affections*.

Active Constituents: These are geranial and neral with caryophyllene oxide, tertenes, glycosides of volatile oil, eugenol glucoside, caffeic acid, flavonoids, cymaroside, cosmosiin, rhamnocitrin, isoquercitrin, triterpene acids (*plant*).

Uses in Literature: Known earlier as antidepressant, antihistaminic, antispasmodic, antiviral; and for Alzheimer diseases, anxiety, headache, hyperthyroidism, nervous excitadility, neurosis, palpitation (Ambasta, 1986; Chopra *et al.*, 1956; Khare, 2007; Jain, 1991).

Mentha piperata L. (Pl. 65C).

Family: Lamiaceae.

Vern.: Pipermint.

Ayurvedic, English and Regional Names:

Ayur.: Vilaayati Pudina;

Eng.: Brandy Mint, Peppermint;

H.: Gamathi Pudina, Paparaminta;

Tel.: Vilayeti, Pudneh.

Distribution: Throughout India; commonly cultivated in native gardens all over plains of India.

Description: Erect or diffuse, perennial herbs with pungent smell and creeping rootstock. Stems hoary pubescent. Leaves smooth, lanceolate, acute, coarsely dentate. Flowers lilac, in loose cylindrical slender spikes. Fruit schizocarpic (carcerulus). Seeds ex-albuminous.

Flowering & Fruiting: July-September.

Habitat Ecology: Cultivated; Bhota, 765m.

Material Examined: HPU-EL-SKS-ABS-15180; 03.7.2008.

Parts Used: Whole Plant. Leaves.

Folk Uses: Leaves or its powder used as a flavouring agent* for preparing 'chutney' and curd. Dried plants kept in storage cabins for repelling insects. 15-20ml decoction of leaves considered good against stomach-ache and vomiting. Poultice of whole plant applied on forehead to cure headache*.

Active Constituents: These are methyl acetate 20%, menthol 9.1%, menthone and hydrocarbon 15%, eriocitrin, luteolin-7-O-rutinoside, hesperidin, isorhoifolin, diosmin, eriodictyol 7-O-glucoside and narirutin, rosmarinic acid, azulenes, cholene, carotenes (*leaves*), pulegone, menthofuran, 1,8-cineole, menthylacetate, isomenthone (*oil*).

Biological activity : Aq. Et OH extr. exhibits antiviral activity against (rinder pest virus) RPV in cattle.

Uses in Literature: Used earlier as antiemetic, antispasmodic, antiseptic, antiviral, carminative, cosmetic, diaphoretic, digestive, diuretic, mild sedative, stimulant; and for bile duct's ailments, bronchitis, cold, cough, diarrhoea, dysentery, dyspepsia, flatulence, gall bladder, rheumatic complaints and vomiting (Ambasta, 1986; Chopra *et al.*, 1956; Jain, 1991; Khare, 2007; Sharma, 2003; Uphof, 1968).

Mentha spicata L. (Pl. 65D).

Syn.:M. *spicata* var. *viridis* L.,M. *viridis* L.

Family: Lamiaceae.

Vern.: Pudina.

Ayurvedic, English and Regional Names:

Ayur.: Podinaka, Pudinaa, Puutihaa, Rochini;

Eng.: Brown Mint, Garden Mint, Lamb Mint, Makaerel Mint, Spearmint;

B.: Pahari Pudina;

Bo.: Pudina;

G.: Pudino;

H.: Paharipudina, Pudina;

Mar.: Pudino;

Tel.: Nagbb, Pudneh, Shahsufian.

Distribution: Throughout India; commonly cultivated in native gardens all over plains of India.

Description: Perennial herbs with pungent smell,

smooth, lanceolate, acute, coarsely dentate leaves and lilac flowers in loose cylindrical slender spikes. Fruit a carcerulus consisting of four nutlets. Seeds ex-albuminous.

Flowering & Fruiting: March-July.

Habitat Ecology: Cultivated; Sujanpur, 567m.

Material Examined: HPU-EL-SKS-ABS-15181; 03.07.2008.

Parts Used: Leaves. Stem.

Folk Uses: 15-20ml decoction of leaves prescribed twice daily as blood purifier*, stomachic* and for vomiting. One tsp of its powdered leaves along with 'ajwain' (*Trachyspermum ammi* L.) and 'harar' (*Terminalia chebula* Retz.) given twice a day for a week to cure gastric problems. Leaves also used as a flavouring agent for preparing 'chutney' and 'kahawa', refreshing drink 'jaljira' which is consumed especially during summers as coolant. Stem and leaf paste applied over body surface of animals to repel exoparasite, also leaves mixed with 'ajwain' (*Carum copticum* Benth.) given in case of dysentery and indigestion in animals.

Active Constituents: Constituents characterized are menthol, carvone (55-75%), limonene (21.4%), phellandrene, dihydrocarvone, diosmin and diosmetin, caffeic acid, rosemarinic acid, pinen, phellandrene, 1-limoene, octyl alcohol, dihydrocarveol, diosmetin, diosmin and volatile oil. (*leaves*).

Uses in Literature: So far known as antispasmodic, carminative, cosmetic, digestive, flavouring agent for making chutney, stimulant; and for bloat (vet.), bronchitis, diarrhoea, diarrhoea (vet.), exoparasite (vet.), indigestion, fever, hysteria, lice (vet.), placental removal (vet.), perfumery, vomiting and wounds (vet.) (Ambasta, 1986; Bhattarai, 1992; Borthakur & Sarma, 1996; Deeba *et al.*, 2009; Khare, 2007; Kirtikar & Basu, 1994; Sharma, 2003; Singh & Kumar, 2000b).

Mesua ferrea L. (Pl. 66A).

Syn.:M. *nagassarium* (Burm. f.) Kosterm.,M. *tedumculata* Wight, M. *roxburghii* Wight,M. *pulchella* Plank & Trian., M. *sclerophylla* Thwaites.,M. *Nagana* Gard.

Family: Clusiaceae.

Vern.: Nagkesar.

Ayurvedic, English and Regional Names:

Ayur.: Ahipushpa, Chaampeya, Naaga, Naagakeshara, Naagakinjalika, Naagapushpa;

Eng.: Iron-wood, Mesu;

Tam.: Sirunagappo, Nagakesaran.

Distribution: Eastern Himalayas; cultivated in other parts of India.

Description: A middle-sized glabrous tree having erect trunk, slender twigs, acute or rounded leaf base which are covered with fine waxy meal beneath. Flowers terminal and solitary or in pairs. Petals 4 spreading. Fruits ovoid, conical-pointed, large-sized.

Flowering & Fruiting: June-October.

Habitat Ecology: Cultivated; Neri, 381m.

Material Examined: HPU-EL-SKS-ABS-15182; 08.07.2008.

Parts Used: Root. Leaves. Flowers. Seed.

Folk Uses: Flowers consumed raw for quenching thirst* and to check perspiration, cough, cold indigestion and dysentery. Poultice of roots applied as antidote against snake bites*. Poultice of leaves applied in forehead against severe cold. Seed oil massaged for rheumatism*, sores*, wounds* and skin affections*.

Active Constituents: These are xanthones, euxanthone, nesuaxanthones A & B (*heartwood*), 4-phenylcoumarin, mesuol, manneigin, mesuagin, mammeisin, mesuone (*seed*) and alpha and beta-amyrin, beta-sitosterol, biflavonoids, mesuaferrones A & B and mesuanic acid (*stamens*).

Biological Activity: Heartwood antimicrobial.

Uses in Literature: Known in published literature as antianaphylaxis, antiasthmatic, antidysentric, antiinflammatory, antimicrobial, astringent, diuretic, haemorrhagic disorders, haemostatic, hypotensive, stomachic; and for bleeding piles, cough, gout, metrorrhagia, renal diseases, urinary bleeding (Ambasta, 1986; Khare, 2007; Kirtikar & Basu, 1994).

Millettia extensa Benth. *ex* Baker (Pl. 66B).

Syn.: *Milletia auriculata* Baker *ex* Brandis, *Otosema extensa* Benth.

Family: Fabaceae.

Vern.: Salangaya. Gauj.

English, Hindi and Sanskrit Names:

Eng.: Daisy, Showy millettia;

H.: Gajoru, Gaujo;

Sans.: Karanja.

Distribution: Foothills of Himalayas, eastwards to Assam,

southwards Central India to north Andhra Pradesh.

Description: Perennial woody climbers having stem and branches green, 9-11, long petioled leaflets which are distantly placed on the rachis. Flowers in very copious axillary racemes. Corolla densely silky, reddish lilac. Pod straight, very hard.

Flowering & Fruiting: October-December.

Habitat Ecology: Common in grasslands; Bhoranj, 950m.

Material Examined: HPU-EL-SKS-ABS-15183; 28.11.2008.

Parts Used: Leaves. Branches. Roots.

Folk Uses: Leaves used as hay* to cattle. Branches beaten for coarse fibre cordage*. Paste of crushed roots applied externally on head to kill lice*; also its paste applied over body of animals to remove ectoparasitic attachment.

Active Constituents: Leaves rich in proteins and calcium, roots; isoflabones, iso-auriculatin, auricularin and rotenoid, sumatrol.

Uses in Literature: Used earlier as pesticidal, vermifuge; and for sore, worm and wounds (vet.) (Agarwal, 2003; Ambasta, 1986; Anonymous, 1998b; Bhatt, 1990; Jain, 1991; Khare, 2007; Kirtikar & Basu, 1994; Manandhar, 1988; Parrotta, 2001; Purohit & Samant, 1995; Raju, 1998; Rama Rao & Henry, 1996; Rastogi & Mehrotra, 1993; Sikarwar, 1996; Varma *et al.*, 1999).

Mimosa pudica L. (Pl. 66C).

Family: Fabaceae.

Vern.: Chhuimui, Lajalu.

Ayurvedic, English, Sanskrit and Regional Names:

Ayur.: Khadirkaa, Laajavanti, Lajjaalu, Namaskaari, Raktapaadi, Shamipatraa;

Eng.: Touch-me-not;

Sans.: Lajjali;

H.: Lajwanti;

Kan.: Laji;

Mal.: Tintarmani;

Tam.: Thottal vadi;

Tel.: Attapati.

Distribution: Throughout India.

Description: Prickly, pubescent, erect or half climbing herbs with bipinnate leaves. Flower heads pink, minute, long-stalked, crowded in globose heads. Pods 3-4 seeded, bristly.

Flowering & Fruiting: July–September.

Habitat Ecology: As an escape; Salasi, 805m.

Material Examined: HPU-EL-SKS-ABS-15184; 20.09.2010.

Parts Used: Plant. Leaves. Seed.

Folk Uses: Planted as an ornamental* and to keep away evil spirits and bad magic*. Children love to play watching the folding of its pinnules. Leaf paste applied to heal fresh cuts and wounds. Seeds soaked and eaten for strength as tonic*. Leaves used as goat fodder*.

Active Constituents: β-amyrin, β-sitosterol, friedelin (*leaves*). 4-O-(beta-d-glucopyranosyl-6-sulphate), gallic acid, C-glycosylflavones, 2''-O-rhamnosylorientin and 2''-O-rhamnosylisoorientin (*plant*).

Uses in Literature: Reported earlier as an antiseptic, astringent, blood purifier, carminative, styptic; and for cuts, diabetes, diarrhoea, dysentery, fistula, glandular swelling, haemophilia, hydrocele, joint pains, leucorrhoea, morbid codition of vagina, piles, scorpion sting, urinary complaints and wounds (Chopra *et al.*, 1956; Chunekar & Pondel, 1999; Khare, 2007; Kirtikar & Basu, 1994; Retnam & Martin, 2006; Sharma, 2003).

Mirabilis jalapa L. (Pl. 66D).

Syn.: *Mirabilis multiflora* (Torr.) Gray

Family: Nyctaginaceae.

Vern.: Gulal.

Ayurvedic, English, Sanskrit and Regional Names:

Ayur : Trisandhi;

Eng.: Four O'Clock Flower, Marvel of Peru;

Sans.: Krishnakeli, Sandhyakali;

B.: Gulabas, Krishnakeli;

Bo.: Gubhaji;

H.: Gulabbas;

Mal.: Antimalari;

P.: Abasi, Gulabas;

Tam.: Antinarulu, Pattarachi;

Tel.: Batharachi, Chandrakanta;

Urdu: Guleabbas.

Distribution: Throughout India.

Description: Erect leafy herbs possessing cordate leaves, pink, long tubular flowers. Fruits elliptical, leathery black.

Flowering & Fruiting: June-September.

Habitat Ecology: Moist localities, wasteplaces, roadsides; Sujanpur, 567m.

Material Examined: HPU-EL-SKS-ABS-15185; 25.05.2011.

Parts Used: Leaves. Flowers. Root.

Folk Uses: Poultice of leaves applied on boils and abcess; its juice applied on skin as coolant*. Flowers used as offering* to appease deities. Root paste fed orally to cattle against sun stroke*.

Active Constituents: Chief constituents of *flowers* are miraxanthins I, II, III and IV, indicaxanthin and vulgaxanthin. triterpenes, alpha-amyrin and its acetate, Mj-AMP-1 and Mj-AMP-2.

Biological Activity: Spasmolytic activity +ve. Mirabilis antiviral protein from tuberous roots (MAP) shows antiproliferative effect on tumour cells.

Uses in Literature: Known in India as an abortifacient, antiproliferative, antitumour, aphrodisiac, ornamental, purgative, tonic, viruses-inhibitory, vegetative; and for boils, blisters, cold at child birth, earache, piles, scabies, scorpion bites and urticaria (Ambasta, 1986; Chopra *et al.*, 1956; Dastur, 1970; Jain, 1991; Khare, 2007; Kirtikar & Basu, 1994; Retnam & Martin, 2006; Sharma, 2003; Upadhye & Kulkarni, 2004).

Momordica charantia L. (Pl. 67A).

Syn.:M. *humilis* Wall.,M. *muricata* DC.,M. *senegalensis* Lamk., *Cucumis africanus* Bot.

Family: Cucurbitaceae.

Vern.: Karela.

Ayurvedic, English, Sanskrit and Regional Names:

Ayur.: Kaaravella, Kaaravellaka, Kaathilla, Sushaavi;

Eng.: Bitter Gourd;

Sans.: Brihadavalli, Karavellah, Karavellaka;

Ass.: Kakiral;

B.: Baramasiya, Potikakkar;

Bo.: Karla;

G.: Karela, Karola;

H.: Karela, Kareli, Karola;

Kan.: Hagal;

Mal.: Kaipavalli, Pavakacheti;

Mar.: Karale;

Oriya: Karena;

Ind. : Pavakkachedi;

Tel.: Mattakakata, Tellakakra.

Distribution: Throughout India.

Description: An annual climbing herb with simple tendrils, long, branched hairy stem, 5-7 lobed leaves which are orbicular, pubescent on both sides and cordate at the base. Flowers yellow, monoecious, solitary and peduncled. Fruits bright green coloured, long, pendulous, fusiform, pointed. Seeds long, compressed, sculptured with corrugate margins.

Flowering & Fruiting: June-October.

Habitat Ecology: Cultivated; Patta, 1,020m.

Material Examined: HPU-EL-SKS-ABS-15186; 22.09.2009.

Part Used: Fruits.

Folk Uses: Fruits cooked as vegetable. 50ml of its juice or 1 tsp of powdered fruit taken daily in morning for diabetes, skin affections*, blood purifier and as anthelmintic. Also, its fruit juice with that of *Azadirachta indica* A. Juss., and *Lycopersicum esculentum* Mill. given (one glass daily, empty stomach) for checking diabetes and digestive disorders*.

Active Constituents: *Fruits* contain acylglucosyl sterols, lipids, charantin, β-sitosterol, glucoside, ascorbigen and a new stigmosterol 5, 25-diene-3 β-O-glucoside. charantin (momordicin M = phytosterolin = β-sitosterol, moriodosides G, F1, F2, I, K and L.

Biological Activity: Positive hypoglycaemic and cytotoxic activity of fruits attributed to charantin-β-sitosterol—glucoside and stigmost 5-25-diene-3 β-O-glucoside.

Uses in Literature: Useful earlier as an abortifacient, anthelmintic, astringent, blood purifier, carminative, colic, emetic, febrifugal, laxative, stomachic, tonic, vegetable; and for arthritis, bilious affection, bleeding piles, blood cancer, burning of the sole of the feet, cooling, cutaneous disease, diabetes, disease of spleen and liver, dysentery, eczema, gout, gummosis, headache, hypertension, jaundice, leprosy, malaria, mouth ulcer, night blindness, odema, peptic ulcer (vet.), poisonous bite (vet.), poultry disease, rheumatic pains, scabies, snake bite, sore, toothache, ulcers, wound maggots, *Momordica* anti-HIV-protein, MAP-30 (seeds & fruits) (Ambasta, 1986; Bajpayee & Dixit, 1996; Bondya & Sharma, 2003; Ghosh *et al.*, 1996; Gill *et al.*, 1993; Jain, 1991; Jain *et al.*, 1994; Kothari & Moorthy, 1996; Kshirsagar *et al.*, 2003; Kulkarni & Kumbhojkar, 1993, 1996; Kumar & Chauhan, 2006; Lalramnghinglova, 1996; Mahato & Mahato, 1996; Negi & Gaur, 1991; Pandey & Verma, 2002; Rama Rao & Henry, 1996; Retnam & Martin, 2006; Sharma, 2003; Sen & Batra, 1997; Sen & Behera, 2003; Sen *et al.*, 2002; Shome *et al.*, 1996; Singh & Pandey, 1996; Watt, 1972).

Major Ayurvedic Formulations: Mahavisagarbha-taila, Karneme.

Trade: Karela.

Moringa oleifera Lam. (Pl. 67B).

Syn.:M. *pterygosperma* Gaertn.,M. *zeylanica* Pers.,M. *polygana* DC., *Hyperanthera moringa* Vahl., *H. Decandra* Willd., *Guilandina moringa* L.

Family: Moringaceae.

Vern.: Suhanjan.

Ayurvedic, English, Sanskrit and Regional Names:

Ayur.: Haritashaaka, Madhushigru, Murandi, Raktaka, Shigru, Shobhaanjana, Sigra;

Eng.: Drum stick Tree, Indian Horse Raddish;

Sans.: Damasamalu, Tikshnandhaka;

B.: Sajina;

Bo.: Mangai, Sanga;

G.: Midhosaragaro, Seyla;

H.: Mungna, Saonjna;

Kan.: Nugge;

Mal.: Muringa, Muringa Elai, Murinna, Tishnagandha;

Mar.: Murinna, Tishagandha;

Oriya: Achajhada, Managachajhada;

P.: Saonjna;

Tel.: Mochakamu, Sajana;

Tam.: Karunjanam, Tavuselam.

Distribution: Widely cultivated or wild, throughout foothills of W. Himalayas: upto 1,000m.

Description: A small tree having usually 3 pinnate leaves, opposite leaflets, corky bark, soft wood and pungent root. Young parts tomentose. Flowers pale white, honey scented. Pods pedunculous. Seeds winged at the angles.

Flowering & Fruiting: February-April.

Habitat Ecology: Hilly slopes, uncommon in forests; Neri, 381m.

Material Examined: HPU-EL-SKS-ABS-15187; 13.04.2008.

Parts Used: Leaves. Flowers. Inflorescence. Pods. Bark. Seed. Gum.

Folk Uses: Young leaves and fruits consumed as vegetable*. 2-3g powdered inflorescence in combination with 25g black pepper and 5g asafoetida given once a day for 2 weeks to check gastric ailments*. 5-10ml decoction of bark treated good for upper respiratory tract infection*, skin infection* and tonsillitis (twice a day for 5-10 days). 1 tablet of dried ginger powder mixed with equal amount of salt, 'hing' (*Ferula assafoetida* L.) and juice of its bark, recommended daily for 21 days for gout, arthritis and joint pains. 1 tsp of powdered gum along with dry fruits given to women with cow's milk for overcoming post labour stress*. Sweet gum given to children for checking dysentery. 3-5g of powdered seeds consumed with milk during night as contraceptive.

Active Constituents: Active properties due to aldotriuronic acid, arabinose, galactose, glumronic acid, rhamnose, mannose, xylose (*gum*), aspartic acid, glutamic acid, glycine, threonine, alanine, valine, leucine, histidine, lysine, tryptophan, cysteine, methimine (*leaves*), threonine, aspartic acid and glutamic (*flowers, fruits*), 4-hydroxymellein, vanillin, octacosanoic acid, b-sitosterol and b-sitosteron (*stem*), benzyl isothiocyanate (*seeds*), and four auxin like factors (fast growing *fruits*).

Biological Activity: 4 (a-L-rhamnosyloxy) benzyl isothiocyanate exhibits antibacterial activity against *Mycobacterium phlei* (40 mM/1) and *Bacillus subtilis* (56mM/1).

Uses in Literature: Recorded so far as an abortifacient, anthelmintic, antilithic, antipyretic, antirheumatic, antiscorbutic, aphrodisiac, appetiser, cardiac stimulant, carminative, cholagogue, cooking oil, diuretic, edible (pods), emetic, flatulent colic, illuminant, lubricants, psychoactive, rubefacient, stimulant, stomachic, symbolic (in folk proverbs), tonic, vesicant; and for acute traumic pain, anthrax (vet.), articular pains, asthma, baldness, blood pressure, boils (vet.), calculus affections, calico printing, catarrhal affections, cholera, circulating stimulants, cold, cosmetics, cough, cuts, debility of nerves, delivery, dental caries, dermatitis, dispersing swellings, dog bite, dropsy, dyeing, dysentery, dyspepsia (vet.), ear complaints, eczema, enlarged spleen or liver, epilepsy, expelling placenta, eye (vet.), eye aliments, fits (vet.), foot sores, glandular swellings, gout, gum boils, headache, hoarseness, hydrocele, hysteria, increasing flow of bile, injuries, intermittent fevers, internal and deep seated inflammation, leprosy, madness, mats, menstrual complaints, otalgia, paper or cordage, paralysis, patches, perfumeries, pneumonia, postnatal, pustules, reducing labour pains in pregnant women, rheumatism (vet.), scabies, scrofula, scurvy, small pox, soreness of mouth and throat, swelling (vet.), syphilis, tetanus, toothache, typhoid, ulcer (vet.), urinary complaints, venomous bites and wounds (Ambasta, 1986; Banerjee, 1996; Chauhan, 1999; Geetha *et al.*, 1996b; Hembrom, 1996; Henry *et al.*, 1996; Hosagoudar & Henry, 1996b; Jain, 1991; Jain *et al.*, 1994; Jha *et al.*, 1996; Khare, 2007; Kulkarni & Kumbhojkar, 1993; Molla

& Roy, 1996; Mukherjee & Namhata, 1990; Negi & Gaur, 1991; Prasad *et al.*, 1996; Rama Rao & Henry, 1996; Saini, 1996a; Shome *et al.*, 1996; Singh, 1996; Siwakoti & Varma, 1996; Srivastava *et al.*, 2003; Vishwanathan, 1997; Vishwanathan & Singh, 1996; Watt, 1972).

Morus alba L. (Pl. 67C).

Syn.:M. *serrata* Wall.,M. *tatarica* L.

Family: Moraceae.

Vern.: Toot.

English, Sanskrit and Regional Names:

Eng.: Mulberg or White Mulbery;

Sans.: Tula;

B.: Tut;

Bo.: Chinni, Satur, Tut;

G.: Shetur;

H.: Chun, Chinni, Tut;

Kan.: Hipnerle, Uppunute;

Kash. : Tut;

Mar.: Ambat,Tut;

Oriya: Tuticoli, Tuto;

P.: Tut,Tutri;

Tam.: Kambali, Pattuppuchi;

Tel.: Pippalipanduchettu, Reshmechattu;

U.**P.:** Shahtoot.

Distribution: Wild or cultivated extensively in drier areas of W. Himalayas upto 3,300m.

Description: A lofty, medium-sized tree having brown rough bark, ovate, toothed cordate leaves and greenish flowers. Male spikes longer than female ones. Peduncles softly tomentose. Fruits ovoid, mucilaginous, black when ripe.

Flowering & Fruiting: March-May.

Habitat Ecology: Common along agricultural fields; Jhaniyari, 900m.

Material Examined: HPU-EL-SKS-ABS-15188; 23.04.2008.

Parts Used: Leaves. Branches. Fruit. Wood.

Folk Uses: Fruits edible and considered refrigerant. Leaves used for rearing silkworm and as fodder for cattle. Branches used for making baskets*, hockey*stides and walking sticks*. Bark used for cordaging* pile of grasses. Wood used for making bed bases* ('pavas'.)

Active Constituents: Main ingredients are N-butanol, β-ã-hexenol, methyl-ethylacetaldehyde, hexaldehyde, isobutyl aldehyde, methyl-ethyl-ketone, methyl-hexyl-ketone, butylamine and acetic, propionic isobutyric acid, astragalin (*leaves*), heptanoic (0.047), caprylic (0.21), pelargonic (0.047), capric (0.07), myristic (0.105), palmitic (11.85), palmitolic (0.188), methylheptadeconic (0.14), stearic (4.46), oleic (12.67), linoleic (68.3), nonodecanoic (0.641), linilenic (0.98%) acids (*fruit*), two phytoalexins-morasin C and moracin D (*shoot*), mulberrin, mulberrochromene and cyclomulberrin (*bark, stem*).

Biological Activity: Mulberrifuran A exhibits antibacterial activity against *Staphylococcus aureus* and *Fusarium roseum* whereas Moran A exhibits hypoglycaemic effect in normal and alloxan induced hyperglycaemic mice.

Uses in Literature: Known in published literature as anthelmintic, diaphoretic, edible, fodder, haemorrhoidal, host for silk worms, purgative, refrigerant, vermifugal; and for agricultural implements, boats, building, diarrhoea, dysentery, dyspepsia, feeding silk worms, fever, furniture, goitre, inducing oestrus cycle, melancholia, paper, sore throat (Ambasta, 1986; Bhardwaj & Gakhar, 2003; Jain, 1991; Khare, 2007; Kshirsagar & Singh, 2000; Maliya, 2004; Negi & Gaur, 1991; Rajwar, 1983; Retnam & Martin, 2006; Saini, 1996a; Shah, 1984; Sharma, 2003; Sharma *et al.*, 2001; Vishwanathan *et al.*, 1981; Watt, 1972).

Trade: Mulberry.

Murraya koenigii Spreng. (Pl. 67D).

Syn.: *Bergera koenigii* L.

Family: Rutaceae.

Vern.: Gandhla, Gandhela.

Ayurvedic, English, Sanskrit and Regional Names:

Ayur.: Surabhini-nimba;

Eng.: Curry Leaf Tree;

Sans : Alakavhaya, Chhandighna, Kadarya, Kalasaka, Krishtna-nimba, Nimba-pattram, Raman, Surabhi, Surabhi-nimba;

Ass.: Bishahari, Narasingha;

B.: Barsungha, Kariaphyll;

Bo.: Goranimb, Karripat, Karrinim;

G.: Kadhi, Limbdo;

H.: Barasunga, Gandhela, Harri, Karipatha, Katnim;

Kan.: Kari-bevu;

Mal.: Karea-pela, Kariveppilei;

Mar.: Jhirang, Karapacha, Kudianim;

Oriya: Barsan, Basan-go, Bhursunga;

P.: Bowala, Gandanim, Gandalu;

Tam.: Karuvembu, Karuvepallai;

Tel.: Karepeka, Kari-vepa, Karivepaku.

Distribution: Throughout India.

Description: Strong smelling umbrageous shrubs having imparipinnate leaflets, white flowers in many-flowered peduncled corymbs. Berries ovoid, apiculate, black rough with glands, two seeded and with whitish pulp.

Flowering & Fruiting: March-June.

Habitat Ecology: Open slopes, wastelands, roadsides; Mattani, 920m.

Material Examined: HPU-EL-SKS-ABS-15189; 27.03.2008.

Parts Used: Leaves. Stem. Fruit.

Folk Uses: Fruits edible. Hot water fomentation of decoction of its leaves with *Cuscuta reflexa* Roxb. and *Dodonaea viscosa* L. Jacq. used for healing internal injuries* and muscular pains. Leaves used for flavouring curries and 'chutney'. Stem considered good for cleaning teeth. ½ tsp infusion of leaves with 'gur' and onion (*Allium cepa* L.) taken with hot water, twice a day, for stomachache and as diuretic*. Also, 2-3g of powdered leaves taken with milk for piles* and diabetes. About 400-500g mixture of its crushed leaves and seeds of 'tikku malu' (*Abrus precatorius* L.) given twice daily for 'patrara' (acute tympany)*, also about 250ml extr. of leaves considered good against dysentery and as blood purifier* in animals.

Active Constituents: Properties mainly due to mahanimbidine, mahanimbine, iso-mahanimbine, mahanimbineine, girinimbine, muconicine, cyclomahanimbine, scopoline, carotene, vitamins (*leaves*), curryangine, curryanine, mukanole, murraycine, murrayacinine, isomurrayazoline, murrayazoline, murrayazolinol, 3 (1,2-dimethylallyl) xanthyletin (stem bark); mahanimbine, koenimbine, koenigicine (*fruit, leaves*) and mahanimbinol (*wood*).

Biological Activity: Root antiameobic, hypotensive whereas aerial parts antiameobic and spasmolytic. However, koenoline exhibits cytotoxic activity against ICB cell culture. Aerial parts-antiprotozoal and spasmolytic.

Uses in Literature: Useful in India as anthelmintic, carminative, edible (vegetable), febrifugal, hair tonic, laxative, stimulant for eruptions, stomachic; and for acute abdominal pains, agricultural implements, bleeding, bone fracture, burning pains, common cold, cough, culinary purposes, dandruff, diabetes, diarrhoea, dysentery, fertility (vet.), fever, flavouring curries, grooms as ornamental, jaundice, joint pains, parasites, renal pains, rinderpest (vet.), snake bite, swelling in various tract of body, tonic, tuberculosis, vomiting and worms (Aminuddin & Girach, 1996; Borthakur, 1996a; Chauhan, 1999; Chinthu *et al.*, 1998; Gill *et al.*, 1997; Gogoi & Borthakur, 2001; Gogoi *et al.*, 2003; Jain, 1991; Khare, 2007; Kurian, 1995; Lal *et al.*, 1996; Maliya & Singh, 2003; Mohanty & Padhy, 1996; Mohanty *et al.*, 1998; Parmar & Kaushal, 1982; Rama Rao & Henry, 1996; Singh, 1996; Siwakoti & Verma, 1996; Sudarsanam *et al.*, 1995; Verma & Chauhan, 2006; Watt, 1972).

Murraya paniculata (L.) Jack. (Pl. 68A).

Syn.:M. *exotica* L.

Family: Rutaceae.

Vern.: Gandhla.

English, Sanskrit and Regional Names:

Eng.: Orange Jesmine;

Sans.: Durabhigraha, Galrabhange;

B.: Kamini;

H.: Kamini, marchula;

Kan.: Angarakana gida, Pandry;

Mar.: Pandari, Kunti, Marchulajuti;

Oriya: Banmallika, harkankali;

Tam.: Konji;

Tel.: Nagagolunga, Karepaku.

Distribution: Throughout India.

Description: Evergreen shrubs with elliptic-lanceolate, shining leaves and white, fragrant flowers. Corymbs terminal and axillary, few or many-flowered. Fruits globose or ovoid berries, red when ripe. Seed with woolly or glabrous testa.

Flowering & Fruiting: June-December.

Habitat Ecology: Open slopes, wastelands, roadsides; Nadaun, 483m.

Material Examined: HPU-EL-SKS-ABS-15190; 25.12.2009.

Parts Used: Stem. Fruit.

Folk Uses: Stem used for scouring teeth*. Fruits edible*.

Active Constituents: Indole, murrangatin, phebalosin, coumarins, characterized from *leaves*.

Uses in Literature: Used so far for agricultural tools, cough, cuts, cosmetics, diarrhoea, dysentery, hysteria, handles, renal pains, rheumatism; and as carminative (Ambasta, 1986; Khare, 2007; Kirtikar & Basu, 1994).

Musa sapientum L. (Pl. 68B).

Syn.: *Musa paradisiaca* L.

Family: Musaceae.

Vern.: Kela.

Ayurvedic, English, Sanskrit and Regional Names:

Ayur.: Ambusaara, Anshumatiphal, Kadali, Mochaa, Rambhaa, Sakrtphala, Vaaranaa;

Eng.: Adam's Fig, Banana, Fig of India, Plantain;

Sans.: Ambusara, Balakapriya, Bhamphala, Mocha, Mous, Nisara, Rambha, Shajratulmouz Sukumara, Tulhtula;

B.: Kela, Khela, Mus;

G.: Kel, Maoz, Mouz;

H.: Amrit, Kach-kula, Kala;

Kan.: Pagnan, Pizang, Vasha, Vazhap, Vellacoi;

Mal.: Ngetpyau, Yakhaing, Ya-thi-lan;

Mar.: Kela;

P.: Kewiro;

Tam.: Anati, Amti, Ariti, Kommuariti, Kadali;

Tel.: Bale, Balenaru.

Distribution: Cultivated throughout India and tropics.

Description: Herbs perennating by underground rhizomes. Stem tall, stoloniferous. Leaf stalk stout. Spike drooping, bract ovate, many fid, deciduous. Fruits oblong, trigonous, sweet, tapering at base and apex, yellowish when ripe.

Flowering & Fruiting: Throughout year.

Habitat Ecology: Planted; Galore (Nalti), 787m.

Material Examined: HPU-EL-SKS-ABS-15190; 15.08.2010.

Parts Used: Leaves. Stem. Fruit.

Folk Uses: Plant considered sacred and a must for all religious and ceremonial occasions like marriage, house warming*, birth and death rituals*. Ripe fruits edible and also offered to appease Lord Satyanaryana; its paste along with sugar and 'ellachi' (fruits of *Elettaria cardamomum* Maton) prescribed to check acidity, nervous complaints* and dysentery. Unripe fruits cooked as vegetable*. Three number of its inflorescence if given by totally naked person believed to check frequent pregnancy loss in animals.

Active Constituents: Constituents isolated are 9-19-cyclotriterpene ketone, a teriterpenoid-24 (R)-4α, 14 α, 24 trimethyl 5 α cholesta-8, 25 (27)-dien3β-ol, dopamine, dopa, noradrenalin, serotonin and caffein, cinnamic, p-coumaric, ferullic, gallic and protocatechuic acid, compesterol, β-sitosterol, stigasterol, cyclomusalenol, cyclomusalenone, steroids, glycosides, terpenes and organic acid (*fruit*).

Uses in Literature: Reported in literature for asthma, black spots, bleeding disorders, boils, burn injuries, cholera, cholera (vet.), cold, cough, diabetes, diarrhoea, diarrhoea (vet.), dysuria, gastric disorder, gastropathies (vet.), haemorrhage (vet.), headache, (vet.) heart pains, heat stroke (vet.), hypertension, infertility (vet.), intestinal disorders, irregular menstrual cycle, jams and jellies, kidney and bladder stone, lactation (vet.), leucorrhoea, polyuria, pooja prasad, pyorrhoea, rheumatism, snake bites, ulcer (vet.), ulcerative colitis, vaginal discharges; and as anthelmintic antibacterial, antifungal, antilithic, edible and laxative (Ambasta, 1986; Anonymous, 1998b; Bhatt *et al.*, 2001; Baruah & Sarma, 1987; Dixit, 1987; Kemparaja *et al.*, 1998; Khanna, 2002; Khare, 2007; Mishra *et al.*, 1996b; Mohanty *et al.*, 1998; Mukherjee *et al.*, 1986; Pandey & Verma, 2002; Ranjan & Sethuraman, 1997; Rastogi & Mehrotra 1993, 1995a,b; Samati, 2004; Sen & Behera, 2003; Sen *et al.*, 2002; Sharma *et al.*, 2002; Sudarsanam *et al.*, 1995; Tiwari & Pande, 2006; Varshney, 1998; Watt, 1972).

Nasturtium officinale R. Br. (Pl. 68C).

Syn.: *N. fontanum* Aschers., *Sisymbrium nasturtium-aquaticum* L., *Rorippa nasturtium-aquaticum* (L.) Hayek

Family: Brassicaceae.

Vern.: Chuch.

English and Regional Names:

Eng.: Water Cress;

P.: Piriya halim.

Distribution: Throughout India, Asia, temperate Europe.

Description: Aquatic herbs having creeping and floating stem, pinnate leaves, sinnuate-lobed leaflets and small white flowers in short racemes. Petals longer than sepals. Lower leaves cut into 3 repand segments. Pods stalked, spreading or bent upwards. Seeds small, 2-seriate.

Flowering & Fruiting: April-September.

Habitat Ecology: Bogs, irrigated channels, streamsides, damp places; Dimmi, 1,006m.

Material Examined: HPU-EL-SKS-ABS-15191; 22.12.2009.

Parts Used: Leaves. Stem. Whole Plant.

Folk Uses: 1 tsp decoction of the plant given once every morning for 3-5 days to expel worms* from the body. 100-200g of its powdered stem and leaves considered good against diabetes in animals. Leaves used as pot herbs; also its decoction (20-30ml, twice daily) good for common cold.

Active Constituents: *Plant* contains vit. A 4720 IU, ascorbic acid 77mg/100g, thiamine, riboflavin, niacin, glucosinolate phenethyl isothiocyanate and biotin, mineral matter, calcium, phosphorous, iron, sulphur, iodine, manganese, zinc, arsenic, copper, protein and amino acids.

Biological Activity: Anticancerous activity confirmed +ive.

Uses in Literature: Recorded in literature as antiscorbutic, appetiser, blood purifier, diuretic, febrifugal, pot herb; and for asthma, blemishes, blotches, catarrh of respiratory organs, chest problems, cold, diabetes (vet.), dry throat, goitre, kidney and bladder detoxification, polypus of nose, spots, stranguary (Anonymous, 1966; Bhargava, 1959; Katewa *et al.*, 2001; Khare, 2007; Uphof, 1968; Watt, 1972).

Nerium indicum Mill. (Pl. 68D).

Syn.: *N. odorum* Soland., *N. odoratum* Lamk., *N. latifolium* Mill.

Family: Apocynaceae.

Vern.: Kaner.

Ayurvedic, English, Sanskrit and Regional Names:

Ayur.: Ashvamaaraka, Divyapushpa, Gauripushpa, Hayamaaraka, Karavira, Shatakumbha, Siddhapushpa, Viraka;

Eng.: India Oleander, Sweet Scented Oleander;

Sans.: Asvamaraka, Pratihasa, Vishavrykshanka;

Ass.: Karavi, Karbira, Karvir;

B.: Karabi;

Bo.: Ganira, Kanhera;

G.: Kanera;

H.: Kaner, Karber, Kuruvira;

Kan.: Kanagalu;

Kash. : Gandeela;

M.**P.:** Arali, Kanave eram;

Mal.: Chuvannarali;

Mar.: Kaneri;

Oriya: Konera, Koniyoro;

P.: Ganhira, Kanhira;

Tam.: Asuvabari;

Tel.: Ettaganneru.

Distribution: N.W. Himalayas: 600-1,500m, C. & N. India.

Description: Erect glabrous shrubs having stout branches, long, pointed, leathery leaves with numerous nerves. Flowers pinkish white, fragrant. Petiole very short. Fruit paired, cylindric.

Flowering & Fruiting: April-November.

Habitat Ecology: Roadsides, wasteplaces; Rangas, 765m.

Material Examined: HPU-EL-SKS-ABS-15192; 22.05.2011.

Parts Used: Leaves. Flowers. Stem.

Folk Uses: Flowers offered* to appease deities, and also for making bouquet*. Its poultice applied on skin as antifungal substance. A pinch of dried leaves snuffed to cure headache and bad cold*. Plant sticks used for making thatched roofs*. Leaves also kept in storage cabinets for repelling worms.

Active Constituents: Leaves possess quercitin, isoquercitin, nerimol, nerifol, oleandrin, ursolic acid, adynerin, a compound similar to rutin, B-D-digitaloside and digitoxigenin.

Biological Activity: Alc. extr. of root, bark, leaf and flowers exhibits antibacterial activity. Oleandrin and oleandrigenin-B-D-diginpyranoside showed cardiopkinetic and diuretic activities.

Uses in Literature: Known in published literature for blisters, eczema, haemorrhoids ulceration, headache, leprosy, ringworms, skin diseases, snake bite; and as antifungal, antiinfLammatory, attenuant, cardiac tonic, diuretic, insecticidal, resolvent, toxic (Katewa *et al.*, 2001; Khare, 2007; Kumar & Chauhan, 2005; Verma & Chauhan, 2006).

Major Ayurvedic Formulations: Karaviradhya-taila, Kasi-sadi-taila, Laghu-visgarbha-taila.

Nicotiana tabacum L. (Pl. 69A).

Family: Solanaceae.

Vern.: Tambaku, Tobacco.

Ayurvedic, English, Sanskrit and Regional Names:

Ayur.: Dhuumrapatraa, Taamraparna;

Eng.: Tobacco;

Sans.: Dhumratatrika, Kalanja, Krimighni, Ksharatatra, Tamakhu, Vajrabhringi;

B.: Tamak;

Bo.: Tambakhu;

G.: Tamak;

H.: Tambaku, Tamaku;

Kan.: Hogesoppu;

Mal.: Pokala;

Tam.: Poghako, Poghu, Pugai-ilai;

Tel.: Pogaku, Dhumrapatramu.

Distribution: Common throughout India, cultivated or as an escape.

Description: Erect viscidly pubescent herbs with decurrent lower leaves and funnel-shaped corolla. Flowers pinkish, pedicelled, bracteate in panicled racemes. Fruit a capsule.

Flowering & Fruiting: April-August.

Habitat Ecology: Edges of cultivated areas, wastelands, dry slopes; Khaya, 787m.

Material Examined: HPU-EL-SKS-ABS-15193; 28.03.2008.

Parts Used: Leaves. Fruits. Seed.

Folk Uses: Poultice of leaves applied on acne*, pimples*, other skin affections* and as muscle relaxant. Its dry leaves mixed with jaggery prescribed for pains, insomnia* and expelling worms* (1/2-1 tsp, twice daily, 3-5 days). Mixture of powdered leaves, fruits and seeds along with a pinch of rock salt and mustard oil massaged on gums for checking toothache. An amulet of its roasted seeds tied to the body for cure of pain. Tobacco chewed, snuffed and smoked (hubble-bubble-'hookah') in various parts of the district.

Active Constituents: Plant posseses nicotine, nicoteine, nicotimine, nabasine (*plant*), oxalic acid (*bark*), tahacinin, tahacilin and 0.25-1.7% isoquercitrin (*leaves*).

Uses in Literature: Described earlier as an antispasmodic, haemostatic, sedative, snuff, vermifugal; and for arthralgia, caterpillar sting, chewing, eczema, eradication of lice and ticks (vet.), food poisoning, gout, itching, local affections, lumbago, muscle relaxation, rheumatism, spongy gums, snake bite, toothache, wounds (Ambasta, 1986; Chopra *et al*., 1956; Jain, 1991; Khare, 2007).

Ocimum basilicum L. (Pl. 69B).

Syn.: *O. pilosum* Willd., *O. album* L., *O. minimum* Burm., *O. hispidum* Lamk., *O. menthaefolium* Benth., *O. caryophyllatum* Roxb., *Plectranthus barrelieri* Spreng.

Family: Lamiaceae.

Vern.: Bhabhri.

Ayurvedic, English, Sanskrit and Regional Names:

Ayur.: Ajgandhikaa, Baabui, Barbari, Kharpushpa, Tulasi, Tungi, Tuvari;

Eng.: Basil, Common Basil, Sweet Basil;

Sans.: Ajagendhika, arjakan, Asuvasa, Barbara, Karahi, Surabhi, Tungi, Varvara;

B.: Babui-tulsi, Debunsha;

G.: Damaro, Nasabo;

H.: Bahari, Kalitulsi;

Kan.: Kam-Kasturi, Sajjebiya;

Mal.: Marva, Sabja;

Oriya: Dhala, Zulasi;

P.: Baburi, Rehan;

Tam.: Tirnupatchie, Tirunitru;

Tel.: Rudrajada,Vipudi-patri.

Distribution: Throughout India.

Description: Erect, softly hairy, strongly-scented annual undershrubs. Leaves elliptic-ovate or lanceolate, entire to serrulate. Flowers small, purplish-white, subsessile in spikes. Nutlets ellipsoid, pitted.

Flowering & Fruiting: June-September.

Habitat Ecology: Commonly cultivated, rarely as an escape in moist slopes; Hamirpur, 787m.

Material Examined: HPU-EL-SKS-ABS-15194; 13.03.2008.

Parts Used: Whole Plant. Leaves. Seeds.

Folk Uses: 2-3g powdered plant given twice a day for 15 days for leucorrhoea*. Green leaves used for culinary purposes and also added to tea to check headache*. Also, 20 ml of its decoction taken twice a day against gastric problems and vomiting*. Decoction of its leaves along with 'basuti' (*Adhatoda vasica* Nees) given to animals for cough. 20-30 ml decoction of seeds prescribed as coolant*.

Active Constituents: Essential oil from *leaves* contains methyl cinnamate, 1-linalool, terpinene, macinnamate, methyl chavicol, linalool, cineole, ocimene, borneol, sambulene and safrole.

Biological Activity: Alc. extr. of leaves and flowers exhibited activity against *Micrococcus pyrogenes* var. *aureus* whereas essential oil antibacterial and antifungal.

Uses in Literature: Useful so far as an anthelmintic, antipyretic, carminative, demulcent, diaphoretic, expectorant, nasal douche, stimulant and stomachic; and for alc. intoxication, boil (vet.), bowl complaints, cholera, cold, cough, cramp, cosmetics, culinary purposes, diarrhoea, dropsy, dysentery, dyspepsia (vet.), earache, epilepsy, fever, flavouring, headache, labour complaints, liver (vet.), mouth ulcer, pus in ears, ring worms, sinuses, sores, sore throat, snake bite, haematuria, congestion, kidney troubles, toothache, tympany (vet.) and wounds (Ambasta, 1986; Borthakur & Sarma, 1996; Jadhav, 2006; Jain, 1991; Khare, 2007; Pal, 1980; Pandey & Verma, 2002; Reddy *et al.*, 1998; Silori & Rana, 2000).

Adulterant or Substitute: Seeds used as an adulterant to *Ficus hederacea* L. and *Ipomoea nil* (L.) Roth.

Ocimum sanctum L. (Pl. 69C).

Syn.: *O. monachorium* L. Mant., *O. tenuiflorum* L., *O. inodorum* Burm.

Family: Lamiaceae.

Vern.: Baramasi Tulsi, Ram Tulsi.

Ayurvedic, English, Sanskrit and Regional Names:

Ayur.: Apet-raakshasi, Bahumanjari, Bhuutaghni, Devadundubhi, Graamya, Manjarikaa, Tulasi, Surasaa, Surasa, Suravalli, Sulabhaa, Shuulaghni, Sulabhaa.

Eng.: Holy Basil, Monk's Basil, Rough Basil, Sacred basil;

Sans.: Ajaka, Arjaka, Amrita, Bharati, Gauri, Laxmi, Madhavi, Punya, Surabhi, Tulasi;

B.: Bantulsi, Tulsi;

Bo.: Tulasa, Tulas-icha-jhada;

G.: Tulsi;

H.: Hural, Kal-tulsi, Tulai, Tulshi;

Kan.: Krishna-tulsi, Nallu-tirtta, Niella-tirtua;

Mal.: Lum;

Mar.: Talasi;

P.: Tulas, Tulasa;

Tam.: Gaggera-chettu, Krishna-tulasi, Tulasi;

Tel.: Tulashi-gida.

Distribution: Throughout India.

Description: Erect, branched, aromatic, hairy perennial herbs. Leaves elliptic-oblanceolate, entire, pubescent on both sides. Flowers greenish, pale yellow and in terminal racemes. Nutlets broad-ellipsoid, smooth, yellow with small black marking.

Flowering & Fruiting: July-November.

Habitat Ecology: Cultivated; Sujanpur, 567m.

Material Examined: HPU-EL-SKS-ABS-15195; 22.11.2010.

Parts Used: Leaves. Inflorescence. Whole Plant.

Folk Uses: Plant considered sacred* and worshipped in 'panch-bhikham'*.

Newly stitched clothes are kept overnight covering the plant before being given to children for wearing as this practice is considered antiseptic. Decoction of inflorescence good for cough. Leaves chewed to arrest foul smell of mouth. Poulitice of leaves applied for suppuration of boils. Mixture of 3g infusion of leaves along with 5g honey and 3g ginger (*Zingiber officinale* Rosc.) prescribed twice a day for asthma, cough and fever. Leaf juice used as nasal drops for checking cold and as expectorant. Leaves form one of the constituents of holy water* which is distributed as 'prasadam' and believed to possess curative* properties. Decoction of its leaves and that of 'basuti' (*Adhatoda vasica* Nees.) and 'ajwain' (*Carum copticum* Benth.) given to check cough* in animals.

Active Constituents: Essential oil contains carracrol, caryophyllene, nerol and camphene, camphor, b-caryophyllene, caryophyllene, camphene, eugenol, methyleclavicol, a pinene, volatile oil and ascorbic acid.

Biological Activity: Hypoglycaemic and spasmolytic activities confirmed.

Uses in Literature: Used earlier as antibacterial, antiseptic, contraceptive, expectorant, insecticidal, stomachic; and for asthma, black patches, boils, bronchitis, burning of feet, conjuctivitis, common cold, cough, cuts, dandruff, earache, eczema, enhancing effect of wine, fertility, fever, flavouring, foul breath, gastric trouble, headache, hepatitis, jaundice, night blindness, perfumery, rat bite, respiratory troubles, ringworm, scabies, skin diseases, spermatorrhoea, sterilization, stomachache, tuberculosis, ulcers, urinary troubles (Ambasta, 1986; Borthakur & Sarma, 1996; Borthakur *et al.*, 2004; Gogoi & Borthakur, 2001; Jadhav, 2006; Khare, 2007; Kshirsagar & Singh, 2000; Kumar & Chauhan, 2006; Maya *et al.*, 2003; Pal, 1980; Reddy *et al.*, 1998; Sen & Batra, 1997; Sen & Behera, 2003; Sen *et al.*, 2002; Sharma & Singh, 2001; Sharma *et al.*, 2002; Varshney, 1998).

Major Ayurvedic Formulations: Tulsiyadi-Kashaya, Honey Tus.

Opuntia dillenii (Ker-Gawl.) Haw. (Pl. 69D).

Syn.: *O. stricta* (Haw.) Haw. var. *dillenii*, *Cactus indicus* Roxb.

Family: Cactaceae.

Vern.: Chu, Thor.

Ayurvedic, English, Sanskrit and Regional Names:

Ayur.: Kanthaari, Naagaphani;

Eng.: Pricklypear, Slipper Thorn;

Sans.: Bhaudagdhika, Kandashaka, Nagaphana, Vidara;

B.: Nagphana, Phenimana, Nug-phana;

G.: Chorhathola, Zhorhatheylo;

H.: Hathathoria, Nagphani, Nag-phana;

Kan.: Papaskalli;

Mal.: Nagmullu, Palakakkalli, Poturiyabiju;

Mar.: Chapal, Nagphana, Samar;

Oriya: Nagophenia, Poturiyabiju;

P.: Chittarthohar;

Tam.: Manjarnagadali, Pattamadugalli;

Tel.: Nagajemudu, Naga-dali, Nagamullu.

Distribution: Planted and naturalised.

Description: Evergreen succulent shrubs with densely elongate-bristly, jointed stem. Areoles raised, glochidiate. Spines 5-7 per areole, yellow. Leaves deciduous, represented by tuft of spines or prickles. Flowers bright yellow, solitary, sessile. Berries pyriform, purple when ripe. Seeds many.

Flowering & Fruiting: April-May.

Habitat Ecology: Wasteplaces, dry slopes, edges of fields; Tikker Khatrian, 908m.

Material Examined: HPU-EL-SKS-ABS-15196; 15.07.2009.

Parts Used: Plant. Fruits.

Folk Uses: Planted outside to keep away black magic* and evil spirits*, also considered as a good hedge plant. 1-2g powdered burnt plant taken with honey for whooping cough. Fruits edible and considered good cardiotonic*, Fruits yield a dye* for dying clothes.

Active Constituents: Glucosides of isorhamnetin, quercetin, isoquercetin, traces of free flavanols, arabinogalacton and betanin have been isolated (*plant*).

Biological Activity: Ether extr. of stem exhibits some antibiotic properties whereas fruit exhibits CNS activity.

Uses in Literature: Recorded in literature as an antipyretic, carminative, diuretic, edible (fruits), expectorant, laxative, refrigerant, stomachic, symbolic in religious; and for asthma, boils, excessive menstruation, eye trouble, fencing, hedge, guinea-worm, leucoderma, ceremonies of 'Santhals' (spines), expelling worms, piles, pimples, swelling, urinary complaints, and whooping cough (Ambasta, 1986; Chopra *et al.*, 1956; Jain, 1991; Jain & Tarafder, 1970, Khare, 2007; Kurian, 1995; Nautiyal, 1981; Parmar & Kaushal, 1982; Singh, 1996, Watt, 1972).

Oroxylum indicum Vent. (Pl. 70A).

Syn.: *Calosanthes indica* Blume, *Bignonia indica* L., *B. pentandra* Lour., *Spathodea indica* Pers.

Family: Bignoniaceae.

Vern.: Tat-palanga.

Ayurvedic, English, Sanskrit and Regional Names:

Ayur.: Bhalluka, Kutannata, Madhukparna, Nata, Patrorna, Prthushimba, Shoshana, Shyonaaka, Tuntuka;

Eng.: Indian Trumpet Flower;

Sans.: Adhvantashatrava, Svarnavalkana, Syonakah;

Ass.: Bhatgila, Dingari, Kering, Toguna;

B.: Nasona;

Bo.: Phalphura, Tetu;

G.: Podval;

H.: Assatsauna, Sauna, Sonapatha;

Kan.: Bunepale, Sonepatta, Tigdu;

Mal.: Palakapayani;

Mar.: Taitu;

Oriya: Phompania;

Tam.: Paluiyudaichi, Vangam;

Tel.: Dundilamu, Mundakaparnamu.

Distribution: Throughout India ascending upto 1,350m.

Description: A small tree having very large, tripinnate, opposite leaves, thick bark, numerous, fetid, purple flowers in large terminal raceme. Calyx and corolla fleshy. Capsule straight, flat, acute, valves semi-woody. Seeds numerous, winged.

Flowering & Fruiting: June-August.

Habitat Ecology: Forest, shrubberies; Tal, 947m.

Material Examined: HPU-EL-SKS-ABS-15197; 05.06.2011.

Parts Used: Leaves. Flowers. Fruits. Bark.

Folk Uses: Poultice of bark applied locally for healing fracture; powdered bark mixed with wheat dough fed to cattle for healing internal injuries. Also, 15ml of bark decoction along with that of fruit, twice a day, considered good against constipation both in human beings and livestock. Leaves and pods given as fodder*. Flowers adorn on the cap*.

Active Constituents: Therapeutic properties attributed to baicalein, its 6-glucuronide and 7-glucuronide, prunetin, scuttellarein, oroxylin A, chrysin and scuttellarein-7-rutinoside (*leaves, stem bark*); glucuronide-oroxidin, synthesis of biacatein-6-glucuronide and 7 glucuronide (*seeds*).

Biological Activity: Spasmolytic (*fruit*) and diuretic (*bark*) activities +ve.

Uses in Literature: Known earlier as anodyne, antidote to snake bite, antiseptic, appetizer, astringent, black dye, carminative, hepatic stimulant, ingredient of 'chyawanprash', lactation, psychoactive, purgative (seed), stomachic, tonic, vermic, vegetable (corolla young shoots, unripe fruits); and for affection of brain, blood dysentery, body pain, bone fracture (vet.), cracks in nipples, dysentery, dysentery (vet.), diseases of chest, indigestion, joint pains, headache, liver ailments, oedema, pain, piles, pimples, rutting (vet.), scabies, scorpion sting, small pox, sore back of herds and bullocks, sores of nails and soles (vet.), tanning, tuberculosis, ulcers, urticaria, vitality and wounds (vet.) (Ambasta, 1986; Borthakur, 1996b; Borthakur & Sarma, 1996; Brahmam & Saxena, 1990; Brahmam *et al.*, 1996; Chhetri, 2005; Chopra *et al.*, 1956; Chowdhery, 1996; Girach, 1992; Girach & Aminuddin, 1995; Jain, 1991; Jain *et al.*, 1994; Khare, 2007; Kothari & Moorthy, 1996; Mehrotra *et al.*, 1995; Mohanty *et al.*, 1998; Satapathy & Brahmam, 1996; Sen & Behera, 2003; Siwakoti & Verma, 1996; Srivastava *et al.*, 2003; Upadhye *et al.*, 1994; Vishwanathan & Singh, 1996; Watt, 1972).

Major Ayurvedic Formulations: Amritarishta, Brhatpanca-mulayadi-kvatha, Chywanprasha, Dashmool-ghrita, Das-hnmul-taila, Mahanarayan-taila, Pushayanug churn, Shyo-naaka-putpale.

Oryza sativa L. (Pl. 70B).

Syn.: *Oryza fatua* Koen. *ex* Trin., *O. palustris* Salisb., *O. plena* (Prain) Chowdhury., *O. repens* Herb. Ham. *ex* Wall., *O. sarghoides* Steud., *O. segetalis* Russ. *ex* Wall.

Family: Poaceae.

Vern.: Dhan.

Ayurvedic, English, Sanskrit and Regional Names:

Ayur.: Shaali, Tandula,Vrihidhaanya;

Eng.: Paddy Rice;

Sans.: Dhanya, Nivara, Syali, Vrithi;

B.: Chal;

Bo.: Bhatta, Dangar;

G.: Choka, Dangar;

H.: Chawal, Dhan;

Kan.: Bhatta, Nellu;

Mal.: Ari, Nellu;

Mar.: Bhat, Bhan, Tandula;

Oriya: Dhan, Rabana;

Tam.: Arisi, Nellu;

Tel.: Vadlu, Ullu.

Distribution: Throughout warmer regions of India.

Description: Leafy tall grasses with creeping or floating stem and flat striated leaves. Spikelets not imbricating, panicle at length drooping. Awn 7-15cm long, yellow or reddish. Grains narrow, free or adnate to the glume and palea.

Flowering & Fruiting: September-December.

Habitat Ecology: Cultivated; Bharari, 920m.

Material Examined: HPU-EL-SKS-ABS-15198; 15.11.2010.

Parts Used: Whole Plant. Leaves. Grains. Straw.

Folk Uses: Grains cooked as staple food. Rice water given to infants as nutritious diet (rice gruel), coolant and for urinary troubles. 1 glass decoction of the whole plant recommended daily for kidney troubles*. Rice flour made into 'chappatis' ('ankalee') and 'idli*'. Powdered roasted grains made into ball by mixing water and given to cure dysentery; also, hot viscous extr. obtained by boiling grains in water applied thrice daily for 10-15 days to check tail gangrene* in animals. Straw and whole plant used as a good fodder.

Active Constituents: These are monoglycosides of cyaniding and delthinidin, diglycosidic anthocyanin, protein, starch, glucotricin (grains).

Uses in Literature: Reported in literature as anthelmintic, anti-inflammatory, diuretic, edible, fodder, gynaecological disorders, refrigerant; and for biliousness, bone fracture, discharges, dysentery, dysuria, epistaxis, haematemesis, lactic disorders, mouth diseases, nausea, prolapsed uterus, throat swelling, wounds and wounds (vet.) (Agarwal, 2003; Ambasta, 1986; Duthie, 1988; Kadel & Jain, 2006; Khare, 2007; Mishra *et al.*, 1996b; Pal & Jain, 1998; Ramadas *et al.*, 2000; Rama Rao & Henry, 1996; Rastogi & Mehrotra, 1990; Watt, 1972).

Oxalis corniculata L. (Pl. 70C).

Syn.: *O. repens* Thunb., *O. pusilla* Salisb., *O. villosa* Bieb., *O. foliosa* Blatt.

Family: Oxalidaceae.

Vern.: Tinpatiyaa, Malori.

Ayurvedic, English, Sanskrit and Regional Names:

Ayur.: Amlapatrikaa, Amlikaa, Chaangeri, Chukraa, Chukrikaa;

Eng.: Indian Sorrel, Yellow Oxalis;

Sans.: Ambashta, Amlika, Amlotaja, Changeri, Shuklika;

Ass.: Chengeritenga;

B.: Amrul, Chalmori;

Bo.: Ambuti, Buti-sarpati;

H.: Amrul, Anboti, Seh;

Mal.: Toliyarala;

Mar.: Ambutibhinsarpali;

P.: Amlika, Amrul;

Tam.: Paliakiri;

Tel.: Ambotikura.

Distribution: Warmer regions, throughout India.

Description: Hairy annual weeds, abundant in cultivated places. Leaves 3-foliate, long petioled. Flowers subumbllate. Capsule yellow, 5 angled, beaked. Seeds several in each cell, ribbed.

Flowering & Fruiting: April-November.

Habitat Ecology: Common in roadsides, wastelands; Bhota, 765m.

Material Examined: HPU-EL-SKS-ABS-15199; 28.06.2008.

Parts Used: Whole Plant. Leaves. Root.

Folk Uses: Powdered plant given to check vomiting and nausea; (½ teaspoon twice a day). Roots eaten like radish. Leaves refreshing and eaten raw as a source of vit C. Decoction of dry or fresh leaves added to 'banefsha' (*Viola odorata* L.), 'tulsi' (*Ocimum sanctum* L.), 'malathi' (*Glycyrrhiza glabra* L.), and given for cough, cold, fever and stomach troubles. Poultice of leaves applied on forehead* to check pains.

Active Constituents: Therapeutic properties attributed to flavonoids, vitexin, isovitexin, vitexin-2'-O-beta-D-glucopyranoside. 1.47% lipids, alpha and beta tocopherol, vitamin C, calcium, oxalates, tartaric, citric and malic acids (*leaves*).

Uses in Literature: Described earlier as an antidote to 'datura' poisoning, antipyretic, antiscorbutic, appetizing, astringent, edible (leaves and fruits), refrigerant, stomachic; and for anaemia, corn, cough, cuts, diarrhoea, dysentery, dyspepsia, epilepsy, eye complaints, fever, indigestion, insect bite, jaundice, piles, rickets, skin diseases, snake bite, swelling, tympanitis and wart (Ambasta, 1986; Drury, 1978; Jain, 1991; Khare, 2007 Watt, 1972).

Papaver rhoeas L. (Pl. 70D).

Syn.: *Papaver atropurpureum* Gilib., *Papaver intermedium* Becker

Family: Papaveraceae.

Vern.: Gulla.

Ayurvedic, English, Sanskrit and Regional Names:

Ayur.: Rakta Posta;

Eng.: Common poppy, Common Red poppy, Corn poppy, Flanders poppy, Red poppy, Shirley poppy;

Sans.: Rhakta-posta-Vrikshah;

H.: Lala, Lal-Post;

Tam.: Sivappu, Kasakasa.

Distribution: Throughout India.

Description: Erect, branched, annual herbs with alternate, petiolate leaves which are hispid on both surfaces. Flowers red, pairs of petals unequal. Capsule stalked, subglobose glabrous.

Flowering & Fruiting: May-October.

Habitat Ecology: Cultivated, roadsides, wastegrounds; Bhira, 780m.

Material Examined: HPU-EL-SKS-ABS-15200; 15.10.2009.

Parts Used: Plant. Flower. Leaves.

Folk Uses: Planted as an ornamental*. Flowers offered to appease deities. 2-3g powdered petals cosumed with honey, twice a day till cure to check cough. Poultice of petals and leaves applied externally on wasp stings*. Latex of capsules considered narcotic*.

Active Constituents: Constituents characterized are α-tocopherol, mecocyanin chloride, cyanidin derivative (*petals*), cyanidin-B and pelargonidin-C (*flowers*); rhoeadine (*leaves, flowers, capsules*); morphine, thebaine; narcotine, meconic acid (*capsules*); protopine and coptisine, phenolic and nonphenolic, crystalline and amorphous bases (*roots, aerial parts*).

Biological Activity: Antioxidant activity and antiradical power have been found to be +ve.

Uses in Literature: Recorded so far as an antibiotic, expectorant, sedative; and for colouring medicines, cough or hoarseness and food materials (Chopra *et al.*,

1956; Chunekar & Pondel, 1999; Khare, 2007; Kirtikar & Basu, 1994; Watt, 1972).

Trade: Anti-Wrinkle Cream.

Parthenium hysterophorus L. (Pl. 71A).

Family: Asteraceae.

Vern.: Chikadu, Congress grass.

English Name:

Eng.: Congress grass.

Distribution: Common weed throughout India.

Description: Profusely branched erect herbs having dissected leaves, angular hairy stem and numerous white heads peduncled in corymb-like cymes. Achenes compressed.

Flowering & Fruiting: August-September.

Habitat Ecology: Wastelands; Chauri, 858m.

Material Examined: HPU-EL-SKS-ABS-15201; 12.07.2009.

Parts Used: Plant. Root.

Folk Uses: Plant before flowering used as fodder. Also, it is used for making manure by uprooting and dumping in a pit in fields; and as fuel. Poultice of root prepared in ghee recommended for slipped naval and neck*. 15-20ml decoction of roots consumed in the morning everyday as tonic.

Active Constituents: These are bornmylacetate, coronopilin, dihydroisoparthenin, hysterin, parthenin, α methylene γ lactone sesquiterpene, hexacosanol, myricyl alcohol, galactose, glucose, 6 hydroxy-kaempferol-3-7-dimethyl ether, tetraneurins A, B, C, D, kaempferol, betulin, ursolic acid and a saponin composed of oleanolic acid and glucose, campesterol, β-sitosterol and stigmasterol (*plant*).

Uses in Literature: So far known as an analgesic, anti-amoebic, antimalarial, emmenagogue, febrifugal, analgesic, emmenagogue, tonic; and for dysentery and nasal blockage (Ambasta, 1986; Bennet, 1985; Chopra *et al.*, 1956; Khare, 2007; Kirtikar & Basu, 1994).

Paspalum scrobiculatum L. (Pl. 71B).

Syn.: *Paspalum commersonni* Lam., *P. scrobiculatum* var. *frumentaceum* Stapf., *P. alternans* Steud., *P. auricularum* Presl., *P. dimidiatum* L., *P. kora* Willd.

Family: Poaceae.

Vern.: Kodri grass.

Ayurvedic, English, Sanskrit and Regional Names:

Ayur.: Kodrava, Korduusha;

Eng.: Kodo millet;

San. : Kodrava;

B.: Kodoa Dhan;

Bo.: Harik, Kodri, Pakodi, Pakod;

G.: Kodro, Menya;

H.: Kodo, kodra;

Mal.: Varagu;

Mar.: Harik, Kodra;

Oriya: Kodus;

Tam.: Varagu.

Distribution: Throughout India; chiefly in Assam & Andamans.

Description: Tufted perennial grasses, upto 1m high with stout rhizome and hairless leaf blades which are angled along the midrib. Infloresence spike-like racemes. Spikelets fringed with hair. Grains free within the hardened glume and pale.

Flowering & Fruiting: July-September.

Habitat Ecology: Cultivated in grasslands, fieldsides; Lambloo, 850m.

Material Examined: HPU-EL-SKS-ABS-15202; 15.06.2009.

Parts Used: Leaves. Whole Plant.

Folk Uses: Plant considered good for phychosis. Leaves used as fodder* and considered good for buffaloes.

Active Constituents: Main ingredients are hentriacontanol, hentriacontanone, sitosterol, arginine, histine, lycine, valine and methionine (*plant*).

Biological Activity: Antioxidant activity found to be +ve.

Uses in Literature: Used so far for psychoses, poisoning in men and animals, schizophrenia (Agarwal, 2003; Ambasta, 1986; Chopra *et al.*, 1956; Chunekar & Pondel, 1999; Duthie, 1988; Islam, 2002; Khare, 2007; Kirtikar & Basu, 1994; Prajapati *et al.*, 2003; Watt, 1972).

Pedilanthes tithymaloides (L.) Poir. (Pl. 71C).

Family: Euphorbiaceae.

Vern.: Nag Bel.

English and Regional Names:

Eng.: Devil backbone, Red bird, Ribbon cactus, Slipper Plant;

B.: Bilati;

Bo.: Vilayti-sher;

H.: Nagdhamni;

Mar.: Vilaayati-sher;

M.P.: Nagdaman, Nag-phani.

Distribution: Afghanistan –Kashmir, N. & W. Asia.

Description: Branched, erect, succulent bush with zig-zag stems. Leaves large, ovate, fleshy, sessile, arranged oppositely on the stem. Flowers red, in dichotomous cyme; bracts bright red. Schizocarpic regma splitting into three cocci with endospermic seeds.

Flowering & Fruiting: August-September.

Habitat Ecology: Grown in Indian gardens; Anu, 900m.

Material Examined: HPU-EL-SKS-ABS-15203; 12.04.2011.

Parts Used: Whole Plant. Latex.

Folk Uses: Planted as an ornamental* with the belief to keep away snakes from the house. Paste of the plant applied to cure snake bites*. Latex used in skin ellergies.

Active Constituents: Azafrin (*root*), octacosanol, cycloartenone, oxime and beta-sitosterol (dried *plant*), enzymes pedilanthain (*latex*), n-hentriacontanol, dehydrodammaronol-A. and diterpene esters (*leaves*) have been isolated.

Uses in Literature: Known earlier as anthelmintic, anti-inflammatory, emetic; and for boils, inflammation, keratoconjunctivitis, latex, leucodermal patches, nausea, pimples, warts, skin irritation, venereal diseases, vomiting (Ambasta, 1986; Katewa & Sharma, 1998; Khare, 2007; Nellis, 1997a, b; Spoerke & Smolinske, 1990).

Pelargonium graveolens L' Hert. (Pl. 71D).

Family: Geraniaceae.

Vern.: Geranium.

English and Hindi Names:

Eng.: Geranium;

H.: Geranium.

Distribution: Hills of Punjab, temperate and subtropical Himalayas upto 2,000m.

Description: A small straggling fragrant and bushy herb upto 100cm. Leaves deeply 5-7 lobed, toothed. Flowers small, rose or pink and purple veined in dense umbels. Fruit capsular, 3-5 lobed, lobes 1-seeded with pendulous or horizontal seeds.

Flowering & Fruiting: March-July.

Habitat Ecology: Grown in Indian gardens; Anu, 944m, Khaggal, 920m.

Material Examined: HPU-EL-SKS-ABS-15204; 03.07.2008.

Parts Used: Whole Plant. Leaves.

Folk Uses: Planted as an ornamental*. Leaves used for flavouring tobacco, desserts, jellies, vinegars, tea* and salads*. Oil good against skin affections.

Active Constituents: Properties attributed to geranium oil, menthone, citronellol, nerol and geraniol, isomenthone, linalool, citronellyl formate.

Uses in Literature: Reported in literature as antifungal, antiinflammatory, antiseptic, antitumour, aromatherapy, astringent, relaxant, sedative; and for acne, bleeding, cancer, bruises, cervical cancer, cosmetics, haemorrhoids, lice, oil, perfumery, poor circulation, eczema, flavouring tobacco, post menstrual and premenstrual problems, massage, nausea, ringworms, skin care, soaps, tonsilitis (Ambasta, 1986; Kirtikar & Basu, 1994; Khare, 2007).

Pennisetum americanum L. (Pl. 72A).

Syn: *P. typhoides* (Burn. F.) Stapf. & C.E. Hubb., *P. typhoideum* Rich. *P. megastachyrum* Steud., *P. giganteum* Ten. *ex* Steud., *P. linnaei* Kunth, *P. alopecuroides* Spreng.

Family: Poaceae.

Vern.: Bajra.

English and Regional Names:

Eng.: Bulrush Millet, Pearl Millet, Spiked millet;

B.: Bajra, Lahra;

G.: Bajri;

H.: Bajra;

Kan.: Sajje;

Mar.: Bajri;

Tam.: Kambu;

Tel.: Ganti, Sajja.

Distribution: Throughout the hotter parts of India.

Description: Annual or perennial erect grasses up to 2m with simple or branched stem and slender or stout root. Leaves lanceolate, spreading or drooping, flat more or less hairy. Spike upto 30cm long, rachis hairy or woolly. Grain oblong, obovoid or pyriform.

Flowering & Fruiting: August-October.

Habitat Ecology: Grown as an escape crop or cultivated; Tauni Devi, 1,050m.

Material Examined: HPU-EL-SKS-ABS-15205; 12.10.2008.

Parts Used: Whole Plant. Grains.

Folk Uses: Plant mainly used as fodder. Powdered grain flour used in making 'chapattis'*, 'sattu' (porridge) and cattle feed*.

Active Constituents: Grains contain proteins 12-15%, starch and gluten.

Uses in Literature: Recorded as coolant, edible, fodder laxative, porridge (Ambasta, 1986; Katewa & Sharma, 1998; Khare, 2007).

Pennisetum purpureum Schum. (Pl. 72B).

Family: Poaceae.

Vern.: Chari.

English Names:

Eng.: Napier grass, Elephant grass.

Distribution: Throughout India.

Description: A robust, deep rooted perennial grass with creeping rhizome.

Leaves dull green. Spikes dense, erect, long and yellow. Grains orbicular or pyriform, smooth, free.

Flowering & Fruiting: September-November.

Habitat Ecology: Cultivated; Dera Parol, 910m.

Material Examined: HPU-EL-SKS-ABS-15206; 18.10.2009.

Part Used: Whole Plant.

Folk Uses: Plant considered good for soil conservation* and as fodder for livestock. Also, used as fuel.

Uses in Literature: Used so far as biofuel, fence, fodder, raw material for paper; and for decontaminator of land, food for elephants, phytoremediation, soil conservation, trapping pests of Maize plant, walls of huts (Duthie, 1888; Ricarda, 2006).

Phoenix sylvestris Roxb. (Pl. 72C).

Syn.: *P. acaulis* (L.) Buch. Ham. *ex* Roxb., *Elate sylvestris* L., *Katou indel* Ham.

Family: Arecaceae.

Vern.: Khajoor.

Ayurvedic, English, Sanskrit and Regional Names:

Ayur.: Kharjuuri;

Eng.: Date Sugar Palm, Dwarf Date Palm, Indian Wine Palm, Sugar Palm, Wild Date Palm;

Sans.: Bhumikhajurika, Haluka, Kharju, Kharjura, Kakakarkati, Skandhappala;

B.: Kajar, Kejur, Khjura;

Bo.: Kajar, Khajuri, Khjura, Sendi;

G.: Khjuri, Kharak;

H.: Jangli Khajur, Kejur, Khaji, Salma, Thakil;

Kan.: Ichalu;

Mal.: Inta, Intappana, Kattinta;

Mar.: Boichand, Sendri, Sindikujuri;

Oriya: Khorjura, Khorjuri, Khorjuro;

P.: Khaji, Khajur;

Tam.: Ichambanai, Periyayinju, Madal;

Tel.: Ita, Peddayita.

Distribution: Throughout India.

Description: Tall dioecious palm trees with stout trunk covered with persistent leaf bases and long, greyish green, glabrous, pinnate leaves upto 4-5m in length. Flower cluster inclined with spreading branches. Spadix compressed, oblong-ellipsoid, golden-orange coloured. Seeds deeply grooved along its whole length, pale brown.

Flowering & Fruiting: May-June.

Habitat Ecology: Open hill slopes, rocky slopes; Bhaleth, 506m.

Material Examined: HPU-EL-SKS-ABS-15207; 14.04.2010.

Parts Used: Leaves. Stem. Branches. Fruits.

Folk Uses: Fresh sap of the stem utilized for making local wine 'sur'. Ripe fruits edible. Leaves employed for making brooms, mats, baskets and thatched roofs*. Fresh leaves used as fodder and branches as fuel.

Active Constituents: Active properties due to protein 0.37%, total sugar 11.01%, reducing sugar 0.97%, mineral matter 0.54%, phosphorus 0.10%, calcium trace, iron 0.04%, nicotonic acid, iso nicotonic and hydrazide, organic acid (fresh *unfermented sap*), amino acids and vitamins.

Biological Activity: Leaves CVS and CNS active.

Uses in Literature: Recorded in Indian literature as cardiotonic, constipative, cooling beverage, diuretic, edible (buds and fruits), intoxicating alcoholic drinks, nervine tonic, psychoactive plant, restorative; and for anaemia, asthma, bakery, brooms, basketry, bronchitis, cold, diabetes, diarrhoea, dysentery, fever, gum trouble, gonorrhoea, heart complaints, loss of consciousness, leprosy, mats and bags, relief in pain during pregnancy, rope, thatching, toothache, venereal diseases and vomiting (Ambasta, 1986; Goud & Pullaiah, 1996a; Jain, 1991; Jain *et al.*, 1994; Katewa *et al.*, 2001; Khare, 2007; Kirtikar & Basu, 1994; Kshirsagar *et al.*, 2003; Mohanty *et al.*, 1998; Patole & Jain, 2002; Ragupathy & Mahadevan, 1991; Rama Rao & Henry, 1996; Ravishankar & Henry, 1992; Sahu, 1996; Singh, 1996; Srivastava *et al.*, 2003; Vishwanathan, 1997; Watt, 1972).

Phragmites karka (Retz.) Trin. *ex* Steud. (Pl. 72D).

Syn.: *Arundo corea* Herb. Rottle., *A. donax* Herb. Ham. *ex* Wall., *A. karka* Retz., *A. roxburghii* Kunth Revis., *A. roxburghiana* Kunth *ex* Steud., *A. tibialis* Roxb., *Oxyanthe japonica* Steud., *Phragmites bifaria* Wight Herb., *P. nepalensis* Nees *ex* Steud., *P. roxburghii* Steud., *Sericura japonica* Steud., *Trichoon karka* Roth

Family: Poaceae.

Vern.: Reed grass.

Ayurvedic, English and Regional Names:

Ayur.: Nala;

Eng.: Common Reed-Grass;

H.: Narakul;

Tam.: Perunanal.

Distribution: Throughout India from Punjab to Myanmar; ascending the Himalayas in Hot Valley upto 1,000m.

Description: Tall perennial grasses with long leaves, very tall stout stems and purple-brown, narrow, slightly flattened spikelets and erect panicals. Branches widely spreading. Glume III rarely more than 4cm. Grains oblong, terete, free within the persistent glumes.

Flowering & Fruiting: July -September.

Habitat Ecology: Moist stony slopes; Khaya, 787m.

Material Examined: HPU-EL-SKS-ABS-15208; 15.05.2010.

Parts Used: Culms. Leaves. Whole Plant.

Folk Uses: Whole plant used as fodder*. Culms used for making paper pulp*. Leaf fibres good for making fish traps, brooms, mats, baskets* etc.

Active Constituents: These are carbohydrates, nitrogenous substances, fats, crude fibres, sucrose, astragalin, ellagic, tannin, quercitin, reducing sugars (*plant*), asparagines, ascorbic acid, polyols betaines, free poline, under ground parts, 13% furfural (*ash*).

Uses in Literature: Described in published literature as an antibacterial, antiemetic, antipyretic, diaphoretic, diuretic, emmenagogue, hypoglycaemic; and for fever, influenza, relief from insect bite (Ambasta, 1986; Chopra *et al.*, 1956; Das *et al.*, 1996; Duthie, 1888, Islam, 2002. Jain, 1991; Khare, 2007; Watt, 1972).

Phyllanthus urinaria L. (Pl. 73A).

Syn.: *P. leprocarpus* Wight, *P. alatus* Blume, *P. cantoniensis* Hornem., *P. mucronatus* Herb., *P. racemosus* Herb., *P. muricatus* Herb., *P. polyphyllus* Herb., *P. echinatus* Herb.

Family: Euphorbiaceae.

Vern.: Bhumi-Amla.

Ayurvedic, Sanskrit and Regional Names:

Ayur.: Bhuumyaamataki;

Sans.: Adhyanda, Ajata, Buhuphala, Jhatamala;

B.: Hazarmani;

G.: Kharsadabonyaanmali;

H.: Hazarmani, Lalbhuinanvalah;

Mal.: Chirukizhukanelli, Chukannakizhanelli;

Mar.: Lalmundajanvalli;

Tam.: Shivappunelli;

Tel.: Ettausirika.

Distribution: Throughout India.

Description: Herbaceous annuals. Leaves oblong, membranous. Flowers yellow, numerous, males solitary, sessile, female with obovate-oblong sepals. Fruits echinate. Capsule depressed-globose, smooth, thinly crustaceous. Seeds with parallel slender ribs and faint cross striae.

Flowering & Fruiting: August-September.

Habitat Ecology: Moist stony slopes; Khaya, 787m.

Material Examined: HPU-EL-SKS-ABS-15209; 05.06.2010.

Parts Used: Whole Plant. Seeds.

Folk Uses: Whole plant used as fodder*. ½ tsp of powdered seeds given once daily for 7 days to cure stomach disorders.

Active Constituents: *Plant* contains astragalin, ellagic, tannin, quercitin, quercitrin, isoquercitrin, rutin, lignan, phyllanthin, hypophyllanthin, niranthin, nirtetralin, and kaempferol.

Uses in Literature: Known so far as an antiinflammatory, appetiser, astringent, diuretic, febrifugal styptic and tonic; and for diarrhoea, dysentery, dyspepsia, gonorrhoea, jaundice, skin eruptions, sleeplessness in children, stomach trouble and urinogenital diseases in children (Ambasta, 1986; Chopra *et al.*, 1956; Jain, 1991; Khare, 2007; Maya *et al.*, 2003; Sharma, 2003, Uphof, 1968; Watt, 1972).

Major Ayurvedic Formulations: Chyawanprash, Tonali tablets, Livomap tablets and syrup.

Trade: Plant traded as Bhumi amla.

Physalis peruviana L. (Pl. 73B).

Syn.: *P. pubescens* Don, *P. edulis* Sims

Family: Solanaceae.

Vern.: Rasbhari.

Ayurvedic, English and Regional Names:

Ayur.: Parpoti;

Eng.: Cape Gooseberry;

Tam.: Perungunni, Pottipallam.

Distribution: Throughout India, cultivated.

Description: Herbaceous annuals. Leaves ovate, angular or scarcely lobed. Flower and fruit calyx larger, usually villous or pubescent. Berries green, globose, loosely enclosed in much enlarged, inflated, 5-angled persistent calyx.

Flowering & Fruiting: April-September.

Habitat Ecology: Cultivated; Jhaniari, 900m.

Material Examined: HPU-EL-SKS-ABS-15210; 23.03.2008.

Parts Used: Leaves. Fruits.

Folk Uses: Fruits edible. Fruits given in anaemia. Infusion of leaves (2-4g) prescribed in stomach troubles*.

Active Constituents: *Plant* contains carotenes, (As Vit. A) thiamine, riboflavin, nicotinic acid, ascorbic acid, mineral matter, phytin phosphorus, Iron, ionizable iron, sodium, potassium, copper, sulphur, pectin, citric acid, mallic acid and tartaric acids, oxygenated ergostane-type of steroids, withanolides and related compounds.

Uses in Literature: So far recorded as an anthelmintic, appetiser, astringent, diuretic, edible as table fruit; and for abdominal disorders, anaemia, vitality (Ambasta, 1986; Chopra *et al.*, 1956; Jain, 1991; Khare, 2007; Maya *et al.*, 2003; Sharma, 2003, Uphof, 1968; Watt, 1972).

Pinus roxburghii Sarg. (Pl. 73C).

Syn.: *P. longifolia* Roxb. *ex* Lamb.

Family: Pinaceae.

Vern.: Chil.

Ayuvedic, English, Sanskrit and Regional Names:

Ayur.: Dhuupavriksha, Ghand Birojaa, Namasu, Oleo resin, Pita-Vriksha, Sarala, Shriveshtaka, Surabhidaaruka;

Eng.: Himalayasn Long Leaved Pine, Three Leaved Pine;

Sans.: Bhadradaru, Drava, Kshire, Puti Kashtha, Srivasa;

Ass.: Telia;

B.: Sarala Kashtha;

G.: Sarala Deodara;

H.: Chir, Dhupasarala, Salla, Saral;

Kash. : Salla;

Mal.: Charalam, Putikashtam;

Mar.: Saraladedara;

Oriya: Sorolokatho;

P.: Anandher, Chir, Drabchi, Nashtar, Saral;

Tam.: Simaidevadari, Suruldevadari;

Tel.: Devadaru, Saralu.

Distribution: Afghanistan, Bhutan, W. & E. Himalayas: 500-2,000m.

Description: A large gregarious tree with whorled branches, naked stem and long needle-like leaves. Bark brown, corky furrowed. Male cones clustered, female cones terminal. Seeds winged. Reproductive cycle : Cone formation from February-May.

Habitat Ecology: Forming extensive forest, roadsides; Hiranagar, 900m.

Material Examined: HPU-EL-SKS-ABS-15211; 03.02.2011.

Parts Used: Needles. Resins. Wood. Female cone.

Folk Uses: Needles used for making brooms*, bed for animals*, and filling matrices. Needles mixed with red soil employed for plastering* walls of mud houses and also mixed with dung for its use as fuel*. Rural populace consume its seed as low quality chilgoza*. Wood employed for making cheap furniture* and building works. Resin collected from stem used for making turpentine, paints* and varnish*. Female cones are burnt for its use as torch light. Fresh air of pine trees considered good for recovery of tuberculotic* patients.

Active Constituents: Various constituents recognized are stilbene, pinosylvin, pinociembrin and pinobanksin (*heartwood*), vit. C, E, carotene α, β pinene (*leaves*), Ä3 carone (*turpentine oil*), pine oil (*needle*), β pinene and Ä 3 carone (*plants*).

Biological Activity: Oil from needles antibacterial against a number of organisms including *Staphyllococcus aureus*.

Uses in Literature: Accounted so far for boils, bone fracture, chest, chronic bronchitis, constipation, construction purposes, cracks in soles of feet, muscle stiffness, respiratory track infection, rheumatism, skin diseases, snake bite, sprain, swelling, tanning, tea, torches, turpentine, typhoid, ulcer and urinary complaints; and also as antipruritic, antiseptic, colic, decongestant, edible, expectorant, flatulent, stimulant (Ambasta, 1986; Chauhan, 1999; Jain, 1968, 1991; Jain *et al.*, 1994; Khare, 2007; Sensarma, 1991; Srivastava *et al.*, 2003; Watt, 1972).

Major Ayurvedic Formulations: Chandanaadi-Vati, Rosin, Turpentine.

Trade: Birozaa, Chil, Chir.

Piper longum L. (Pl. 73D).

Syn.: *P. latifolium* Hunter., *P. sarmentosum* Wall., *Chavica Roxburghi* Miq., *C. sarmentosa* Miq.

Family: Piperaceae.

Vern.: Mugh.

Ayuvedic, English and Regional Names:

Ayur.: Maagadha, Maagadhaa, Maagadhi, Maagadhikaa, Pippali;

Eng.: Indian Long Pepper, Joborandi;

Tam.: Thippili, Arisi thippili, Thippiliver.

Distribution: Throughout hotter provinces of India; cultivated as spices.

Description: Glabrous shrubs having soft angular branches which are grooved when dry. Lower leaves long petioled, ovate-cordate whereas upper ones narrower, sessile, amplexicaul. Fruiting spikes short, sub-erect. Fruits greyish green or darker grey berries.

Flowering & Fruiting: July-January.

Habitat Ecology: Cultivated; Bijhari, 823m.

Material Examined: HPU-EL-SKS-ABS-15212; 03.07.2008.

Parts Used: Spikes. Roots. Berries.

Folk Uses: Berries edible* and consumed as mouth freshner* and for checking ulcers*. Powdered fruits given with honey for appetite, indigestion, cough, asthma, bronchitis, tonic and fever. Decoction of dried flowers and roots recommended for abdominal pains* (15-20ml, twice a day for 3-5 days).

Active Constituents: Chief constituents are aristolactams, dioxoaporphines, isobutylamide, longamide, guineensine lignans, pluviatilol, methyl pluviatilol, sesamin and asarinine, piperine, N-isobutyl-deca-trans-2-trans-4-dienamide (*fruits*).

Uses in Literature: Recorded in India as analeptic, antifertility, antipyretic, antitubercular, appetizer, carminative, cholagogaue, CNS stimulant, digestive, emmenagogue, hypotensive, sedative; and for asthma, bronchitis cold, cough (Ambasta, 1986; Jain, 1968, 1991; Khare, 2007; Watt, 1972).

Pistacia integerrima Stewart *ex* Brandis (Pl. 74A).

Syn.: *P. khinjuk* Stew. *ex* Brandis, *P. chinensis* Bunge, *P. chinensis* var. *integerrima* (Stewart) Zohary, *Rhus integerima* Wall., *R. kakra-singee* Royle

Family: Anacardiaceae.

Vern.: Kakad-singi, Kakarsingi.

Ayurvedic, English, Sanskrit and Regional Names:

Ayur.: Karkatashringi, Karkatashringikaa, Shringi;

Eng.: Crab's Claw;

San.: Chakra, Chakrangi, Chandrasapada, Ghosha, Karkali, Kasavinashini, Mahaghosa, Navanga, Vakra;

Ass.: Kakaisrnghi;

B.: Kakra, Kakrsingi;

G.: Kakra;

H.: Kakra, Kakra singi;

Kan.: Dushtapuchattu, Kakadsasingi, Karkatasrini;

Mal.: Gurgu, Kakkar;

Mar.: Kakra, Kakrasingi;

Oriya: Kakadashringi, Kakadashrungi;

P.: Kakkar, Kakkeram, Kaklashne, Tungu;

Tam.: Karkata-shingi (galls);

Tel.: Kakar-shingi, Karkatashrungi.

Distribution: W. Himalayas, W. Nepal: 600-2,400m.

Description: Medium-sized glabrous trees having pinnate leaves and 4-5 pairs of lanceolate, caudate, acuminate leaflets. Flowers small, red coloured, in axillary racemes. Drupes globose, glabrous, grey when ripe and wrinkled.Wood very hard and good.

Flowering & Fruiting: March-May.

Habitat Ecology: Open slopes, grasslands, roadsides; Tal Mahal, 947m.

Material Examined: HPU-EL-SKS-ABS-15213;15.09.2009.

Parts Used: Leaves. Fruit. Bacterial Galls. Wood.

Folk Uses: Leaves useful as fodder* and also chewed against toothache*. 3g ash of burnt powdered galls mixed with honey considered as good expectorant and for checking asthma. Curved fruits smoked for whooping cough. Wood utilized for making furniture and as fuel*.

Active Constituents: Properties mainly attributed to α-pinene, camphene, dl-limonene, cineole, α-terpenol, aromadendrene, β-sitosterol, gummastic, pistacin and pistacinin and two riterpene acids, pistacienonic acid A and B (*galls, drupes*).

Biological Activity: Gall oil antibacterial, antiprotozoal and antispasmodic.

Uses in Literature: So far employed as an antidote to scorpion sting and snake venom, expectorant, fodder and tonic; and for asthma, cough, dysentery, fever, furniture, ingredient of 'Chawanprash', phthisis, ploughs, respiratory trouble, roping, spinning wheel, tanning, turnery and want of appetite, wood for construction and furniture works (Ambasta, 1986; Chauhan, 1999; Gupta, 1981; Jain, 1991; Kapur & Nanda, 1996; Khare, 2007; Mehrotra *et al.*, 1995; Negi *et al.*, 1985; Watt, 1972).

Major Ayurvedic Formulations: Ingredient of 'Chawanprash', Sringynadi churn, Dashmularista Kantakari, Kantakari, Avaleha, Nokufcough, Kasni.

Trade: Kakarsinghi.

Plumbago zeylanica L. (Pl. 74B).

Syn.: *P. auriculata* Blume, *Thela alba* Lour.

Family: Plumbaginaceae.

Vern.: Chitra, Chitrak.

Ayurvedic, English, Sanskrit and Regional Names:

Ayur.: Agni Vahni, Chitraka, Dahana, Hutaasha, Jvalanaakhya, Krishaanu, Sikhi;

Eng.: Ceylon Lead Wort, White Flowered Lead Wort;

Sans.: Agni, Agnisikha, Analanania, Bhali, Chitraka, Jataveda, Vaishavan-Ara;

B.: Chita, Chitruk;

Bo.: Chitra, Chitrack;

G.: Chitaro,Gholochitro;

H.: Chita, Chitarak, Chitra, Chiti;

Kan.: Chitramula, Vahni;

Mal.: Tumpokotuveli;

Mar.: Chitraka, Chitramula;

Oriya: Chitramula, Krishanu;

P.: Chitrak;

Tam.: Adigarradi, Tigana, Vellaikkodiveli;

Tel.: Agnimata, Chitramulamu, Tellachitramulamu.

Distribution: Throughout India ascending to 1,650m.

Description: A rambling sub-scandent perennial herb possessing diffuse branches, elliptic ovate, truncate to attenuate leaves with apiculate apex. Flowers white and in long spikes having glandular rachis. Corolla tube long. Capsule glandular hairy, globose. Seeds solitary.

Flowering & Fruiting: April-November.

Habitat Ecology: Moist slopes, wastelands, roadsides; Bhira, 816m.

Material Examined: HPU-EL-SKS-ABS-15214;13.03.2008.

Parts Used: Plant. Roots. Stem. Leaves.

Folk Uses: Young plant edible as pot herb*. Stem used for scouring teeth*. Extr. of leaves and its roots mixed with mustard oil applied for relieving rheumatic pain. Paste of roots with milk, vinegar or salt and water good against skin infections; if mixed with latex of *Euphorbia royleana* Boiss. checks pus* formation in livestock.

Active Constituents: Constituents characterised are plumbagin, 3-choloroplumbagin, droserone, 3, 31-bi plumbagin, zeylanone, iso-zeylanone and ellipitinone, new bi-napthaquinone-chitranone with zeylinone, elliptinone drosene and plumbagic acid (*roots*), isoschinandone, sitosterol, plumbazeylone, a new dihydronaptho-quinone-dihydroserone-tyrosine (4.5), threonine (4.1), alanine (3.5), histidine (3.4), glycine (1.7), methionine(0.5), hydroxyproline (0.02%) (*aerial parts*).

Biological Activity: Root extr. antibacterial and antimicrobial. Plumbagin (0.75mg/kg) active against P 338 lymphocytic leukaemia and at 4mg/kg exhibits antibacterial and antifungal activity against wide variety of bacteria and fungi. Bonded plumbagin induces abortifacient, anticoagulant antiimplantation, antiovulatory activities.

Uses in Literature: Useful earlier as an abortifacient, antiperiodic, appetiser, diuretic, expectorant, vesicant; and for abdominal pain, antifertility, aphthae, body pains (roots worn on arm), cancer of lungs, breast and oral cavity, diarrhoea, dysentery, dyspepsia, eczema, fever, haemorrhoids, headache, hydrocele, indigestion, leprosy, leucoderma, malaria, muscular pain, paralysis, piles, rheumatism, skin diseases, skin diseases (vet.), spleen complaints, stomach disorder, sexually transmitted diseases, wounds (vet.) (Ambasta, 1986; Balasubramaniam & Prasad, 1996; Bhandary *et al.*, 1996; Borthakur, 1993; Brahmam & Saxena, 1990; Brahmam *et al.*, 1996; Goud & Pullaiah, 1996b; Hembrom, 1996; Hembrom & Goel, 2005; Jain, 1991; Jain & Patole, 2001; Jain & Sharma, 1996; Jain *et al.*, 1994; Jha & Varma, 1996; Khare, 2007; Manadhar, 1987; Ranjan, 1996; Reddy *et al.*, 1996; Sharma & Borthakur, 2003; Srivastava *et al.*, 2003; Tirkey, 2004; Varma, 1997; Venkataratnam & Venkata Raju, 2004; Verma & Chauhan, 2006; Watt, 1972).

Major Ayurvedic Formulations: Agni-tundi-vati, Chandraprabha–vati, Chitraka, Chitrak-ghirta, Devaradi-kwath, Mahasank-vati, Mahamritunjay ras, Navayas-lauha, Panchkola-churna, Sarvajwar-lauha, Sudarshan-churna, Trimalla-churna, Yog-raj-gugglu.

Trade: Chitrak (roots).

Pogostemon benghalensis Burm. f. (Pl. 74C).

Syn.: *P. plectranthoides* Desf., *Origanum benghalensis* Burm.f., *O. indicum* Roth, *Mentha secunda* Roxb.,M. *fruticulosa* Roxb.

Family: Lamiaceae.

Vern.: Dusyan, Kali Basuti.

Regional Names:

B.: Bakoha, Jin, Jui-Lata;

Bo.: Pangla;

Gar.: Lujra;

Oriya: Dumobadottoka, poksunga;

Tel.: Gondripula, Kasurijang.

Distribution: H.P., E.Nepal upto 1,300m, W. Himalayas.

Description: Pubescent shrubs. Leaves membranous, irregularly toothed. Flowers whitish purple, crowded in spikes. Nutlets smooth.

Flowering & Fruiting: February-April.

Habitat Ecology: Wastelands, roadside; Nalti, 787m.

Material Examined: HPU-EL-SKS-ABS-15215; 28.03.2008.

Parts Used: Roots. Leaves.

Folk Uses: ½ teaspoon of root ash with honey or sugar pellets prescribed for 15 days to cure whooping cough. Leaves used as fodder*, its poultice after heating applied on head to check headache. Also, its decoction (25-30ml) good for cough* (twice daily till cure). Powdered leaves with onion (*Allium cepa* L.), 'ajwain' (*Trachyspermum ammi* L.) and 'gur' in ratio of 2:1:1 respectively (1 tsp, twice daily) checks stomachache*. Moreover, pellets prepared by heating, 'dhakh' (dehydrated grapes) in earthen 'chullah' (fire place), wrapped in leaves of this plant, are given for quick relief from cough.

Active Constituents: These are 0.07% essential oil (fresh *leaves*), sesquiterpene lactone, caryophyllen 9-beta-10-olide (*whole plant*).

Biological Activity: Leaves show antifungal activity against *Helminthosporium sativum.*

Uses in Literature: Known earlier as an antidote to scorpion sting and snake bite, antihaemorrhagic, cleaning wounds, edible (eaten in time of scarcity), febrifugal, green manure, insecticidal, styptic; and for body pains, boils, cuts, detergent, diarrhoea, diarrhoea (vet.), drink, wounds and wounds in animals (Ambasta, 1986; Brahmam & Saxena, 1990; Chopra *et al.*, 1956; Datt, 1996; Jain, 1991; Khare, 2007; Kothari & Moorthy, 1996; Mokat & Deokule, 2004; Rama Rao & Henry, 1996; Siwakoti & Verma, 1996).

Polyalthia longifolia (Sonn.) Thw. (Pl. 74D).

Family: Annonaceae.

Vern.: Ashok.

Ayurvedic and English Names:

Ayur.: Devadaari;

Eng.: False Ashoka, Green Champa, Indian Mast Tree.

Distribution: Throughout India.

Description: Trees with pendulous branches and long narrow lanceolate leaves having undulate margins. Flowers subumbelled, numerous, dense yellow green, star-like or pale green. Fruits in cluster of 10-20 and purple or black when ripe.

Flowering & Fruiting: February-April.

Habitat Ecology: Along roadsides; Bhota, 765m.

Material Examined: HPU-EL-SKS-ABS-15216; 22.05.2011.

Parts Used: Plant. Leaves.

Folk Uses: Plant known for its ornamental* value; festoons of its leaves are strung across doors* alongwith that of mango on ceremonial occasions*.

Active Constituents: Active constituents are (3S, 4R)-3,4,5-trihydroxypentanoic acid-1,4-lactone (1). Polyalthialdoic acid, Kolavenic acid, aporphine alkaloid, liriodenine, and Azafluorene (*stem bark*).

Biological Activity: EtOH extr. of stem showed promising antibacterial activity against thirteen Gram +ve and nine Gram –ve organisms.

Uses in Literature: Reportedly described in India for arches of festoons of its leaves, barrels, carriage shafts, kidney diseases, matches, packing cases, pencils, scaffolding, strengthening of gums, toothache; and as anthelmintic, analgesic, anticancerous, antihyperglycaemic, antiinflammatory, hepatoprotective (Ambasta, 1986; Chopra *et al.*, 1956; Khare, 2007; Kirtikar & Basu, 1994).

Pongamia pinnata Pierre. (Pl. 75A).

Syn.: *P. glabra* Vent., *Derris indica* (Lam.) Benn., *Galedupa indica* Lam., *G. arborea* Roxb., *Robinia mitis* L., *Dalbergia arborea* Willd.

Family: Fabaceae.

Vern.: Sukhchain.

Ayurvedic, English and Regional Names:

Ayur.: Karanja;

Eng.: Pongam oil tree, Indian beech;

Tam.: Pungam.

Distribution: Common in sea coast throughout India, sporadically in Shivalik range in H.P., also in gardens upto 1,200m.

Description: Trees with drooping branches, unequally pinnated with 2-3 pairs of ovate, acuminate shining green leaves, laden with rose tinted white flowers in axillary racemes. Pods woody with a short decurved point, turgid, 1-seeded.

Flowering & Fruiting: April-June.

Habitat Ecology: Cultivated in gardens; Awah Devi, 1,232m.

Material Examined: HPU-EL-SKS-ABS-15217; 22.05.2011.

Parts Used: Leaves. Flowers. Bark. Seeds.

Folk Uses: Hot bath with extr. of flowers and leaves used for skin affections. 2-4g powdered bark consumed with luke warm water against stomach disorders; also its decoction prescribed after child birth as antigastric*. Poultice of leaves good for sores and ulcers. Powdered seeds consumed 10g daily in the morning with cow's milk to check diabetes*. Seed oil considered antiseptic and used against skin infections and leucoderma. Paste of its seeds and roots used locally for fish poisoning*.

Active Constituents: Active constituents are flavonoids, flvones, furanoflavonoids, chromenoflavones, chromenochalcones, coumarones, flavone glucosides, sterols, triterpenes and phenyl alanine dipeptide (*bark*), karanjin, pongamol, pongapin kanjone (*seed oil*).

Uses in Literature: Useful earlier for bleeding piles, bronchitis, cold, cough, cutaneous infections, diarrhoea, dyspepsia, flatulence, foul ulcers, gonorrhoea, herpes, leprosy, leucoderma, liver ailments, pneumonia, scabies, scrofulous enlargements, whooping cough; and as antileprotic, antipyretic, antirheumatic, antispasmodic (Chopra *et al.*, 1956; Khare, 2007; Kirtikar & Basu, 1994).

Populus nigra L. (Pl. 75B).

Syn.: *P. nigra* var. *italica* (Munchh) Koehne

Family: Salicaceae.

Vern.: Poplar.

English and Regional Names:

Eng.: Black Poplar, Lombardy poplar;

Kash. : Frasht;

P.: Buins, Do, Frast, Kramadi, Pakshubut.

Distribution: N. W. Himalayas: 900-3,700m.

Description: Large-sized trees with upright branches forming a narrow cylindrical crown. Leaves broad and stalked. Dioecious plant; flowers in pendulous catkins, each with a cup-shaped lobed disc, which represents remains of perianth. Capsules one-celled with many seeds, each with basal tuft of white, silky hair.

Flowering & Fruiting: April-July.

Habitat Ecology: Planted along roadsides, Dimmi, 1,006m.

Material Examined: HPU-EL-SKS-ABS-15218; 25.05.2009.

Parts Used: Whole Plant. Tender Shoots. Wood.

Folk Uses: Planted on wetlands for reclaimation* of soil. Tender shoots used as fodder. Wood used in building works, plywood* industries and as fuel*.

Active Constituents: Active constituents are salicin, salicortin, d, new phenol, glycoside-nigracin (*bark*), 2-methylpropyl, 2-methyl-2-propenyl 2-methylbutyl, 3-methylbutyl-, 2-methyl-2-butenyl-, 4-methylpeatyl, 5-methylhexyl and 3,7-dimethyl-2,6-octadienyl caffeates (*bud exudates*).

Uses in Literature: Reported so far as antiinflammatory, antiseptic, coacher, depurative, fodder, substitute of quinine; and for bone fracture, carving, cold, cross-beams, furniture, haemorrhoids, hair growth, matches, poles, small boxes, timber, uric acid discharge (Ambasta, 1986; Jain, 1991; Khare, 2007; Singh *et al.*, 1996a; Watt, 1972).

Premna latifolia Roxb. (Pl. 75C).

Syn.: *P. latifolia* var. *mucronata* C.B. Clarke, *P. mucronata* Roxb.

Family: Verbenaceae.

Vern.: Bankar.

Ayurvedic, English and Regional Names :

Ayur.: Agnimantha;

Eng.: Dusky Fire Brand Bark.

B.: Goharo;

H.: Agnium, Bakarcha, Jhatela;

Mal.- : Knappa;

Oriya: Gondhona;

P.- : Bankar, Ganhila;

Tam.: Erumaimunai, Munnai, Panchumullai;

Tel.: Koora, Peddanell kure.

Distribution: Temperate Himalayas upto 1,400m.

Description: Small trees with subcoriaceous leaves and broad-ovate, greenish-white globose, dark purple flowers in panicles. Drupes bright crimson.

Flowering & Fruiting: July-September.

Habitat Ecology: Common on Roadsides; Masayana, 900m.

Material Examined: HPU-EL-SKS-ABS-15219; 05.06.2011.

Parts Used: Leaves. Bark.

Folk Uses: Decoction of the plant with bark used for bath against skin infection*. Leaves used as fodder. Juice of leaves rubbed as an emollient* on affected parts. 20 ml decoction of leaves with 'malathi' (*Glycyrrhiza glabra* L.) and 'dhansoa' (*Anethum sowa* Roxb. *ex* Flem.) in equal quantity given to cure diabetes*. Paste of bark applied on abscess in animals. Wood used as fuel*.

Active Constituents: Properties attributed to β-sitosterol and its glucoside, 5-hydroxy-4'-methoxyflavone-7-O-arabinopyranosyl-rhamnoside,.two iridoside-7-deoxyloganic acid (bisdesoxydi-hydromonotropein) and geniposidic acid (*leaves*).

Uses in Literature: Known so far as colic, colic (vet.), diuretic, firewood, fodder, vegetable (leaves, tender shoots); and for blisters, boils, charcoal, dropsy, headache, flavouring, gonorrhoea, indigestion, ring worms, syphilis and ulcers (Ambasta, 1986; Khanna & Kumar 2000; Khare, 2007; Venkataratnam & Venkata Raju, 2004).

Prunus amygdalus Baill. (Pl. 75D).

Syn.: *P. communis* Fritsch, *Amygdalus communis* L., *P. dulcis* (Mill.) D. A. Webb.

Family: Rosaceae.

Vern.: Badam.

Ayurvedic, English, Sanskrit and Regional Names:

Ayur.: Vaataada, Vaataama;

Eng.: Almonds;

Sans.: Badama, Netroamaphala;

B.: Bilati-badam;

G.: Badam;

H.: Badam;

Mal.: Vatam-kotta;

Mar.: Badam;

Tam.: Vadamkottai;

Tel.: Adam vittulu.

Distribution: Throughout India.

Description: Unarmed glabrous trees with oblong-lanceolate serrated leaves and solitary, pink flowers. Drupes, velvety. Stone slightly flattened covered with shallow wrinkled and minute holes. Seed long, oval.

Flowering & Fruiting: April-August.

Habitat Ecology: Planted; Sujanpur, 567m.

Material Examined: HPU-EL-SKS-ABS-15220; 15.04.2011.

Parts Used: Twigs. Fruits. Seeds.

Folk Uses: Ripe fruits edible. Dried branches used as fuel wood*. Seeds oil given to check constipation*. 5-10 seeds soaked in water prescribed empty stomach early morning as brain tonic and for vitality. Decoction of its seeds prepared in milk along with grates said to be good aphrodisiac preparation; paste used as lotion for burns, seed oil massaged in skin ailments.

Uses in Literature: Useful earlier as astringent, demulcent, detergent, diuretic, edible (kernel), laxative, nervine tonic, stimulant, symbolic; and for confectionary, cosmetics, cough, diabetes, peptic ulcer, perfumery, skin eruptions, sores and turnery (Ambasta, 1986; Jain, 1991; Khare, 2007; Shome *et al.*, 1996).

Prunus domestica L. (Pl. 76A).

Syn.: *P. insititia* L., *P. communis* Huds. var. *insititia* (L.) Hook. f.

Family: Rosaceae.

Vern.: Aloo Bukhara, Plum.

Ayurvedic, English and Regional Names:

Ayur.: Aaruka, Aaluubukhaaraa;

Eng.: Plum Tree, Prum Tree, Common Plum;

B.: Alubukhara;

G.: Alubukhara;

H.: Alubukhara.

Distribution: W. Himalayas.

Description: Cultivated trees without spines. Receptacle hollow. Leaves obovate-ovate or ovate-lanceolate. Peduncles solitary or in pairs. Flowers stalked, pink. Drupes globose or ovoid drooping with 1-seeded, smooth stone.

Flowering & Fruiting: April-August.

Habitat Ecology: Planted; Kuthera, 900m.

Material Examined: HPU-EL-SKS-ABS-15221, 15.05.2011.

Parts Used: Fruits. Branches. Leaves.

Folk Uses: Ripe fruits edible; mashed fruits given twice a day for jaundice*, also employed for making jams. Fruit juice prescribed after every two hours to cure dysentery. Dried branches used as fuel wood*. Leaves used as fodder*.

Active Constituents: Ingredients comprise phloracetophenone, 4-O-methyl-phloracetophenone, and its 6-O-glucoside (I) and frakinol (*bark*), galacturmic acid, arabinose, galactose, xylose and rhamnose (*fruits*), naringenin, quercitin, chrysin (*heartwood*).

Biological Activity: Naringenin possesses significant antiulcer activity in experimentally induced ulcer in pylorus ligated rats (132 mg/kg).

Uses in Literature: Described in literature as aperient, astringent, cabinet work, cathartic, demulcent, desert, digestive, edible, laxative, refrigerant; and for bilious states of the system, cold, confectionary, decoction, fever, jams, lay and turnery, match boxes, nutritious, refreshing, skeleton of paper (Ambasta, 1986; Khare, 2007; Samati, 2004; Watt, 1972).

Prunus persica Benth. & Hook. f. (Pl. 76B).

Syn.: *Amygdalus persica* L., *A. colinus* Wall., *Persica vulgaris* Mill., *P. saligna* Royle

Family: Rosaceae.

Vern.: Adu, Aru.

Ayurvedic, English and Regional Names:

Ayur.: Aaluka, Aaruka Aru, Pichuka;

Eng.: Peach Tree, Nectarine;

H.: Aru;

Oriya: Pishu;

P.: Aru, Bembi, Chirmaru;

Urdu: Adud.

Distribution: N. W. Himalayas, cultivated.

Description: Small deciduous trees. Leaves conduplicate. Flowers pink, solitary or in small cymose clusters, crowded towards the end of dwarf branches. Drupes fleshy and downy, irregularly furrowed single-seeded stones.

Flowering & Fruiting: February-May.

Habitat Ecology: Planted; Bhaleth, Sujanpur, 567m.

Material Examined: HPU-EL-SKS-ABS-15222; 17.03.2009.

Parts Used: Roots. Leaves. Fruit. Gum. Seeds.

Folk Uses: Fruits edible as well as pickled. Seeds also relished. Ash of burnt leaves considered good against burn injuries*. Infusion of leaves and roots good for cold*. Leaf juice rubbed for healing wounds (vet.); hot extr. of young leaves considered good for expelling intestinal worms (given twice daily for 7 days) in animals. Gum used as an adhesive*.

Active Constituents: Chief constituents are kaempferol (*leaves*), 6-o-(β glucoronopyranosyl)-D-galactose, 6-o-(4-o-methyl β-D-glucoronopyranosyl)-d-galactose and 2-o-(β-glucorono pyranosyl), D-mannose, rhamnose, xylose, arabinose and galactose (*gum*), ã-hexalactone, ã-heptalactone, ã-octalactone, ã-nanolactone, ã-decalactone, ethanol, hexanol, benzylalcohol, acetaldehyde, benzaldehyde, acetic acid, pentanoic acid, hexanoic acid, hexyl formate, methyl acetate and ethylacetate, vit.A, protein 1.2, fat 0.03%, fibre 1.2, C15, Mg 21, oxalic acid, iron 2.4, total P 41, Na 2, K 453.0, Cu 0.06, S 26.0, thiamine 0.02, riboflavin 0.03, nicotinic acid 0.05 and ascorbic acid 6.0 mg /100g (*fruits*).

Biological Activity: Aerial parts diuretic. Alcoholic extr. of embryo and immature seeds exhibits analgesic and antiinflammatory activities.

Uses in Literature: Used in India as an abortifacient, antiemetic, antiscorbutic, ascaricide, cattle feed, demulcent, diuretic, edible, peach brandy, stomachic, table fruit, tonic, vegetable; and for building purposes, canning, cough, dessert, eczema, fertilizer, foot and mouth (vet.), headache, hook worms cure (vet.) pickling, piles, removing bad smell of mouth and sputum, scabies, stomach trouble, whooping cough (Ambasta, 1986; Chinthu *et al.*, 1998; Jain *et al.*, 1994; Kapur & Singh, 1996; Khare, 2007; Lal *et al.*, 1996; Malik *et al.*, 2009; Parmar & Kaushal, 1982; Samati, 2004; Sharma, 1996; Sinha, 1987; Tiwari & Pande, 2006).

Psidium guajava L. (Pl. 76C).

Family: Myrtaceae.

Vern.: Amrood.

Ayurvedic, English, Sanskrit and Regional Names:

Ayur.: Amaruuda, Peruka;

Eng.: Common Guava, Guava tree;

Sans.: Aprithaktvacha, Ovidhavija;

Ass.: Madhuriam;

B.: Goaachni, Peyara;

Bo.: Jam, Perala;

G.: Jamrud, Peru;

H.: Amrud, Safed safari;

Kan.: Pera, Koyera;

Mal.: Koyya, Malakkaperu;

Mar.: Jamba;

Oriya: Bodojambo, Julabojamb;

P.: Amrud, Anjirizarod;

Tam.: Segappugnoyya, Vellaikoyya;

Tel.: Ettajama, Tellajama.

Distribution: Widely cultivated throughout India.

Description: Small cultivated trees. Leaves opposite, ovate-oblong, on very short petioles. Flowers white, solitary with quincuncial sepals. Petals broad. Berries globose, green, pulp yellow with white or pink seed.

Flowering & Fruiting: July-September.

Habitat Ecology: Cultivated; Nadaun, 483m.

Material Examined: HPU-EL-SKS-ABS-15223; 05.05.2011.

Parts Used: Leaves. Fruits.

Folk Uses: Fruits edible and considered good for anaemia* and constipation. Fruits also cooked as vegetable*. Leaves used for scouring teeth and tongue and its paste rubbed on gums to cure pyorrhoea. Chewing of leaves along with 'katha' (*Acacia catechu* Willd.) cures mouth ulceration*. Poultice of leaves with jaggery and 'ajwain' (*Carum copticum* Hiern.), used as medicine for both mankind and animals as stomachic. About 1 litre extr. of its leaves and that of mint (*Mentha longifolia* (L.) Huds.) in equal ratio given thrice daily till cure to control dysentery (vet.)*.

Active Constituents: Properties attributed to sesquiguavaene, ellagic acid, ellagitanin, guavin β, guavin A, C and D, galaturonic acid, galactose, arabinose with traces of xylose, rhamnose (*fruits*) and quercitin-3-α arabino pyranoside (*leaves*).

Biological Activity: Spasmolytic activity of leaves mainly due to quercitin. (+). Gallocatechin exhibits antimutagenic activity against UV-induced mutation in *Escherchia coli.*

Uses in Literature: Known earlier as an antiemetic, antinarcotics to overcome the influence of liquor, astringent, colic, colouration material in 'Mithila' paintings, edible, emetic, laxative, mouth wash for swollen gums, spear handle and instrument of various kinds, tongue cleaner, tonic; and for carbuncle, cholera, chutneys, cold, cough, diarrhoea, dysentery, headache, inflammation, itches, jam, ketchup, loose motions, malaria, marmalades, piles, sprains, stomachache, stones, tanning, toothache, ulcer and wound (vet.) (Ambasta, 1986; Augustine & Sivadasan, 2004; Bhandary *et al.*, 1996; Bhardwaj & Gakhar, 2003; Borthakur, 1993; Chowdhury *et al.*, 1996; Gill *et al.*, 1993; Girach & Aminuddin, 1995; Gogoi & Borthakur, 2001; Hosagoudar & Henry, 1996c; Jain, 1991; Jain *et al.*, 1991; Jha & Basak, 1994; Khare, 2007; Lalramnghinglova, 1996; Mohanty *et al.*, 1998; Pal & Jain, 1998; Pandey & Verma, 2002; Samati, 2004; Sen & Behera, 2003; Shome *et al.*, 1996; Singh & Maheshwari, 1992; Siwakoti & Verma, 1996; Tirkey *et al.*, 2003; Upadhyay & Singh, 2005; Verma *et al.*, 1995; Vishwanathan & Singh, 1996; Viswanathan *et al.*, 2001).

Punica granatum L. (Pl. 76D).

Syn.: *P. nana* L.

Family: Punicaceae.

Vern.: Daadu, Daardn.

Ayurvedic, English, Sanskrit and Regional Names:

Ayur.: Daadima, Daadimba, Lohitpushpaka, Raktakusumaa;

Eng.: Pomegranate Tree;

Sans.: Bijapura, Dadima, Dantabijaka, Granat, Pindapushpa, Vrittaphala;
Ass.: Dalim;
B.: Dalimgachh, Dalimphal;
Bo.: Anara, Dalimba;
G.: Dadam, Daramphal;
H.: Anar, Dhalim, Dharimb;
Kan.: Dalimbe, Darrim;
Mal.: Pumatalam, Uruyapalam;
Mar.: Dalimba;
Oriya: Dalim, Dalimba;
P.: Anar, Daan, Daru, Daruna;
Tam.: Kalumal-adalai, Tusagam;
Tel.: Dadimba, Dadimanu, Dalimma, Purvudanimma;
U.P.: Anar, Rumman;
Urdu: Anamitha.

Distribution: Cultivated throughout India; W. Himalayas 700-2,700m.

Description: A small deciduous tree with thorny branches and dark grey bark. Leaves opposite, oblong-lanceolate, glabrous. Flowers sessile, bright red, terminal. Berries globose. Seeds angled, aril filled with an acidic sweet juice. Rind coriaceous.

Flowering & Fruiting: May-September.

Habitat Ecology: Stony slopes, roadsides, edges of agricultural fields; Brahmni village, 858m.

Material Examined: HPU-EL-SKS-ABS-15224; 03.07.2008.

Parts Used: Plant. Rind. Fruits. Seeds.

Folk Uses: Plant considered sacred and worshipped. Fruits edible and considered good for hyperacidity*, anaemia*, fatigue and rich source of vit. C. Dried seeds 'anardana'* used for 'chutney'* preparation by grinding it with mint leaves and chillies. One tsp of powdered rind taken with honey for seven days to cure cough. Rind added in vegetable preparation of 'jimikand' (*Amorphophallus campanulatus* (Roxb.) Blume *ex* Decne.) for removing bitterness*. Dried rind with curd also given to pregnant ladies to cure dysentery and as coolant. 2-3g mixture of powdered seeds (20g) in combination with 5g 'soa' (*Anethum sowa* Roxb.), 5g 'kadu' (*Picorrhiza kurroa* Royle *ex* Benth.), 5g 'patish' (*Aconitum heterophyllum* Wall. *ex* Royle), 5g 'magan' (*Piper longum* L.) taken with honey twice a day till cure for cough, stomachache and giddiness*. 2-3g mixture of its powdered seeds (100g) and those of 10g 'soa' (*Anethum sowa* Roxb.), 10g 'black pepper' (*Piper nigrum* L.) and 10g rock salt prescribed thrice a day for 15-20 days for curing liver ailments. 'Chutney' of its seeds grinded along with coriander, 'soa' and leaves of 'pudina' (*Mentha spicata* L.) prescribed for persistent cough.

Active Constituents: Properties attributed to tannin-punicalagin and puncalin, granatin A and B (*rind*), punicalin, punicalagin, punicafolin, granatins A and B, corilagin, strictinin, 1, 2, 3, 4, 6-Penta-o-galloyl-β-D-glucose, 1, 2, 4, 6-tetra-o-galloyl-β-D-glucose and new hydrolysable tannin 2-o-galloyl-4, 6 (ä, I)-galloyl-D-glucose (I) (*leaves*); cyanidin-3-glucoside, delphindin-3-glucoside, and three, 5-diglucoside (*seed coat*), elagitannins, punicacorteins A-D, punigluconin, causuariin, punicalin, puicalagin and pullestierine; 28% gallotanic acid together with yellow colouring matter (*rind*), glucose, fructose, maltose, starch, carotene, thiamine, riboflavin, nicotinic acid, vitamin, pectin and amino acid (edible portion of *fruit*), elagitannins, punicacorteins A, D, punigluconin, casuarin, casuarinin, punicalin, punicalagin and pullestierine (*bark*).

Biological Activity: Aerial parts CNS depressant and diuretic whereas flower buds antifungal and antibacterial. Fruit extr. exhibits high activity against *Micrococcus pyogenes* var. *aureus, Escherichia coli, Salmonella typhosa, S. monterideo.* Aq. extr. of roots completely inhibits activity of *Mycobacterium tuberculosis.*

Uses in Literature: Useful as an anthelmintic, antivomiting, astringent, cardiac, coloration material in 'Mithila' paintings, coolant, condiments, edible, emetic, expectorant, good source of vit. C, laxative, refreshing drink, souring agent, stomachic, symbolic, tonic and wine; and for brain affection, bronchitis, cholera, congestion of liver, cough, diarrhoea, dysentery, diabetes, eye complaints, fertility, gastric, intestinal worms, jaundice, kidney stones, pimples, pyorrhoea, sore throat, toothache, urinary troubles, uterine disease (Ambasta, 1986; Anonymous, 1990, 2008; Bhardwaj & Gakhar, 2003; Bondya & Sharma, 2003; Chopra *et al.*, 1956; Dagar & Dagar, 1996; Ghosh *et al.*, 1996; Gogoi & Borthakur, 2001; Jain *et al.*, 1991; Jha & Basak, 1994; Kapur & Singh, 1996; Katewa *et al.*, 2001; Khanna *et al.*, 1996b; Khare, 2007; Mahato & Mahato, 1996; Mohanty & Padhy, 1996; Mohanty *et al.*, 1998; Pandey & Verma, 2002; Rana *et al.*, 1996; Singh, 1996; Tirkey *et al.*, 2003a,b; Verma & Chauhan, 2006; Viswanathan *et al.*, 2001).

Major Ayurvedic Formulations: Bhaskara-lavana, Dadimadi-ghrit, Dadimastak-churn, Lavanbhaskar-churn, Eladi-churn, Sukra-matrka-vati.

Trade: Fruits and dried seeds marketed as 'anardana'.

Putranjiva roxburghii Wall. (Pl. 77A).

Family: Euphorbiaceae.

Vern.: Putranjan.

Ayurvedic, English, Sanskrit and Regional Names:

Ayur.: Arthsaadhana, Putrajiva, Putrajivaka, Putranjiva, Yashtipushpa;

Eng.: Life of the child;

Sans.: Putranjiva;

B.: Putranjiva;

H.: Putranjiva;

Mal.: Pongalam;

Mar.: Jewanputr;

P.: Jiyaputra;

Tam.: Karupale;

Tel.: Kadrajuvi, Putrajivi.

Distribution: Throughout tropical India.

Description: A moderate-sized evergreen tree. Leaves dark green, elliptic-oblong. Flowers yellowish, males in dense axillary clusters, female ones solitary. Drupes tomentose, globose to ovoid, size of the cherry or smaller.

Flowering & Fruiting: March-April.

Habitat Ecology: Cultivated in temple courtyards; Krishna Nagar, 858m.

Material Examined: HPU-EL-SKS-ABS-15225; 24.04.2011.

Parts Used: Whole Plant. Seeds. Wood.

Folk Uses: Planted for its shade; also considered sacred and worshipped for long life of male child in the family. Wood used in construction* and furniture*. Childern love to play* with its seeds.

Uses in Literature: Used so far as sacred (Ambasta, 1986, Khare, 2007; Sharma, 2003).

Pyrus communis L. (Pl. 77B).

Syn.: *P. pyrifolia* Nakai var. *culta* Nakai, *P. sinensis* Hort.

Family: Rosaceae.

Vern.: Nashpati.

English and Regional Names:

Eng.: Pear, Wild Pear;

H.: Naspati;

Kash.: Batang, Kistabahira;

P.: Batank, Naspati, Tangi;

Tam.: Perrikaya;

Tel.: Berikaya;

U.P.: Bagugohsa.

Distribution: Cultivated in N.W. Himalayas.

Description: Trees with woolly young parts and inflorescence. Leaves long green. Flowers white, a few flowered corymbs terminating on short shoots. Fruit pyriform, yellow. Seeds obovoid-ellipsoid, smooth.

Flowering & Fruiting: June-August.

Habitat Ecology: Cultivated; Tauni Devi, 1,050m.

Material Examined: HPU-EL-SKS-ABS-15226; 02.06.2010.

Part Used: Fruit.

Folk Uses: Fruits edible and considered good for checking dysentery* and diarrhoea*; also, its juice regarded as refreshing* and given in fever and for promoting appetite*.

Active Constituents: Fruits rich in lupeol, β-sitosterol, betulin, betulinic acid, hydroguinone and arbutin.

Biological Activity: Lupeol, β-sitosterol, betulin, betulinic acid, hydroguinone and arbutin possess bacteriostatic activity against *Escherchia coli, Salmonella typhii, Shigella flexneri* and *Staphylcoccus aureus.*

Uses in Literature: Useful as beverages, edible, psychoactive; and for acute fever, carving, engraving purposes, malaria, mathematical and drawing instrument, turnery and wines (Ambasta, 1986; Gill *et al.*, 1993; Jain *et al.*, 1994; Khare, 2007; Singh, 1999a).

Pyrus pashia Ham. (Pl. 77C).

Syn.: *P. variolosa* Wall., *P. verruculosa* Bertol., *P. nepalensis* Herb. Ham. *ex* Hort.

Family: Rosaceae.

Vern.: Kainth, Kaenth.

Regional Names:

Ass.: Chalthei, Soh-jhur, Soh-shur;

H.: Mehal Mol;

P.: Batangi, Kaenth, Keitha, Kiat, Ku, Shehgal, Tang.

Distribution: Kashmir -Bhutan (Temperate Himalayas).

Description: Small deciduous trees having young woolly parts, acuminate leaves, spinescent branches, white flowers in corymbs. Fruits globose, rough due to small white spots, dark yellow brown and scurfy.

Flowering & Fruiting: March-May.

Habitat Ecology: Open grassy slopes, moist and drier areas, edges of fields; Tauni Devi, 1,050m.

Material Examined: HPU-EL-SKS-ABS-15227; 13.04.2010.

Parts Used: Leaves. Stem. Fruit. Bark. Whole Plant.

Folk Uses: Plant used for grafting* of 'nashpati' (*Pyrus communis* L.). Fruits edible and its extr. effective against dandruff* and mouth ulcers*. Pulp of fruits taken with black pepper, twice a day for three days to check dysentery*. Crushed fruits considered good for making local wine. Poultice of leaves applied on head for conditioning of hair and as dye*. Leaves used as an excellent fodder. One tsp of powdered bark and leaves (1:2 ratio) with a pinch of black pepper (*Piper nigrum* L.) prescribed till relief for rheumatic pains*. Also, fruit pulp applied (two times a day for 3-4 days) for suppuration of blisters on tongue of animals. Paste of unripe fruits applied on foot sores* of animals. Wood used as fuel and for making agricultural implements*. Seeds* sold at high price.

Active Constituents: These are n-triacontane, myricyl alcohol as chloride and sitosterol (*leaves*), friedelin, β-sitosterol (*bark*).

Biological Activity: Aerial parts anticancerous and CNS depressant.

Uses in Literature: Considered useful as bobbins, catarrh (vet.), edible, fodder, fuel; and for blisters on tongue (vet.), eye complaints, eye infection (vet.), combs, injury (vet) and making mortar and walking sticks (Ambasta, 1986; Gaur *et al*, 1992; Joseph & Kharkonger, 1981; Singh & Kumar, 2000b; Sinha, 1996; Tiwari & Pande, 2006).

Randia dumetorum (Retz.) Lam. (Pl. 77D).

Syn.: *R. nutans* DC. & W. & A., *R. brandisii* Gamble, *R. tomentosa* W. & A. non Blume, *R. longispina* DC. & W. & A., *R. floribunda* DC. & W. & A., *R. rotteri* W. & A., *R. stipulosa* Miq., *R. spinosa* Blume, *R. malabarica* Wall., *Gardenia nutans* Roxb., *G. spinosa* L., *G. longispina* Roxb., *G. floribunda* Roxb., *G. dumetorum* Retz., *G. stipularis* Rottl. & Willd., *G. glabra* Br., *G. propinqua* Br., *G. latifolia* Schelecht., *Posoqueria dumetorum* Roxb., *P. nutans* Roxb., *P. longispina* Roxb., *P. floribunda* Roxb., *Ceriscus malabaricus* Gaertn., *Catuneregum nutans* (DC.) Tiruv., *Xeromphis spinosa* (Thunb.) Keay.

Family: Rubiaceae.

Vern.: Rada.

Ayurvedic, English, Sanskrit and Regional Names:

Ayur.: Chhardana, Madana, Pindi, Shalayaka, Vishapushpaka;

Eng.: Common Emetic Nut;

Sans.: Bastishodhana, Ramachhardanaka;

Ass.: Bahmona, Gural, Mon;

B.: Madan;

Bo.: Gehala, Gelaphal;

G.: Medhola;

H.: Arar, Madanphala

Kan.: Banegara, Kara, Mangri;

Kash. : Kirkala, Kokoa;

Mal.: Karalikkaya, Kattumaranna;

Oriya: Potova;

P.: Mandkolla, mindhal;

Tam.: Karadam, Murkka-rai;

Tel.: Madanamu, Mangara.

Distribution: Subtropical Himalayas upto 1,200m.

Description: A small tree armed with strong, straight, opposite decussate spines. Leaves obovate-obtuse, wrinkled, shining above. Flowers creamish-yellow, fragrant, sessile. Fruits ovoid, yellowish-brown. Seeds numerous, compressed, smooth, closely packed in pulp, pericarp thick.

Flowering & Fruiting: April-June.

Habitat Ecology: Wasteplaces, drier areas; Khaya, 787m.

Material Examined: HPU-EL-SKS-ABS-15228; 22.05.2011.

Parts Used: Roots. Fruits. Seeds. Spine.

Folk Uses: Fresh fruits eaten and also used for washing woollen clothes. Poultice of fresh root good against foot infection in animals. Also, its paste rubbed on boils as an antiseptic*. It is believed that milch animals tied to its trunk start yielding milk*. One tsp of powdered fruit and 'ajwain' (*Trachyspermum ammi* L.) taken with 'gur' (jaggery) to cure stomachache* (twice a day) and constipation*. Tail of animals punctured with its spine in case of 'gahli' (tail gangrene)*.

Active Constituents: Isolated constituents are dumetoronins A, B, C, D, E & F, β-D-galactopyran-osyl (1→3) deanolic acid, lupeol and two new triterpenoid glycosides (*fruits*).

Uses in Literature: Known earlier as an abortifacient, anodyne, anthelmintic, astringent, edible, emetic, expectorant, fish poison, fodder, insect repellent, intensifier, sacred tree; and boils, bodyache during fever, diarrhoea, dyeing, dysentery, jaundice, lactation, pimples, soap industry, umbrella handles and wood for walking sticks (Ambasta, 1986; Chauhan, 1999; Chopra *et al.*, 1956; Jain, 1991; Khare, 2007; Lalramnghinglova, 1996; Singh & Prakash, 1994; Watt, 1972).

Major Ayurvedic Formulations: Madnaphala-churn, Pipalayadi-churn.

Trade: Madanphal.

Ranunculus muricatus L. (Pl. 78A).

Syn.: *R. cabulicus* Boiss.

Family: Ranunculaceae.

Vern.: Buttercup.

Ayurvedic and English Names:

Ayur.: Kaandira, Kaandakatuka;

Eng.: Bilster Buttercup, Celery-leaved crowfoot.

Distribution: Swampy places at low elevations.

Description: Erect or diffuse, glabrous, rarely hairy annual herbs. Leaves irregularly cut, base rounded or cordate. Flowers solitary, leaf apposed, terminal and panicled. Achenes in large globose heads.

Flowering & Fruiting: February-April.

Habitat Ecology: Common weed in swampy places; Kuthera, 900m.

Material Examined: HPU-EL-SKS-ABS-15229; 03.03.2011.

Parts Used: Whole Plant. Leaves. Twigs.

Folk Uses: Poultice of dried plant used in skin diseases*. Leaves used as fodder*. Powdered dried aerial plant parts (¼ tsp) prescribed twice a day for a week with hot water against throat infection*.

Active Constituents: Contains anemonin, proptoanemonin, ranunculine, serotonin, tryptamine derivatives, Serotonin (5-hydroxytryptamine) (*plant*).

Biological Activity: Effective against gram positive and gram negative bacteria; also inhibits growth of *E. coli, Staphylococcus aureus, Candida albicans.*

Uses in Literature: Considered earlier as antibacterial, antifungal, antiviral, cytopathogenic, diuretic, emmenagogue, galactagogue, highly acrid, rubefacient, stimulant, tonic, toxic, vermicidal, vesicant and poisonous; and for asthma, dysuria, eczema, inactivating diphtheria toxins, rheumatism and sores (Ambasta, 1986; Baruah & Sarma, 1984; Chopra *et al.*, 1956; Khare, 2007; Saxena & Vyas, 1981; Sinha, 1987).

Raphanus sativus L. (Pl. 78B).

Syn.: *R. sativus* var. *caudatus* (L.) Vilm, *R. sativus* var. *mougri* Helm., *R. raphanistrum* subsp. *caudatus* (L.) Thell.

Family: Brassicaceae.

Vern.: Mooli.

English, Sanskrit and Regional Names:

Eng.: Raddish;

Sans.: Mulaka;

B.: Mula, Mura, Muli, Muri;

H.: Mula, Mura, Muli, Muri;

Mar.: Mula, Mura, Muli, Muri;

Tam.: Malangi;

Tel.: Malangi.

Distribution: Cultivated throughout India.

Description: Tall coarse rough or glabrous annual or biennial herbs. Leaves pinnatifid. Flowers large usually white or lilac in corymbose raceme. Pods indehiscent, terete, thick, constricted between the seeds. Seeds separated by pith.

Flowering & Fruiting: November-March.

Habitat Ecology: Cultivated; Kohli, 851m.

Material Examined: HPU-EL-SKS-ABS-15230; 25.01.2010.

Parts Used: Roots. Leaves. Fruits.

Folk Uses: 'Mooli' worshipped on 'Hoi' festival for long life of children. Juice of fresh leaves checks jaundice* and constipation*; leaves also cooked as 'saag' (a type of vegetable preparation). Fruits ('mongara') cooked as vegetable. Fresh roots eaten raw and also cooked as vegetable. Roots considered diuretic and good for piles, jaundice, kidney trouble and gastrodynia.

Active Constituents: *Root* possesses seven new b-carboline derivatives I which are characterised as 1-(2-pyrrolidinethion-3-yl) -1, 2, 3, 4-tetrahydro-B–carbolin-3-carboxylic acid.

Biological Activity: A purified basic protein homologus to non specific lipid transfer protein from seed showed antifungal activity.

Uses in Literature: Considered useful as carminative, diuretic, laxative, vegetable; and for flatulence, impotency, kidney stones, piles, ringworm and reducing acidity (Ambasta, 1986; Chandra Sekar & Srivastava, 2003; Khanna & Kumar 2000; Khare, 2007; Kumar & Chauhan, 2005; Samati, 2004).

Rauwolfia serpentina **Benth. ex Kurz. (Pl. 78C).**

Syn.: *Ophioxylon serpentinum* L., *O. trifoliatum* Gaertn., *O. obversum* Miq.

Family: Apocynaceae.

Vern.: Sarpgandha.

Ayurvedic, English, Sanskrit and Regional Names:

Ayur.: Sarpagandhaa;

Eng.: Indian Snakeroot, Rauvolfia root, Serpentina root;

Sans.: Chandrika, Sarpagandha;

B.: Chandra;

G.: Sarpgandha;

H.: Chotachand;

Mal.: Chuvannayilpuri;

Tam.: Charpagandhi;

Tel.: Patala-gandhi.

Distribution: Almost all parts of India upto 1,000m.

Description: Small shrubs upto 1m tall having white bark and leaves narrowing into a short petiole. Flowers white or pinkish, many flowered cymes. Drupes black small rounded dark purple or blackish when dry.

Flowering & Fruiting: May-September.

Habitat Ecology: Cultivated for medicinal properties; Bramhni, 858m, Khaggal, 1,000m.

Material Examined: HPU-EL-SKS-ABS-15231; 13.08.2008.

Parts Used: Roots. Leaves.

Folk Uses: Juice of leaves used as drops for eye affections*; also its 20-25ml decoction prescribed as sedative and for lowering blood pressure and in insanity (twice daily, 10-15 days). 1 tsp of root extr. given on empty stomach for deworming and as antidote for snake or insect bites* whereas its (30-40ml) decoction considered good for better uterine contraction during labour pains.

Active Constituents: Contains alkaloids reserpine, ajmaline, ajmalicine, ajmalicine, yohimgine, coryanthine, iso-ajmaline, neo-ajmaline, papaverine, raubasine, rauwolscine, rescinnamine, reserpine, sarpagine, serpentine, serpentinine, serpinine and deserpidine (*root, root bark*).

Biological Activity: Tranquilizing and antiarrhythmic activities found positive.

Uses in Literature: Useful earlier for anxiety, bleeding disorders, corneal diseases, depression, easy pregnancy, gastric and duodenal ulcers, hypertension, mental disorders, psychomotor irritation, sinustachycardia, tension, uterine contraction; and as abortifacient, anthelmintic, antidote for snake venom, diuretic, hypotensive, sedative and tranquilizer (Ambasta, 1986; Khare, 2007).

Ayurvedic Formulations : Sarpagandhadi-churna, sarpagandhayoga, sarpagandha-vati, various patent medicines as serpina tablets etc.

Reinwardtia indica Dumort. (Pl. 78D).

Syn.: *R. trigyna* (Roxb.) Planch., *R. tetragyna* Planch., *R. repens* Planch., *Linum trigynum* Roxb., *L. repens* Don, *Macrolinum trigynum* Reichb., *Kittelocharis trigyna* Alef.

Family: Linaceae.

Vern.: Basant, Piyan.

Ayurvedic, English and Regional Names:

Ayur.: Baasanti;

Eng.: Winter flax, Yellow flax;

H.: Basanti;

Bo.: Abai;

P.: Balbasant, Gudbatal, Karkum.

Distribution: Pakistan, S.W. China, India, Myanmar, S.E. Asia, upto 1,800m.

Description: Erect-spreading shrubs with alternate, ovate-lanceolate leaves and minute, subulate, caducous stipules. Flowers yellow, in axillary and terminal cymose. Capsule globose.

Flowering & Fruiting: November-May.

Habitat Ecology: Edges of terraced fields, grazing grounds, wastelands; Badhani, 908m.

Material Examined: HPU-EL-SKS-ABS-15232; 05.01.2009.

Parts Used: Leaves. Flowers.

Folk Uses: Poultice of leaves applied for healing cuts* and injuries* and in animals for suppuration of boils*. Flowers offered for worshipping* various deities, also its paste considered good for cuts and wounds.

Uses in Literature: Known to be used as an edible (flowers), poison (vet); and for founder diseases of cattle, maggot infested wounds, paralysis, typhoid, whitlow (Ambasta, 1986; Bhatt, 1990; Jain, 1991; Joshi, 1993; Khare, 2007; Lal *et al.*, 1996; Manandhar, 1990; Watt, 1972).

Rhynchostylis retusa Blume (Pl. 79A).

Syn.: *R. praemorsa* Blume, *R. guttata* Reichb., *R. garwalica* Reichb., *Saccolabium guttatum* Lindl., *S. praemorsum* Lindl., *S. rheedii* Wight, *Sarcanthus guttatus* Lindl., *Aerides retusum* Sw., *A. guttatum* Roxb., *A. spicatum* Don, *Epidendrum retusum* L.

Family: Orchidaceae.

Vern.: Bhangaru.

English and Regional Names:

Eng.: Orchid, Fox tail Orchid;

Asm.: Kopou Phool;

H.: Draupadi mala;

Man.: Samjirei;

Mar.: Gajara;

Kon: Panas;

Tel.: Chintaranamu.

Distribution: Tropical Himalayass: Garwal-Assam.

Description: Epiphytes with stout stem and thick linear leaves upto 60cm, with pendulous spikes of 38-58cm bearing many tiny white but heavily spotted with magenta-purple flowers, each upto 2cm in diameter hook-shaped. Column short stout, rostellum shortly beaked. Capsules pyriform, pedicelled. Seeds numerous, minute, testa drawn out at the end or winged.

Flowering & Fruiting: May-June.

Habitat Ecology: Grows as an epiphyte preferably on Mango tree; Anu, 900m.

Material Examined: HPU-EL-SKS-ABS-15233; 05.06.2010.

Parts Used: Leaves. Flowers.

Folk Uses: Poultice of leaves applied for healing fractured bones; also 3-5g of its powder with honey given for the same. In veterinary the larger dose of 25-50g is given for bone related ailments. Flowers cooked as porridge ('halwa') and given for internal injuries*, joint pains* and bone fractures.

Uses in Literature: Recorded in literature as antirheumatic, state flower of Arunachal and Assam; and for bone injury, bruises, cuts, fracture, internal injury, strength, swelling and wounds (Ambasta, 1986; Khare, 2007).

Ricinus communis L. (Pl. 79B).

Syn.: *R. inermis* Jacq., *R. lividus* Jacq., *R. speciosus* Burm., *R. spectabilis* Blume, *R. viridis* Willd., *Croton spinosus* L.

Family: Euphorbiaceae.

Vern.: Airand, Arind, Erand.

Ayurvedic, English, Sanskrit and Regional Names:

Ayur.: Chanchu, Chitraa, Chitrabija, Eranda, Gandharva-hastaa, Manda, Panchaan-gula, Tribija, Triputi, Uruvaka, Uttaanpatraka, Vaataari, Vardhamaana, Vyaaghrapuchha;

Eng.: Castor Oil Plant, Palma Christi;

Sans.: Dirghadantaka, Triputiphala, Vyadatvaka;

Ass.: Eri;

B.: Bheranda;

Bo.: Rendi;

G.: Diveligo;

H.: Arand, Arend, Arangi, Erand, Ind, Rand;

Kan.: Avanakka, Chittavanaku;

Mar.: Yorandicha;

Oriya: Bheronta, Chitroka, Tindoko;

P.: Arand, Aneru, Bedanjir, Harnauli;

Tam.: Amanakku, Muttukkottilai, Sittamunuk;

Tel.: Amadam, Chittamudamu, Eramudapu;

U.**P.:** Rend.

Distribution: Planted and naturalised in Himalayan villages upto 2,400m.

Description: Evergreen soft wooded shrubs. Leaves peltate, large palmately lobed, serrated. Flowers reddish or yellowish, in large terminal branches. Capsules prickly. Seeds shining black with an aril at its base.

Flowering & Fruiting: March-August.

Habitat Ecology: Wastelands, dry areas, open slopes, roadsides; Jhaniyari, 900m.

Material Examined: HPU-EL-SKS-ABS-15234; 28.05.2008.

Parts Used: Leaves. Flower. Stem. Root. Seed Oil.

Folk Uses: Root extr. given for expelling intestinal worms. Stem used for scouring teeth*. Warm leaves applied as bandage on sprains. Flowers hung at the main door to ward off evil spirits and also for offering to deities. Seed paste mixed with butter applied for skin infection; massage with its seed oil alone or in combination of warm mustard oil considered useful for rheumatic pains and as hair tonic*. Also, 2 tsp of castor oil with hot milk given at bed time to cure constipation.

Active Constituents: These are ricinine and 1-methyl-3-cyano-4-methoxy-2-pyridone (*seeds, leaves*), β-amyrin, stigmasterol, β-sitosterol, ricinine, quercetin, rutin, hyperoside, n–dimethylricinine, 3-o-β-d-xylopyranosides, 3-o-β-d glucopyranosides and 3-o-β-d-rutinosides of kaempferol and quercetin (*leaves*), lysine, palmitic, stearic, hexadecenoic, oleic, linoleic, linoleni, ricinoleic acids, arachidic, chlorogenic, oleic, palmitic, ricinoleic, stearic and dihydrostearic acid, hexadecanoic, hydrocyanic and uric acid (*oil*).

Biological Activity: Activities exhibited are antiameobic, anticancerous, hypoglycaemic (*root, stem*), antiviral hypoglycaemic (*leaf*), diuretic. anticholestatic and hepatoprotective (*seed*).

Uses in Literature: Described earlier as an antidote to food poisoning, carminative, contraceptive, fish poison, edible, fodder, laxative, lubricant, luminent, psychoactive, purgative, spice, symbolic to ward off evil, vermicidal; and for abortion, aching joints, anasarca, boils, bone sprain of horses, burns, cholera, conjunctivitis, constipation, cough (vet.), delirium, dropsy, dysentery, earache, gout, headache, heat stroke, hydrocele, increasing milk yield, injury, intestinal worms, irregular menstruation, lumbago, migraine, muscular pain, night blindness, pneumonia, rearing silk worms, rheumatism, sciatica, scorpion bites, skin disease, skin ulcer, sores, swelling, throat problems (vet.), wounds (Ambasta, 1986; Bhandary *et al.*, 1996; Bhardwaj & Gakhar, 2003; Bhatt *et al.*, 2001; Borthakur, 1993; Hembrom, 1996; Hembrom & Goel, 2005; Jadhav, 2006; Jain, 1968, 1991; Jain & Saklani, 1992; Jha *et al.*, 1991; Kadel & Jain, 2006; Katewa & Arora, 1997; Khare, 2007; Kharkonger & Joseph, 1981; Khanna *et al.*, 1996a; Kirtikar & Basu, 1994; Kumar & Chauhan,

2005; Kumar *et al.*, 2004; Lal *et al.*, 1996; Mahato & Mahato, 1996; Mokat & Deokule, 2004; Panda, 1996; Pawar & Patil, 2004; Rama Rao & Henry, 1996; Sen & Behera, 2003; Sharma & Borthakur, 2003; Tirkey *et al.*, 2003a, b; Tiwari & Tiwari, 1996; Verma & Chauhan, 2006; Vihari, 1995; Watt, 1972).

Major Ayurvedic Formulations: Balarista, Eranda-pak, Eranda-saptaka-kvatha, Eranda tel, Erandimubhi-kwath, Gangharvahastadi -kvatha -churna, Laghuvisgaravataila, Rasnasaptak-kwath, Sadangaguggulu, Ratari-guggula.

Trade: Erandi, Eranda-bija, Manjeeth (roots). Castor seeds and oil are marketed.

Rosa damascena Mill. (Pl. 79C).

Family: Rosaceae.

Vern.: Gulab.

Ayurvedic, English, Sanskrit and Regional Names:

Ayur.: Taruni;

Eng.: Damask rose, Otto of roses;

Sans.: Fasli gulab;

H.: Shataputri;

Tam.: Isora.

Distribution: Throughout India.

Description: Perennial shrubs with unequal large-sized prickles. Sepals reflexed, moderately hooked stem. Leaf stipulate, compound, imparipinnate. Flowers pink, sweet scented in terminal axillary corymbs. Achenes hard.

Flowering & Fruiting: Greater part of the year.

Habitat Ecology: Cultivated; Muthan, 900m.

Material Examined: HPU-EL-SKS-ABS-15235; 15.04.2008.

Parts Used: Flowers. Fruit.

Folk Uses: Planted as an ornamental*; also considered sacred* and its flowers offered to deities. Petals eaten for tonsillitis, and sore throat*. 'Gulkhand' prepared from its petals mixed with sugar pellets prescribed with milk for constipation*, also, fresh or dried flowers mixed with raw sugar given for constipation to animals. Rose water (extr. of flowers) used in preparing face packs* and also used as coolant* for sore eyes*. Eye drops prepared by mixing rose water with 3 fruits of 'laung' (*Syzygium aromaticum* (L.) Merr. & Perry.), 1 black pepper, 1g 'fatkari' (crystals of potassium aluminium sulphate), 1g white salt, 1g camphor, ¼ tsp honey, 1 tsp water good for checking cataract of eyes* (1 drop twice daily, for 15 days).

Active Constituents: Constituents characterized are quercetin, kaempferol, cyaniding (*whole plant*) and cyaniding-3,5-diglucoside (*petals*).

Uses in Literature: Used as beverages, confectionery, drugs, food and perfume oil (Ambasta, 1986, Khare, 2007; Kirtikar & Basu, 1994; Sharma, 2003).

Rosmarinus officinalis L. (Pl. 79D).

Family: Lamiaceae.

Vern.: Rosemary.

English and Folk Names:

Eng.: Rosemary;

Folk.: Rusmari.

Distribution: Cultivated in hilly terrain of subtropical areas.

Description: Evergreen shrubs with lavender-like leaves, erect stem and sessile, opposite, smooth green leaves. Flowers in clusters. Fruit oval-shaped.

Flowering & Fruiting: February-June.

Habitat Ecology: Cultivated; Khaggal, 920m.

Material Examined: HPU-EL-SKS-ABS-15236; 03.07.2008.

Parts Used: Whole Plant. Leaf. Petal Oil.

Folk Uses: 25-30ml decoction of the plant consumed thrice daily for 10-15 days for urinary disorders, rheumatism and as carminative*. Poultice of leaves applied for joints pains*, skin ailments* and stomachache*. Infusion of petals used in hair wash. Oil added to food stuff for flavouring*.

Active Constituents: Alpha-pinene, apigenin, borneol, camphene, camphor, 1,8-cineole, diosmetin, diosmin, flavonoids, isobutyl acetate, limonene, linalool, 3-octanone, oleanolic acid, rosmarinic acid, terpineol, verbenol, and other ursolic acids (*volatile oil*).

Biological Activity: Leaf juice shows antibacterial activity, kills *Staphylococcus aureus, E. coli.,* Bacillus subtilis.

Uses in Literature: Reported to be useful as antiinflammatory, antiseptic, astringent, hair tonic, hair wash; and for bath, dyspeptic complaints, hepatic and biliary functions, incipient catarrh, muscular affections, and rheumatism (Ambasta, 1986; Chopra *et al.*, 1956; Khare, 2007; Kirtikar & Basu, 1994).

Roylea cinerea (D. Don) Baill. (PI. 80A).

Syn.: *R. elegans* Wall., *R. calycina* (Roxb.) Briq.,*Ballota cinerea* Don, *Phlomis calycina* Roxb.

Family: Lamiaceae.

Vern.: Karkoi.

Hindi and Regional Names:

H.: Patkarru;

P.: Kaur, Kauri.

Distribution: Himalayas (Kashmir-Nepal): 1,200-3,700m.

Description: Shrubs upto 1.6m with shortly stalked, coarsely toothed leaves. Branches slender terete, tomentose. Flowers white often tinged with pink, in whorl. Nutlets obovoid-oblong, smooth.

Flowering & Fruiting: March-November.

Habitat Ecology: Frequent on open rocky slopes; Patnaun, 1,196m.

Material Examined: HPU-EL-SKS-ABS-15237; 17.10.2008.

Parts Used: Roots. Flowers. Leaves.

Folk Uses: Poultice of roots useful against joint pains*. Paste of leaves rubbed against insect bites*, boils, dermatitis* and expelling ectoparasites. In veterinary, juice of leaves applied as an emollient for healing open wounds, mouth sores and killing lice. Flowers used in perfumery* and held sacred for offerings; its extr. (1 tsp twice daily, till cure) considered antidiarrhoeal*, emetic, refrigerant* and as drops for eye troubles*.

Active Constituents: Principle activites due to betulin, β-amyrin, β-sitosterol, stigmasterol, cetyl alcohol, glucose, fructose, arabinose and palmitic, stearic, oleic, oxalic and tartaric acids (*leaves*).

Biological Activity: Plant extr. has depressing effect on CNS and relaxed skeletal muscle activities.

Uses in Literature: Used in India as blood purifier, febrifuge, snuff in tonsils, tonic; and for contusion produced by winds, fever and pimples (Ambasta, 1986; Jain, 1991; Khare, 2007).

Rubus ellipticus Sm. (PI. 80B).

Syn. : *R. rotundifolius* Wall., *R. gownj-phul* Roxb., *R. gowree-phul* Roxb., *R. flavus* Ham., *R. sessilifolius* Miq., *R. hirsutus* Roxb., *R. wallichianus* W. & A, *R. affinis* Maiden

Family: Rosaceae.

Vern.: Aakhe, Kharyadi.

English and Regional Names:

Eng.: Himalayasn Yellow-Raspberry;

Ass.: Jotelupoka;

H.: Anehhu, Hinsalu;

P.: Akhi.

Distribution: Throughout India upto 2,300m.

Description: Shrubs with bristly branches and compound leaves. Leaflets-3. Flowers white. Fruits globose, yellow, succulent.

Flowering & Fruiting: January-May.

Habitat Ecology: Shrubberies, cultivated areas; Hira Nagar, 900m.

Material Examined: HPU-EL-SKS-ABS-15238; 17.04.2010.

Parts Used: Whole Plant. Roots. Stem. Leaves. Fruit. Wood.

Folk Uses: Plant used as fuel wood*; and its leaves relished by sheep and goats. A gargle of its crushed roots along with a pinch of black pepper useful against tonsillitis*. Fruits edible; its decoction prescribed as a mild laxative*. Wood suitable for gun-powder*.

Active Constituents: Principle constituents are octacosanol, β-sitosterol, its glucoside, octacosanoic, ursolic and rubitic acids (*roots*) and deoxysericoside (*leaves*).

Biological Activity: Effect on CNS and antifertility activity found positive; spasmolytic activity confirmed.

Uses in Literature: Recorded earlier for cracked heals, dyeing, dysentery, malaria, stomachache, worms; and as edible (fruits) (Jain, 1991; Khare, 2007; Lal *et al.*, 1996; Mao 1993; Megoneitso & Rao, 1983; Panda, 1996; Watt, 1972).

Rumex hastatus Don (Pl. 80C).

Family: Polygonaceae.

Vern.: Maroli.

Regional Names:

Kum. : Amlora, Chulmora;

P.: Amla, Amlora, Ami, Khattimal, Malori-gha.

Distribution: Common in Himalayas from Kashmir-Kumaon.

Description: Bushy shrubs having broadly triangular, stalked leaves, woody rootstock and tiny, greenish-white flowers. Fruits pinkish, wings papery. Nuts brown.

Flowering & Fruiting: November-April.

Habitat Ecology: Stony slopes, common on walls of cultivated terraces; Krishna nagar, 858m.

Material Examined: HPU-EL-SKS-ABS-15239; 25.12.2010.

Parts Used: Whole Plant. Leaves.

Folk Uses: Planted for conservation of soil*. Leaves eaten raw as a source of vit. C*, also used in culinary* preparations. Decoction of leaves used as mouth wash* and astringent*. Poultice of leaves applied on skin affections*.

Active Constituents: Vit. A, calcium, β-carotene (*leaves*), 21-23% tannin (*root, bark*).

Uses in Literature: Regarded useful for burns, cuts, cooling, chutneys, fever; and as pot herb (Ambasta, 1986; Chandra Sekar & Srivastava, 2003; Khare, 2007).

Ruta graveolens L. (Pl. 80D).

Syn. : *R. angustifolia* Pers., *R. chalepensis* Wall.

Family: Rutaceae.

Vern.: Rue, Ruta.

English and Regional Names:

Eng.: Garden Rue;

Tam.: Aruvada.

Distribution: Cultivated in India.

Description: Strong smelling herbs possessing petioled triangular leaves which are decompound and with various segments. Corymbs spreading. Sepals triangular, acute. Petals ciliate. Capsule obtuse, shortly pedicelled.

Flowering & Fruiting: July-September.

Habitat Ecology: Planted in India; Neri, 381m.

Material Examined: HPU-EL-SKS-ABS-15240; 03.07.2008.

Parts Used: Plant. Leaves.

Folk Uses: Leaves eaten raw for increasing appetite* and to check gastric trouble and stomachache. 15-20ml decoction of the plant taken twice daily for 5-10 days for menstrual disorders and uterine problems*.

Active Constituents: These are 2-undecanone, 2-nonanone, 2-nonyl acetate, psoralen, bergapten, xanthotoxin, rutin, flavonoids: quercetin, coumarins, bergapten, daphnoretin, isoimperatorin, naphthoherniarin, rutamarin, rutarin, scopoletin, umbelliferone (*plant*).

Uses in Literature: Known to be useful as abortifacient, antiepileptic, antispasmodic, antitumourous, colic, emmenagogue, irritant, rubefacient stomachic, tonic; and for ammenorrhoea, arthritis, chest pain, cough, croupy affections, flatulence, headache, hysterical conditions, menorrhoea, muscular bronchitis and sciatica (Ambasta, 1986; Khare, 2007).

Saccharum arundinaceum Retz. (Pl. 81A).

Syn.: *Erianthus arundinaceus* (Retz.) Jesw. *ex* Hyene, *S. cillare* Anderss, *S. munja* Roxb., *S. procerum* Roxb.

Family: Poaceae.

Vern.: Sarphar.

Ayurvedic, Sanskrit and Regional Names:

Ayur.: Bhadramuja, Munja, Raamshara, Sara, Shara, Vaana;

Sans.: Munja;

B.: Teng;

Mal.: Munja;

P.: Sarkanda;

Tam.: Munji;

Tel.: Adavicheruk.

Distribution: Throughout the plains and low hills of India.

Description: A gigantic tufted grass with glabrous stem having very long internodes and long upper leaves which are broad at the middle and taper to a slender point. Midrib white. Leaf sheath glabrous. Panicle long, clothed with soft cream-coloured, rarely purplish Hair. Grain oblong.

Flowering & Fruiting: October-December.

Habitat Ecology: Wastelands, drier areas, rocks; Tikker, 800m.

Material Examined: HPU-EL-SKS-ABS-15241; 27.06.2010.

Parts Used: Whole Plant. Stem. Flowering Scape.

Folk Uses: Plant employed for roofing huts; also its pieces of stem used as writing pen*. Stems used for making 'chicks', (door/ window curtains) and 'moorahs', (sitting armless chair). Flowering scape woven into winnowing trays* ('chhaj'), and as cover* for grain heaps and carts and strings for weaving bed steads*. Dried cane used as fuel*.

Uses in Literature: Described in Indian literature as aphrodisiac, demulcent, diuretic, refrigerant; and for blood troubles, burning sensation, delivery, paint brushes, thatched houses, thirst, urinary complaints (Ambasta, 1986; Khare, 2007).

Saccharum officinarum L. (Pl. 81B).

Syn.: *S. sinense* Roxb.

Family: Poaceae.

Vern.: Ganna, Kamandi.

Ayurvedic, English, Sanskrit and Regional Names:

Ayur.: Asipatra, Bhuurirasa, Dirgha-chhada, Gudamuula, Ikshu, Madhutrna, Morata, Trnarasa;

Eng.: Sugar Cane, Noble Cane;

Sans.: Adhipatra, Gudakastha, Rasola;

Ass.: Pura-Kuhiyar;

B.: Ak, Kushiar, Puri;

Bo.: Gol, Serdi;

G.: Naisakar, Sheradi, Uns;

H.: Ganna, Ikh, Kumad;

Mal.: Kantarakam, Madhutrianam;

Mar.: Aos, Kabbo;

Oriya: Aku, Gudodaru;

P.: Ganna, Khaud, Paunda;

Tam.: Angarigai, Madudirunam;

Tel.: Arukanupulakranugu, Tellacheraku.

Distribution: Cultivated throughout India.

Description: Tall perennials possessing solid stem and linear, lanceolate, scabrid leaves. Spikelets lanceolate, surrounded by dense silky white hair. Seeds endospermic with shield shaped scutellum.

Flowering & Fruiting: January-April.

Habitat Ecology: Cultivated; Bijari, Cha kmoh, 736m.

Material Examined: HPU-EL-SKS-ABS-15242; 15.05.2010.

Parts Used: Plant. Stem. Leaves.

Folk Uses: Plant considered sacred and worshipped* on 'Hoi' (a day when mothers in N. India fast for good health and long life of their sons). Stems chewed by children and elders for sweet cane juice; also extracted cane juice considered good for jaundice, refrigerant and refreshing; and employed for making 'gur' (jaggery)*, 'shakkar' and local wine 'sur'. Left out stems used as fuel and cattle feed. Dried cane used as torches and thatching material for huts. Liquid jaggery boiled with pods of 'amaltas' (*Cassia fistula* L.) and given twice daily till cure for constipation* in animals. If mixed with 'ajwain' (*Carum copticum* Benth.) it is considered good for abdominal pain* in cattle. For other cattle diseases like 'saridan' (upward fixation of patella)*, crushed powder obtained by heating liquid jaggery mixed with desi ghee is given twice daily, whereas for ' padjibhi' (glossitis)*, hot liquid jagerry applied on tongue with the help of a brush. For enhancing milk* yield and to check constipation, hot liquid jaggery considered useful. Sugarcane molasses 'mali' or its leaves given orally for easy placental evagination* in animals.

Active Constituents: Saccharans A-F, neocarlinoside, neoisoschaftoside, orientin, swerti-ajaponin, vicenin-2, p-coumaric, ferulic, caffeic and 3, 4 di-o-methyl caffeic acids (*stalk juice*) have been isolated.

Biological Activity: Saccharans A-F exhibited hypoglycaemic activity in normal and alloxan produced hyperglycaemic rats.

Uses in Literature: Reported to be useful as anthelmintic, antidote to arsenical poisoning, antiseptic, diuretic, emollient, fodder, laxative, manure, nutritious, refrigerant, sweetener, symbolic in sculpture (bow of Kamdev-God of love made from canes), thatching material for huts, tonic, torches (dried cane); and for bagasse or fibre, beverages, disorder of bile and gall bladder, hepatitis, inflammation, jaundice (juice as drink) lactation (vet.), rough cordage, soothing and protecting the alimentary tract, strengthening the gum, sugarcane wax, tying, ulcer, wax papers and wounds (vet.) (Ambasta, 1986; Borthakur, 1996a; Borthakur *et al.*, 2004; Dagar & Dagar, 1996; Gogoi *et al.*, 2003; Jain, 1991; Khare, 2007; Kurian, 1995; Lalramnghinglova, 1996; Mishra *et al.*, 1996b; Pal & Jain, 1998; Sahu, 1996; Siwakoti & Verma, 1996; Tirkey *et al.*, 2003a; Watt, 1972).

Major Ayurvedic Formulations: Bala-taila, Navaratnaralamrganka-rasa.

Saccharum spontaneum L. (Pl. 81C).

Family: Poaceae.

Vern.: Kans.

Ayurvedic, Sanskrit and Regional Names:

Ayur.: Kaasha, Kandekshu, Shvetachaamara;

Sans.: Kasa;

B.: Kagara, Kanga, Kans, Kas, Kash;

G.: Kans, Kansado;

H.: Kans, Kas;

Mal.: Nan-nana;

Mar.: Kagara;

P.: Kahi, Kanh, Kans, Sarkara;

Tam.: Nanal;

Tel.: Kakicheraku.

Distribution: Throughout warmer parts of India; ascending upto 2,000m in Himalayas.

Description: A tall erect grass having narrowly linear, acuminated, rigid, coriaceous, glabrous leaves and lanceolate, silky hairy panicles. Rachis slender. Spikelet lanceolated, bearded with spreading silky hair. Glumes 4.

Flowering & Fruiting: Greater part of the year.

Habitat Ecology: Drier areas; Nalti, 787m.

Material Examined: HPU-EL-SKS-ABS-15243; 25.06.2010.

Parts Used: Whole Plant. Stem. Leaves.

Folk Uses: Plant considered sacred*. Plant also employed for thatching roof of huts as it keeps hut cool during summers and warm in winters. Stems utilized for making winnowing trays and 'kalam' (writing pens)*. Àlso its pieces kept near corpse for the last rites*. Young leaves used as fodder.

Uses in Literature: Known in India as an aphrodiasic, fodder, haemorrhagic, laxative; and for biliousness, burning sensation, diatheris, paper manufacturing, phthisis, stranguary, thatching, urinary troubles, vesical calculi diseases of blood (Ambasta, 1986; Chhetri, 2005; Khare, 2007).

Salvia coccinea L. (Pl. 81D).

Syn.: *S. gerardiana* Benth.

Family: Lamiaceae.

Vern.: Phultulsi.

Ayurvedic and English Names:

Ayur.: Samudrashosha;

Eng.: Sage.

Distribution: More or less cultivated throughout India.

Description: Erect pubescent herbs with scarlet red flowers borne in terminal racemes. Flowers arranged in verticillaster inflorescence. Nutlets smooth.

Flowering & Fruiting: June-August.

Habitat Ecology: Dry slopes;Tarkawari, 980m.

Material Examined: HPU-EL-SKS-ABS-15244; 19.03.2010.

Parts Used: Leaves. Flowers.

Folk Uses: Leaves used as fodder*. Flowers used for offering* to various deities.

Uses in Literature: Considered useful for lumbago, renal troubles and tubercular cough (Ambasta, 1986; Khare, 2007).

Santalum album L. (Pl. 82A).

Family: Santalaceae.

Vern.: Chandan.

Ayurvedic and English Names:

Ayur.: Bhadra-Shree, Chandana, Ekaangi, Gandhsaara, Hima, Malayaja, Shvetachandana, Shrikhanda;

Eng.: White Sandalwood.

Distribution: Cultivated in India.

Description: A semiparasitic tree having woody stem and simple, entire, exstipulate leaves. Flowers minute, regular, bisexual and in racemes or spikes. Fruits a drupe.

Flowering & Fruiting: April-August.

Habitat Ecology: Cultivated as novelties; Brahmni, 858m.

Material Examined: HPU-EL-SKS-ABS-15245; 29.05.2011.

Parts Used: Twigs. Leaves. Wood.

Folk Uses: Leaves used as fodder*. Wood considered sacred and employed in 'pooja', 'hawan', funeral pyre* and sporting 'tilak'. Paste of heartwood applied for checking inflammation, skin acnes*, pimples*, rashes, headaches* and as coolant. Heartwood used in cosmetics and 'batana' (a mixture of its powder along with 'basin' and rose water applied to bride and groom in the marriage ceremony).

Active Constituents: Main ingredients are triterpene-urs 12-en-3 butyl-palmitate (*bark*), alpha & beta santol, sesquiterpene hydrocarbons-alpha & beta, epibeta-santalene, alpha & beta curcumene and beta farnesene, dihydrogarofuran (*heartwood*).

Uses in Literature: Considered useful as antibacterial, antifungal, anti-inflammatory, antipyretic, antiseptic, aphrodiasic, diaphoretic, diuretic, expectorant; and for biliousness, bronchitis, burning sensation, diseases of gallbladder, dysuria, heart troubles, thirst, urethral discharges, urinary chronic cystitis and vaginal discharges (Ambasta, 1986; Khare, 2007).

Sapindus mukorossi Gaertn. (Pl. 82B).

Syn.: *S. abruptus* Lour.

Family: Sapindaceae.

Vern.: Doddae, Doda, Reetha, Ritha.

Ayurvedic, English, Sanskrit and Regional Names:

Ayur.: Arishtaka, Phenila;

Eng.: Soap Nut, Tree of North India;

Sans.: Arishta, Phenila, Urista;

Ass.: Haithaguti;

B.: Chotta-Ritha, Ritha;

Bo.: Kanmar, Ritha;

H.: Aritha, Dodam, Kanmar, Ritha;

Oriya: Ita;

P.: Aritha, Dodan, Haritha (the fruit), Ritha.

Distribution: Cultivated throughout N. India, ascending to 1,650m.

Description: Large deciduous trees possessing long leaves, opposite, lanceolate, glabrous leaflets and small, pale brown, globose flowers. Drupes yellow, globose. Seeds black, loosely contained in fleshy pericarp.

Flowering & Fruiting: May-August.

Habitat Ecology: Open slopes, edges of agricultural fields; Tal, 947m.

Material Examined: HPU-EL-SKS-ABS-15246; 22.05.2011.

Parts Used: Leaves. Fruits. Seeds.

Folk Uses: Poultice of leaves applied on body part affected by dog's bite in animals. Powdered fruits used as a substitute for bath soap. Water boiled with its fruits used for washing woollens and silken clothes; and as shampoo for hair. Seeds worshipped by natives as God 'Bharoin' and in magico-religious* ceremonies. Powdered seeds with red chillies applied on rabied dog's bite. Necklace* of seeds born to ward off evil eyes*.

Active Constituents: Active properties due to sapindoside V, sapindoside A & B, heedragenin as a common glyconel, sapindoside V, mukorosside, a new saponin-emarginatoside (*fruits*), protein (31.87), carbohydrate (2.21) and starch (11.94%) (*kernel*).

Biological Activity: Saponins show high spermicidal activity in both spot and IPPF tests with human semen, with MED of 0.05 and 0.51, respectively. Saponins from fruits when injected in rabbits lower cholesterols and brings down B.P. by 25%.

Uses in Literature: Recorded earlier as an anthelmintic, cattle feed, contraceptive, detergent, emetic, expectorant, febrifuge, fish poison, hair wash, oil seeds, purgative; and for charcoal making, chlorosis, epilepsy, polishing jewellery, pneumonia, salvation, sinusitis, washing and bleaching cardamoms, ulcer (vet.) (Ambasta, 1986; Banerjee & Pal, 1996; Borthakur, 1993; Chauhan, 1999; Chhetri, 2005; Chopra *et al.*, 1956; Hajra & Baishya, 1981; Jain, 1965, 1991; Jain & Borthakur, 1980; Karnick *et al.*, 1981; Khare, 2007; Rao, 1996; Rao *et al.*, 1996; Singh, 1999b, Siwakoti & Verma, 1996; Venkataratnam & Venkata Raju, 2004; Vihari, 1995).

Major Ayurvedic Formulations: Mahalaxadi-Tel.

Trade: Dried fruits sold in the local and national markets as Reetha.

Saraca asoca (Roxb.) De Wild. (Pl. 82C).

Syn.: *S. indica* auct. non L., *S. arborescens* Burm., *S. minor* Miq., *S. zollingeriana* Miq., *Jonesia asoca* Roxb. *J. pinnata* Willd.

Family: Fabaceae.

Vern.: Ashoka.

Ayurvedic, English, Sanskrit and Regional Names:

Ayur.: Ashoka, Ashoku, Ghandhapushpa, Hempushpa, Pindapushpa, Taamrapallava;

Eng.: Ashok Tree, Ashoka Tree;

Sans.: Karnapuraka;

Ass.: Ashok, Jasundi;

B.: Asok, Asoka;

Bo.: Ashok, Jasundi;

G.: Ashopalava;

H.: Ashok;

Kan.: Aksunkar, Asokadamara, Kenkali;

Mal.: Hemapushpam, Vanjulam;

Mar.: Asoka, Jasundi;

Oriya: Osoka;

Tam.: Asogam, Malaikkarunai;

Tel.: Asokamu, Vanjalamu.

Distribution: W. Ghats; often cultivated.

Description: Large evergreen trees with rigidly coriaceous paripinnate leaves. Leaflets oblong-lanceolate, drooping. Flowers orange-yellow, in dense sessile corymbose axillary panicles. Pods glabrous, truncate, flat, indehiscent, coriaceous with 4-6 exalbuminous seeds.

Flowering & Fruiting: March-September.

Habitat Ecology: Edges of fields, open slopes; Khaggal, 920m.

Material Examined: HPU-EL-SKS-ABS-15247; 03.07.2008.

Parts Used: Plant. Bark.

Folk Uses: Planted as an ornamental* shadow tree; and also considered sacred. ½ tsp of powdered bark with luke warm water prescribed once a day for 15 days to cure menorrhagia, leucorrhoea and chest infection*. Wood used for agricultural implements, construction and furniture.

Active Constituents: Therapeutic properties attributed to leucopelarogonidin-3-o-β-D glucoside, leucopelarogonodin, leucocyanidin, β-sitosterol (*stem bark*).

Biological Activity: Flower bud inhibits *Escherichia coli, Salmonella viballerup, Vibrio cholerae, Shigella dysenteriae* and *Shigella boydii.*

Uses in Literature: Reported earlier as abode of god (sacred), astringent, blood purifier, diuretic, edible, haemorrhagic, masticatory, psychoactive plant, substitute of milk in tea, tonic to prevent miscarriage, uterine sedative; and for diabetes, heart disease, leucorrhoea, menorrhagia, menstrual irregularity, plough, shaft, treatment of excessive menstruation, urinogenital diseases and uterine affections (Ambasta, 1986; Bhandary *et al.*, 1996; Gogoi & Borthakur, 2001; Jain *et al.*, 1991; Khare, 2007; Lalramnghinglova, 1996; Maya *et al.*, 2003; Sharma *et al.*, 2002; Siwakoti & Verma, 1996; Srivastava *et al.*, 2003).

Major Ayurvedic Formulations: Ashoka-dikvatha, Ashoka-vasi, Ashoka-ghrita, Ashoka-rishta.

Sechium edule Sw. (Pl. 82D).

Family: Cucurbitaceae.

Vern.: Lonku.

English and Regional Names:

Eng.: Chayote, Chow: Chow;

Kan.: Badane, Seema;

Tam.: Kattirikkai, Seeme.

Distribution: Cultivated.

Description: Perennial climbers. Tendril tri -or penta-fid. Leaves hirsute on both sides. Flowers greenish-yellow. Fruits echinate, one-seeded.

Flowering & Fruiting: June-September.

Habitat Ecology: Cultivated; Muthan, 900m.

Material Examined: HPU-EL-SKS-ABS-15248; 05.06.2011.

Part Used: Fruit.

Folk Uses: Fruit cooked as vegetable and considered good for gastro-intestinal disorders* and sick people*.

Active Constituents: These are α-amino ã-ureidobutyric acid (*seeds*), glucosides of sitosterol and stigmasterol (*fruits*).

Biological Activity: β-sitosterol and stigmasterol-β-D-glucosides exhibit antiinflammatory activity in rat paw odema.

Uses in Literature: Used earlier for its edible fruits, fibre, fodder, diabetes; and as vegetable (Ambasta, 1986 Khare, 2007; Negi & Gaur, 1991; Samati, 2004).

Sesamum indicum DC. (Pl. 83A).

Syn.: *S. luteum* Retz., *S. occidentales* Heer & Regel, *S. orientale* L.

Family: Pedaliaceae.

Ayurvedic, English, Sanskrit and Regional Names:

Ayur.: Snehphala, Tila;

Eng.: Gingelly, Sesame;

Sans.: Homadhamya, Jattle, Pavitra, Puraphal;

Ass.: Tila;

B.: Bhadutil, Kalatil, Katil;

Bo.: Ashadital, Bariktil, Krishnatil, Tal;

G.: Mithutel, Tal, Tel, Til;

H.: Bariktel, Gingli, Krishnatel, Mithatel, Til;

Kan.: Ellu;

Mal.: Bijan;

Mar.: Til, Tila, Zilechatil;

Oriya: Khasa, Rasi;

P.: Kanjad, Til, Tili;

Tam.: Ellu, Nuvvulu:yelluchedi;

Tel.: Nuvvu, Nuvvula, Pollanuvvulu.

Distribution: Throughout warmer parts of India.

Description: A branched erect hairy annual with linear leaves, axillary, solitary and light pink or purplish flowers and tetragonal 2- or 4-valved capsule. Seeds numerous.

Flowering & Fruiting: July-September.

Habitat Ecology: Cultivated, wastelands, damp places; Nadaun, 483m.

Material Examined: HPU-EL-SKS-ABS-15249; 04.10.2010.

Parts Used: Leaves. Seeds.

Folk Uses: Leaf paste applied externally for painful swellings*. Seeds considered sacred and used in performing various religious ceremonies; also constitute an important component of 'hawan'. Seed oil used for cooking purposes; and massaged externally for rheumatic pains*. Infusion of seeds given with butter twice daily till relief for bleeding piles. 2ml of its oil added to milk taken once a day to check body pains and for reducing cholesterol*. Seeds used for making 'panjiri', 'laddu' (sweet balls) or roasted and eaten* as such to keep the body warm especially during winters. Seed oil also given to animals to increase milk production.

Active Constituents: Active constituents are globulin, thiamine and niacine (*seeds*), pedaliin (*leaves*).

Uses in Literature: Known in Indian literature as an abortifacient, demulcent, diuretic, edible, emollient, galactagogue, sacred, source of protein thiamine and niacine, tonic; and for amenorrhoea, boils, bone fracture (vet.), broken horns (vet.), burn (vet.), contusions (vet.), dysentery, dysentery (vet.), dysmenorrhoea, eye and spleen complaints, expel placenta (vet.), general horse care (vet.), kidney and bladder disorder, lactation (vet.), mouth diseases (vet.), paste for burns, piles, retained placenta (vet.), scalds, ulcer and wounds (vet.) (Ambasta, 1986; Bhatt *et al.*, 2001; Bhattacharjee *et al.*, 1980; Chopra *et al.*, 1956; Goud & Pullaiah, 1996b; Jain, 1991, Kumar *et al.*, 2004; Ramadas *et al.*, 2000; Reddy *et al.*, 1998; Sharma, 2003; Sikarwar *et al.*, 1994; Sudarsanam *et al.*, 1995).

Setaria viridis Beauv. (Pl. 83B).

Syn.: *S. affinis* Schult., *S. ambigua* Guss., *S. imberbis* Roem. & Schult., *S. nana* Dum., *S. penicillata* Wall., *S. villosa* Beauv., *Panicum bicolor* Moench., *P. laevigatum*

Lam., *P. psilocaulum* Steud., *P. purpurascens* Opiz, *P. reclinatum* Vill., *P. viresceus* Salzm., *P. viride* L.

Family: Poaceae.

Vern.: Hathi grass.

Ayurvedic and English Names:

Ayur.: Kangu, Kanguni, Kangunikaa;

Eng.: Green bristle-grass.

Distribution: Throughout India, temperate and sub-tropical regions.

Description: Grasses having stems upto 100cm. Leaves usually pale green, glabrous or with few scattered hair, margins rough. Spikelets nearly sessile, crowded in dense spike-like panicles. Bristles unequal, green or tinged with red. Fruiting glume smooth. Grain pale, free and enclosed within the hardened glume.

Flowering & Fruiting: September-December.

Habitat Ecology: Wild grass; Kakkar, 850m.

Material Examined: HPU-EL-SKS-ABS-15250; 26.12.2009.

Parts Used: Whole Plant. Leaves.

Folk Uses: Plant considered nutritious green fodder for cattle. Straw useful for making hay. Decoction of leaves given for easy evagination of placenta* in animals.

Active Constituents: Proteins, fats, minerals, crude fibre, carbohydrates, prolamin, albumin, globulin (*grains*).

Uses in Literature: Described earlier for its use as fodder, hay, veterinary medicines; and for gravid uterus, pain after parturition, rheumatism (Agarwal, 2003; Ambasta, 1986; Duthie, 1888; Khanna, 2002; Khare, 2007).

Solanum erianthum Don (Pl. 83C).

Syn.: *S. verbascifolium* L., *S. pubescens* Roxb.

Family: Solanaceae.

Vern.: Rada.

English, Sanskrit and Regional Names:

Eng.: Potato Tree;

Sans.: Gandira, Pathi, Priyamkari, Vidari;

Gar. : Asedu;

Kan.: Kadusinde, Sow-dangigida;

M.P.: Chichora;

Mal.: Chunta, Erichunta, Malanjunta;

Mar.: Kutri;

P.: Kalamewa, Tiari, Ola;

Tam.: Anaichundai, Malaichundai.

Distribution: Throughout India.

Description: A shrub covered with dense, stellate tomentum. Flowers greenish-white. Berries globose, yellow.

Flowering & Fruiting: July-October.

Habitat Ecology: Wastelands, roadsides; Tal, 947m.

Material Examined: HPU-EL-SKS-ABS-15251; 24.07.2008.

Part Used: Leaves.

Folk Uses: Poultice of leaves good for healing wounds*.

Active Constituents: Solasonine (*leaves, fruits*).

Uses in Literature: Known to be useful for body pains, burning sensation, edible purposes, expelling worm, horses affected with glanders, inflammations, urinary troubles, vaginal discharge and vertigo (Ambasta, 1986; Banerjee, 1996; Bhargava, 1981; Henry *et al.*, 1996; Hosagoudar & Henry, 1996b; Jain, 1991; Khare, 2007; Siwakoti & Verma, 1996).

Solanum melongena L. (Pl. 83D).

Syn.: *S. incarnum* L., *S. insanum* L., *S. zeilanicum* Scop., *S. undulatum* Lamk., *S. trongum* Lamk., *S. longum* Roxb., *S. pseudoundulatum* Blume, *S. esculentum* Dunal, *S. heteracanthum* Dunal, *S. melanocarpum* Dunal, *S. sativum* Dunal, *S. pressum* Dunal, *S. cumingii* Dunal, *S. torrum* var. *inerme* Dalz. & Gibbs

Family: Solanaceae.

Vern.: Baingan.

Ayurvedic, English, Sanskrit and Regional Names:

Ayur.: Bhantaa, Bhantaki, Vaartaaka, Vaartaakini, Vaartaaku, Vrintaaka;

Eng.: Brinjal, Egg Plant;

Sans.: Drighavartaki, Nilakantaka;

Ass.: Jatibengani;

B.: Bartaku, Kalibegun;

Bo.: Waingi;

G.: Rihgni;

H.: Baigun, Baingan;

Kan.: Badanekayi, Doddabadane;

Kash.: Tu kmibadanjanurmi, Vangun;

Mal.: Hinguli, Nilavalutina;

Mar.: Vangi;

Oriya: Bartakki;

P.: Bengan, Vataun;

Tam.: Kalliyam, Mullukkattari;

Tel.: Mettavanga, Nidupuvanga.

Distribution: Widely cultivated in warmer regions of India.

Description: Erect herbs with shoot prickles. Leaves elliptic with rounded lobes which are woolly beneath. Flowers solitary blue. Berries elongated, dark purple.

Flowering & Fruiting: June-September.

Habitat Ecology: Cultivated; Ghodiwala pind, 930m.

Material Examined: HPU-EL-SKS-ABS-15252; 25.09.2008.

Part Used: Fruit.

Folk Uses: Fruits cooked as vegetable and considered good for heart*, gastric problems* and to ladies after child birth as a source of iron*. Ash of burnt fruits with honey heals blisters on tongue* and mouth ulcers and piles*. Infusion of plant applied on hands and feet to check sweating*.

Active Constituents: Solasodine and diosegin characterized from *fruits.*

Uses in Literature: Known earlier as an alternative, antiasthmatic, diuretic, laxative, stimulant, tonic, vegetable; and for abscess, asthma, burns, cold sores, dropsy, enlargment of liver, epitaxis (vet.), fever, hydrophobia, inflammation of kidney and bladder, otitis producing uncomfortable pregnancy, sleep, skin disease, toothache, ulcers, virulent gonorrhoea (Ambasta, 1986; Gill *et al.*, 1993; Jain, 1991; Khare, 2007; Rama Rao & Henry, 1996; Sen & Behera, 2003).

Adulterant or Substitute: Roots used for that of 'sadabahar' (*Catharanthus roseus* G. Don).

Solanum nigrum L. (Pl. 84A).

Syn.: *S. rubrum* Mill., *S. triangular* Lamk., *A. villsoum* Lamk., *S. incertum* Dunal, *S. nodiflorum* Jacq., *S. uliginosum* Blume, *S. roxburghii* Dunal, *S. fistulosum* Rich., *S. miniatum* Bernh., *S. paludosum* Dunal, *S. erithraeum* Dunal

Family: Solanaceae.

Vern.: Makko.

Ayurvedic, English, Sanskrit and Regional Names:

Ayur.: Dhvankshamaachi, Kaakaahya, Kaakamaachi, Kaakamaataa;

Eng.: Black Nightshade, Common Nightshade, Garden Nightshade, Hounds Berry, Morelle, Petty Morel;

Sans.: Bahuphala, Bahutikta, Ghanaghana, Jahenephala, Kaka, Rasayani, Saravatik-ata, Sundari, Tiktika, Vayasi;

Ass.: Pichakati;

B.: Ghati, Kamuni;

Bo.: Ghati, Kamnni, Mako;

H.: Gurkamai, Kabaiya, Mekoi;

P.: Kachmach, Kambei, Meko;

Tam.: Manattakkali;

Tel.: Gajuchettu, Kachi, Kakamachi, Kamanchi, Kasaka;

Urdu. : Makoya.

Distribution: Throughout India.

Description: Annual herbs. Leaves ovate-lanceolate, acuminate, glabrous, thin. Flowers white, small. Calyx saucer-shaped. Berries globose, purplish, smooth, shining. Seed discoid, yellow.

Flowering & Fruiting: January-July.

Habitat Ecology: Wastelands, moist areas; Dhoh, 1,034m.

Material Examined: HPU-EL-SKS-ABS-15253;10.07.2008.

Parts Used: Plant. Leaves. Flower. Fruit.

Folk Uses: 15-20ml extr. of plant prescribed for inflammatory swellings* (once daily for 3-5 days). Poultice of leaves checks swellings. Ripe berries considered good for fever, diarrhoea and as laxative. Powdered berries and flowers taken with honey for cough.

Active Constituents: Active properties of the *plant* due to 12-hydroxysolasodine, N-methyl solasodine, 23-o-acetyl-12-β-hydroxy solasodine, solanocapsine and solamargine.

Biological Activity: Plant CNS depressant, hypotensive and spasmolytic.

Uses in Literature: Known in India as an antidote to opium toxicity, antiseptic, aphrodisiac, diaphoretic, diuretic, emollient, expectorant, hydragogue, laxative, pot herb, sedative, stomachic; and for body swelling, boils, colitis, conjuctivitis, cough, diarrhoea, dysentery, ear complaints, easy expulsion of foetus, eye complaints, fever, facilitate delivery, goitre, heart ailments, inflammation of scortum, jaundice, kidney bladder, liver complaints, nostril complaints, piles, psoriasis, rheumatism, ringworm, skin diseases, sores, sprains, swelling, testicles, throat troubles, ulcers in mouth and urinary complaints (Ambasta, 1986; Chatterjee & Pakrashi, 1997; Chauhan, 1999; Chopra *et al.*, 1956; Dey, 1973; Dhiman, 1976; Jain, 1991; Khanna & Kumar, 2000; Khare, 2007; Kurian, 1995; Pawar & Patil, 2004; Upadhyay & Singh, 2005).

Solanum surattense Burm. f. (Pl. 84B).

Syn.: *S. armatum* Br., *S. jacquinii* Willd., *S. diffusum* Roxb., *S. virginiamum* Jacq., *S. xanthocarpum* Schrad. & Wendl., *S. maccanni* Sant.

Family: Solanaceae.

Vern.: Choti-Kandayai.

Ayurvedic, English, Sanskrit and Regional Names:

Ayur.: Dhaavani, Duhsparshaa, Kantakaari, Kantakaarikaa, Keteri, Kshudraa, Nidigdhaa, Nidigdhikaa, Vyaaghri;

Eng.: Dwarf Lid Brinjal, Yellow Berried Nightshade;

Sans.: Kantkaree, Nidigadhika;

B.: Kantikari;

G.: Bhoyaringani;

H.: Katai, Kateli, Ringni;

Mar.: Ankranti, Bheji-behun;

P.: Kandyali, Mahori, Warumba;

Tam.: Cundung katric, Kandankattiri;

Tel.: Nela-mulaka, Pinna-mulaka, Vankuda.

Distribution: Throughout India.

Description: Prickly diffuse perennial herbs having zig-zag stem with numerous branches, compressed prickles, hairy leaves which are armed on midrib with long yellow sharp prickles. Flowers blue, in a few flowered extra-axillary cyme. Berries globose yellow when ripe.

Flowering & Fruiting: April-October.

HabitatEcolog : Dry localities, roadsides, wastelands; Tal, 947m.

Material Examined: HPU-EL-SKS-ABS-15254; 03.07.2008.

Parts Used: Plant. Stem. Fruits. Seeds.

Folk Uses: One tsp of powdered stem and fruits with water recommended twice a day till relief to cure fever. Smoke of seeds and 'ajwain' (*Trachyspermum ammi* L.) inhaled to cure dental problems*. A pinch of powdered seeds kept in nasal passage to check sneezing*. Plant given as fodder* after removing prickles to increase milk* content of animals.

Active Constituents: Active principles attributed to solasodine, solmargine, β-solamargine, solasonine, diosgenin, solasurine, β-sitosterol, stigmasteryl glucoside (*fruits*), arachidic, linoleic, oleic, palmitic and stearic acids and solanines (*fruit oil*), isochlorogenic, neochlorogenic, chlorogenic acid and caffeic acid.

Biological Activity: Aq. and alc. extr. of the plant hypotensive.

Uses in Literature: Described in literature as an appetiser, carminative, diuretic, edible (seeds), laxative; and for asthma, blister, boils, bronchitis, chest pain, cold, cough, curries, diarrhoea, dropsy, ear and eye complaints, fever, gonorrhoea, gum trouble, heart diseases, lumbago, migraine, muscular pain, pain in jaw, paralysis, piles, pyorrhoea, rheumatism, scorpion bite, snake bite, sore throat, sterility in women, swelling and watery eyes (Ambasta, 1986; Chopra *et al.*, 1956; Jain, 1991; Kalpesh *et al.*, 2004; Khare, 2007; Kumar & Chauhan, 2005; Kurian, 1995; Maliya & Singh, 2003; Mohanty *et al.*, 1998; Uphof, 1968).

Major Ayurvedic Formulations: Chyawanprash, Dasamoolarist, Dasamool Kwath Kantakariavleha, Kantakari-ghrit,

Vyaghiharitaki.

Trade: Sold as 'Kantakari' in crude drug markets of India.

Solanum tuberosum L. (Pl 84C).

Family: Solanaceae.

Vern.: Aloo.

Ayurvedic, English and Regional Names:

Ayur.: Aaluka, Aaruka, Golaalu;

Eng.: Potato;

B.: Alu;

G.: Batata;

H.: Alu;

Kan.: Alu:gidde, Batate;

Mal.: Kizhangu, Urulam;

Mar.: Batata;

Oriya: Alu;

P.: Alu;

Tam.: Urala kilangu, Wallaraikilangu;

Tel.: Bangaladumpa, Uralagadda.

Distribution: Commonly cultivated from sea level upto 2,300m.

Description: Herbs with pinnately dissected leaves. Flowers regular white or violet in cymose, helicoid clusters. Capsules with minute seeds having abundant endosperm.

Flowering & Fruiting: May-September.

Habitat Ecology: Cultivated; Lamloo, 850m.

Material Examined: HPU-EL-SKS-ABS-15255; 05.12.2010.

Part Used: Tuber.

Folk Uses: Tubers relished as vegetable. Boiled tubers eaten on religious fast and believed to check hypertension; its paste applied on face to remove dark circles*. Also, these are given as sacred offering* ('sidha') with wheat flour and sugar to Brahmins on religious days. Tuber paste applied over burn injuries in animals.

Active Constituents: Constituents isolated are cycloartenol, stigmasterol, β-sitosterol, campesterol, cholesterol, a mixture of 4-mono-methyl sterol, chlorogenic, neochlorogenic and iso chlorogenic, 1-caffeoylquinic acid and glycosides of quercitin (*leaves*).

Uses in Literature: Used earlier as antacid, antiscorbutic, antispasmodic, aperient, diuretic, galactagogue; and for adjuvant, bloat (vet.), burns, burn injuries (vet.) cough, ethyl alcohol, food, feed, hypertension, medium for microorganisms, potato chips, starch, vegetable (Ambasta, 1986; Deeba *et al.*, 2009; Khare, 2007; Samati, 2004; Tiwari & Pande, 2006).

Sonchus asper Hill (Pl. 84D).

Syn.: *S. ferox* Wall., *S. oleraceus* Wall.

Family: Asteraceae.

Vern.: Bhursalae.

Ayurvedic, English and Regional Names:

Ayur.: Sahadevi;

Eng.: Spiny Leaves Sow, Thistle;

H.: Didhi;

Mar.: Mhatara.

Distribution: Throughout India and in Himalayas upto 4,000m.

Description: Milky herbs with clasping stem. Flowers yellow. Achenes compressed. Pappus copious, white.

Flowering & Fruiting: March-April.

Habitat Ecology: Common along roadsides, wastelands; Jhaniker, 1,050m.

Material Examined: HPU-EL-SKS-ABS-15256; 14.04.2011.

Parts Used: Aerial plant parts. Root.

Folk Uses: ½ teaspoon of powdered aerial plant parts taken twice a day to check headache*. Paste of roots massaged on gums to check toothache* and dental caries*.

Active Constituents: These are α and β lactucerol, ceryl alcohol, mannitol and traces of acetic acid (*latex*), sesquiterpene glycosides, ionone glycosides and ascorbic acid (*plant*).

Uses in Literature: Considered useful as an emollient, salad (young shoots), vegetable (whole plant); and for boils, cut, injuries and stomach pain (Anonymous, 1948-1976; Jain & Patole, 2001; Khare, 2007).

Soymida febrifuga A. Juss. (Pl 85A).

Syn.: *S. soymida,* Duncan., *S. rubra* Rottl., *Swietenia febrifuga* Roxb.

Family: Meliaceae.

Vern.: Rehana.

Ayurvedic, English, Sanskrit and Regional Names:

Ayur.: Maansrohini, Prahaarvalli, Rohinaa, Rohini;

Eng.: Bastard Cedar, Indian Red-wood;

Sans.: Paranga, Rohuna;

B.: Rohan, Rohina;

Bom. : Rohan, Rohing;

G.: Rohina;

H.: Rakat, Ruhun, Ruhunna, Rohan;

Kan.: Some, Suani;

Mar.: Ruhin, Ruhan;

Oriya: Karwai;

Tam.: Shem, Wond;

Tel.: Sonidsa manu, Sumi.

Distribution: Throughout India.

Description: Small glabrous trees possessing bitter bark, paripinnate leaves, opposite, entire obtuse leaflets, greenish white flowers in axillary or terminal panicles and dull red hard wood. Capsule smooth, black when ripe. Seeds numerous, flattish, winged at both ends.

Flowering & Fruiting: August-December.

Habitat Ecology: Occurs in forest, hill slopes; Panjot, 1,120m.

Material Examined: HPU-EL-SKS-ABS-15257; 07.05.2011.

Parts Used: Bark. Wood. Gum.

Folk Uses: ¼ tsp of powdered gum given to ladies along with dry fruits to overcome post labour complications*. 1 tsp of powdered bark given twice a day for joint pains*. Poultice of bark mixed with horse dung prescribed as plaster for cracked bones. Paste of its bark, rock salt and leaves of 'basuti' (*Colebrookia oppositifolia* Smith) applied as poultice to check joint pains. Moreover, hot fomentation of decoction of its bark, along with *Cuscuta reflexa* Roxb., *Vitex negundo* L., *Dodonaea viscosa* (Jacq.) L. in 2:1:1:1 ratio recommended for internal injuries and rheumatic pains, arthritis*, and bone injuries. Stem bark cooked with jaggery given as 'halwa' in case of fracture, also its paste tied as such with bamboo pieces over fractured part or with leaves of 'mehandru' (*Dodonaea viscosa* L.), rhizome of turmeric and leaves of 'bhangru' (*Vanda roxburghii* R. Br.), in animals.

Active Constituents: Isolated constituents are tetranor-triterpenoids and flavonoids febrifugine, A and B, febrinins A & B. Flavonoids naringenin, quercetin, myricetin dehydromyricetin (*plant*), sitosterol, obtusifoliol, syringetin, dihydrosyringetin, tetranor-triterpenoids (*root bark*), steroids, and methyl angolensate (*stem bark*).

Uses in Literature: Known in literature as antipyretic, astringent, tonic; and for bone fracture (vet.), debility, rheumatic swellings, stomatitis, sore and ulcer (vet.) substitute of oak bark and cinchona bark (Ambasta, 1986; Borthakur, 1980; Chopra *et al.*, 1956; Khare, 2007; Kirtikar & Basu, 1994; Verma *et al.*, 1999; Watt, 1972).

Spilanthes acmella Murr. (Pl. 85B).

Syn.: *S. calva* DC., *S. mauritiana* DC., *S. paniculata* auct. non DC., *S. pseudoacmella* auct. non (L.) Murr.

Family: Asteraceae.

Vern.: Karkara.

Ayurvedic and Regional Names:

Ayur.: Marahattikaa;

Ass.: Piraza;

Mal.: Kuppa manjel;

Mar.: Akkalkara, Pipulka;

P.: Akarkarha, Pokarmul;

Tel.: Maratimogga, Maratiteega.

Distribution: Throughout India.

Description: Annual herbs. Leaves opposite, stalked, ovate-lanceolate, toothed. Flower heads conical, solitary on long stalk and with yellow flowers. Heads long-peduncled axillary or terminal. Achenes dorsally compressed, enclosed in a scale.

Flowering & Fruiting: May-August.

Habitat Ecology: Grassy meadows, moist areas; Sujanpur, 567m.

Material Examined: HPU-EL-SKS-ABS-15258; 13.04.2008.

Part Used: Flower.

Folk Uses: Chewing of ray florets recommended as mouth freshner*, mouth wash* and for mouth ulceration*, toothache, throat infection and increasing flow of saliva*. Poultice of florets applied to check headache* and stammering in infants.

Active Constituents: Attributed properties due to myricyl alcohol, β-sitosterol its glucosides, stigmasterol glucoside, α and β-amyrin and amyrin acetate, palmitic and stearic acid, a eudesmanolide I (*flower heads*) and spiloanthol.

Biological Activity: Ether extr. of fresh flowering tops lethal to Anophelin larvae. Spilanthol shows strong sialagogic action, and acts as local anaesthetic.

Uses in Literature: Recorded to be useful for caries, dysentery, inflammation of jaw bones, paralysis of tongue, scabies, stammering of children, throat infection, toothache; and as diuretic, fish poison, local anaesthesia, lithotropic, purgative, silagogue and stimulant (Ambasta, 1986; Khare, 2007; Upadhyay & Singh, 2005).

Spinacia oleracea L. (Pl. 85C).

Syn.: *S. tetrandra* Roxb.

Family: Chenopodiaceae.

Vern.: Palak.

Ayurvedic, English and Regional Names:

Ayur.: Paalankikaa, Paalankya, Paalakyaa;

Eng.: Garden Spinach;

Tam.: Vasaiyila-keerai.

Distribution: Cultivated throughout India.

Description: Annual herbs having deltoid, ovate acuminate, acute, broadly pinnatifidly lobed leaves. Flowers dioecious, ebracteate; males in terminal leafless spikes, females in axillary clusters. Utricles hard, compressed, adanate to the perianth. Seed vertical, testa thin, albumen floury.

Flowering & Fruiting: January-April.

Habitat Ecology: Cultivated; Bhira, 780m.

Material Examined: HPU-EL-SKS-ABS-15259; 22.12.2010.

Parts Used: Whole Plant. Leaves.

Folk Uses: Plant cooked as pot herb* and enters in various preparations like soups*, pakora etc.; also its extr. recommended to anaemic patients.

Active Constituents: These are rutin, hyperoside, astragalin, caffeic, chlorogenic, neochlorogenic, protocatechuic acids (*aerial parts*), glycoprotein-bound hexosamine (*seed*), spirasaponins (*root*).

Uses in Literature: Reported for cure of anaemia, jaundice, nitrogenous substances; and as coolant, hydrocarbons, iron sesqui-oxide and laxative (Ambasta, 1986; Jain, 1999; Khare, 2007).

Stephania glabra (Roxb.) Miers (Pl. 85D).

Syn.: *S. rotuda* Hook.f., *Cissampelos glabra* Roxb., *C. longa* Miers, *C. japonica* Miers, *C. roxburghiana* Miers, *C. glabra* Miers, *C. glandulifera* Miers, *Cocculus glabra* Roxb., *C. roxburghianus* Wall. Cat., *C. finlaysoniansus* Wall., *Clypea wightii* Arn.

Family: Menispermaceae.

Vern.: Bis-khappar.

Ayurvedic and Regional Names:

Ayur.: Paathaa;

Gar. : Gindaru;

U.**P.:** Gindaru, Purha.

Distribution: Tropical and temperate Himalayas.

Description: Large climbers having broad ovate or suborbicular, membranous leaves which are pale-glaucescent below. Umbels in lax cymes. Petals shorter. Drupes obovate, compressed and pisiform. Roots subglobose.

Flowering & Fruiting: June-August.

Habitat Ecology: Dry places, open places; Balooni, 858m.

Material Examined: HPU-EL-SKS-ABS-15260; 24.04.2011.

Parts Used: Roots. Tubers. Fruits.

Folk Uses: Fresh roots mixed with a pinch of rock salt prescribed for flatulence* in cattle. Smoke of fruits cure inflamed teats in animals. Pieces of roots tied as an amulet to the neck of animals for treatment of pus*, 'basaihar' (mastitis)*. Pieces of corm also given orally to check cough* in animals. Root paste prepared in mustard oil applied for inflamed* parts. Tuber paste applied for mouth ulceration* both in human beings and livestock. Also, a garland of pieces of its tubers hung on the door to keep evil spirits away*.

Active Constituents: *Tubers* possess gindarine, gindarinine with tetrahydro-palmatine, palmatine, dehydrocorydalmine, palmatrubine, stepharanine, crydalmine, stepholidine.

Biological Activity: Hypoglycaemic, antiinflammatory, CNS depressant, anticholin esterase and spasmolytic activities found +ve.

Uses in Literature: Useful for asthma, bodyaches, diarrhoea, dysentery, eye disease, fever, headache, intestinal complaints, pulmonary tuberculosis, skin diseases, tumour in stomach and veterinary ailments; and as psychoactive plant, vermifuge (vet.) (Ambasta, 1986; Gaur *et al.*, 1992; Jain, 1986; Jain *et al.*, 1991; Khare, 2007; Mohanty *et al.*, 1998; Shah & Joshi, 1971).

Tabernaemontana divaricata R. Br. *ex* Roem. & Sch. (Pl. 86A).

Syn.: *T. divaricata* Blume, *T. coronaria* R. Br., *Nerium divaricatum* L., *N. coronarium* Jacq., *Jasminum zelanicum* Burm., *Ervatamia coronaria* Stapf.

Family: Apocynaceae.

Vern.: Tagar.

Ayurvedic, English, Sanskrit and Regional Names:

Ayur.: Nandi Pushpa, Nandivriksha, Tagar;

Eng.: East India Rosebay;

Sans.: Nandivirksha, Tagara;

Ass.: Tagar;

G.: Tagar;

H.: Chameli, Chandni, Tagar;

Kan.: Kotuhalle, Nantiyavattam;

Mal.: Kuttampale, Takaram;

Mar.: Ananta, Gondetagara, Tagar;

Tam.: Adukkundiyavattai, Nandiyavattam;

Tel.: Gadhitagarpu, Nandivardhamu.

Distribution: Cultivated Throughout India.

Description: Evergreen, dichotomously branched, glabrous, small trees having oblong-lanceolate leaves. Flowers white, fragrant, in axillary terminal cymes. Fruits oblong or reniform, follicles dry or fleshy. Seeds linear oblong, smooth striate or rugose.

Flowering & Fruiting: July-October.

Habitat Ecology: Cultivated; Naduan, 483m.

Material Examined: HPU-EL-SKS-ABS-15261; 27.07.2008.

Parts Used: Whole Plant. Flowers.

Folk Uses: Plant valued for ornamental purposes. Flowers offered* to deities; its 5ml decoction prescribed on empty stomach in the morning as anthelmintic.

Active Constituents: Indole alkaloids, kaempferol, ibogamine, janetine, voaphyllene, voacangine, voacamine and loroneridine dregamine, tabernaemontanine (*plant*), tabersonine (*flowers*).

Biological Activity: Indole alkaloid (I) inhibits HCl induced ulcer in mice by 48.8%. Tabersonine shows hypotensive effect on anaesthetized cats.

Uses in Literature: Described earlier as anodyne, anthelmintic, antidote to scorpion bite, incense, refrigerant; and for boils, decorative purposes, dental caries, eye complaints, fever, headache, perfumery and sores (Ambasta, 1986; Gogoi & Borthakur, 1991; Jain, 1991; Khanna *et al.*, 1996a; Khare, 2007; Pandey *et al.*, 1996; Rao *et al.*, 1996).

Tagetes erecta L. (Pl. 86B).

Family: Asteraceae.

Vern.: Ghatak.Genda.

Ayurvediv, English, Sanskrit and Regional Names:

Ayur.: Gendaa, Jhandu;

Eng.: African Marigold, Marigold;

Sans.: Gandugai, Sandu, Stahulapushpa;

B.: Genda;

Bo.: Guljaharo, Gul-Jafari, Makhmal;

G.: Chaphul, Rojia;

H.: Genda, Gultera;

Kan.: Chandumallige, Seemes hamantige;

Mal.: Chendumalli;

Mar.: Gulgoto, Rajia-cha-Phul, Zendu;

Oriya: Gendu;

P.: Genda, Gul jafari, Makhmal, Mentok.

Distribution: Naturalized in India; extensively cultivated as a garden plant.

Description: Shrubby herbs. Leaves pinnately dissected, lobes lanceolate, sharply toothed. Ray florets golden-yellow, disc florets tubular and perfect. Head or capitulum consisting of sessile flowers. Achenial cypsela bearing persistent calyx. Seeds albuminous with straight embryo.

Flowering & Fruiting: October-January.

Habitat Ecology: Cultivated; Tikker Didwin, 800m.

Material Examined: HPU-EL-SKS-ABS-15262; 28.10.2009.

Parts Used: Leaves. Flower. Receptacle.

Folk Uses: Receptacle eaten* by children. Infusion of leaves rubbed to subside pain and swelling caused due to bee or wasp stings*. Paste of florets applied for henna effect* on hands. Flowers offered to appease deities*; also its garland hung at entrance of the main door and employed for decorative* purposes.

Active Constituents: Constituents isolated are syringic acid, methyl 3, 5-dihydroxy-4-methoxybenzonate, quercetin, tagetus oil (*flowers*), piperitone (20.0), (E) -β -ocimene (13.1), terpinolene (12.4), limonene (11.0 %) (*leaf essential oil*).

Biological Activity: Essential oil exhibited antimicrobial activity against fungi, gram+ve and gram –ve microorganism.

Uses in Literature: Reported in earlier literature for boils, bleeding piles, bronchitis, carbuncle, caterpillar bite, cold, cuts, eye troubles, muscular pains, renal trouble, rheumatism, ringworms, scorpion bite, wounds; and as carminative, chicken feed, diuretic, emmenagogue, laxative, vermifuge (Ambasta, 1986; Khare, 2007; Mokat & Deokule, 2004; Sen & Behera, 2003; Upadhye & Kulkarni, 2004).

Tamarindus indica L. (Pl. 86C).

Syn.: *T. occidentalis* Gaertn., *T. officinalis* Hook.

Family: Fabaceae.

Vern.: Imli.

Ayurvedic, English, Sanskrit and Regional Names:

Ayur.: Amli, Amlikaa, Chandikaa, Chinchaa, Chukaa, Chukrikaa, Suktaa, Tintidika;

Eng.: Tamarind Tree;

Sans.: Amlika, Bhukta, Charitra, Sukha, Tintrini;

Ass.: Tetuli;

B.: Anbli, Tentul;

G.: Ambli, Amli;

H.: Amli, Amlika, Anbli, Imli, Nuli, Teter;

Kan.: Amli, Huli;

Mal.: Amlam, Puli;

Mar.: Chicha,Chinchu;

Oriya: Konya, Tentuli;

P.: Imbli;

Tam.: Amliam, Puli;

Tel.: Chintachettu, Chintapandu, Sintachettu.

Distribution: Most parts of India.

Description: Spineless trees having even pinnate, oblong, glabrous, entire leaves, yellow flowers with pink striations, and are lax racemed at the end of branchlets. Pods 3-10 seeded with thin crustaceous epicarp and thick pulpy mesocarp.

Flowering & Fruiting: May-August.

Habitat Ecology: Open fields; Khya Village, 787m.

Material Examined: HPU-EL-SKS-ABS-15263; 26.06.2010.

Parts Used: Fruits. Seeds.

Folk Uses: Pulp of fruit used for preparing 'chutney', sauce and also added to pulses and vegetables to give acidic taste. Seeds used by children for playing various games*. 25-30ml pulp water checks vomiting and nausea*. 20-30ml docoction of bark prescribed in fever* and as mouth wash*.

Active Constituents: *Fruit pulp* rich in 10% free tartaric acid, 8% acid potassium tartarate, 25-40% invert sugar, tamarindienal, alkaloids, glycosides, tannins, organic acid and amino acids.

Biological Activity: Tamarindienal showed fungicidal activity against *Aspergillus niger* and *Candida albicans* and strong bacterial activity against *Bacillus subtilis, Staphylococcus aureus, Escherichia coli* and *Pseudomonas aeruginosa*.

Uses in Literature: Useful so far as antiscorbutic, carminative, fodder, green manure, laxative, refrigerant; and for agricultural implements, beverages, biliousness, boils, chutney, common cold, diarrhoea, dyeing, dysentery, indigestion, jams, jellies, kibs, menorrhoea, polishing and clearing metalware, scorpion stings, souring agent for sauces, tanning, tool handles, vomiting and wounds (Ambasta, 1986; Bhardwaj & Gakhar, 2003; Bondya *et al.*, 2006; Gogoi & Borthakur, 2001; Hembrom & Goel, 2005; Kadel & Jain, 2006; Katewa *et al.*, 2001; Khare, 2007; Kumar & Jain, 1998; Mohanty *et al.*, 1998; Pandey & Rout, 2006; Ram *et al.*, 2002; Sen & Behera, 2003; Sharma & Singh, 2001; Upadhye & Kulkarni, 2004).

Taraxacum officinale Wigg. (Pl. 86D).

Syn.: *T. dens-leonic* Desf., *Leontodon taraxacum* L.

Family: Asteraceae.

Vern.: Dudhli, Dulal.

Ayurvedic, English and Regional Names:

Ayur.: Dugdh-pheni, Luutaari, Payaswani;

Eng.: Dandelion, Fortune teller, Milk Gowan;

B.: Pitachumki;

G.: Pathardi;

H.: Barau, Dulal, Kanphul;

Kan.: Kaadu, Servent;

Kash. : Hamd;

Mar.: Undarkani;

P.: Baran, Bathur, Dudli, Kanphul, Shamukei.

Distribution: Throughout temperate Himalayas.

Description: Perennial scapigerous milky herbs. Leaves radical, entire, pinnatisect and basal. Rootstocks stout. Heads solitary on leafless scapes. Achenes ribbed.

Flowering & Fruiting: March-November.

Habitat Ecology: Grassy meadows, roadsides, damp places; Kanjiyan, 910m.

Material Examined: HPU-EL-SKS-ABS-15264; 18.03.2010.

Part Used: Leaves.

Folk Uses: 2-3 fresh leaves or ¼ teaspoon of powdered leaves taken twice a day for two days to cure stomachache*. Leaves used as fodder* for milching animals.

Active Constituents: These are taraxin, taraxacin, taraxacerin and inulin, saponin, cerylalcohol, lactuce-roltaraxacin (*rhizome, roots*).

Biological Activity: Plant exhibits slight antibiotic activity against *Mycobacterium tuberculosis.*

Uses in Literature: Described in Indian literature as an antiscorbutic, antiseptic, aperient, blood purifier, detoxicant, diuretic, laxative, liver tonic, vegetable; and for blisters, blood complaints, dislocation of joints, dysentery, gastric ulcers, headache, infective hepatitis, kidney disorder, liver complaints, preparation of beer, urinary rheumatism, vertigo, wines and wounds (Ambasta, 1986; Chhetri, 2005; Jain, 1991 Khare, 2007).

Adulterant or Substitute: Leaves used as an adulterant and substitute to Henbane leaves and flower tops of *Hyoscyamus niger* L.

Tecoma grandiflora Loisel. (Pl. 87A).

Syn.: *Campsis grandiflora* (Thunb.) K. Schum., *Bignonia grandiflora* Thunb.

Family: Bignoniaceae.

Vern.: Shehnai Phool.

English Name:

Eng.: Chinese Trumpet Creeper.

Distribution: Planted.

Description: Handsome shrubs having 7-9 leaflets. Flowers large, in drooping clusters. Corolla tubular, orange coloured. Capsule linear or narrow, loculicidally 2-valved, compressed at right angles to the septum. Seeds thinly discoid; wing hyaline, broad.

Flowering & Fruiting: April-August.

Habitat Ecology: Cultivated; Loharari, 858m.

Material Examined: HPU-EL-SKS-ABS-15265; 07.04.2011.

Parts Used: Plant. Flowers.

Folk Uses: Plant grown for ornamental* purposes; its flowers considered sacred* for offerings.

Active Constituents: These are furfural, 5-methylfural, furfurlyl alcohol and 2-acetylfuran (essential oil of *flowers*).

Use in Literature: So far used in diabetes (Ambasta, 1986).

Tecoma stans (L.) H.B. & K. (Pl. 87B).

Syn.: *Stenolobium stans* (L.) D. Don, *Bignonia stans* L.

Family: Bignoniaceae.

Vern.: Shehnai Phool.

English and Regional Names:

Eng.: Yellow Elder, Yellow Bells;

Tam.: Sonapaati, Thanga Arali, Naga Sambagam.

Distribution: Cultivated in gardens.

Description: Shrubs or small trees possessing opposite leaves and flowers in racemes or panicles. Calyx tubular. Corolla yellow. Capsule linear, two-valved septicidal. Seeds non-endospermic.

Flowering & Fruiting: April-August.

Habitat Ecology: Cultivated; Hamirpur (Pakka Bharo), 840m.

Material Examined: HPU-EL-SKS-ABS-15266; 19.05.2010.

Parts Used: Plant. Flowers.

Folk Uses: Plant grown for ornamental* purposes; and its flowers considered sacred* for offerings.

Active Constituents: These are β-carotene, zeaxanthin (*flower*), phenolic acids, β-sitosteron, triterpenoids, ursolic acid, oleanolic acid, alpha-anarine (*plant*), indole-metabolizing enzyme, indole-oxygenase (*leaves*).

Uses in Literature: Used earlier for diabetes; and as diuretic, vermifuge (Ambasta, 1986; Khare, 2007).

Terminalia arjuna Wt. & Arn. (Pl. 87C).

Syn.: *T. berryi* W. & A., *T. glabra* W.& A., *T. ovalifolia* Rottl., *Pentaptera glabra* Roxb., *P. angustifolia* Roxb.

Family: Combretaceae.

Vern.: Arjun.

Ayurvedic, English, Sanskrit and Regional Names:

Ayur.: Aartagala, Arjuna, Dhananjaya, Indravriksha, Kaakubha, Kakubha, Paartha, Virataru, Viravriksha;

Eng.: Arjun, White Murdah;

Sans.: Arjunah, Dhananjaya, Karaviraka, Savya Sachi;

Ass.: Orjun;

B.: Arjun, Kahu;

Bo.: Arjunasadadra, Jamla;

G.: Arjunasadada, Dhaulasadr;

H.: Arjun, Arjun saadadaa, Kangach, Kowa;

Kan.: Maddi;

Mal.: Vellamarutu;

Mar.: Arjunladada, Saumadat;

Oriya: Orjuna, Ponoyosahajo;

P.: Arjan, Jumla;

Tam.: Attumarudu, Kulamatrudu;

Tel.: Kakubhamu, Tittumaddi;

U.P.: Arjan, Kaucha.

Distribution: Commonly along riverian areas in the sub-Himalayans tract.

Description: Large-sized trees with huge buttressed trunk, glabrous, oblong-elliptic leaves and green flowers in panicle spikes. Fruits penta-winged, wings straight.

Flowering & Fruiting: May-September.

Habitat Ecology: Roadsides, open slopes; Anu, Hamirpur, 900m.

Material Examined: HPU-EL-SKS-ABS-15267; 06.05.2010.

Parts Used: Bark. Leaves. Wood.

Folk Uses: One tsp of powdered bark or its decoction (twice a day till relief) prescribed for heart palpitation and as cardiotonic. Leaves used as fodder*. Wood used for agricultural implements and building works. Poultice of stem bark prepared in coconut oil considered good against burn injuries* to cattle.

Active Constituents: Properties associated due to cevasidine, β-sitosterol, friedelin, methyl deanolate and gallic, ellegic and arjunic acid, (+) catehol, (+)

gallocatenol, epicateenol, epigalco, catechol, arjunolone, arjunglucoside III, arjunoside I and II, arjunflucoside I, arjunetin, leucoganidin, terminoic acid and arjunin, arjungenin (*bark* extr.).

Biological Activity: Stem bark abortifacient.

Uses in Literature: Known to be useful as an antidote to poison, antidysenteric, astringent, blood purifier, detergent, expectorant, hair wash, psychoactive, purgative, sacred, sex stimulant; and for agricultural implements, anaemia, anthrax, blood diseases, blood dysentery, blood pressure, boats and canoes, boils, carbuncle, chest pain and heart palpitation, cardiac complaints, colic pain, cuts and wounds, earache, ephemeral fever (vet.), fever, fractures, gastric troubles, heart burn, high blood pressure, house construction, impaction (vet.), insect bite (vet.), leucorrhoea, menstrual disorder, musical instrument, painful urination, plywood manufacture, pneumonia, sores, tanning, tumours, ulcers, water traps, water trough, wheels of country carts, wounds, wounds (vet.) (Ambasta, 1986; Chauhan, 1999; Datt, 1996; Girach *et al.*, 1998; Gogoi & Borthakur, 2001; Hembrom, 1991; Hosagoudar & Henry, 1996a; Jain, 1968, 1991; Jain *et al.*, 1994; Khanna, 2002; Khare, 2007; Kothari & Moorthy, 1996; Kumar & Jain, 1998; Kumar & Pullaiah, 1998; Kurian, 1995; Lalramnghinglova, 1996; Masih, 2003; Ramadas *et al.*, 2000; Rama Rao & Henry, 1996; Reddy *et al.*, 1998; Satapathy & Brahmam, 1996; Sen & Behera, 2003; Sharma & Sood, 1997; Singh & Prakash, 1994; Srivastava *et al.*, 2003).

Major Ayurvedic Formulations: Arjunaghrita, Arjunakasivapalla, Arjunarishta, Arjunatwal-churna, Arjun-kwath, Arvindaasava, Devadarista, Hridayarnavaasa, Kakubhaadi-churna, Nagarjunabhra-rasa, Parthadyarista.

Trade: Arjun.

Terminalia bellirica (Gaertn.) Roxb. (Pl. 87D).

Syn.: *Myrobalamus bellirica* Gaertn.

Family: Combretaceae.

Vern.: Behra, Bahera.

Ayurvedic, English, Sanskrit and Regional Names:

Ayur.: Aksha, Alshaka, Baibhita, Bibhita, Bibhitaka, Bibhitaki, Kaamaghna, Kalidru Kali, Karshaphala, Vibhitaka;

Eng.: Bastard Myrobalan, Beddanuts, Belleric Myrobalan;

Sans.: Anilaghnaka, Kaliyugalaya, Karshaphala;

Ass.: Banri, Hulluch;

B.: Bahera, Buhura;

Bo.: Buhudda, Bherda, Yella;

G.: Bahedamunjad, Bahedo;

H.: Bahera, Bhairach, Sagona;

Kan.: Tare-kai, Shanti Kayi;

Kash..: Babelo, Balali;

Mal.: Tanni, Tusham;

Oriya: Horinda, Jonghihorida;

P.: Bahira, Bayrah;

Tam.: Akkam, Kalitturuman;

Tel.: Karakkaya, Nallakaroka.

Distribution: Throughout the plains & sub-Himalayasn tract.

Description: Large deciduous trees with ashy grey bark, alternate, obovate-elliptic, coriaceous leaves and yellow flowers in spikes. Upper flowers of the spike males whereas lower ones hermaphrodite. Fruit long, ovoid, velvety.

Flowering & Fruiting: May-September.

Habitat Ecology: Open slopes, roadsides; Jhanikar, 1,050m.

Material Examined: HPU-EL-SKS-ABS-15268; 07.05.2010.

Parts Used: Fruit. Seed.

Folk Uses: ½ tsp fruit powder taken with honey to cure cough*, bronchitis epilepsy, stomachache and for vitality; also its powder soaked overnight used for washing eyes. Fruit forms one of the constituents of 'triphala' powder by grinding 'harar' (*Terminalia chebula* Retz.): 'behra' (*Terminalia bellirica* (Gaertn.) Roxb.) and 'amla' (*Emblica officinalis* Gaertn.) in ratio of 1:2:3 and considered good appetizer*, laxative, liver tonic and for gastrointestinal problems and piles. Seeds eaten like almonds*. Outer covering given to animals for stomach trouble. Powdered dried fruits in combination with 'harad' (*Terminalia chebula* Retz.) and 'amla' (*Emblica officinalis* Gaertn.) in ratio of 1:2:1 given after mixing in wheat dough or decoction prepared from this combination given twice daily for 3-5 days for cough*, dysentery, constipation, intestinal obstruction* and bloating* in animals. The 'triphala-churna' mixed with 'ajwain' (*Carum copticum* Benth.) 'kodi saunf' (*Anethum sowa* Roxb.), ginger powder, asafoetida, considered good for bloating* in cattle.

Active Constituents: Fruit pulp comprises hexahydroxydiphenic acid ester, β-sitosterol, gallic acid, ellegic acid, ethyl gallate, galloyl glucose, chebulegic acid, and cardioglycoside bellericanin.

Biological Activity: Anticancerous activity +ve.

Uses in Literature: Useful earlier as anthelmintic, antiseptic, aperient, astringent, brain tonic, colic, diuretic, expectorant, fish poison, fodder best for milching cows, important constituent of Triphala, laxative, narcotic, ornamental around temple yard, psychoactive plant, vegetable and writing inks; and for acidity, agricultural tools, asthma, black hair, boards, boats, carts, cholera, coffee boxes, cold, colic (vet.), constipation, diabetics, diarrhoea (vet.), dropsy, enlargement of spleen, fever, foot and mouth diseases, gastric complaints, gastropathies (vet.), headache, hoarseness, inflammation, inflammatory disease of eyes, jaundice, leprosy, liver complaints, muscular pain, piles, plough, skin disease, slate frames, small crafts, snake bite, soap manufacture, stomach pain, tannin, throat trouble, traditional dwelling, turning, ulcer (vet.), vigour, vitality, vomiting, weakness and wounds (vet.) (Ambasta, 1986; Aminuddin & Girach, 1993; Anonymous, 1998b; Borthakur, 1996b;

Borthakur & Gogoi, 1994; Chauhan, 1999; Dixit, 1987; Dwivedi, 1993; Goud & Pullaiah, 1996b; Khanna, 2002; Khare, 2007; Kothari & Moorthy, 1996; Kumar & Pullaiah, 2003; Lalramnghinglova, 1996; Masih, 2003; Mishra *et al.*, 1997; Mohanty *et al.*, 1998; Nazarudeen *et al.*, 1996; Pal & Jain, 1998; Patel *et al.*, 2003; Patil & Merat, 2003; Prasad *et al.*, 1996; Pullaiah & Kumar, 1996; Ramadas *et al.*, 2000; Rama Rao & Henry, 1996; Rao, 1996; Sharma & Borthakur, 2003; Sharma & Singh, 2001; Sharma & Sood, 1997; Singh, 1999b; Siwakoti & Verma, 1996; Sudarsanam *et al.*, 1995; Tirkey *et al.*, 2003; Venkataratnam & Venkata Raju, 2004; Verma *et al.*, 1995).

Major Ayurvedic Formulations: Agnitundi-vati, Aragyavardhini-vati, Bibhitak-taila, Chandra-prabha-vati, Gokshuvadiguggula, Herbotone, Karanjadiyoga, Khadiraista, Kutajavaleha, Lavangudi-vati,-Navayasloaha, Unarnava-mandur, Sanjivani-vati, Triphaditaila, Triphala-churna, Triphala-kwath, Vidangadi-louha.

Trade: Bahera, Belleric myrobalam.

Terminalia chebula Retz. (Pl. 88A).

Syn.: *T. aruta* Ham., *T. reticulata* Roth., *Myrobalanus chebula* Gaertn., *Embryogonia arborea* Teys & Binn.

Family: Combretaceae.

Vern.: Harar.

Ayurvedic, English, Sanskrti and Regional Names:

Ayur.: Haritaki, Kaayasthaa, Pathyaa, Shivaa, Shreyasi;

Eng.: The Chebulic or Black Myrobalan;

Sans.: Haritaki, Harra;

Ass.: Haritaki, Silikha;

B.: Hirda, Haritaki;

G.: Hardo;

H.: Harna, Har, Harar, Harra;

Kan.: Hirda;

Mal.: Kadukka;

Mar.: Hirao, Hirda;

Oriya: Karedha, Haridra;

P.: Har, Harar;

Tam.: Kadakai, Tadukkai;

Tel.: Karaka;

U.P.: Harr.

Distribution: Throughout India.

Description: A moderate-sized tree with glabrous, alternate, elliptic-oblong leaves having acute rounded base. Flowers sessile, yellow with an offensive smell. Drupes pendulous, ellipsoid, glabrous, when dry yellowish-green. Stone bony, very thick, obscurely angled.

Flowering & Fruiting: April-June.

Habitat Ecology: Open dry slopes, edges of agricultural fields; Near Shukkar Khad, 381m.

Material Examined: HPU-EL-SKS-ABS-15269; 29.05.2011.

Part Used: Fruit.

Folk Uses: ½ tsp of its powdered fruits preserved in sugary syrup with milk before bed time considered a good laxative. ½ tsp of powdered fried 'harar' with honey given twice a day to cure cough*. Also, 1 tsp of powdered 'harar' and 'saunf' (*Foeniculum vulgare* Mill.) fried in ghee taken once daily with hot water to check dysentery. Mixture comprising 'harar' (*Terminalia chebula* Retz.), 'behra' (*Terminalia bellirica* (Gaertn.) Roxb.), 'amla' (*Emblica officinalis* Gaertn.) and 'mehendi' (*Lawsonia alba* Lam.) considered good dye for greying hair. Fruit in combination with almond (*Prunus amygdalus* Baill.), belliric myrobolan (*Terminalia bellirica* (Gaertn.) Roxb.), and Indian gooseberry given to cure diabetes. Powdered fruits act as one of the important constituents of 'triphala' powder. Hot extr. of its fruit given with seeds of 'bhera' (*Terminalia bellirica* Roxb.) and 'amla' (*Emblica officinalis* Gaertn.) to check cough* in animals (2-3 litres daily till cure).

Active Constituents: *Fruits* rich in ellagitannin, terchebulin, punicalagin, terflavin A, B, C and D, shilimic acid, gallic triacontanoic and palmitic acid, β-sitosterol, daucosterol, triethyl ester of chebulic acid and ethyl ester of gallic acid, arjungenin, terminoic acid, arjunolic acid and terchebin.

Biological Activity: Antibacterial and spasmolytic activity +ve.

Uses in Literature: Used so far as an alternative, astringent, blood purifier, cardiac tonic, colic, dye for hair, fodder, important constituent of Triphala, ornamental, psychoactive plant, sacred for making ink; and for acidity, agricultural tools, anthrax (vet.), asthma, boards, boats, body pains, carts, cholera, coffee boxes, cold, constipation, constipation (vet.), diabetes, diarrhoea, diarrhoea (vet.), dropsy, dysentery (vet.), ephemeral fever (vet.), enlargement of spleen, fever, gargles in inflammation of mucous membrane of mouth, gastric complaints, gastropathies (vet.), headache, hoarseness, inflammation, inflammatory disease of eyes, jaundice, leprosy, liver complaints, liver infection (vet.), muscular pain, piles, plough, skin disease, slate frames, small crafts, snake bite, soap manufacture, stomach pain, sweating, tannin, throat trouble, turning, ulcer (vet.), ulceration of gums, vaginal discharge, vigour and vitality, vomiting, worms, wounds and wounds (vet.) (Ambasta, 1986; Aminuddin & Girach, 1993; Borthakur, 1996 a,b; Borthakur & Gogoi, 1994; Chowdhery, 1996; Dixit, 1987; Dwivedi, 1993; Girach, 1992; Gogoi & Borthakur, 1991, 2001; Hosagoudar & Henry, 1996a; Jain, 1986; Jain & Patole, 2001; Jain *et al.*, 1994; Khanna & Kumar 2000; Khare, 2007; Kshirsagar & Singh, 2000; Kumar, 1996; Lalramnghinglova, 1996; Masih, 2003; Mohanty & Padhy, 1996; Mohanty *et al.*, 1998; Pal & Jain 1998; Ramadas *et al.*, 2000; Rama Rao & Henry, 1996; Reddy & Raju, 2000; Reddy *et al.*, 1998; Sarangi & Sahu, 2004; Sharma & Borthakur, 2003; Sharma & Sood, 1997; Srivastava *et al.*, 2003; Sudarsanam *et al.*, 1995; Tirkey *et al.*, 2003a, b; Varshney, 1998; Verma & Chauhan, 2006; Verma *et al.*, 1995; Watt, 1972).

Major Ayurvedic Formulations: Abhaya-lavana, Abhaya-mo-dak, Abhayarista, Agastya-hairtaki-rasayana, Amrita-haritaki, Avleha, Brahma-rasayana, Chandraprabha vati, Dantiharitaki, Gokshuraadi-gugglu, Hairtaki-churna, Karanjaadi-yoga, Medigatone, Panchraska churna, Sanjivani-vati, Sudarshan-churna, Triphala, Triphala-churna, Triphaladi-taila, Triphala-ghrit.

Trade: Myrobalam, Chebulic-myrobalam (tree, fruit), Harada, Jangli Harad (for small variety).

Thevetia neriifolia Juss. *ex* Steud. (Pl. 88B).

Syn.: *T. peruviana* (Pers.) K. Schum., *Cerbera peruviana* Pers., *Cascabela thevetia* (L.) Lipp.

Family: Apocynaceae.

Vern.: Kaner.

Ayurvedic, English, Sanskrit and Regional Names:

Ayur.: Ashvaghna, Divyapushpa, Pita-Karavira;

Eng.: Bastard Oleander, Exile Oleander, Lucky Nut Tree, Yellow Oleander;

Sans.: Ashvamaraka, Nakharatlia, Sidhapushpa;

B.: China-Karab, Kokil:phul;

Bo.: Pilakaner, Zardakunel;

G.: Pilakanir;

H.: Pila, Pila kananir;

Kan.: Kadukasi; Kanogalu;

Mal.: Pachchaarali;

Mar.: Pivalakanhera;

Oriya: Konyar-phul;

Tam.: Pachaiyalari, Tiruvachippu;

Tel.: Pila Kaner, Pit kaner.

Distribution: Throughout India.

Description: Shrubs. Leaves linear-lanceolate, sessile. Flowers yellow, infundibuliform corolla, in dichasial cymes. Stigma dumbell-shaped. Drupes angular, poisonous. Seeds small with glucoside.

Flowering & Fruiting: July-September.

Habitat Ecology: Forest, shrubberies; Dimmi, 1,006m.

Material Examined: HPU-EL-SKS-ABS-15270; 10.04.2010.

Parts Used: Plant. Leaves.

Folk Uses: Planted in temple courtyards as an ornamental. Powdered leaves with honey (1 tsp thrice daily till cure) prescribed as blood purifier*.

Active Constituents: Important constituents of *fruits* are arabinose, galactose, glues, galacturonic acid and xylose, thevetine (*seed*), neriifolin, cerberin, peruvosise (*leaves, plant*).

Biological Activity: Cardiotonic and isotonic effects and musculotropic activity confirmed.

Uses in Literature: Known in published ethnobotanical literature as abortifacient, aeronarcotic poison, alexetric, bitter cathartic, cardiotonic against pentobarbital induced heart faliure, emetic, laxative, ornamental (beads), poison (vet.), vegetable; and for carotene disease, dropsy, impotency, skin disease, snake bite, swelling (vet.), swelling of foot, toothache, tumours and worms (Ambasta, 1986; Bhandary *et al.*, 1996; Chauhan, 1999; Huidrom, 1996; Jain, 1991; Jha *et al.*, 1996; Khare, 2007; Kurian, 1995; Mishra *et al.*, 1996a; Pandey & Verma, 2002; Saini, 1996a; Sharma & Sood, 1997; Singh & Pandey, 1996; Siwakoti & Verma, 1996).

Warning : All parts of the plant contain cardiac glycosides and considered poisonous particularly seed kernels and milky juice from all parts of plant.

Thuja orientalis L. (Pl. 88C).

Syn.: *Biota orientalis* (L.) Endl.

Family: Cupressaceae.

Vern.: Morpankh.

English and Regional Names:

Eng.: Oriental Arbor-Vitae;

H.: Mayur Pankh, Morphankhi.

Distribution: Cultivated.

Description: Aromatic, evergreen gymnospermous shrubs having scale-like, overlapping leaves. Cones compressed, pyriform, 2-5 sporangia per microsporangiophore and 2-3 ovules per megasporophyll.

Flowering & Fruiting: July-October.

Habitat Ecology: Cultivated; Sameerpur, 1,196m.

Material Examined: HPU-EL-SKS-ABS-15271; 10.04.2010.

Parts Used: Plant. Leaves.

Folk Uses: Planted for ornamental* purposes. Leaves kept in books by children as 'vidya patri'*, i.e. for gift of knowledge from goddess 'Saraswati'.

Active Constituents: Pimeric and isopiameric acid, Rhodoxanthin, amentoflavone, hinokiflavone, quercetin, myricetin, carotene, thujone, iso-thujone, borneol characterized from *leaves.*

Biological Activity: Quercitin exhibits haemostatic activity in mice.

Uses in Literature: So far used for barrels and casks, dysentery, fencing posts, furniture, house building and tannin; and also as antipyretic, diuretic, insecticidal plant, psychoactive, tonic (Ambasta, 1986; Jain *et al.*, 1994; Khare, 2007; Singh & Prakash, 1994).

Tinospora cordifolia (Willd.) Miers (Pl. 88D).

Syn.: *T. palminervis* Miers, *Cocculus cordifolius* DC., *C. convolvulaceus* DC., *C. verrucosus* Wall., *Menispermum cordifolium* Willd.

Family: Menispermaceae.

Vern.: Galon, Gulja, Giloy.

Ayurvedic, English, Sanskrit and Regional Names:

Ayur.: Amritaa, Amrita, Amritalataa, Amritavalli, Chinnaruuhaa, Chinnodbhavaa, Guduuchi, Guduuchikaa, Guluuchi, Kundalini, Madhuparni, Tantrikaa, Vatsaadani;

Eng.: Gulancha Tinospora, Heart leaved Mvonseed;

Sans.: Amrita, Amritalata, Chakrangi, Dhira, Jwaranashini, Guluchi, Guduchi, Jirantiha, Madhuparnika, Nirajara, Tantri, Vara, Vishalya;

Ass.: Amarlata, Gulancha, Siddhilata;

B.: Gadancha, Giloe, Gulancha;

Bo.: Ambarrel, Gharol, Giroli, Guloe, Gulwel;

G.: Gado, Galo, Gulo;

H.: Giloe, Gulancha, Gulbel, Gurach, Gurcha;

Kan.: Amrita balli, Amruta balli;

Kash.: Amrita Gilo;

Mal.: Amrytu, Peyamrytam, Sittamrytu;

Mar.: Ambarvel, Gharol, Giroli, Gulavela, Gulvel;

Oriya: Gulancha, Gulochi;

P : Batindu, Gilogularich, Garum, Gilo;

Tam.: Amudam, Amridavalli, Kaippuchindil, Kunale;

Tel.: Duyutige, Guduchi, Jivatika, Somida;

U.P.: Gurch.

Distribution: Throughout tropical India.

Description: A large glabrous climber possessing succulent corky stem and branches sending down slender pendulous fleshy roots. Leaves membranous, roundish, deeply cordate. Flowers yellowish-green, in axillary or terminal racemes. Drupes size of a pea or small cherry, red.

Flowering & Fruiting: May-July.

Habitat Ecology: Wastelands, common in hedges; Mattani, 920m.

Material Examined: HPU-EL-SKS-ABS-15272; 27.03.2008.

Parts Used: Plant. Stem. Leaves.

Folk Uses: 20-30ml decoction of the plant given on empty stomach for increasing platelet count* in blood. ½ tsp of dry powdered stem taken with luke warm water daily to lower blood pressure*, and for 15 days twice daily to cure abdominal diseases, congestion of liver and persistent fever. Pieces of dried stem used in 'hawan'* for purification of surroundings. Poultice of fresh stem applied against rheumatic ailments* and joint pains. Stem fed to cattle orally as an appetiser and for checking abortions* ('taruna'), also stem and leaves mixed with fodder given to improve lactation in animals.

Active Constituents: Active properties due to giloin, giloinin, gilosterol and berberine (*stem bark*), tinosporin, tinosporon, tinosporol, tinosporic acid, tinosporoside, cordifolide and a furanoid diterpene identical with columbin, clerodane, clerodane furanoditerpene I, II, III, IV (*plant*).

Biological Activity: Antiinflammatory, analgesic, anticancerous, antipyretic, antiviral, CNS depressant, and hypoglycaemic activity found +ve.

Uses in Literature: Described earlier as anthelmintic (vet.), antiperiodic, antipyretic, antidote to snake bites, aphrodisiac, colic (vet.), expectorant, febrifugal, fibre for tying bundle, fodder, psychoactive plant, tonic; and for anthrax, anthrax (vet.), appetiser (vet.), asthma, bloat, bone fracture, bone fracture (vet.), cough, cough (vet.), debility, diarrhoea, diptheria, dysentery, dysentery (vet.), exoparasite (vet.), fever, fever (vet.), gastric disorder, gonorrhoea, haematuria (vet.), headache, impaction (vet.), indigestion (vet.), insect bite (vet.), jaundice, jaundice (vet.), ingredient of 'Chyawanprash', liver disorder, liver infection (vet.), malaria, mastitis, nocturnal emission, panting, piles, pimples, poultry diseases, pulmonary tuberculosis, sexual strength, stomach ailments in animal, stomachic (vet.), ulcer (vet.), swelling in mammary glands (vet.), tonic (vet.), urinary diseases, vein injuries, venereal complaints, vermifuge (vet.), visceral obstruction, wound (vet.) (Ambasta, 1986; Anonymous, 1998b; Bhatt, 1990; Chinthu *et al.*, 1998; Gaur *et al.*, 1992; Goud & Pullaiah, 1996b; Jain, 1968, 1991; Jain & Saklani, 1992; Jain *et al.*, 1994; Jha *et al.*, 1991; Joshi, 1995; Katewa & Arora, 1997; Khare, 2007; Kiruba *et al.*, 2006; Kshirsagar *et al.*, 2003; Kumar & Pullaiah, 1998; Lalramnghinglova, 1996; Mahato & Mahato, 1996; Malik *et al.*, 2009; Manandhar, 1988; Mehrotra *et al.*, 1995; Pandey & Verma, 2003; Patil & Merat, 2003; Ranade, 1998; Reddy & Raju, 2000; Reddy *et al.*, 1996, 1998; Rekha *et al.*, 1998; Sebastian & Bhandari, 1984; Sen & Behera, 2003; Sharma, 1996; Silori & Rana, 2000; Singh & Pandey, 1996; Srivastava *et al.*, 2003; Tiwari & Pande, 2006; Uniyal, 1968; Varshney, 1998; Verma & Chauhan, 2006; Verma *et al.*, 1995).

Major Ayurvedic Formulations: Amritaastaka-kwatha, Amritasava, Chandra-prabha-vati, Giloyasatva, Guduchyadi-kwath, Kaishore-guggulu, Rasna-panchak-kwath, Sanjivani-vati, Sudarshan-churna.

Trade: Gilory or Gulancha (dried stem).

Trichodesma indicum R. Br. (Pl. 89A).

Syn.: *T. perfoliatum* Wall., *T. hirsutum* Edgew., *Borago spinulosa* Roxb., *B. indica* L.

Family: Boraginaceae.

Vern.: Chhota Kalpa, Rukhabi.

Ayurvedic, Sanskrit and Regional Names:

Ayur.: Adah-pushpi, Adhomukha, Andhaka Gandhapushpika;

Sans.: Adah-pushpi, Adhomukha, Avakpushpi, Darvika, Dhenujivha, Gandhapushpika, Golomi, Romalu, Sarasa;

B.: Chhota-kulpha;

H.: Chhota-kulpha, Ratmundiya;

Kash.: Nilakrai, Ratisurkh;

Mar.: Lahanakalpa;

P : Kallri-buti, Ratmandu;

Tam.: Kazuthai-tumdal;

Tel.: Guvva-gutti;

U.P.: Ratmandi.

Distribution: Throughout tropical India.

Description: Coarsely branched annual herbs, 15-45cm tall with bulbous based Hair. Stem erect or diffused. Leaves stem-clasping. Flowers pale blue turning to pink and finally to white in terminal a few flowered cymes. Fruits pyramidal with the persistent style, 4-ribbed. Nutlets 4.

Flowering & Fruiting: September-December.

Habitat Ecology: Open slopes, dry areas; Kehdru, 842m.

Material Examined: HPU-EL-SKS-ABS-15273; 03.04.2008.

Part Used: Whole Plant.

Folk Uses: 1 tsp decoction of the plant given twice a day for stomachache. Paste of the plant applied for swelling of joints, inflammation*, healing wounds* and internal injuries*.

Active Constituents: Active ingredients are linoleic, linolenic, oleic, palmitic and stearic acids (*seeds*), hexacosane, ethylhexacosanoate and 21, 24-hexacosadienoic acid ethylester (*leaves*).

Uses in Literature: Known earlier as a blood purifier, depurative, diuretic, edible (leaves are eaten as such), emollient, tonic for brain; and for eczema, expelling of worms in animals, stomachache and swelling (Ambasta, 1986; Drury, 1978; Jain, 1991; Khare, 2007).

Tridax procumbens L. (Pl. 89B).

Family: Asteraceae.

Vern.: Dhautha, Lamb, Gorakhmundi.

Ayurvedic, English and Regional Names:

Ayur.: Jayanti;

Eng.: Coatbuttons;

H.: Pardesi Langri;

Kan.: Gabbu, Sanna Savanthi;

Tam.: Vettukkaaya,-thalai;

Tel.: Raavanaasuruditalkaai.

Distribution: Throughout tropical India.

Description: A straggling perennial hirsute herb, having a few opposite, pinnatisect leaves and heterogamous, rayed heads borne on a long, slender, solitary peduncle. Achenes turbinate or oblong, silky, brown, pappus of feathery bristle, shining.

Flowering & Fruiting: July-August.

Habitat Ecology: Common on moist rocky slopes; Kuthera, 900m.

Material Examined: HPU-EL-SKS-ABS-15274; 22.07.2010.

Parts Used: Whole Plant. Leaves. Stem.

Folk Uses: Leaves and stems used as fodder*, ½ tsp infusion of the plant prescribed with water for stomachache.

Active Constituents: Principle constituents are fumaric acids, dotriacontanol, beta-amyrin, beta-amyron, lupeol, fucosterole, beta-setosterol, glucoluteolin, isoquercetin, quercetin, lauric, myristic, palmitic, stearic, arachidic, behenic, palmitoleic, linoleic and lionlenic acid (*flowering tops*).

Uses in Literature: Known to be used as an antidiarrhoeal, antidysenteric, antiseptic, haemostatic, hair tonic, insecticidal, piscicidal, stomachic; and for bleeding, blisters, boils bronchial catarrh, cuts, eczema, eye diseases, fever, leprosy, scorpion bite, sores, stones in urinary bladder, toothache, ulcers, wounds and wounds (vet.) (Ambasta, 1986; Jain, 1991; Khare, 2007; Kirtikar & Baṣu, 1994; Kurian, 1995).

Trifolium alexandrinum L. (Pl. 89C).

Family: Fabaceae.

Vern.: Barseem.

English Names:

Eng.: Barseem, Egyptian Clover.

Distribution: Cultivated for green manure.

Description: Annual herbs. Stem trailing. Leaves trifoliate. Leaflets obovate, toothed, veins prominent. Flowers white or tinged with pink, shortly stalked, clustered in dense heads. Pods 3-4 seeded.

Flowering & Fruiting: March-June.

Habitat Ecology: Cultivated; Kangoo, 649m.

Material Examined: HPU-EL-SKS-ABS-15275; 25.04.2008.

Part Used: Whole Plant.

Folk Uses: Planted as a major fodder crop to increase milk production..

Active Constituents: Contains protein, calcium and phosphorus (*plant*).

Uses in Literature: Known so far as green fodder, silage, hay, pasture (Ambasta, 1986).

Trigonella foenum-graecum L. (Pl. 89D).

Family: Fabaceae.

Vern.: Methi.

Ayurvedic, English, Sanskrit and Regional Names:

Ayur.: Bodhaini, Bahupatrikaa, Dipani, Gandhaphala, Methi, Methikaa, Methini, Vastikaa.

Eng.: Fenugreek, Greek Hayes;

Sans.: Methi;

B.: Methi;

G.: Methi;

H.: Methi, Muth;

Kan.: Menthya;

P.: Methi;

Tam.: Vendayam;

Tel.: Mentulle.

Distribution: Kashmir-Punjab, Upper Gangetic plains.

Description: An erect herb having fragrant leaflets which are shortly serrate from the middle upward. Flowers sessile, yellow in racemose inflorescence. Pods solitary, hairy, glabrous, curved, 5-7.5cm long.

Flowering & Fruiting: May-August.

Habitat Ecology: Cultivated; Sujanpur, 567m.

Material Examined: HPU-EL-SKS-ABS-15276; 10.02.2010.

Parts Used: Leaves. Seed.

Folk Uses: Leaves used in preparation of 'saag'* which is relished along with 'makki-ki-roti' (a thick 'chappatti' prepared from maize flour) by one and all. Powdered leaves used as a spice*. Fresh leaves stuffed in wheat or maize flour to make 'parantha'*. One tsp of powdered seeds taken daily with cold water as an easy remedy for diabetes and with hot water for curing gastric problems. Seeds soaked in mustard oil overnight massaged for checking bodyaches* and on scalp for maintaining black and long hair. Seeds in powdered form given to animals for bloating*.

Active Constituents: Attributed properties due to β-sitosterol, kaemferol, quercetin, saponins-graecumins A-F and (*leaves*), tigogenin, neotigogenin, diosgenin, yamogenin, gitogenin and 4-hydroxy isoleucine, vitexin, iso-vitexin, saponaeretin, homoorientin, vicenin-1 and vicenin-2 (*seeds*).

Uses in Literature: Reported to be useful for acne, baldness, boils, controling blood cholesterol, blood pressure, diarrhoea, dyspepsia, flatulence, enhancing lactation (vet.), lipids, loose motion (vet.), enlargement of liver and spleen, gastric trouble, gonorrhoea, hair growth, loss of appetite, odema, piles, rheumatism, sores, syphilis and ulcers; and as an aphrodisiac, appetiser (vet.), carminative, colic, demulcent, emollient in intestinal inflammation, fodder, galctagogue, pot herb, spice, tonic and urine discharge (vet.) (Ambasta, 1986; Anonymous, 1998b; Jain, 1991; Khare, 2007; Kumar *et al.*, 2004; Sen & Behera, 2003; Sudarsanam *et al.*, 1995; Uphof, 1968; Varshney, 1998).

Adulterant or Substitute: Seeds considered a good alternative source of *Dioscorea deltoidea* Wall. *ex* Kuntz.

Triticum aestivum L. (Pl. 90A).

Syn.: *T. sativum* Lam., *T. vulgare* Vill.

Family: Poaceae.

Vern.: Gehun.

Ayurvedic, English, Sanskrit and Regional Names:

Ayur.: Godhuuma;

Eng.: Common Wheat, Bread Wheat;

Sans.: Godhuma, Mahagodhuma, Madhuli, Mihsuki;

B.: Gam, Giun,Gom;

G.: Ghavum, Govum, Gawn;

H.: Gehun, Giun, Gandham, Kanak;

Kan.: Godhi;

Mal.: Gendum, Godamba, Kotanpam;

Mar.: Gahun, Gahung;

P.: Gehun, Kanak;

Tam.: Godumai, Godumbayarisi;

Tel.: Goodhumalu.

Distribution: Throughout India, cultivated as major food crop.

Description: Tall cultivated herbs. Leaves broadly linear, leaf-base sheathing. Stem clasping. Spikelets awned. Flowers with feathery stigmas. Fruits achenial, caryopsis and endospermic seeds with single cotyledon.

Flowering & Fruiting: November-April.

Habitat Ecology: Cultivated; Cha kmoh, 736m.

Material Examined: HPU-EL-SKS-ABS-15277; 07.01.2010.

Parts Used: Whole Plant. Grain.

Folk Uses: Plant considered sacred*. Its grains boiled eaten or flour used for making 'chappati' and 'prasadam for offering' to deities on every festive and religious occasion. Wheat straw used as fodder and bedding for cattle to rest comfortably. Milky extr. 'seera'* of soaked grains cooked in ghee and sugar prescribed for strengthening body as a tonic, coolant, and memory enhancer. Noodles prepared from wheat served as pudding*. Dried grains or soaked ones given to animals for 'taruna' (abortion)* and for enhancing lactation*. Sweet pudding ('halwa') made from its 'rawa' along with ginger powder and coconut oil given for 'mastitis'*. Powder of its roasted grains in 'desi' ghee mixed with wheat dough given for upward fixation of patella* ('saridan').

Active Constituents: Gluten (*grain*).

Uses in Literature: Reported in India for alcohol, anaemia, bedding for cattle, biscuit, bone fracture (vet.), constipation (vet.), chappaties, dysentery (vet.), fattening, gastric problems (vet.), increasing appetite, malted wheat; and as coolant, laxative, loaf bread, tonic (Ambasta, 1986; Anonymous, 1998b; Khare, 2007; Kumar *et al.*, 2004; Mishra *et al.*, 1996b).

Tylophora indica (Brum. f.) Merr. (Pl. 90B).

Syn.: *Tylophora asthmatica* Wight & Arn.

Family: Asclepiadaceae.

Vern.: Damabuti.

Ayurvedic, English, Sanskrit and Regional Names:

Ayur.: Antamuula, Arkaparni, Muulini;

Eng.: Emetic, Indian or Country-Ipecacuanha, Swallow Wort;

Sans.: Antrapachaka, Arkaparini, Lataksiri;

B.: Antomul, Ananthamul;

G.: Daminivel;

H.: Jangli Pikvam;

Kan.: Adumuttaha, Nepala;

Mal.: Villipalla;

Mar.: Khodiki Raasna, Pitkari;

Ori. : Mendi, Mulini;

Tam.: Nach-Churuppam, Naja-Murich-Chaan;

Tel.: Kakapalla, Kaakapalla.

Distribution: Upto 1,000m in plains and hills of India.

Description: A twinning shrub. Leaves ovate or very narrow, 5-10cm long, usually pointed at the tip and in opposite pairs. Flowers large, dull yellow, in short umbellate cymes. Fruits 5-10cm long, pointed at tips, ridged with many fine ridges and in pairs.

Flowering & Fruiting: May-October.

Habitat Ecology: Found in peninsular India, cultivated in other parts; Khaggal, 920m.

Material Examined: HPU-EL-SKS-ABS-15278; 03.07.2008.

Parts Used: Whole Plant. Root. Leaves.

Folk Uses: Infusion or decoction of the plant given in asthma and bronchitis. 2-4g powdered roots prescribed with honey against dysentery*. Leaves also eaten raw for curing asthma and bronchitis. Plant considered a good fibre plant for its fine fabrics*.

Active Constituents: Therapeutic properties due to tylophorine, tylphorinine, desmethyl tylophorine, desmethyl-tylophorinine, kaemferol (*plant*), tylophorinidine, gammafagarine (*root*), d-septicine, d-iso-tylocrebrine, triterpenoids, alpha & beta amyrin, beta-sitosterol, stigmasterol, campesterol, phenylalanine and quercetin (*leaves*).

Biological Activity: Plant exhibited antiamoebic activity against axenic and polyaxenic strains of *Entamoeba histolytica.*

Uses in Literature: Used so far for allergic rhinitis, bronchial asthma, hepatic and intestinal amoebiasis, leukaemia; and as anti-tumourous and anti-inflammatory (Agarwal, 2003; Ambasta, 1986; Jain, 1991; Khare, 2007; Kumar *et al.*, 2004; Parrotta, 2001; Prajapati *et al.*, 2003; Rastogi & Mehrotra, 1990).

Urtica dioica L. (Pl. 90C).

Family: Urticaceae.

Vern.: Bichhu-Buti, Choti Kogsi.

Ayurvedic, English and Hindi Names:

Ayur.: Vrishchhiyaa-shaaka;

Eng.: Stinging Nettle;

H.: Bhichoo Buti.

Distribution: N.W. Himalayass upto 3,500m: Kashmir-Shimla.

Description: Erect or stout grooved herbs with stinging hair, cordate, long, stalked, toothed leaves and green, dioecious flowers clustered on drooping axillary panicles. Achenes flat, included in persistent perianth.

Flowering & Fruiting: June-October.

Habitat Ecology: Moist slopes, wastelands; Dhaneta Dhaar, 1,070m.

Material Examined: HPU-EL-SKS-ABS-15279; 05.06.2011.

Parts Used: Whole Plant. Leaves. Shoot. Roots.

Folk Uses: Poultice of plant efficacious in chronic painful joints. Plant also utilized as a good source for fibres*. Necklace of roots worn to keep away black magic and evil spirits*. Leaves considered a source of fodder* and considered good for enhancing lactation* in milch animals. Tender leaves and shoots also consumed as pot herb and good against asthma*, biting cold* and rheumatic pains.

Active Constituents: Properties attributed to acetylcholine, histamine, 5-hydroxytryptamine (5-HT) (*hair*). flavonoids rutin, sterols, carotenoids, vitamins (C, B & K), minerals, phenolic acids, and coumarin scopoletin (*leaves*).

Uses in Literature: Recorded in l iterature as antihaemorrhagic, astringent, diuretic, eliminates uric acid, lithotropic; and for boils, constipation, dandruff, diarrhoea, gout, haemorrhages, nephritis, rheumatism, sprain, swelling, throat infection, wounds (Agarwal, 2003; Ambasta, 1986, Khanna, 2002; Khare, 2007; Manandhar, 2000; Sood & Thakur, 2004).

Vanda roxburghii R. Br. (Pl. 90D).

Syn.: *V. tessellate* Lodd. *ex* Loud.

Family: Orchidaceae.

Vern.: Bhangaru, Rasna.

Ayurvedic and English Names:

Ayur.: Raasnaa;

Eng.: An Orchid.

Distribution: Throughout India, U.P.-W. Bengal extending southwards to Kerala; found growing on mango trees in H.P.

Description: Epiphytic orchids 30-60cm high with leafy stem and thick coriacious, recurved, plicate, obtusely keeled leaves. Flowers greenish-yellow, mottled with brown on the mid-lobe of lip with purple caruncles in long dense drooping cylindric racemes. Column short, stout; rostellum shortly-beaked. Seeds numerous, minute; testa drawn out or winged.

Flowering & Fruiting: May-June.

Habitat Ecology: Grows as an epiphyte on mango trees; Sujanpur, 567m.

Material Examined: HPU-EL-SKS-ABS-15280; 27.01.2009.

Parts Used: Whole Plant. Leaves. Flowers.

Folk Uses: Poultice of tender leaves applied against rheumatism and for healing bones* and internal injuries*. Leaves also added to animal fodder* for bone healing. Flowers cooked as 'halwa' (pudding) and prescribed for cracked bones. Whole plant along with inforescence given for curing 'lalu' (foot and mouth disease)* of domestic animals.

Active Constituents: Properties attributed to tetracosylferulate, β-sitosterol, D-glucoside, saponin, β-sitostrol, tannins, heptacosane, octacosanol (*root, plant*).

Uses in Literature: Known so far as anti-inflammatory, antipyretic, laxative, livertonic, tranquilizer; and for abdomen, diseases of nervous system, earache (vet.), fractures, inflammation, lumbago, neurological affections, rheumatism, swelling (vet.) (Agarwal, 2003; Ambasta, 1986, Chauhan, 1988a, b; Khare, 2007; Sikarwar *et al.,* 1994; Singh & Kumar, 2000b).

Ayurvedic Formulations: Rasnaptak-kwath, Rasna-guggal, Rasnadasmula-kwatha, Rasna-panchak-kwath, Maharasnadi-kwatha, Mahamash-taila.

Verbascum thapsus L. (Pl. 91A).

Syn.: *V. indicum* Wall., *V. blattaria* Schmidt.

Family: Scrophulariaceae.

Vern.: Ban tambakhoo.

Ayurvedic, English and Regional Names:

Ayur.: Ban Tambaaku, Gidar Tambaaku, Phullaa;

Eng.: Common Mullein, Cow's Lungwort;

H.: Gidar tamaku;

Kan.: Kaduhogesooppina gida;

P.: Gidar tamaku, Gurganna, Karathri, Phasruk, Vulr.

Distribution: Temperate Himalayas: Kashmir-Bhutan.

Description: Erect herbs densely clothed with soft, yellow grey stellate hair. Stem robust, winged with prolonged leaf bases. Leaves entire, radical; lower ovate stalked, upper ones sessile, oblong-lanceolate. Flowers sessile, yellow, crowded in terminal spikes. Capsule tomentose, ovoid.

Flowering & Fruiting: January-April.

Habitat Ecology: Wastelands, open slopes, rocky slopes, cultivated areas; Kehdru, 842m.

Material Examined: HPU-EL-SKS-ABS-15281; 08.05.2008.

Parts Used: Aerial Plant parts. Leaves.

Folk Uses: Hot poultice of leaves good for joint pains*. Powdered aerial plant parts mixed with wheat dough given to cure constipation* in animals and stem alongwith leaves and inforescence given to check 'ban' (intestinal obstruction)*.

Active Constituents: Isolated principles are rotenone and coumarin (*leaves*) and crocetin (*flowers*).

Uses in Literature: Already known as an antispasmodic, aphrodisiac, demulcent in pectoral complaints, emollient, febrifugal, fish poison, heart stimulant, sedative, symbolic to keep evils away; and for asthma, cough, cramps (vet.), diarrhoea, inflammation, leucorrhoea, migraine (vet.), pulmonary disease of cattle and pyrexia (vet.) (Ambasta, 1986; Chandra Sekar & Srivastava, 2003; Jain, 1991; Khare, 2007).

Adulterant or Substitute: Dry leaves substituted for that of tobacco leaves.

Vernonia anthelmintica Willd. (Pl. 91B).

Syn.: *Conyza anthelmintica* L., *Serratula anthelmintica* Roxb., *Ascaricida indica* Cass., *Centratherum anthelminticum* Kuntze

Family: Asteraceae.

Vern.: Brahmjiri.

Ayurvedic, English, Sanskrit and Regional Names:

Ayur.: Dandotpalaa, Sahadevi, Uttamkanyaka;

Eng.: Iron weed, Purple flebane;

Sans.: Aranyajirak, Sonaraji;

H.: Buckshi, Kalijiri, Somraj;

Tam.: Naichotte poonde.

Distribution: Throughout India ascending up to 1,500m.

Description: Erect annual herbs possessing rough, ovate lanceolate, coarsely toothed, long pointed, narrowed into a short stalk, alternate leaves and robust stems glandular-pubescent near the top. Flowers in heads, involucral bracts narrowly oblong, tips dilated. Outer bracts longer than the inner. Pappus downy. Corollas often unequal. Achenes ribbed.

Flowering & Fruiting: July-August.

Habitat Ecology: Wastelands, slopes near the fields; Patnaun, 1,000m.

Material Examined: HPU-EL-SKS-ABS-15282; 11.08.2008.

Parts Used: Plant. Leaves. Seeds.

Folk Uses: Plant used as fodder. Powdered seeds given in case of 'basaihar' (mastitis), 20-50ml decoction of leaves given in case of asthma*, spasm in bladders and as anthelmintic (vet.)*.

Active Constituents: Isolated principles are luteolin-7-mono-beta-D-glucopyranoside. Triterpene compounds-beta-amyrin acetate, lupeol; sterol-beta-amyrin (*plant*).

Uses in Literature: Already known as anthelmintic, blood purifier, diaphoretic, febrifugal, fodder, hay; and for abdominal pains, dysuria, haematological disorders, leucorrhoea, mastitis (vet.), spasm of bladder and udder (Ambasta, 1986; Deeba *et al.*, 2009; Iqbal *et al.*, 2005; Jain, 1991; Joshi, 1995; Khare, 2007; Kumar *et al.*, 2004; Singh & Pandey, 1998).

Vicia pallida Facquem. (Pl. 91C).

Syn.: *Vicia bobartii* Forst., *V. Sativa* L.

Family: Fabaceae.

Vern.: Bada Kaer, Roothi.

English and Regional Names:

Eng.: Common Vetch, Spring Vetch;

H.: Akra, Ankara;

B.: Ankari;

Oriya.: Choni, Rothi.

Distribution: Throughout India.

Description: Annual herbaceous climbers with a weak stem, pinnate leaves, the rachis ending in a tendril. Leaflets many, opposite or nearly so. Flowers red-blue, or solitary, sometimes in pairs. Pods glabrous, 10 seeded.

Flowering & Fruiting: December-April.

Habitat Ecology: Wastelands, rocky slopes, cultivated areas; Bhareri, 920m.

Material Examined: HPU-EL-SKS-ABS-15283; 15.03.2009.

Parts Used: Whole Plant. Pods. Seeds.

Folk Uses: Mainly planted for its fodder value. Pods used for vegetable by locals. Kids eat* its seeds while playing.

Active Constituents: Pods rich in proteins, gamma-L-glutamyl-beta-alanine and gamma-L-glutamyl-gamma-aminobutyric acid.

Uses in Literature: Known to be used in Indian literature as edible, fodder and pot herb (Ambasta, 1986; Dwivedi, 1993; Jain, 1991; Khare, 2007; Khanna, 2002; Rastogi & Mehrotra, 1990; Singh & Kumar, 2000b; Sood & Thakur, 2004).

Viola canescens Wall. (Pl. 91D).

Family: Violaceae.

Vern.: Banaksha, Banfshah.

English and Regional Names:

Eng.: Sweet Violet;

H.: Banaksha.

Distribution: Throughout the temperate Himalayass upto 2,000m.

Description: A small perennial, softly dense pubescent herb with tufted rootstocks. Stems none or very short, nearly always producing long, leafy runners. Leaves tufted, ovate, deeply cordate, crenate, stipules fringed. Flowers lilac, solitary. Stigma terminal, truncate not beaked. Capsule ovoid, opening horizontally by boat-shaped valves. Seeds ovoid, ejected with force.

Flowering & Fruiting: April-November.

Habitat Ecology: Wastelands, hilly slopes, field areas after harvesting of maize; Dhoh, 1,034m.

Material Examined: HPU-EL-SKS-ABS-15284; 25.15.2009.

Parts Used: Plant. Flower. Roots.

Folk Uses: 30-50ml decoction of plant and flowers used as wonderful remedy for cough, cold and fever (twice daily till cure). Also ¼ glass decoction of its plant along with that of *Ocimum sanctum* L., *Elettaria cardamomum* Maton., *Glycyrrhiza glabra* L. and a pinch of black salt, given twice a day for urinary troubles, bleeding piles and asthma.

Active Constituents: Main ingredients are glucoside methyle salicylate, violine, a glycoside, viola quercitrin, saponins (*root*), methyl-salicylate glucoside (*leaves, flowers*).

Uses in Literature: So far recorded as an aperient, antipyretic, coolant, demulcent, diaphoretic, diuretic, emetic, emollient, expectorant, febrifuge, purgative; and for asthma, bleeding piles, cancer of throat, constipation, headache and skin diseases. (Ambasta, 1986; Chauhan, 1984; Dey, 1980; Dwivedi, 1993; Dymock *et al.*, 1889-1891; Jain, 1991; Khare, 2007; Khanna, 2002; Rastogi & Mehrotra, 1990; Singh & Kumar, 2000b).

Vitex negundo L. (Pl. 92A).

Syn.: *V. bicolor* Willd., *V. arborea* Desf., *V. panniculata* Lamk., *V. incisa* Lamk.

Family: Verbenaceae.

Vern.: Bana.

Ayurvedic, English, Sanskrit and Regional Names:

Ayur.: Nirgundi, Shephaalikaa, Sindhuka, Sindhuvaara, Sugandhikaa, Suhavaa;

Eng.: Common Chaste tree, Indian Privet, Nigandi;

Sans.: Indrani, Nilapushpa, Shepkali, Svetasuraso;

Ass.: Aggla-chita, Pasutia;

B.: Nishinda, Samalu;

Bo.: Katri, Lingur, Nirgur, Shiwari;

G.: Nagaol, Nigod, Nirgari;

H.: Mewri, Panikisambhalu, Nirgori, Nirgunda, Sanbhalu, Shiwali, Sindhuca, Sinduari;

Kan.: Lakkigida, Nakkilu, Nekki;

M.P.: Nagad;

Mal.: Indrani, Nochi, Vellanochi, Vennochi;

Mar.: Lingur, Nirguda, Nirgundi, Nirgur;

Oriya: Begundia,Indrani;

P.: Bankahu, Moraum;

Tam.: Nirkkundi, Nochi, Siduvaram, Tiributamerirran, Vennocho;

Tel.: Nellavavili, Sinduvaramu, Valili;

U.P.: Tarvan, Kali-Nirgundi.

Distribution: Throughout India.

Description: A large shrub or small tree having foliolate leaves which are opposite, long-stalked. Leaflets lanceolate-acute. Flowers bluish-purplish, in branched cymes. Drupes black when ripe. Fruits succulent, black ovoid containing four 1-seeded nutlets.

Flowering & Fruiting: March-June.

Habitat Ecology: Wastelands, roadsides, dry areas; Aghar, Bhota, 765m.

Material Examined: HPU-EL-SKS-ABS-15285; 05.06.2008.

Parts Used: Buds. Branches. Leaves. Twigs.

Folk Uses: A garland of its branches and leaves hung on main door to ward off communicable diseases*. Stem used for scouring teeth* and also chewed for healing mouth sores. Mustard oil burnt with wood rotting fungi growing on this plant used after filtering as drops for ear aches. Juice of fresh leaves mixed with water utilized for washing eyes (1-2 drops daily). Buds with 'gur' (jaggery) given to cure stomach disorders. Twigs used in magico religious purposes ('jhhada'). Leaves burnt to repel insects. Inhalation of steam emanating from hot water containing its leaves good for pains due to inflammation of joints. Paste of leaves used as antidote for snake bite*, honey bee and scorpion bites*. Fresh leaves given to animals for mastitis whereas roasted ones mixed with rock salt added to water in earthen pot as a good appetiser* for domestic cattle.

Active Constituents: Various constituents are iridoid glycosides, cucubin, agnuside, nishindaside, negundoside (*leaves*), vanitin and as mixture of unidentified alkaloids (*seeds*).

Biological Activity: Extr. of leaves exhibits anticancerous activity against *Ehrlich ascites* tumour cells. Also, leaf possesses anti-inflammatory, analgesic and CNS depressant activity.

Uses in Literature: Also known earlier as an abortifacient, astringent, cephalic, colic, diuretic, expectorant, insect repellent, sterilizer, stomachic, symbolic to keep evils away, tonic, vermifuge; and for anasarca, anthrax, basketry, blister, bodyache, boils, bone fracture, cold, cold (vet.), cough, cough (vet.), cut (vet.), diarrhoea, dropsy, dysentery, earache, eczema, endoparasite (vet.), ephemeral fever (vet.), eye complaints, fever, foetid discharge, foot diseases, gonorrhoea, epididymitis, giddiness, gout, gum trouble, headache, inflammation, itching, leprosy, liver disorder, mastitis, mental disorder, mosquito repellent (vet.), paralysis, paralysis (vet.), piles, pneumonia, post-natal delibity, rabies, reducing sex desire, rheumatism, scabies, sciatica, skin diseases, sores, spleen enlargement, sprain (vet.), swelling, swelling (vet.), syphilis, testies disorder, toothache, udder wounds (vet.), ulcers, waist pain (vet.), wounds (vet.) (Ambasta, 1986; Aminuddin & Girach, 1996; Bhandary *et al.*, 1996; Bhatt *et al.*, 2001; Borthakur & Sarma, 1996; Borthakur *et al.*, 2004; Chhetri, 2005; Geetha *et al.*, 1996b; Gogoi & Borthakur, 2001; Gogoi *et al.*, 2003; Hembrom & Goel, 2005; Jain *et al.*, 1994; Katewa & Arora, 1997; Khare, 2007; Kiruba *et al.*, 2006; Kshirsagar & Singh, 2000; Lal *et al.*, 1996; Mahato & Mahato, 1996; Malik *et al.*, 2009; Manandhar, 1988; Painuli & Maheswari, 1996; Pal & Jain, 1998; Pandey & Rout, 2006; Ragupathy & Mahadevan, 1991; Ram *et al.*, 2002; Ramadas *et al.*, 2000; Rama Rao & Henry, 1996; Reddy *et al.*, 1998; Rekha *et al.*, 1998; Saini, 1996a; Sharma, 1996; Sharma & Sood, 1997; Singh *et al.*, 1996b, Srivastava *et al.*, 2003, Sudarsanam *et al.*, 1995; Vir Jee *et al.*, 1984).

Major Ayurvedic Formulations: Himacolin cream, Nirgundi-ghrit, Nirgundi-kwath, Pilex, Rumalayas (cream and tablets), Varnasodhana-taila.

Trade: Nirgundi-phal, Samhaalu-phal.

Vitis jacquemontii Parker (Pl. 92B).

Syn.: *Ampelocissus jacquemontii* (Parker) Raizada

Family: Vitaceae.

Vern.: Geedar-Daakh.

Distribution: Subtropical regions of India.

Description: A stout woody climber with branched tendrils. Leaves ovate-cordate toothed, woolly hairy beneath. Flowers in leaf opposed dense cluster. Berries globose, succulent, 1-4 seeded.

Flowering & Fruiting: June-August.

Habitat Ecology: Forests, along roadsides; Chauri, 858m.

Material Examined: HPU-EL-SKS-ABS-15286; 17.08.2009.

Part Used: Fruit.

Folk Uses: Wild grapes edible* and considered as laxative*.

Uses in Literature: There is no mention of its use in literature.

Vitis vinifera L. (Pl. 92C).

Family: Vitaceae.

Vern.: Angoor.

Ayurvedic, English and Hindi Names:

Ayur.: Draakshaa, Go-stani, Mrdvikaa;

Eng.: Wine grapes, European Grape;

H.: Angoor.

Distribution: Throughout India, tropical and sub-tropical parts.

Description: A woody shrubby climbing vine. Leaves simple or digitately or pedately compound. Tendrils leaf opposed, yellow green, simple or branched. Flowers often in racemes or cymes. Berries globose, succulent, 1-4 seeded.

Flowering & Fruiting: March-June.

Habitat Ecology: Cultivated for its fruits; Anu Khurd, 858m.

Material Examined: HPU-EL-SKS-ABS-15287; 25.05.2011.

Parts Used: Leaves, Fruits.

Folk Uses: Fruits edible as such. One glass of its fruit juice prescribed daily for anaemia, jaundice, weakness; and also as coolant and source of vitamins. Dried fruits used in confectionary* and preparations of sweet dishes. Also, its heated dried fruit after wrapping in leaf of *Colebrookia oppositifolia* Smith. given for cure of cough, tonsillitis*, constipation and liver ailments.

Active Constituents: Active ingredients are flavonoids tannins, tartrates, inositol, carotenes, choline (*grape vine*), sugars, pectins, tartaric acid, malic acids, flavone glycosides, vitamin A, B, B1, B2, C, minerals and anthocyanins (*fruit*).

Uses in Literature: Known to be used earlier as invigorating, nourishing; and for alcoholism, anaemia, antherosclerosis, constipation, cough, decreased hepatic injury from alcoholism, diarrhoea, dyspepsia, enlarged liver, gout, haemorrhoids, heavy menstrual bleeding, haemorrhagic diseases, jaundice, night vision, oedema due to injury, post surgery oedema, respiratory track catarrh, spleen, uterine haemorrhage, varicose veins and venous insufficiency (Ambasta, 1986; Dwivedi, 1993; Jain, 1991; Khare, 2007; Khanna, 2002; Rastogi & Mehrotra, 1990; Singh & Kumar, 2000b).

Wendlandia heynei (Roem. & Schult.) Sant. & Merch. (Pl. 92D).

Syn.: *W. exserta* (Roxb.) DC., *W. cinerea* Wall., *Rondeletia exserta* Roxb., *R. cinerea* Wall., *R. thrysiflora* Roth., *R. orissensis* Roth., *R. heynei* Roem. & Sch.

Family: Rubiaceae.

Vern.: Panseera.

Ayurvedic and Regional Names:

Ayur.: Tilaka;

Ass.: Kadam;

B.: Mimri, Tilki;

H.: Birsa, Chaulai, Chila, Chilkiya, Tila, Tilki;

Kan.: Tilai;

Mar.: Padgira;

P.: Chilkiya, Pansara, Pudhara.

Distribution: Abundant in dry forest of tropical Himalayas ascending to 1,350m.

Description: Small trees with terete branches, coriaceous leaves, and ovate, acute, recurved stipules. Flowers white, fragrant, sessile in panicles. Capsule globose, many seeded, white.

Flowering & Fruiting: April-June.

Habitat Ecology: Wasteplaces, edges of cultivated fields, slopes, dry forests; Man Khad, 483m.

Material Examined: HPU-EL-SKS-ABS-15288; 22.06.2010.

Parts Used: Leaves. Twigs.

Folk Uses: Twigs used as tooth brush*. Leaves used as fodder and its juice helps in expelling leeches* from nostrils of livestock.

Uses in Literature: Described in India as fodder and for agricultural implements, building works, skin diseases, snake bites, urinary complaints (Ambasta, 1986; Jain, 1991; Khare, 2007; Murthy *et al.*, 1986; Saxena, 1986).

Withania somnifera (L.) Dunal (Pl. 93A).

Syn.: *W. ashwagandha* Kaul., *Physalis somnifera* Link, *P. flexuosa* L., *P. arborescens* Thunb., *P. tomentosa* Thunb.

Family: Solanaceae.

Vern.: Ashvagandha, Asgandh, Asugandha, Indian ginseng.

Ayurvedic, English, Sanskrit and Regional Names:

Ayur.: Ashwagandhaa, Ashwakanda, Gandharvagandhaa, Gokarnaa, Hayagandhaa, Turaga;

Eng.: Winter Cherry;

Sans.: Aswaganda;

B.: Ashuagandh;

Bo.: Asgandh;

G.: Asan;

H.: Asgandh, Ashwagandha;

Kan.: Aswagandhi;

Mal.: Ammukirram;

Mar.: Askanda, Tilli;

Oriya: Asua gandha;

Tam.: Amukurra;

Tel.: Pannerugudda.

Distribution: Drier parts of India, ascending upto 1,700m in the Himalayas.

Description: Unarmed perennial shrubs upto 1.5m tall. Leaves entire, thinly woolly, ovate. Branches flexuose, terete and densely tomentose. Flowers greenish or lurid yellow and in sub sessile umbelliform cymes. Berries enclosed in inflated calyx, red when ripe. Seeds enclosed in pulp.

Flowering & Fruiting: June-November.

Habitat Ecology: Wasteplaces as well as cultivated; Bramni village, 858m.

Material Examined: HPU-EL-SKS-ABS-15289; 13.03.2008.

Parts Used: Roots. Leaves.

Folk Uses: Leaves valued for preventing greying of hair* and inflammation. 20-30ml decoction of roots given once daily for asthma*, cough*, urinary troubles and debility from old age. 1-3g powdered roots prescribed with honey before bed time as tonic for providing physical and mental health*.

Active Constituents: Main ingredients are ipuranol, a crystalline alcohol, withaniol, withanine, withaferin-A, hentriacontane, phytosterol, fatty oil (*plant*), alkaloids withanine and somniferine, somniferinine (*root*).

Uses in Literature: Used as antiarthritis, antibacterial, anti-inflammatory, anti-tumourous, diuretic, hepatoprotective, hypnotic, sedative; and for hypotoxicity, rheumatism, scrofula, swellings (Ambasta, 1986; Jain, 1991; Khare, 2007; Murthy *et al.*, 1986; Saxena, 1986).

Ayurvedic Formulations : Asvagandha-ghrita, Asvangandharishta, Ashvagandhahadi-pak, Ashwagandha-rasayana.

Woodfordia floribunda Salisb. (Pl. 93B).

Syn.: *W. fruticosa* (L.), Kurz, *W. tomentosa* Bedd., *Lythrum fruticosum* L., *Grislea tomentosa* Roxb., *G. uniflora* Rich., *G. punctuata* Ham., *G. micropetala* Hocsht. *ex* Steud.

Family: Lythraceae.

Vern.: Dhai, Dhavae, Dhavi, Dhaun.

Ayurvedic, English, Sanskrit and Regional Names:

Ayur.: Bahupushpi, Dhaatri, Dhaataki, Kunjaraa, Taamrapushpi, Vahnijwaalaa;

Eng.: Fire Flame Bush, Red Bell Bush, Shiranjitea Woodfordia;

Sans.: Agnijvala, Bahupusphika, Dhataki, Reyya-manu, Rodhrapushpini;

Ass.: Eda, Era;

B.: Dhadki, Dhavanji, Dawa;

Bo.: Dhavri, Dowari;

G.: Dhawadina;

H.: Dawi, Dhai Dhaura, Thawi;

Kan.: Dhataaki, Huvvu, Tamrapushpi;

Kash. : Thavi;

Mal.: Tattivipu, Tatire;

Mar.: Dhaiphal, Dhazatichi, Phulsatti;

Oriya: Datubiphala, Horibadi;

P.: Dawi, Gulbahar, Khurd;

Tam.: Thatarijargi, Velakkai;

Tel.: Gaddapisniki, Gaji-Godari, Kusumamu Reyyamanu.

Distribution: Throughout India.

Description: Evergreen shrubs having opposite, lanceolate, pointed leaves and bright red, tubular flowers in axillary cluster. Capsule rounded or cordate at the base, included in the calyx.

Flowering & Fruiting: March-June.

Habitat Ecology: Forests, open slopes; Pucca Bharo, 840m.

Material Examined: HPU-EL-SKS-ABS-15290; 25.03.2008.

Parts Used: Flowers. Leaves. Wood.

Folk Uses: Hot poultice of flowers tied to the stomach for curing stomachache* and constipation*. Powdered dried flowers sprinkled for quick healing of open wounds. Application of flower poultice on fore-head relieves headache*. Also, 1 tsp decoction of flowers prescribed twice a day for 7-10 days to regulate liver disorders. Leaves form a good fodder*. Wood used as fuel.

Active Constituents: Active principles associated with cyaniding-3, 5-diglucoside, pelargonidin-3, 5-diglucoside, myricetin-3-diglucoside, polystachoside, quercitin-3-rhamnoside, naringenin-7-glucoside, kaempferol-3-glucoside, chrysophanol-8-D-β-glucopyranoside (*flowers*), a new C-glucoside-norbergenin (*stem*), ellagic acid, polystachoside, myricetin-3-galactoside and pelar-gonidine-3, 5-diglucoside (*leaves, flowers*).

Biological Activity: Flowers abortifacient and CNS depressant.

Uses in Literature: Useful earlier as an astringent, blood purifier, drink, febrifugal, fuel wood, red dye, stimulant, tonic; and for bleeding in pregnancy and menstruation, bone fracture, bowel complaints, burns, cancer, carbuncle, cholera, conception, complaints of liver, conjunctivitis, cough, diarrhoea, dysentery, edible purposes (raw or cooked), erysipelas, haemetemesis, injuries, loss of appetite in pregnancy, menorrhoea, miscarriage, muscular pain, nausea, night blindness, rheumatism, seminal weakness, skin diseases, sores and ulcers (vet.), uterus pain, vomiting, wounds (Ambasta, 1986; Aminuddin & Girach, 1996; Borthakur, 1993; Chhetri, 2005; Hembrom & Goel, 2005; Jain, 1991; Khare, 2007; Kothari & Moorthy, 1996; Kulkarni & Kumbhojkar, 1996; Masih, 2003; Mohanty *et al.*, 1998; Pandey & Rout, 2006; Patil, 2004; Patole & Jain, 2002; Rama Rao & Henry, 1996; Sarangi & Sahu, 2004; Singh, 1999b; Verma & Chauhan, 2006).

Major Ayurvedic Formulations: Arvindasava, Brhatgangadhara-churna, Dhatakyadi-taila, Gangadhar-churna, Kutajarista, Laghugangadhar-churna, Pushayanug-churn.

Trade: Dhaataki-phul, Dhai phool.

Xanthium strumarium L. (Pl. 93C).

Syn.: *X. indicum* DC., *X. roxburghii* Discolor, *X. orientale* Blume

Family: Asteraceae.

Vern.: Bhakhadda.

Ayurvedic, English, Sanskrit and Regional Names:

Ayur.: Aartagala, Arishta, Shankheshwara;

Eng.: Bur-weed, Cockle bur;

Sans.: Arishta, Chanda, Hara, Pitapushpi;

Ass.: Agara;

B.: Banokra;

Bo.: Shankeshvara;

G.: Gadriyun;

H.: Banokra, Chhotgokhru;

Kash. : Lanetsuru;

Mal.: Buah anjang;

Mar.: Dumundi, Sankeshwara;

P.: Chirru, Gudal;

Tam.: Marlumutta;

Tel.: Marulamatangi, Parsvapu.

Distribution: Throughout the hotter parts of India.

Description: Coarse annual, unarmed herbs with 3 fid spines. Leaves petioled, triangular, 3-lobed and greenish-white. Heads in terminal and axillary racemes. Achenes clothed with strong hooked spines.

Flowering & Fruiting: September-November.

Habitat Ecology: Open slopes, wastelands, roadsides; Chhoti Kashi, Nadaun, 483m.

Material Examined: HPU-EL-SKS-ABS-15291; 27.09.2010.

Part Used: Seeds.

Folk Use : Seeds eaten raw to check headache.

Active Constituents: Active properties due to sesquiterpene lactone-xanthinin, its stereoisomers-xanthanol and iso-xanthanol, xanthumin, isohexacosane, xanthatin, β-sitosterol (*leaves*).

Biological Activity: Hypoglycaemic activity found to be positive.

Uses in Literature: Recorded in published literature as a bitter tonic, demulcent, diaphoretic, diuretic, drink, emollient, pot-herb, sedative, sudorific; and for boils, cancer wounds, cooling, eye diseases, fencing, headache, herpes, inflammatory swellings, killing leucorrhoea, maggots, malaria, menorrhoea, night blindness, piles, rheumatism, ringworm, scrofula, snake-bite, toothache, ulcers and urinary complaints (Ambasta, 1986; Jain, 1991; Khare, 2007; Kirtikar & Basu, 1994; Patil & Merat, 2003; Punjani, 1998; Venkataratnam & Venkata Raju, 2004).

Yucca aloifolia L. (Pl. 93D).

Family: Liliaceae.

Vern.: Ramban.

English and Hindi Names:

Eng.: Beer Grass;

H.: Bhalu Ghas.

Distribution: Common in gardens in warm tropical regions.

Description: Perennial rhizomatous shrubs with rosette fleshy leaves and thorny edges. Roots adventitious. Stem woody, erect, branched. Flowers in racemes. Fruit a capsule or berry. Seeds endospermic with black testa.

Flowering & Fruiting: June-September.

Habitat Ecology: Cultivated for its flowers; NIT, Anu, 900m.

Material Examined: HPU-EL-SKS-ABS-15292; 02.06.2009.

Parts Used: Plant. Leaves. Flowers.

Folk Uses: Planted as an ornamental plant for hedge* purposes. Leaves used for tieing grass* bundles. Flowers used for making bouquet*. Fibres obtained from leaves used for making ropes, cordages and strings.

Active Constituents: Main ingredients are aloifoline (*flower*), indole melanins (*seeds*), tigogenin, sarsasapogenin, gitogenin, hecogenin, smilagenin, neotigogenin and samogenin (*leaves*).

Uses in Literature: Used earlier as decorative, purgative; and for cordages, ropes and strings (Agarwal, 2003; Ambasta, 1986; Chopra *et al.*, 1956; Dwivedi, 1993; Jain, 1991; Khare, 2007; Khanna, 2002; Rastogi & Mehrotra, 1990; Singh & Kumar, 2000b).

Zanthoxylum armatum DC. (Pl. 94A).

Syn.: *Z. alatum* Roxb., *Z. planispinum* Sieb & Zucc., *Z. hostile* Wall.

Family: Rutaceae.

Vern.: Tirmer.

Ayurvedic, English, Sanskrit and Regional Names:

Ayur.: Tejabala, Tejaswani, Tejohva, Tejovati, Tumburu;

Eng.: Wing leaf prickly ash;

Sans.: Andhaka, Dhira, Gandhalu, Mahamuni, Sanuja, Saurabh, Sauravanaja, Sugandhi, Tumbaru, Tumburn;

B.: Gaira, Nepali-dhamia, Tambul, Tun;

G.: Tejabala, Tejbal;

H.: Darmar, Tejbal, Tirmur, Tezmal, Timbur, Tumru;

Kan.: Jimmi, Tejovanti, Tejpatri, Tumbura, Tumburudra;

Mal.: Thumboonal, Thunbooni, Valiyavaluzhavam;

Mar.: Tejabal;

Oriya: Tundopoda;

P.: Tezbal, Timmal, Timru, Kababa;

Tam.: Thejyovathi;

Tel.: Tejovathi.

Distribution: Hot valleys of subtropical Himalayas.

Description: Shrubs or small trees with corky bark and numerous long straight spines on branchlet and leaf stalk. Branches armed; prickles flattened, pungent aromatic taste and smell. Leaves pinnate. Flowers yellow in short lateral clusters. Drupes pale red, globose, splitting in two valves. Seeds aromatic, solitary, black shining.

Flowering & Fruiting: April-June.

Habitat Ecology: Open slopes, dry areas; Anu Khurd, 900m.

Material Examined: HPU-EL-SKS-ABS-15293; 15.05.2011.

Parts Used: Twigs. Carpels. Seed. Bark. Wood.

Folk Uses: Twigs used for scouring teeth. Paste of its seeds and bark mixed with a pinch of rock salt and black pepper (*Piper nigrum* L.) massaged on gums for dental problems. Gargles with decoction of bark good for mouth ulcer*. Carpels used as condiments*. Long thick stick of its wood used for grinding* purposes in kitchen. Powdered seeds and fruits (25g for 3-5 days) given with milk or water twice daily to cure indigestion in animals*.

Active Constituents: Active ingredients are dictamnine (*stem-bark*), 1.5% essential oil (wartara oil), a source of linalool (*fruit*), ketone free fraction of essential oil (*leaves*) linalyl acetate, sesquiterpene hydrocarbon and tricabane, β-sitosterol, pinoresionol, dimethyl ether (*aerial parts*).

Biological Activity: Stem exhibits hypoglycaemic and spasmolytic properties. Essential oil from fruits shows antifungal activity.

Uses in Literature: Known to Indians as an anthelmintic, anthelmintic (vet.), carminative, deodorant, disinfectant, fish poison, spice, stomachic and tonic; and for cholera, cough, dental troubles, diarrhoea, eczema, fever, flatulence, house building, indigestion, itching, leucoderma, madness, making walking sticks and clubs, piles, religious ceremonies, rheumatism, scouring teeth, small pox, snake bite, soap making, toothache and warding off house flies (Ambasta, 1986; Ansari, 1991; Bhattarai, 1992; Borthakur & Sarma, 1996; Chauhan, 1999; Chhetri, 2005; Chowdhery, 1996; Iqbal *et al.*, 2005; Jain, 1991; Kapur & Singh, 1996; Khare, 2007; Kirtikar & Basu, 1994; Lal *et al.*, 1996; Manandhar, 1996; Mohanty *et al.*, 1998; Negi & Gaur, 1991; Ravisankar & Henry, 1992; Siwakoti & Verma, 1996; Verma & Chauhan, 2006).

Major Ayurvedic Formulations: Brahmi-vati, Dantamanjan, Hingodi-taila, Kalaka churna (Lepa), Tumbarwadi-churna.

Trade: Timroo or Tejbel (dried fruits), Tumbru.

Zea mays L. (Pl. 94B).

Family: Poaceae.

Vern.: Makki, Bhuta, Chhali.

Ayurvedic, English, Sanskrit and Regional Names:

Ayur.: Mahaa-Kaaya;

Eng.: Maize;

Sans.: Yavanala;

B.: Bhutta;

Bo.: Makai;

H.: Makai;

Kan.: Mekkejola;

Mal.: Cholam;

P.: Makai;

Tam.: Malsbasholam;

Tel.: Mokhajanna.

Distribution: Cultivated.

Description: Tall, stout annual herbs. Leaves large. Spikelets unisexual. Female flowers in spadix. Glumes membranous, lodicules 2. Stamen 3. Female spikelet lodicules zero. Grains sub-globose.

Flowering & Fruiting: July-August.

Habitat Ecology: Planted; Bhira, 780m.

Material Examined: HPU-EL-SKS-ABS-15294; 17.07.2010.

Parts Used: Grains. Stem. Cobs.

Folk Uses: Grains roasted and consumed as popcorns; also its flour used for making 'makki-ki-roti' which is relished along with 'saag' prepared from mustard leaves and butter. Stem used as fuel* and fodder. Dry deseeded mature cobs used as fuel whereas youngones used for making soups and roasted and eaten ('bhutta'). Maize flour employed as an adulterant* to 'basen' (flour of *Cicer arietinum* L.). Decoction of its inflorescence prescribed once daily on empty stomach for constipation*, also soaked pieces of 'chappati' made from maize flour given to animals for cough*.

Active Constituents: These are 3 galactoside-cyanidol coumarate (*corn* pigment) and a-b-D-glucoside (*roots*).

Biological Activity: Plant-nephro protector.

Uses in Literature: Used earlier as astringent, diuretic, fodder, food; and for antifertility, antifertility (vet.), corn flakes, diabetes, lactation (vet.), popcorns, weakness (vet.) (Ambasta, 1986; Hembrom & Goel, 2005; Kadel & Jain, 2006; Khare, 2007; Mishra *et al.*, 1996b; Samati, 2004; Sensarma, 1989).

Zingiber officinale Rosc. (Pl. 94C).

Syn.: *Amomum zingiber* L.

Family: Zingiberaceae.

Vern.: Adrak.

Ayurvedic, English, Sanskrit and Regional Names:

Ayur.: Aadraka, Aadrikaa, Katubhadra, Shrngavera, Shrngibera;

Eng.: Ginger;

Sans.: Anupama, Ardraka, Ardrashaka, Kandara, Machhaka, Mahija, Mulaja;

B.: Ada;

H.: Ada, Adraka;

Mal.: Andrakam, Inchi;

Mar.: Ale;

Tam.: Allam, Inji;

Tel.: Allamu, Sonthi.

Distribution: Cultivated in hot and humid areas of S. India, Assam, H.P.

Description: Herbs. Leaves linear-lanceolate, sessile. Rootstock horizontal, tuberous, Flowers in spike. Corolla segments unequal. Capsule with globose, arillate seeds.

Flowering & Fruiting: July-October.

Habitat Ecology: Cultivated; Khaya, 787m.

Material Examined: HPU-EL-SKS-ABS-15295; 18.01.2009.

Part Used: Rhizome.

Folk Uses: 5-10g of powdered rhizome fried in ghee prescribed with milk for rheumatic pains. Also, powdered dried rhizomes ('sund') used for making 'panjiri' given to ladies after deliveries*. Rhizome also used as condiments* and for pickling*. Pieces of rhizome added to tea for flavouring* and to cooked vegetables for checking flatulence. 1 tsp of rhizome juice prescribed with honey and pepper, twice a day till cure for cough and asthma. 500ml decoction of dried rhizome mixed with 'taramira' (*Eruca sativa* Mill.), water and 'hing' (*Ferula asafoetida* L.) given for mastitis*, whereas dried rhizome given for prolapsed uterus* and indigestion to cattle.

Active Constituents: Active properties attributed to gingerol, shagols, dihydrogingerol, hexahydro-curcumin, gingerdiols, zingerone, 4 new diarylheptenones-gingerenones A, B, C and isogingerenone-B, two more diarylheptanoids, meso-3, 5-diacetoxy-1, 7-bis (4-hydroxy-3, 5-dimethoxyphenyl)-7-(4-hydroxy-3-metho-xyphenyl heptane) (*rhizome*).

Biological Activity: Gingerenone A exhibited anticoccidium activity *in vitro* in cultural kidney cells of chicken and strong antifungal effect against *Pyricularia oryzae*.

Uses in Literature: Known earlier for arthritis, cough, flatulence, indigestion (vet.), snake and insect bite, sore throat, soup, throatache, vomiting; and as anthelmintic, anthrax, appetiser (vet.), blood purifier (vet.), bone fracture (vet.), carminative, cold and constipation (vet.), diarrohoea (vet.), dropsy (vet.), dyspepsia (vet.), expectorant, fever (vet.), gastric (vet.), indigestion (vet.), rabies, stimulant, strength (vet.), tympany (vet.), vegetable and warmth (vet.) (Ambasta, 1986; Chinthu *et al.*, 1998; Choondal, 1988; Deeba *et al.*, 2009; Gogoi & Borthakur, 2001; Iqbal *et al.*, 2005; Jain, 1995; Jha *et al.*, 1991; Kemparaja *et al.*, 1998; Khare, 2007; Kumar *et al.*, 2004; Mishra *et al.*, 1996b; Pal, 1981; Ranjan & Sethuraman, 1997; Reddy *et al.*, 1998; Samati, 2004; Sharma & Singh, 2001; Singh, 1998; Varshney, 1998; Yadav & Gupta, 1998).

Zizyphus jujuba Mill. (Pl. 94D).

Syn.: *Z. mauritiana* Herb., *Z. sororia* Scult., *Z. trinervis* Roth., *Z. sativa* Gaertn DC., *Z. vulgaris* Lamk., *Rhamnus jujuba* L.

Family: Rhamnaceae.

Vern.: Ber.

Ayurvedic, English, Sanskrit and Regional Names:

Ayur.: Badar, Kola;

Eng.: Chinese date, Indian Cherry, Indian jujube, Indian plum;

Sans.: Ajapriya, Badari, Dridhabija, Koli, Prithukoli, Rajakoli, Svaccha, Tanubija;

B.: Bogri;

Bo.: Bhurmi, Bordi;

G.: Ber, Boreli, Boyedi;

H.: Baer, Ber, Beri;

Kan.: Elanji, Yalachi;

Mal.: Badaram, Badari, Perinutulali;

Mar.: Baher, Bor;

Oriya: Barkoli, Bodokoli;

P.: Ber, Beri, Unab;

Tam.: Attiram, Ilandai, Iratti, Koli, Kulvali, Sivagam;

Tel.: Badari, Badaramu, Gangaregu, Karkhanduvu;

Urdu: Ber.

Distribution: Throughout India.

Description: Small glabrous bushy shrubs having divaricate, zig-zag branches. Leaves elliptic, spinous dentate. Flowers greenish, in axillary cymes. Drupes globose, glabrous, fleshy 2-celled and mealy, red when ripe.

Flowering & Fruiting: June-August.

Habitat Ecology: Wastelands, dry areas; Tikker Khatrian, 908m.

Material Examined: HPU-EL-SKS-ABS-15296; 03.02.2010.

Parts Used: Leaves. Twigs. Fruits.

Folk Uses: Leaves used as cattle feed*. Poultice of leaves applied on head overnight to check hairfall or head bath taken with its decoction (for a week) to check dandruff*. Dried twigs used as fuel*. Fruits edible and considered sacred for offering to appease Lord Shiva on day of 'Mahashivratri. Dried fruits also added to decoction of 'benafsha' (*Viola canescens* Wall.) and 'Malathi' (*Glycyrrhiza glabra* L.) for cough* and cold*. Dried fruits also prescribed for phlegm* and whooping cough (25-30ml decoction, once a day for 10 days).

Active Constituents: Fruits rich in fatty acids, phytosterol, organic acid, reducing sugar, pectic mucilage, catechols, tannins, catechins, p-coumaraylates of alphitolic acid, gunosine-3'-5' monophosphate, saponins I, II, III and jujubuside B.

Biological Activity: Aerial parts hypotensive.

Uses in Literature: Described in published literature as an anodyne, antiemetic, astringent, blood purifier, colic, diaphoretic, edible, febrifugal, purgative, sedative, stomachic, symbolic worshipping and tonic (vet.); and for bleeding from nose, cholera, cooling, conjunctivitis, constipation, diarrhoea, dysentery, fall of hair, fencing, fever, foot and mouth diseases, headache, inflammation, odema, rheumatism, scorpion stings, sores, spleen complaints, ulcer, wounds, whooping cough, yolk galls (Bhatt *et al.*, 2001; Chatterjee & Pakrashi, 1997; Gogoi & Borthakur, 1991; Jain, 1991; Kadel & Jain, 2006; Khare, 2007; Kumar & Chauhan, 2006; Kumar *et al.*, 2004; Lalramnghinglova, 1996; Mohanty & Padhy, 1996; Mohanty *et al.*, 1998; Punjani, 1998; Ragupathy & Mahedevan, 1991; Ram *et al.*, 2002; Rama Rao & Henry, 1996; Sahu, 1996; Sharma *et al.* 2002; Shome *et al.*, 1996; Shukla & Verma, 1996; Sikarwar & Kaushik, 1992; Singh, 1996; Srivastava *et al.*, 2003; Upadhye & Kulkarni, 2004).

Major Ayurvedic Formulations: Madhusdudi-kshay.

Chapter 3
Epilogue

Ethnobotanical Diversity

The study highlights the ethnic uses of 296 plant species belonging to 243 genera under 86 families (medicinal 263 species, edible 133 species, fodder 108 species, ornamental 67 species, sacred 52 species, magico-religious 34 species, fibre 29 species, veterinary ailments 123 species, miscellaneous purposes 84 species; Figs. 3.13, 3.14) by the rural communities of district Hamirpur, H.P. Of the plants utilized for their sustenance, there are 116 herbs representing 39.18% of the overall diversity, 59 shrubs (19.93%), 88 trees (29.72%), 10 undershrubs (3.37%), and 6 lianas (2.02%) (Figs. 3.3, 3.4; Tables 3.1, 3.6) with floristic composition dominated by dicots constituting bulk share of 84.45% representation (250 species, 205 genera, 71 families; Figs. 3.5, 3.6), followed by monocots 13.51% (40 species, 33 genera, 10 families), gymnosperms 1.01% (3 species, 3 genera, 3 families: *Cycas, Pinus Thuja*) and pteridophytes 1.01% (3 species, 2 genera, 2 families: *Adiantum, Equisetum*) (Figs. 3.1, 3.2, 3.5, 3.6; Table 3.8; Appendix I). Amongst these, the study identifies the use of 175 wild plants, 108 species under cultivation and 13 species as wild and under cultivation such as *Alstonia scholaris, Bryophyllum calycinum, Cynodon dactylon, Duranta repens, Ficus religiosa, Gossypium arboreum, Helianthus annuus, Jasminum humile, Lawsonia alba, Mimosa pudica, Musa sapientum, Physalis peruviana, Prunus persica* (Table 3.7) with predominant usage of aerial parts (766; 92.62%; Figs. 3.9, 3.10) and of members belonging to polypetalous families (126 species, 106 genera, 39 families; Tables 3.9, 3.10). Phenologically, a large majority of the taxa used ethnically produce flowers and fruits predominantly during summers (45%), and spring (23%) seasons (Figs. 3.20, 3.21). In comparison, the proportion of species and genera belonging to monocots to dicots is 1: 6.2 and of families 1: 7.1 (Figs. 3.7, 3.8) thereby indicating that usage of dicotyledonous families by the local populace are over 7 times more than the monocotyledonous ones. This ratio for the region is also similar to information for whole India, i.e., 1: 7 (Hooker, 1872-1897). With regard to species richness, the high used families pertain to Fabaceae (26 species), Poaceae (18 species), Asteraceae (17 species), Solanaceae (13 species), Moraceae (12 species), Lamiaceae (11 species), Rosaceae (10 species), Rutaceae (10 species), Euphorbiaceae (9 species), Apocynaceae (9 species), Cucurbitaceae (7 species), Brassicaceae (7 species), Amaranthaceae (7 species), Liliaceae (7 species), etc (Figs. 3.15-3.17; Tables 3.12, 3.13) with top 10 ranking families accounting for 45.27% of the recorded species diversity from the study area (Table 3.16). Statistically, the relative percentage in taxa for predominant Fabaceae amongst dicots accounts for 8.78% whereas for monocots it is 6.08 (Poaceae) (Table 3.11). For organ-wise analysis, the data reveals predominant usage of leaves (206), whole plant (128), fruits (91), seeds (82), rhizomes/bulbs/tuber/root (48), flower (53), wood (48), stem (47), twig/aerial plant parts (41) whereas inflorescence (4), oil cake (4), rind (3), receptacle (2) and spine (1) do not find much favour in the region (Fig. 3.11, 3.12). From the view point of utilitarian genera *Ficus* (7 species), *Citrus* (5 species), *Solanum* (5 species), *Brassica* (3 species), *Terminalia* (3 species), *Acacia* (3 species), *Jasminum* (3 species), *Prunus* (3 species) and *Saccharum* (3 species), etc. predominate (Figs. 3.18, 3.19) in their day-to-day life. Undeniably the rural inhabitants of the district have good knowledge about the bioresources in their area since time immemorial and are using these resources very carefully to ensure year-round supply of food, fuel, fodder, medicine to them and their cattle.

Table 3.1: Ethnobotanically Important Herbs of District Hamirpur

S.No.	Name of the Plant	Family	Vernacular Names
1.	*Abelmoschus esculentus*	Malvaceae	Bhindi
2.	*Abelmoschus moschatus*	Malvaceae	Kasturi Bhindi
3.	*Acorus calamus*	Araceae	Baryan
4.	*Adiantum capillus-veneris*	Adiantaceae	Hansraj
5.	*Adiantum lunulatum*	Adiantaceae	Hansraj
6.	*Ageratum conyzoides*	Asteraceae	Neela Fulunu
7.	*Allium cepa*	Liliaceae	Piaz
8.	*Allium sativum*	Liliaceae	Lahsun
9.	*Amaranthus paniculatus*	Amaranthaceae	Cholai
10.	*Amaranthus spinosus*	Amaranthaceae	Cholaee
11.	*Amaranthus viridis*	Amaranthaceae	Cholaee
12.	*Amorphophallus campanulatus*	Araceae	Zimmikand
13.	*Andrographis paniculata*	Acanthaceae	Kalmegh
14.	*Anetham sowa*	Apiaceae	Kaudi saunf
15.	*Arachis hypogaea*	Fabaceae	Moongphali
16.	*Argemone mexicana*	Papaveraceae	Lea
17.	*Arisaema tortuosum*	Araceae	Kirr-Ri-Chhhali
18.	*Asparagus adscendens*	Liliaceae	Sanspai
19.	*Avena sativa*	Poaceae	Joi, Javi
20.	*Bacopa monnieri*	Scrophulariaceae	Jal Brahmi
21.	*Boerhavia diffusa*	Nyctaginaceae	Itsat, Utsata
22.	*Brassica campestris*	Brassicaceae	Sarson
23.	*Brassica juncea*	Brassicaceae	Rai
24.	*Brassica nigra*	Brassicaceae	Rai, Ohari
25.	*Bryophyllum calycinum*	Crassulaceae	Patharchat, Gillar pattar
26.	*Cannabis sativa*	Moraceae	Bhang
27.	*Canscora diffusa*	Gentianaceae	Shankhpushpi
28.	*Capsicum annuum*	Solanaceae	Mirch
29.	*Carduus nutans*	Asteraceae	Lea
30.	*Carthamus oxycantha*	Asteraceae	Pohli

contd...

Table 3.1: *contd...*

S.No.	Name of the Plant	Family	Vernacular Names
31.	*Carum copticum*	Apiaceae	Ajwain
32.	*Catharanthus roseus*	Apocynaceae	Sadabahar
33.	*Celosia argentea*	Amaranthaceae	Kukkar-phul
34.	*Centella asiatica*	Apiaceae	Brahmi
35.	*Centipeda minima*	Asteraceae	Nakchikdu
36.	*Chenopodium album*	Chenopodiaceae	Kunnah, Bathu
37.	*Chenopodium ambrosioides*	Chenopodiaceae	Kah-ajwain
38.	*Cicer arietinum*	Fabaceae	Chollae
39.	*Colocasia esculenta*	Araceae	Kachalu, Geju
40.	*Coriandrum sativum*	Apiaceae	Dhania
41.	*Curcuma domestica*	Zingiberaceae	Haldi, Haldan
42.	*Cymbopogon citratus*	Poaceae	Lemongrass
43.	*Cynodon dactylon*	Poaceae	Doob, Dhruv, Doorva
44.	*Daucus carota*	Apiaceae	Gajjar
45.	*Eclipta prostrata*	Asteraceae	Bhringraj
46.	*Elettaria cardamomum*	Zingiberaceae	Ellachi, Elaichi
47.	*Eruca sativa*	Brassicaceae	Taramira
48.	*Eupatorium adenophorum*	Asteraceae	Kali Basuti
49.	*Foeniculum vulgare*	Apiaceae	Saunf, Soay
50.	*Fragaria chiloensis*	Rosaceae	Strawberry
51.	*Fumaria indica*	Fumariaceae	Pitpapara, Papara, Pitpapada
52.	*Glycine max*	Fabaceae	Soyabean
53.	*Glycyrrhiza glabra*	Fabaceae	Mallathi
54.	*Habenaria pectinata*	Orchidaceae	Orchid
55.	*Helianthus annuus*	Asteraceae	Sooraj-Mukhi
56.	*Hordeum vulgare*	Poaceae	Jau, Satu
57.	*Impatiens balsamina*	Balsaminaceae	Teur, Balsam
58.	*Lagenaria siceraria*	Cucurbitaceae	Bitter Gourd
59.	*Lens culinaris*	Fabaceae	Masar
60.	*Lepidium sativum*	Brassicaceae	Hallon
61.	*Linum usitatissimum*	Linaceae	Alsi
62.	*Lycopersicum esculentum*	Solanaceae	Tamatar
63.	*Martynia annua*	Martyniaceae	Kaun
64.	*Melissa officinalis*	Lamiaceae	Balm
65.	*Mentha piperata*	Lamiaceae	Pipermint
66.	*Mentha spicata*	Lamiaceae	Pudina
67.	*Mimosa pudica*	Fabaceae	Chhui Muui, Lajwanti
68.	*Mirabilis jalapa*	Nyctaginaceae	Gulal, 4 o'clock
69	*Musa sapientum*	Musaceae	Kela
70.	*Nasturtium officinale*	Brassicaceae	Chuch

contd...

Table 3.1: *contd...*

S.No.	Name of the Plant	Family	Vernacular Names
71.	*Nicotiana tabacum*	Solanaceae	Tambaku, Tammaku
72.	*Ocimum basilicum*	Lamiaceae	Bhabhri
73.	*Ocimum sanctum*	Lamiaceae	Ram-Tulsi, Baramasi
74.	*Oryza sativa*	Poaceae	Dhan, Jari
75.	*Oxalis corniculata*	Oxalidaceae	Maroli
76.	*Papaver rhoeus*	Papaveraceae	Gulla
77.	*Parthenium hysterophorus*	Asteraceae	Congress grass
78.	*Paspalum scrobiculatum*	Poaceae	Kodro grass
79.	*Pelargonium graveolens*	Geraniaceae	Geranium
80.	*Pennisetum americanum*	Poaceae	Bajra, Barley
81.	*Pennisetum purpureum*	Poaceae	Chari
82.	*Phragmites karka*	Poaceae	Elephant grass
83.	*Phyllanthus urinaria*	Euphorbiaceae	Bhumi- Amla
84.	*Physalis peruviana*	Solanaceae	Rasbhari
85.	*Plumbago zeylanica*	Plumbaginaceae	Chitrak, Chitra
86.	*Ranunculus muricatus*	Ranunculaceae	Buttercup
87.	*Raphanus sativus*	Brassicaceae	Mooli
88.	*Rhynchostylis retusa*	Orchidaceae	Bhangru
89.	*Ruta graveolens*	Rutaceae	Sudab
90.	*Saccharum arundinaceum*	Poaceae	Sarphar
91.	*Saccharum spontaneum*	Poaceae	Kans
92.	*Salvia coccinea*	Lamiaceae	Phultulsi
93.	*Sesamum indicum*	Pedaliaceae	Til
94.	*Setaria viridis*	Poaceae	Hathi grass
95.	*Solanum melongena*	Solanaceae	Baingan
96.	*Solanum nigrum*	Solanaceae	Makko
97.	*Solanum surattense*	Solanaceae	Choti-Kandayali
98.	*Solanum tuberosum*	Solanaceae	Aloo
99.	*Sonchus asper*	Asteraceae	Bhursalae
100.	*Spilanthes acmella*	Asteraceae	Karkara
101.	*Spinacia oleracea*	Chenopodiaceae	Palak, sag
102.	*Tagetes erecta*	Asteraceae	Ghatak
103.	*Taraxacum officinale*	Asteraceae	Dudhli, Dulal
104.	*Trichodesma indicum*	Boraginaceae	Chhota Kalpa, Rukhabi
105.	*Tridax procumbens*	Asteraceae	Lamb

contd...

Table 3.1: *contd...*

S.No.	Name of the Plant	Family	Vernacular Names
106.	*Trifolium alexandrinum*	Fabaceae	Barseem
107.	*Trigonella foenum-graecum*	Fabaceae	Methe, Methi
108.	*Triticum aestivum*	Poaceae	Gehun
109.	*Urtica dioica*	Urticaceae	Bichhubuti, Choti kogsi
110.	*Vanda roxburghii*	Orchidaceae	Bhangru
111.	*Verbascum thapsus*	Scrophulariaceae	Ban Tambakhoo
112.	*Vernonia anthelmintica*	Asteraceae	Brahmjiri
113.	*Viola canescens*	Violaceae	Banfsha, Banaksha
114.	*Xanthium strumarium*	Asteraceae	Bhakhadda
115.	*Zea mays*	Poaceae	Makki, Bhuta
116.	*Zingiber officinale*	Zingiberaceae	Adrak

Table 3.2: Ethnobotanically Important Shrubs of Hamirpur District

S.No.	Name of the Plant	Family	Vernacular Names
1.	*Abroma augusta*	Sterculiaceae	Ulat Kamal
2.	*Adhatoda vasica*	Acanthaceae	Basuti
3.	*Agave americana*	Agavaceae	Ramban
4.	*Aloe barbedensis*	Liliaceae	Ghi-kawer
5.	*Aloe vera*	Liliaceae	Ghi-kawer
6.	*Artemisia vulgaris*	Asteraceae	Charmad
7.	*Bambusa arundinacea*	Poaceae	Bainjh
8.	*Bambusa nutans*	Poaceae	Bans
9.	*Berberis asiatica*	Berberidaceae	Kasmal
10.	*Bougainvillea spectabilis*	Nyctaginaceae	Garden glory
11.	*Calotropis procera*	Asclepiadaceae	Ak, Madar
12.	*Carissa carandas*	Apocynaceae	Karonda, Bada Garna
13.	*Carissa spinarum*	Apocynaceae	Chhota Garna
14.	*Cestrum nocturnum*	Solanaceae	Rat-ki-Rani
15.	*Colebrookia oppositifolia*	Lamiaceae	Dushane, Dus
16.	*Dendrocalamus hamiltonii*	Poaceae	Baingh, Maggar
17.	*Dodonaea viscosa*	Sapindaceae	Mehandru
18.	*Duranta repens*	Verbenaceae	Neelkanthi, Neel Kanta
19.	*Equisetum debile*	Equisetaceae	Ghornali
20.	*Euphorbia pulcherrima*	Euphorbiaceae	Lalpatta

contd...

Table 3.2: *contd...*

S.No.	Name of the Plant	Family	Vernacular Names
21.	*Euphorbia royleana*	Euphorbiaceae	Chu, Chui
22.	*Gossypium arboreum*	Malvaceae	Kapah, Kapas
23.	*Helicteres isora*	Sterculiaceae	Marorphali
24.	*Hibiscus rosa-sinensis*	Malvaceae	China-Rose, Shoe Flower
25.	*Hypericum cernuum*	Hypericaceae	Pinli
26.	*Ipomoea carnea*	Convolvulaceae	Badi Basuti
27.	*Jasminum humile*	Oleaceae	Chameli
28.	*Jasminum multiflorum*	Oleaceae	Malti
29.	*Jasminum sambac*	Oleaceae	Motia
30.	*Lantana camara*	Verbenaceae	Phulnu
31.	*Lavandula angustifolia*	Lamiaceae	Lavender
32.	*Lawsonia alba*	Lythraceae	Mehendi
33.	*Malvaviscus conzatii*	Malvaceae	Gudhal
34.	*Murraya koenigii*	Rutaceae	Gandhla, Curry patta
35.	*Murraya paniculata*	Rutaceae	Gandhla
36.	*Nerium indicum*	Apocynaceae	Kaner
37.	*Opuntia dillenii*	Cactaceae	Chu, Thor
38.	*Pedilanthus tithymaloides*	Euphorbiaceae	Nagbel
39.	*Piper longum*	Piperaceae	Mugh
40.	*Pogostemon benghalensis*	Lamiaceae	Kali Basuti, Dusyan
41.	*Rauwolfia serpentina*	Apocynaceae	Sarpgandha
42.	*Reinwardtia indica*	Linaceae	Basant, Piyan
43.	*Ricinus communis*	Euphorbiaceae	Airand, Arind, Erand
44.	*Rosa damascena*	Rosaceae	Gulab
45.	*Rosmarinus officinalis*	Lamiaceae	Rosemary
46.	*Roylea cinerea*	Lamiaceae	Korru
47.	*Rubus ellipticus*	Rosaceae	Akhe
48.	*Rumex hastatus*	Polygonaceae	Maroli
49.	*Saccharum officinarum*	Poaceae	Ganna
50.	*Solanum erianthum*	Solanaceae	Rada
51.	*Tecoma grandiflora*	Bignoniaceae	Shehnai Phool
52.	*Tecoma stans*	Bignoniaceae	Tecoma, Shehnai Phool
53.	*Thevetia neriifolia*	Apocynaceae	Kaner
54.	*Thuja orientalis*	Cupressaceae	Saru , Morphankh
55.	*Vitex negundo*	Verbenaceae	Banna
56.	*Withania somnifera*	Solanaceae	Asgandh, Ashwagandha
57.	*Woodfordia floribunda*	Lythraceae	Dhai, Dhavae, Dhaun
58.	*Yucca aloifolia*	Liliaceae	Ramban
59.	*Zizyphus jujuba*	Rhamnaceae	Ber

Table 3.3: Ethnobotanically Important Trees of District Hamirpur

S.No.	Name of the Plant	Family	Vernacular Names
1.	*Acacia catechu*	Fabaceae	Khair
2.	*Acacia concinna*	Fabaceae	Shika- kai
3.	*Acacia nilotica*	Fabaceae	Kikker, Shami
4.	*Achras zapota*	Sapotaceae	Chikku
5.	*Aegle marmelos*	Rutaceae	Bil
6.	*Albizia lebbeck*	Fabaceae	Sirin
7.	*Albizia stipulata*	Fabaceae	Oayee
8.	*Alstonia scholaris*	Apocynaceae	Shaitan-Ka-Per
9.	*Annona squamosa*	Annonaceae	Sitaphal
10.	*Artocarpus integrifolia*	Moraceae	Kathal
11.	*Artocarpus lakoocha*	Moraceae	Dheun
12.	*Azadirachta indica*	Meliaceae	Neem
13.	*Bauhinia variegata*	Fabaceae	Kachnar, Karale, Karayalae
14.	*Bombax ceiba*	Boraginaceae	Simbal, Simal
15.	*Butea monosperma*	Fabaceae	Palah, Palash, Plah
16.	*Callistemon lanceolatus*	Myrtaceae	Bottle brush
17.	*Carica papaya*	Caricaceae	Papita
18.	*Cassia fistula*	Fabaceae	Allia, Aahali, Kanyar
19.	*Cedrela toona*	Meliaceae	Tooni
20.	*Cinnamomum camphora*	Lauraceae	Kapur, Kapoor
21.	*Cinnamomum tamala*	Lauraceae	Tejpatra
22.	*Citrus aurantium*	Rutaceae	Santara, Santra
23.	*Citrus grandis*	Rutaceae	Chakotra-Khatta
24.	*Citrus jambhiri*	Rutaceae	Jhambiri, Jambirdi, Kimb
25.	*Citrus limon*	Rutaceae	Khatta, Galgal
26.	*Citrus medica*	Rutaceae	Nimbu
27.	*Cordia dichotoma*	Cordiaceae	Lasura
28.	*Cycas revoluta*	Cycadaceae	Sago Palm
29.	*Dalbergia sissoo*	Fabaceae	Tahali, Shisham
30.	*Elaeocarpus ganitrus*	Elaeocarpaceae	Rudraksha
31.	*Emblica officinalis*	Euphorbiaceae	Ambla, Amla
32.	*Eriobotrya japonica*	Rosaceae	Lukat, Laukhat
33.	*Eucalyptus umbellata*	Myrtaceae	Safeda
34.	*Eugenia cerasoides*	Myrtaceae	Kathman
35.	*Eugenia jambolana*	Myrtaceae	Jamun
36.	*Ficus benghalensis*	Moraceae	Bar, Bad, Barh
37.	*Ficus carica*	Moraceae	Anjir
38.	*Ficus glomerata*	Moraceae	Umreyan, Umara
39.	*Ficus hispida*	Moraceae	Devryan
40.	*Ficus palmata*	Moraceae	Phalgu, Anjhir
41.	*Ficus religiosa*	Moraceae	Peepal
42.	*Ficus roxburghii*	Moraceae	Triamble, Trayambalu
43.	*Flacourtia ramontchi*	Flacourtiaceae	Kangu
44.	*Grevillea robusta*	Proteaceae	Silver oak

contd...

Table 3.3: *contd...*

S.No.	Name of the Plant	Family	Vernacular Names
45.	*Grewia oppositifolia*	Tiliaceae	Beul, Bhuel
46.	*Holarrhena pubescens*	Apocynaceae	Kurchi, Keur, Inderjaun
47.	*Jacaranda acutifolia*	Bignoniaceae	Nili-Gulmohar
48.	*Jatropha curcus*	Euphorbiaceae	Rattanjot, Jamnotta
49.	*Juglans regia*	Juglandaceae	Akharot
50.	*Lagerstroemia indica*	Lythraceae	Har-shingar
51.	*Lannea coromandelica*	Anacardiaceae	Kehmble
52.	*Leucaena leucocephala*	Fabaceae	Alseenia
53.	*Litchi chinensis*	Sapindaceae	Lichi
54.	*Madhuca indica*	Sapotaceae	Mahua
55.	*Mallotus philippinensis*	Euphorbiaceae	Kambhal, Kambal
56.	*Mangifera indica*	Anacardiaceae	Amb, Aam
57.	*Melia azedarach*	Meliaceae	Draek
58.	*Mesua ferrea*	Clusiaceae	Nagkesar
59.	*Moringa oleifera*	Moringaceae	Suhanjan, Sungna
60.	*Morus alba*	Moraceae	Toot
61.	*Oroxylum indicum*	Bignoniaceae	Tat-palanga
62.	*Phoenix sylvestris*	Arecaceae	Jangali Khajoor
63.	*Pinus roxburghii*	Pinaceae	Chil
64.	*Pistacia integerrima*	Anacardiaceae	Kakarsinghi, Kakad-singhi
65.	*Polyalthia longifolia*	Annonaceae	Ashok
66.	*Pongamia pinnata*	Fabaceae	Sukh- Chain
67.	*Populus nigra*	Moraceae	Poplar
68.	*Premna latifolia*	Verbenaceae	Bhankar
69.	*Prunus amygdalus*	Rosaceae	Badam
70.	*Prunus domestica*	Rosaceae	Aloo Bukhara, Plum
71.	*Prunus persica*	Rosaceae	Aru, Adu
72.	*Psidium guajava*	Rosaceae	Amrood
73.	*Punica granatum*	Punicaceae	Daaru, Daadu
74.	*Putranjiva roxburghii*	Euphorbiaceae	Putranjan
75.	*Pyrus communis*	Rosaceae	Nakh, Nashpati
76.	*Pyrus pashia*	Rosaceae	Kainth, Kaenth
77.	*Randia dumetorum*	Rubiaceae	Rada
78.	*Santalum album*	Santalaceae	Chandan
79.	*Sapindus mukorossi*	Sapindaceae	Reetha, Doda, Doddae
80.	*Saraca asoca*	Fabaceae	Ashoka
81.	*Soymida febrifuga*	Meliaceae	Ruhan, Rehana
82.	*Tabernaemontana divaricata*	Apocynaceae	Tagar
83.	*Tamarindus indica*	Fabaceae	Imali
84.	*Terminalia arjuna*	Combretaceae	Arjun Tree
85.	*Terminalia bellirica*	Combretaceae	Behra, Bahera
86.	*Terminalia chebula*	Combretaceae	Harar
87.	*Wendlandia heynei*	Rubiaceae	Panseera
88.	*Zanthoxylum armatum*	Rutaceae	Tirmer

Table 3.4: Ethnobotanically Important Undershrubs of District Hamirpur

S.No.	Name of the Plant	Family	Vernacular Names
1.	*Achyranthes aspera*	Amaranthaceae	Puthkanda
2.	*Achyranthes bidentata*	Amaranthaceae	Puthkanda
3.	*Aerva lanata*	Amaranthaceae	Bhadra
4.	*Artemisia japonica*	Asteraceae	Chamber
5.	*Asclepias curassavica*	Asclepiadaceae	Ghanera
6.	*Asparagus racemosus*	Liliaceae	Satavari
7.	*Canna indica*	Cannaceae	Sudarshan, Hardarshan
8.	*Cassia occidentalis*	Fabaceae	Bada-relu, Alliuna
9.	*Datura metel*	Solanaceae	Datura
10.	*Datura stramonium*	Solanaceae	Datura, Dhatura

Table 3.5: Ethnobotanically Important Climbers of District Hamirpur

S.No.	Name of the Plant	Family	Vernacular Names
1.	*Abrus precatorius*	Fabaceae	Ratti, Raktgunja
2.	*Bauhinia vahlii*	Fabaceae	Tor
3.	*Cardiospermum halicacabum*	Sapindaceae	Tikku-malu
4.	*Cissampelos pareira*	Menispermaceae	Patindu, Bhatindu
5.	*Clematis grata*	Ranunculaceae	Dhagali Bel
6.	*Cucumis sativus*	Cucurbitaceae	Kheera, Kakri
7.	*Cucurbita hispida*	Cucurbitaceae	Petha, Dhuda Petha
8.	*Cucurbita pepo*	Cucurbitaceae	Kaddu
9.	*Cuscuta reflexa*	Cuscutaceae	Akashbel, Amarvela, Halwan
10.	*Dioscorea bulbifera*	Dioscoreaceae	Tardi
11.	*Ipomoea cairica*	Convolvulaceae	Railway Creeper
12.	*Luffa aegyptiaca*	Cucurbitaceae	Ghiya-Tori
13.	*Momordica charantia*	Cucurbitaceae	Karela
14.	*Sechium edule*	Cucurbitaceae	Lonku
15.	*Stephania glabra*	Menispermaceae	Bis-Khappar
16.	*Tylophora indica*	Asclepiadaceae	Damabuti
17.	*Vicia pallida*	Fabaceae	BadaKaer, Roothi

Table 3.6: Ethnobotanically Important Lianas of Hamirpur District

S.No.	Name of the Plant	Family	Vernacular Names
1.	*Actinidia chinensis* var. *deliciosa*	Actinidiaceae	Kiwi
2.	*Celastrus paniculatus*	Celastraceae	Sankhiryan, Sankheere
3.	*Millettia extensa*	Fabaceae	Salangaya
4.	*Tinospora cordifolia*	Menispermaceae	Gulaja, Giloy, Galon
5.	*Vitis jacquemontii*	Vitaceae	Geedar Dhakh
6.	*Vitis vinifera*	Vitaceae	Angoor

Table 3.7: Plant Resource Categories of District Hamirpur (H.P.)

S.No.	Wild	Cultivated	Wild as Well as Cultivated
1.	*Abroma augusta*	*Abelmoschus esculentus*	*Alstonia scholaris*
2.	*Abrus precatorius*	*Abelmoschus moschatus*	*Bryophyllum calycinum*
3.	*Acacia catechu*	*Achras zapota*	*Cynodon dactylon*
4.	*Acacia concinna*	*Actinidia chinensis* var. *deliciosa*	*Duranta repens*
5.	*Acacia nilotica*	*Allium cepa*	*Ficus religiosa*
6.	*Achyranthes aspera*	*Allium sativum*	*Gossypium arboreum*
7.	*Achyranthes bidentata*	*Amaranthus paniculatus*	*Helianthus annuus*
8.	*Acorus calamus*	*Amorphophallus campanulatus*	*Jasminum humile*
9.	*Adhatoda vasica*	*Anetham sowa*	*Lawsonia alba*
10.	*Adiantum capillus-veneris*	*Annona squamosa*	*Mimosa pudica*
11.	*Adiantum lunulatum*	*Arachis hypogaea*	*Musa sapientum*
12.	*Aegle marmelos*	*Artocarpus integrifolia*	*Physalis peruviana*
13.	*Aerva lanata*	*Avena sativa*	*Prunus persica*
14.	*Agave americana*	*Azadirachta indica*	
15.	*Ageratum conyzoides*	*Bougainvillea spectabilis*	
16.	*Albizia lebbeck*	*Brassica campestris*	
17.	*Albizia stipulata*	*Brassica juncea*	
18.	*Aloe barbedensis*	*Brassica nigra*	
19.	*Aloe vera*	*Callistemon lanceolatus*	

contd...

Table 3.7: *contd...*

S.No.	Wild	Cultivated	Wild as Well as Cultivated
20.	*Amaranthus spinosus*	*Capsicum annuum*	
21.	*Amaranthus viridis*	*Carica papaya*	
22.	*Andrographis paniculata*	*Carum copticum*	
23.	*Argemone mexicana*	*Celosia argentea*	
24.	*Arisaema tortuosum*	*Cestrum nocturnum*	
25.	*Artemisia japonica*	*Cicer arietinum*	
26.	*Artemisia vulgaris*	*Cinnamomum camphora*	
27.	*Artocarpus lakoocha*	*Cinnamomum tamala*	
28.	*Asclepias curassavica*	*Citrus aurantium*	
29.	*Asparagus adscendens*	*Citrus medica*	
30.	*Asparagus racemosus*	*Colocasia esculenta*	
31.	*Bacopa monnieri*	*Coriandrum sativum*	
32.	*Bambusa arundinacea*	*Cucumis sativus*	
33.	*Bambusa nutans*	*Cucurbita hispida*	
34.	*Bauhinia vahlii*	*Cucurbita pepo*	
35.	*Bauhinia variegata*	*Curcuma domestica*	
36.	*Berberis asiatica*	*Cycas revoluta*	
37.	*Boerhavia diffusa*	*Cymbopogon citratus*	
38.	*Bombax ceiba*	*Daucus carota*	
39.	*Butea monosperma*	*Elaeocarpus ganitrus*	
40.	*Calotropis procera*	*Elettaria cardamomum*	
41.	*Canna indica*	*Eriobotrya japonica*	
42.	*Cannabis sativa*	*Euphorbia pulcherrima*	
43.	*Canscora diffusa*	*Foeniculum vulgare*	
44.	*Cardiospermum halicacabum*	*Fragaria chiloensis*	
45.	*Carduus nutans*	*Glycine max*	
46.	*Carissa carandas var. congesta*	*Grevillea robusta*	
47.	*Carissa spinarum*	*Hordeum vulgare*	
48.	*Carthamus oxycantha*	*Jacaranda acutifolia*	
49.	*Cassia fistula*	*Juglans regia*	
50.	*Cassia occidentalis*	*Lagenaria siceraria*	
51.	*Catharanthus roseus*	*Lagerstroemia indica*	
52.	*Cedrela toona*	*Lavandula angustifolia*	
53.	*Celastrus paniculatus*	*Lens culinaris*	
54.	*Centella asiatica*	*Lepidium sativum*	
55.	*Centipeda minima*	*Linum usitatissimum*	
56.	*Chenopodium album*	*Litchi chinensis*	
57.	*Chenopodium ambrosioides*	*Luffa aegyptiaca*	
58.	*Cissampelos pareira*	*Lycopersicum eculentum*	
59.	*Citrus grandis*	*Madhuca indica*	
60.	*Citrus jambhiri*	*Malvaviscus conzatii*	
61.	*Citrus limon*	*Melissa officinalis*	
62.	*Clematis grata*	*Mentha piperata*	
63.	*Colebrookia oppositifolia*	*Mesua ferrea*	
64.	*Cordia dichotoma*	*Momordica charantia*	
65.	*Cuscuta reflexa*	*Moringa oleifera*	

contd...

Table 3.7: *contd...*

S.No.	Wild	Cultivated	Wild as Well as Cultivated
66.	*Dalbergia sissoo*	*Ocimum basilicum*	
67.	*Datura metel* auct. non	*Ocimum sanctum*	
68.	*Datura stramonium*	*Oryza sativa*	
69.	*Dendrocalamus hamiltonii*	*Paspalum scrobiculatum*	
70.	*Dioscorea bulbifera*	*Pedilanthus tithymaloides*	
71.	*Dodonaea viscosa*	*Pelargonium graveolens*	
72.	*Eclipta prostrata*	*Pennisetum americanum*	
73.	*Emblica officinalis*	*Pennisetum purpureum*	
74.	*Equisetum debile*	*Piper longum*	
75.	*Eruca sativa*	*Polyalthia longifolia*	
76.	*Eucalyptus umbellata*	*Pongamia pinnata*	
77.	*Eugenia cerasoides*	*Populus nigra*	
78.	*Eugenia jambolana*	*Prunus amygdalus*	
79.	*Eupatorium adenophorum*	*Prunus domestica*	
80.	*Euphorbia royleana*	*Psidium guajava*	
81.	*Ficus benghalensis*	*Putranjiva roxburghii*	
82.	*Ficus carica*	*Pyrus communis*	
83.	*Ficus glomerata*	*Raphanus sativus*	
84.	*Ficus hispida*	*Rauwolfia serpentina*	
85.	*Ficus palmata*	*Rosa damascena*	
86.	*Ficus roxburghii*	*Rosmarinus officinalis*	
87.	*Flacourtia ramontchi*	*Ruta graveolens*	
88.	*Fumaria indica*	*Saccharum officinarum*	
89.	*Glycyrrhiza glabra*	*Santalum album*	
90.	*Grewia oppositifolia*	*Saraca asoca*	
91.	*Habenaria pectinata*	*Sechium edule*	
92.	*Helicteres isora*	*Sesamum indicum*	
93.	*Hibiscus rosa-sinensis*	*Solanum melongena*	
94.	*Holarrhena pubescens*	*Solanum tuberosum*	
95.	*Hypericum cernuum*	*Spinacia oleracea*	
96.	*Impatiens balsamina*	*Tabernaemontana divaricata*	
97.	*Ipomoea cairica*	*Tagetes erecta*	
98.	*Ipomoea carnea*	*Tecoma grandiflora*	
99.	*Jasminum multiflorum*	*Tecoma stans*	
100.	*Jasminum sambac*	*Thevetia neriifolia*	
101.	*Jatropha curcus*	*Thuja orientalis*	
102.	*Lannea coromandelica*	*Trigonella foenum- graecum*	
103.	*Lantana camara*	*Triticum aestivum*	
104.	*Leucaena leucocephala*	*Tylophora indica*	
105.	*Mallotus philippinensis*	*Vitis vinifera*	
106.	*Mangifera indica*	*Yucca aloifolia*	
107.	*Martynia annua*	*Zea mays*	
108.	*Melia azedarach*	*Zingiber officinale*	
109.	*Mentha spicata*		
110.	*Millettia extensa*		
111.	*Mirabilis jalapa*		

contd...

Table 3.7: *contd...*

S.No.	Wild	Cultivated	Wild as Well as Cultivated
112.	*Morus alba*		
113.	*Murraya koenigii*		
114.	*Murraya paniculata*		
115.	*Nasturtium officinale*		
116.	*Nerium indicum*		
117.	*Nicotiana tabacum*		
118.	*Opuntia dillenii*		
119.	*Oroxylum indicum*		
120.	*Oxalis corniculata*		
121.	*Papaver rhoeus*		
122.	*Parthenium hysterophorus*		
123.	*Phoenix sylvestris*		
124.	*Phragmites karka*		
125.	*Phyllanthus urinaria*		
126.	*Pinus roxburghii*		
127.	*Pistacia integerrima*		
128.	*Plumbago zeylanica*		
129.	*Pogostemon benghalensis*		
130.	*Premna latifolia*		
131.	*Punica granatum*		
132.	*Pyrus pashia*		
133.	*Randia dumetorum*		
134.	*Ranunculus muricatus*		
135.	*Reinwardtia indica*		
136.	*Rhynchostylis retusa*		
137.	*Ricinus communis*		
138.	*Roylea cinerea*		
139.	*Rubus ellipticus*		
140.	*Rumex hastatus*		
141.	*Saccharum arundinaceum*		
142.	*Saccharum spontaneum*		
143.	*Salvia coccinea*		
144.	*Sapindus mukorossi*		
145.	*Setaria viridis*		
146.	*Solanum erianthum*		
147.	*Solanum nigrum*		
148.	*Solanum surattense*		
149.	*Sonchus asper*		
150.	*Soymida febrifuga*		
151.	*Spilanthes acmella*		
152.	*Stephania glabra*		
153.	*Tamarindus indica*		
154.	*Taraxacum officinale*		
155.	*Terminalia arjuna*		
156.	*Terminalia bellirica*		

contd...

Table 3.7: *contd...*

S.No.	Wild	Cultivated	Wild as Well as Cultivated
157.	*Terminalia chebula*		
158.	*Tinospora cordifolia*		
159.	*Trichodesma indicum*		
160.	*Tridax procumbens*		
161.	*Trifolium alexandrinum*		
162.	*Urtica dioica*		
163.	*Vanda roxburghii*		
164.	*Verbascum thapsus*		
165.	*Vernonia anthelmintica*		
166.	*Vicia pallida*		
167.	*Viola canescens*		
168.	*Vitex negundo*		
169.	*Vitis jacquemontii*		
170.	*Wendlandia heynei*		
171.	*Withania somnifera*		
172.	*Woodfordia floribunda*		
173.	*Xanthium strumarium*		
174.	*Zanthoxylum armatum*		
175.	*Zizyphus jujuba*		

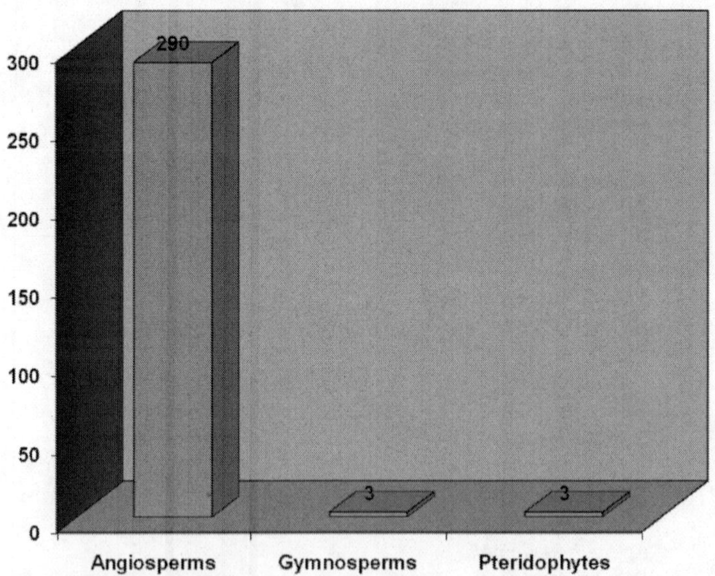

Fig. 3.1. Histogram Showing Total Number of Plants Under Various Divisions of Plant Kingdom Used in District Hamirpur.

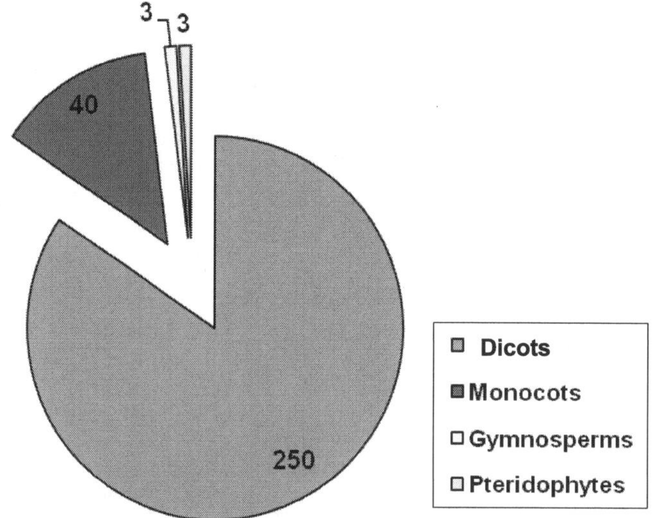

Fig. 3.2. Pie Diagram Showing Relative Number of Dicots, Monocots, Gymnosperms and Pteridophytes of Hamirpur.

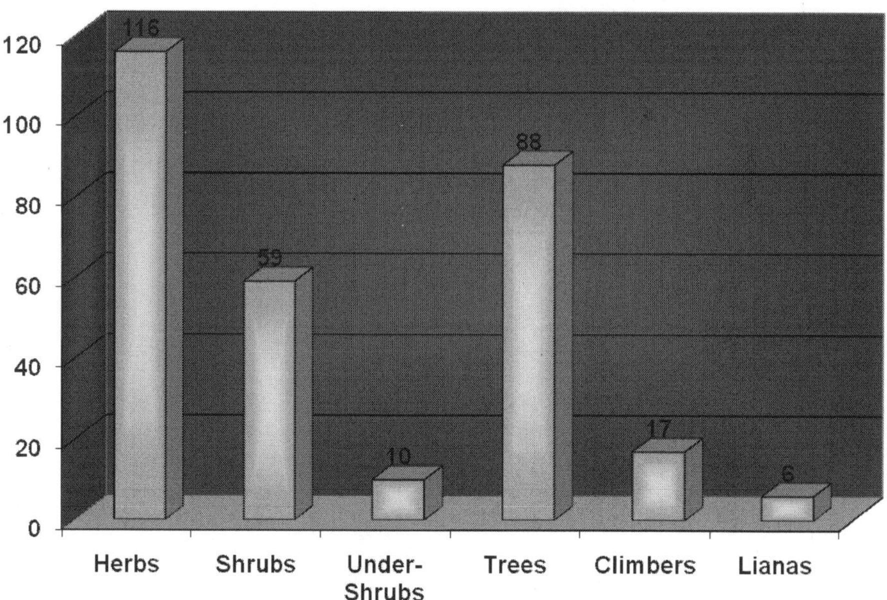

Fig. 3.3. Total Number of Herbs, Shrubs, Undershrubs, Trees, Climbers and Lianas Used by Rural Populace of District Hamirpur.

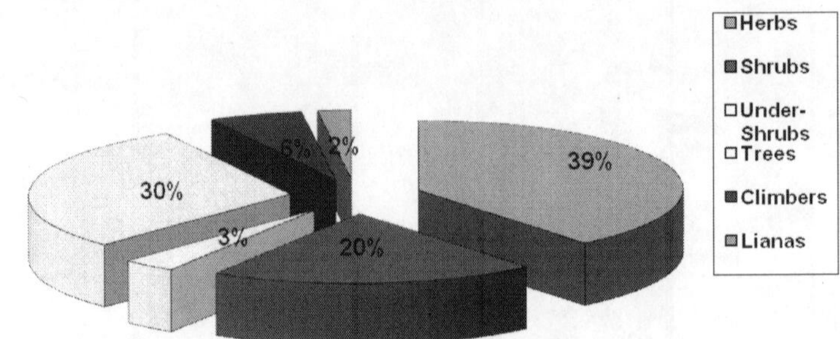

Fig. 3.4. Relative Percentage of Ethnobotanically Collected Herbs, Shrubs, Undershrubs, Trees, Climbers and Lianas.

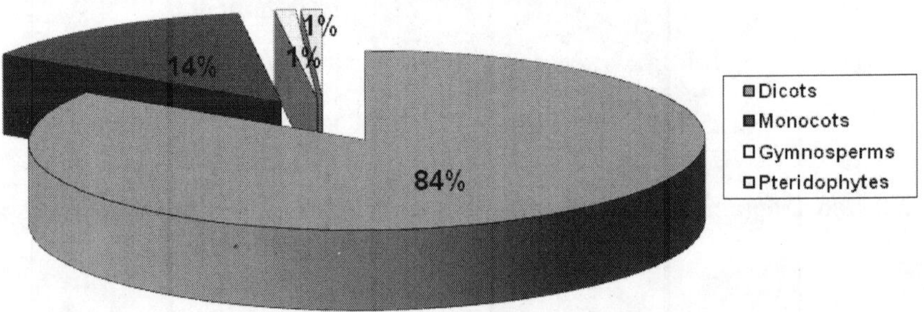

Fig. 3.5. Relative Percentage of Ethnobotanically Collected Dicots, Monocots, Gymnosperms and Pteridophytes

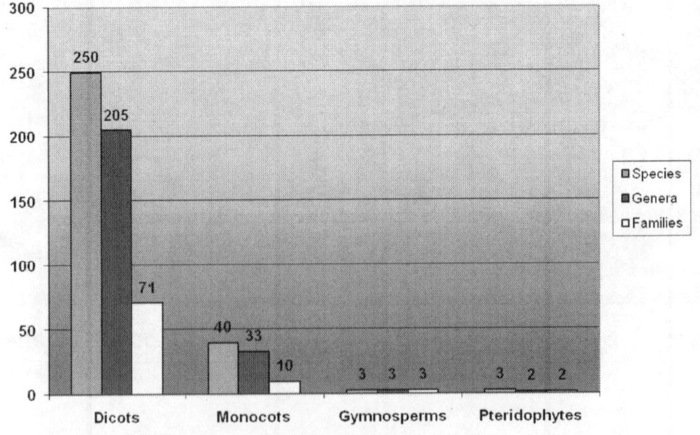

Fig. 3.6. Total Number of Ethnobotanically Collected Species, Genera and Families Under Various Divisions of Plant Kingdom.

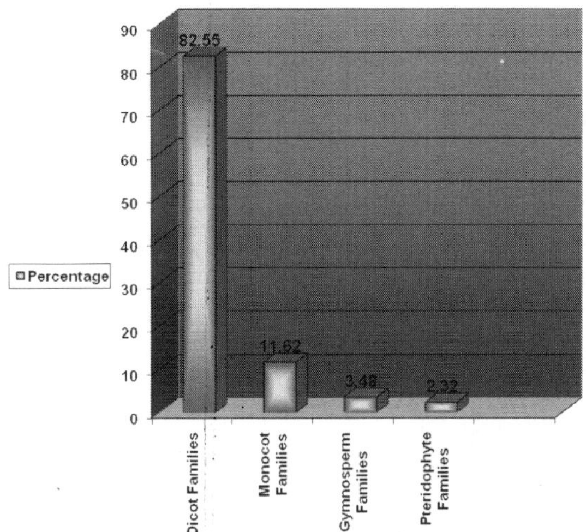

Fig. 3.7. Relative Percentage of Angiospermous Gymnospermous and Pteridophytic Families.

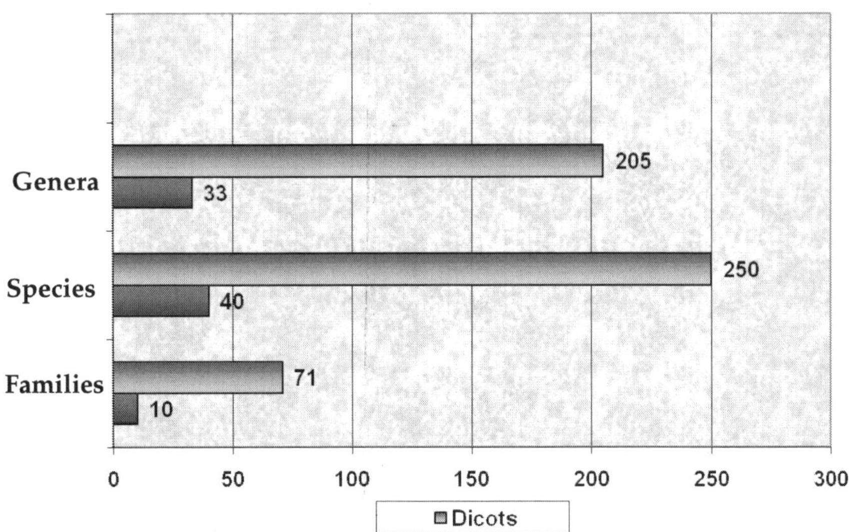

Fig. 3.8. Histogram Showing Comparative Number of Genera, Species and Families of Monocots and Dicots Used Ethnobotanically in Hamirpur.

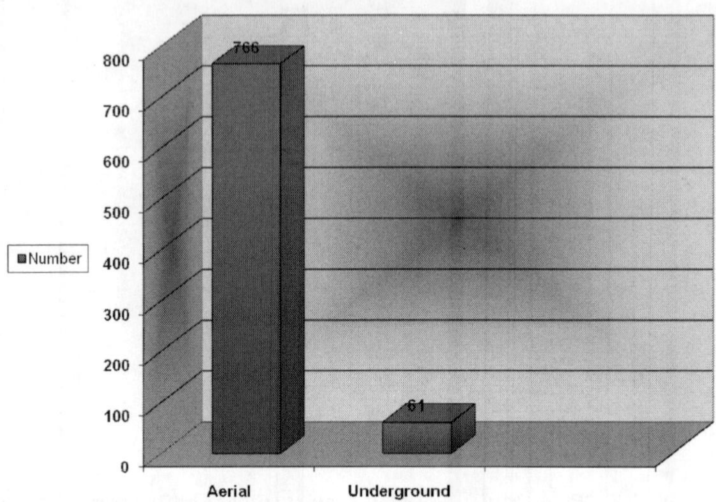

Fig. 3.9. Relative Number of Aerial and Underground Plant Parts Used by the Rural Populace of Hamirpur.

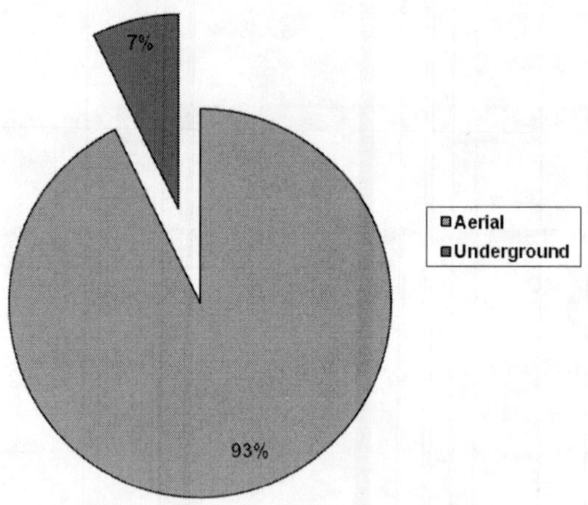

Fig. 3.10. Pie Diagram Showing Relative Disposition of Underground and Aerial Plant Parts Consumed in the Study Area.

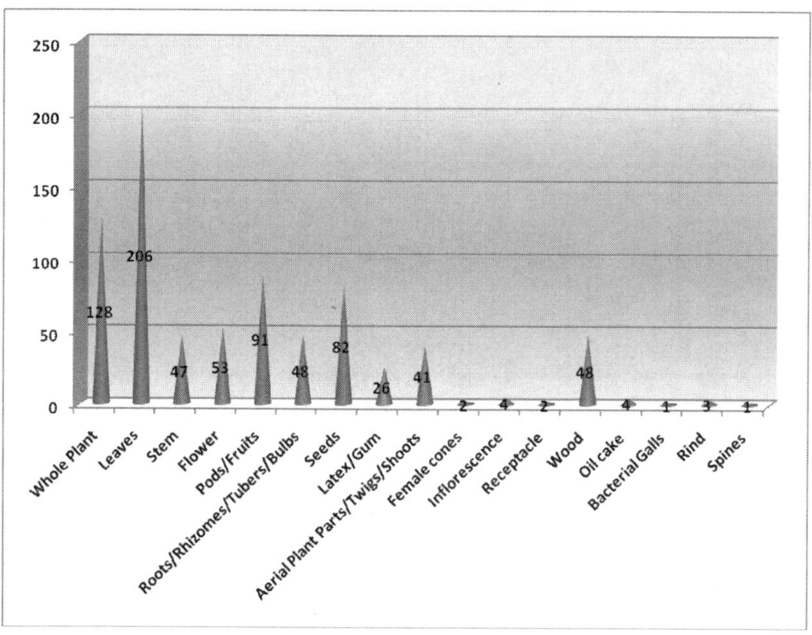

Fig. 3.11. Quantitative Reflection of Various Plant Parts Consumed by the Rural Populace of District Hamirpur.

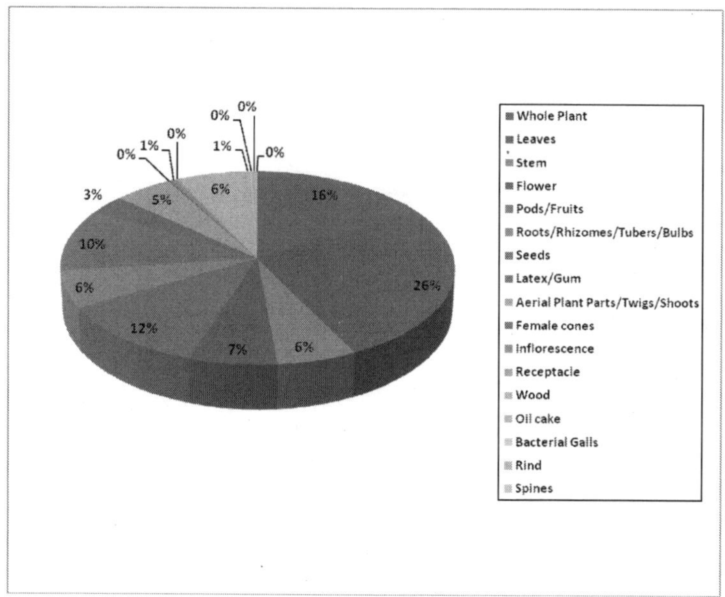

Fig. 3.12. Relative Percentage Use of the Various Plant Parts (the Study Area).

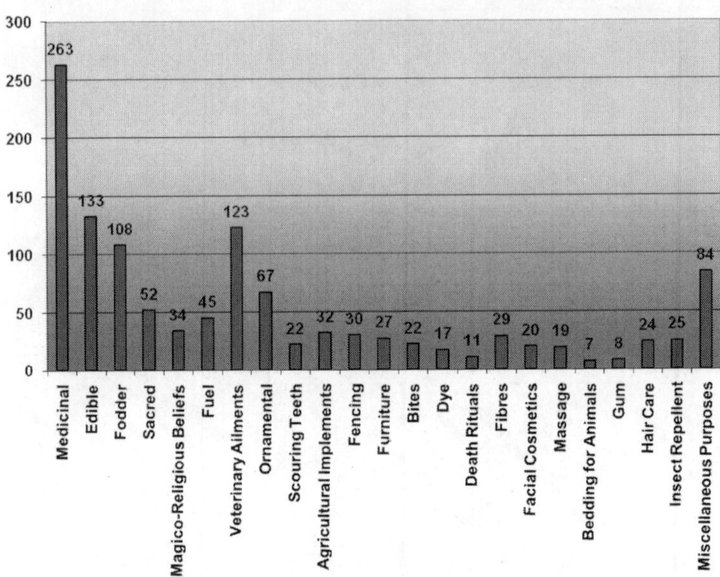

Fig. 3.13. Quantitative Uses of Plants for Various Purposes by the Rural Populace of District Hamirpur.

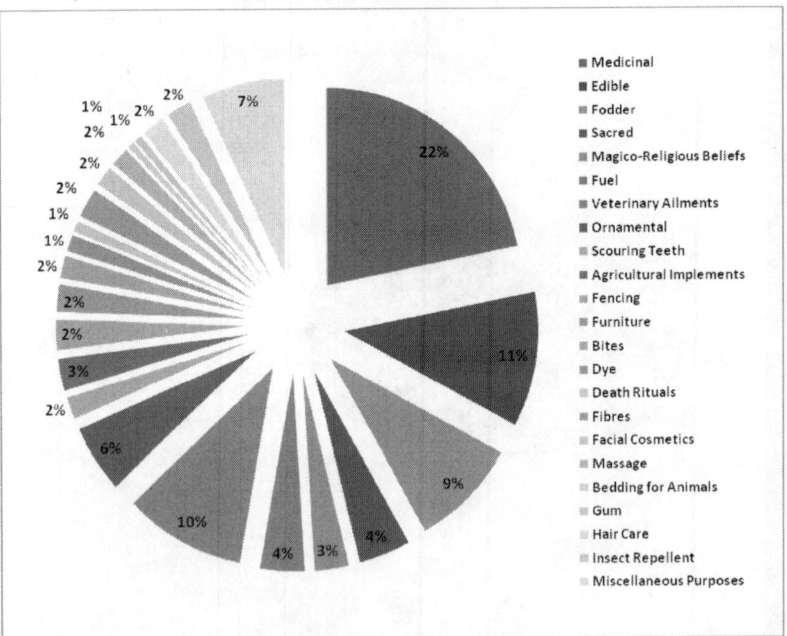

Fig. 3.14. Percentage Uses of Plants Under Various Categories by the Rural Populace of District Hamirpur.

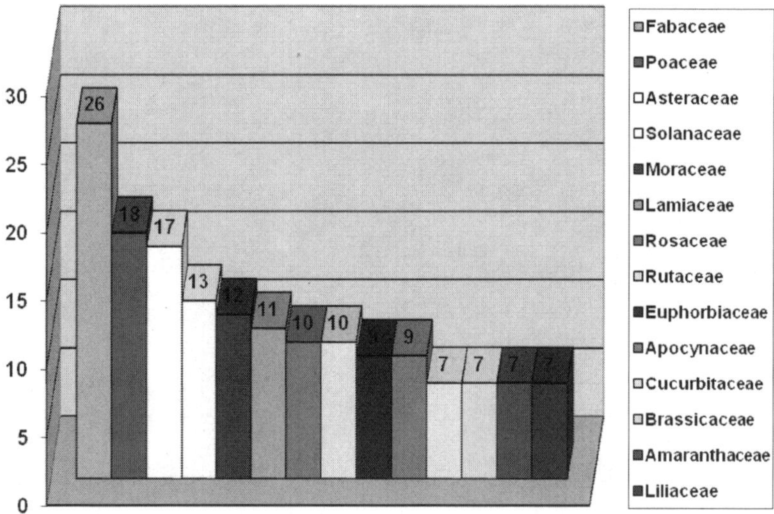

Fig. 3.15. Relative Distribution of Taxa Under Predominant Ethnobotanical Families of District Hamirpur.

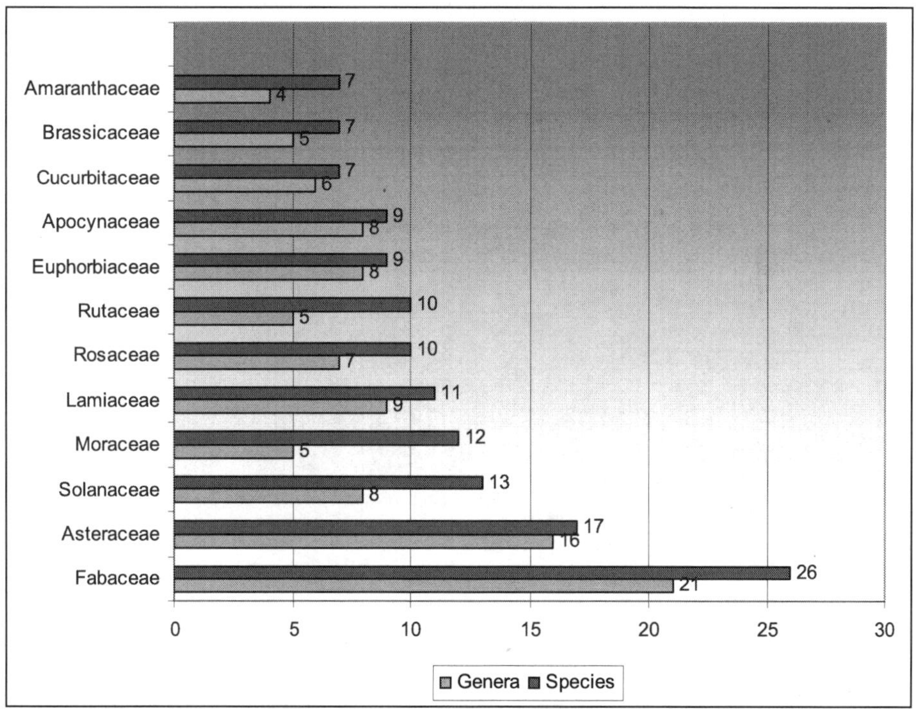

Fig. 3.16. Relative Disposition of Species and Genera Under Predominant Ethnobotanical Dicot Families of District Hamirpur.

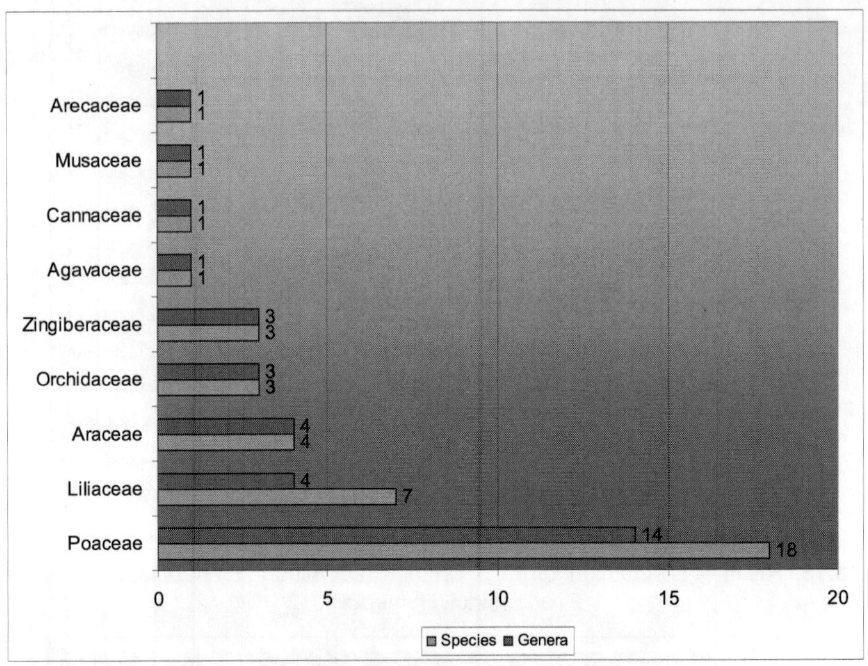

Fig. 3.17. Predominant Ethnobotanical Monocot Families of the Study Area Showing Relative Disposition of Genera and Species.

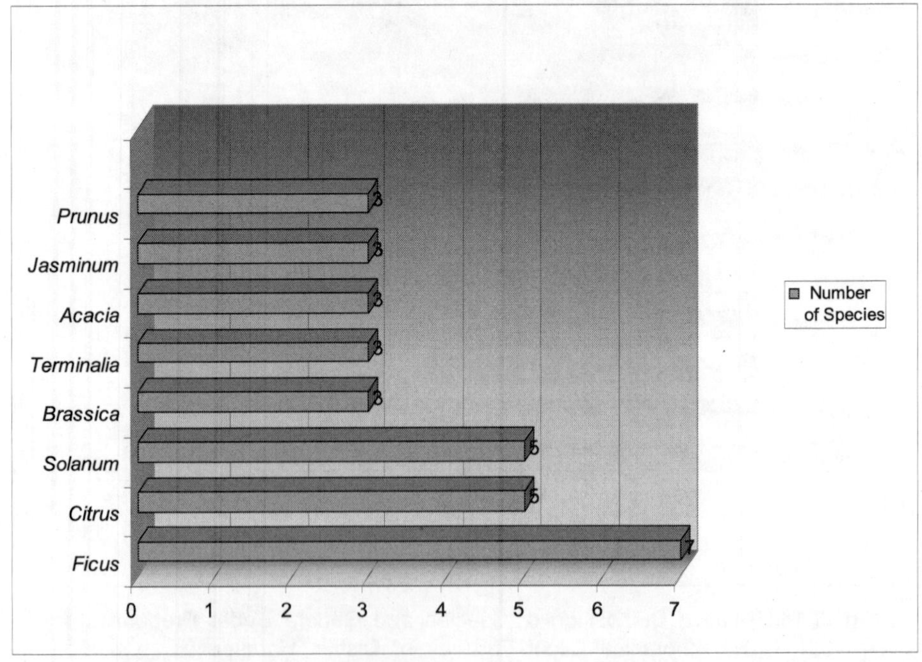

3.18. Predominant Ethnobotanical Dicot Genera of District Hamirpur.

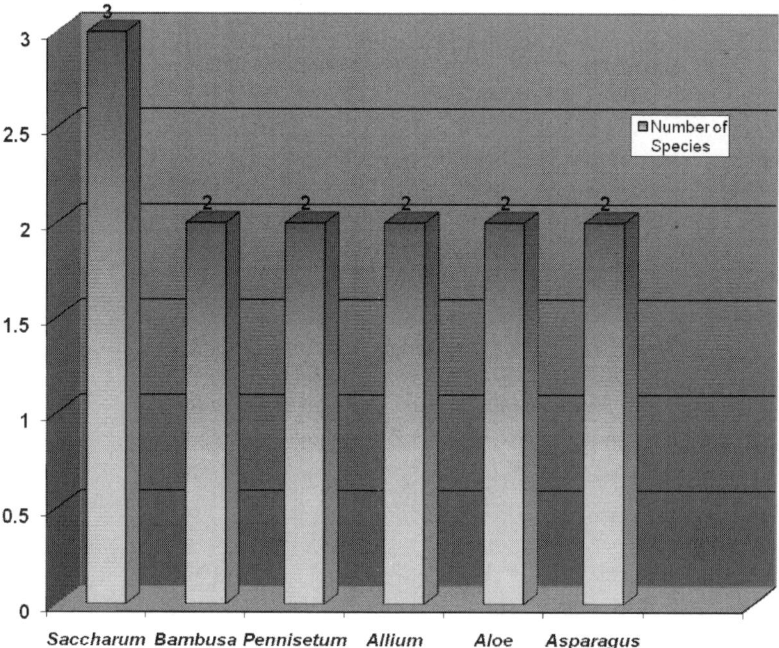

Fig. 3.19. Predominant Ethnobotanical Monocot Genera of District Hamirpur.

Table 3.8: Families, Genera and Species of Ethnobotanically Important Plants Under Various Divisions of Plant Kingdom

S.No.	Division	Total No. of Families	Total No. of Genera	Total No. of Species
1.	Dicots	71	205	250
2.	Monocots	10	33	40
3.	Gymnosperms	03	03	03
4.	Pteridophytes	02	02	03

Table 3.9: Comparative Index of Ethnobotanically Important Plants of District Hamirpur Under Various Groups of Plant Kingdom

Plant Group	Families	Genera	Species
Pteridophytes	02	02	03
Gymnosperms	03	03	03
Monocots	10	33	40
Polypetalae	39	106	126
Gamopetalae	20	70	82
Monochlamydae	12	29	42

Table 3.10: Layout of Ethnobotanical Families of Hamirpur District as per Bentham & Hooker System of Classification (1862-1883)

Polypetalous Families	Gamopetalous Families	Mono-chlamydous Families	Monocot Families	Gymno-spermous Families	Pteridophytic Families
Actinidiaceae	Acanthaceae	Amaranthaceae	Agavaceae	Cupressaceae	Adiantaceae
Anacardiaceae	Apocynaceae	Chenopodiaceae	Araceae	Cycadaceae	Equisetaceae
Annonaceae	Asclepiadaceae	Euphorbiaceae	Arecaceae	Pinaceae	
Apiaceae	Asteraceae	Juglandaceae	Cannaceae		
Balsaminaceae	Bignoniaceae	Lauraceae	Dioscoreaceae		
Berberidaceae	Boraginaceae	Moraceae	Liliaceae		
Brassicaceae	Convolvulaceae	Nyctaginaceae	Musaceae		
Cactaceae	Cordiaceae	Piperaceae	Orchidaceae		
Caricaceae	Cuscutaceae	Polygonaceae	Poaceae		
Celastraceae	Gentianaceae	Proteaceae	Zingiberaceae		
Clusiaceae	Lamiaceae	Santalaceae			
Combretaceae	Martyniaceae	Urticaceae			
Crassulaceae	Oleaceae				
Cucurbitaceae	Pedaliaceae				
Elaeocarpaceae	Plumbaginaceae				
Fabaceae	Rubiaceae				
Flacourtiaceae	Sapotaceae				
Fumariaceae	Scrophulariaceae				
Geraniaceae	Solanaceae				
Hypericaceae	Verbenaceae				
Linaceae					
Lythraceae					
Malvaceae					
Meliaceae					
Menispermaceae					
Moringaceae					
Myrtaceae					
Oxalidaceae					
Papaveraceae					
Punicaceae					
Ranunculaceae					
Rhamnaceae					
Rosaceae					
Rutaceae					
Sapindaceae					
Sterculiaceae					
Tiliaceae					
Violaceae					
Vitaceae					

Table 3.11: Respective Percentage Distribution of Taxa Under Predominant Ethnobotanical Families of District Hamirpur (H.P.)

Family	**Species of Ethnic Use**	***Percentage in Taxa***
Fabaceae	26	*8.78*
Poaceae	18	*6.08*
Asteraceae	17	*5.74*
Solanaceae	13	*4.39*
Moraceae	12	*4.05*
Lamiaceae	11	*3.71*
Rosaceae	10	*3.37*
Rutaceae	10	*3.37*
Euphorbiaceae	9	*3.04*
Apocynaceae	9	*3.04*
Cucurbitaceae	7	*2.36*
Brassicaceae	7	*2.36*
Amaranthaceae	7	*2.36*
Liliaceae	7	*2.36*
Apiaceae	6	*2.02*
Malvaceae	5	*1.68*
Araceae	4	*1.35*
Meliaceae	4	*1.35*
Sapindaceae	4	*1.35*
Verbenaceae	4	*1.35*
Bignoniaceae	4	*1.35*
Myrtaceae	4	*1.35*
Anacardiaceae	3	*1.01*
Asclepiadaceae	3	*1.01*
Lythraceae	3	*1.01*
Menispermaceae	3	*1.01*
Nyctaginaceae	3	*1.01*
Orchidaceae	3	*1.01*
Zingiberaceae	3	*1.01*
Chenopodiaceae	3	*1.01*
Combretaceae	3	*1.01*
Oleaceae	3	*1.01*
Acanthaceae	2	*0.67*
Adiantaceae	2	*0.67*
Annonaceae	2	*0.67*
Boraginaceae	2	*0.67*
Convolvulaceae	2	*0.67*
Lauraceae	2	*0.67*
Linaceae	2	*0.67*
Papaveraceae	2	*0.67*
Ranunculaceae	2	*0.67*
Rubiaceae	2	*0.67*
Sapotaceae	2	*0.67*
Scrophulariaceae	2	*0.67*
Sterculiaceae	2	*0.67*
Vitaceae	2	*0.67*

Table 3.12: Relative Disposition of Species and Genera under Predominant Ethnobotanical Families of District Hamirpur

Families	Total No. of Genera	Total No. of Species
Fabaceae	21	26
Poaceae	14	18
Asteraceae	16	17
Solanaceae	08	13
Moraceae	05	12
Lamiaceae	09	11
Apocynaceae	08	09
Rosaceae	07	10
Rutaceae	05	10
Euphorbiaceae	08	09
Cucurbitaceae	06	07
Brassicaceae	05	07
Amaranthaceae	04	07
Liliaceae	04	07
Apiaceae	06	06
Malvaceae	04	05
Araceae	04	04
Meliaceae	04	04
Sapindaceae	04	04
Verbenaceae	04	04

Table 3.13: Predominant Ethnobotanical Dicot Families of Hamirpur

Name of the Family	Genera	Species
Fabaceae	21	26
Asteraceae	16	17
Solanaceae	8	13
Moraceae	5	12
Lamiaceae	9	11
Rosaceae	7	10
Rutaceae	5	10
Euphorbiaceae	8	9
Apocynaceae	8	9
Cucurbitaceae	6	7
Brassicaceae	5	7
Amaranthaceae	4	7

Table 3.14: Predominant Ethnobotanical Monocot Families of Hamirpur

Name of the Family	Genera	Species
Poaceae	14	18
Liliaceae	4	7
Araceae	4	4
Orchidaceae	3	3
Arecaceae	1	1
Zingiberaceae	1	1
Agavaceae	1	1
Musaceae	1	1
Dioscoreaceae	1	1
Cannaceae	1	1

Table 3.15 Predominant Ethnobotanical Monocot Genera of District Hamirpur

Genus	Family	Number of Species
Saccharum	Poaceae	03
Bambusa	Poaceae	02
Pennisetum	Poaceae	02
Allium	Liliaceae	02
Aloe	Liliaceae	02
Asparagus	Liliaceae	02
Acorus	Araceae	01
Agave	Agavaceae	01
Amorphophallus	Araceae	01
Arisaema	Araceae	01
Avena	Poaceae	01
Canna	Cannaceae	01
Colocasia	Araceae	01
Curcuma	Zingiberaceae	01
Cymopogon	Poaceae	01
Cynodon	Poaceae	01
Dendrocalamus	Poaceae	01
Dioscorea	Dioscoreaceae	01
Elettaria	Zingiberaceae	01
Habenaria	Orchidaceae	01
Hordeum	Poaceae	01
Musa	Musaceae	01
Oryza	Poaceae	01
Paspalum	Poaceae	01
Phoenix	Arecaceae	01
Phragmites	Poaceae	01
Rhynchostylis	Orchidaceae	01
Setaria	Poaceae	01
Triticum	Poaceae	01
Vanda	Orchidaceae	01
Yucca	Liliaceae	01
Zea	Poaceae	01
Zingiber	Zingiberaceae	01

Table 3.16: Predominant Ethnobotanical Families of District Hamirpur with Percentage in Taxa

Family	Genera of Ethnic Use	Percentage in Taxa
Fabaceae	21	8.64
Asteraceae	16	6.58
Poaceae	14	5.76
Lamiaceae	9	3.70
Euphorbiaceae	8	3.29
Solanaceae	8	3.29
Apocynaceae	8	3.29
Rosaceae	7	2.88
Apiaceae	6	2.46
Cucurbitaceae	6	2.46
Brassicaceae	5	2.05
Moraceae	5	2.05
Rutaceae	5	2.05
Amaranthaceae	4	1.64
Araceae	4	1.64
Liliaceae	4	1.64
Malvaceae	4	1.64
Meliaceae	4	1.64
Sapindaceae	4	1.64
Verbenaceae	4	1.64
Anacardiaceae	3	3.4
Asclepiadaceae	3	3.4
Bignoniaceae	3	3.4
Lythraceae	3	3.4
Menispermaceae	3	3.4
Myrtaceae	3	3.4
Nyctaginaceae	3	3.4
Orchidaceae	3	3.4
Zingiberaceae	3	3.4

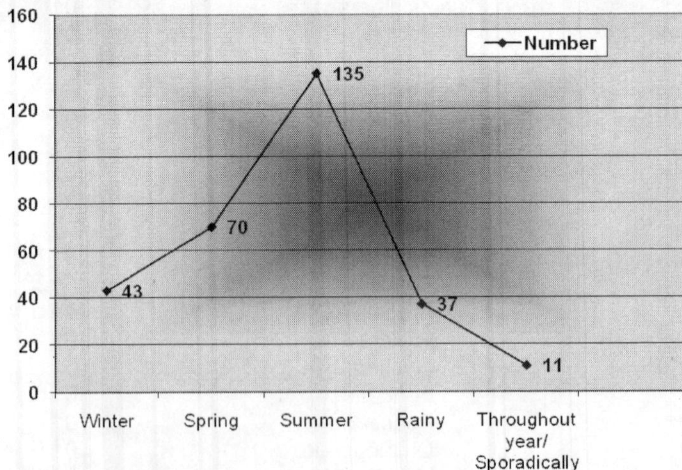

Fig. 3.20. Flowering and Fruiting Seasons of Ethnobotanically Important Plants of District Hamirpur.

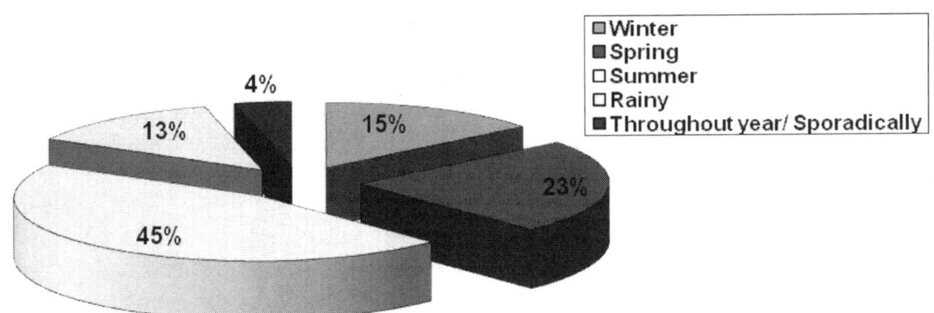

Fig. 3.21. Relative Percentage of Flowering and Fruiting Seasons of Ethnobotanically Important Plants (Present Study).

By and large, therapeutic usage of 263 plants belonging to 218 genera under 82 families have been documented singly (130) or in combination (133), for curing 67 types of ailments affecting the rural populace of district Hamirpur (Fig. 3.22; Table 3.17). Most of these herbal drugs are prescribed orally. As evidenced from use, the most prevalent ailments in the region are skin ailments (pimples, boils, acne, cracked skin, rashes, dermatitis, eczema: 70 species), stomach disorders (54 species), joint diseases (50 species), gastrointestinal disorders (49 species), cough (42 species), backache/ pains (38 species), followed by febrifuge (29 species), cold, liver disorders, mouth sores, throat/chest/respiratory tract infection (26 each), asthma (25), hair care (24), anaemia, bites (23 each), coolant/refrigerant (28 species), diabetes (19 species), constipation (18), blood purifier, expelling worms, massage (17 each), diuretic, laxative, swelling/ inflammation (16 each), appetiser, fractures/ sprains (14 each), burns, delivery, neuraligic disorders, menstrual disorders (13 each), astringent, bronchitis, heart ailments, piles, vomiting/ nausea (12 each), antilithic, eyesight (11 each), antiemetic, flatulence, sex diseases (10 each), antiseptic, clotting, kidney stones (9 each), ulcer (8), body heat, fear psychosis (7 each), burning soles, ear complaints, sweating (6 each), leucorrhoea, uterine disorders (5 each), tuberculosis (4), dark circles (3), bandage, communicable diseases, excessive saliva, rabies (2 each), hernia, impotency, nose bleeding, and slipped neck (1 each) thereby indicating their less prevalence in the area (Table 3.17). Their relative percentage disposition are also delineated in Fig. 3.23. Nevertheless, the maximum number of plant species used in treatment of one disease shows lesser consensus whereas less number of plant species used in cure of a single disease indicates highest consensus. Various plant part/s used in order of preference for alleviating prevalent health problems are: leaves (190 species), whole plant (108 species), fruits (78 species), seeds (76 species), roots (48 species), flower (46 species), stem (41 species), twig/ branches/ aerial parts/shoots (39 species), wood (31 species), stem bark (28 species), latex/ gum (26 species), rhizomes/ tubers/ bulbs (13 species), oil cake (4 species), receptacle (2 species), rind (2 species), spine, galls, inflorescence/ female cones (1 species each), etc. Among the genera, *Citrus* (5 species) and *Solanum* (5 species) showed highest species diversity, followed by

Ficus (4 species), *Acacia, Amaranthus, Brassica, Jasminum, Prunus, Terminalia,* (3 species each), *Abelmoschus, Achyranches, Adiantum, Albizia, Allium, Aloe, Artemisia, Artocarpus, Asparagus, Bambusa, Bauhinia, Carissa, Cassia, Chenopodium, Cinnamomum, Datura, Eugenia, Euphorbia, Mentha, Ocimum, Pyrus and Vitis* (2 species each), and rest of 187 genera represented by one species each. Out of the 82 families employed for healthcare needs in the region, the top ranking families in order of preference are: Fabaceae (23 species), Asteraceae (16 species), Solanaceae (13 species), Poaceae (12 species), Lamiaceae (10 species), Apocynaceae (9 species), Rutaceae (9 species), Moraceae (9 species), Rosaceae (9 species), Euphorbiaceae (8 species), Amaranthaceae (7 species), Brassicaceae (7 species), Liliaceae (7 species), Apiaceae (6 species), Cucurbitaceae (6 species), Araceae (4 species), Malvaceae (4 species), Meliaceae (4 species), Sapindaceae (4 species), Verbenaceae (4 species), Anacardiaceae (3 species), Chenopodiaceae (3 species), Combretaceae (3 species), Lythraceae (3 species), Menispermaceae (3 species), Myrtaceae (3 species), Oleaceae (3 species), Orchidaceae (3 species), Zingiberaceae (3 species) and rest of 53 families are represented with one or two species each. Also, it is pertinent to mention that as many as 375 additional medicinal usages of plants indicated by asterisk against their names in Table 3.17 are new to Indian ethnopharmacopoeia, suggesting ample need for their critical scientific evaluation and practical utility as perspective medicinal plants for the future. No doubt attempts will help integrate the folk medicines with the modern system of medicine, thereby making science of it.

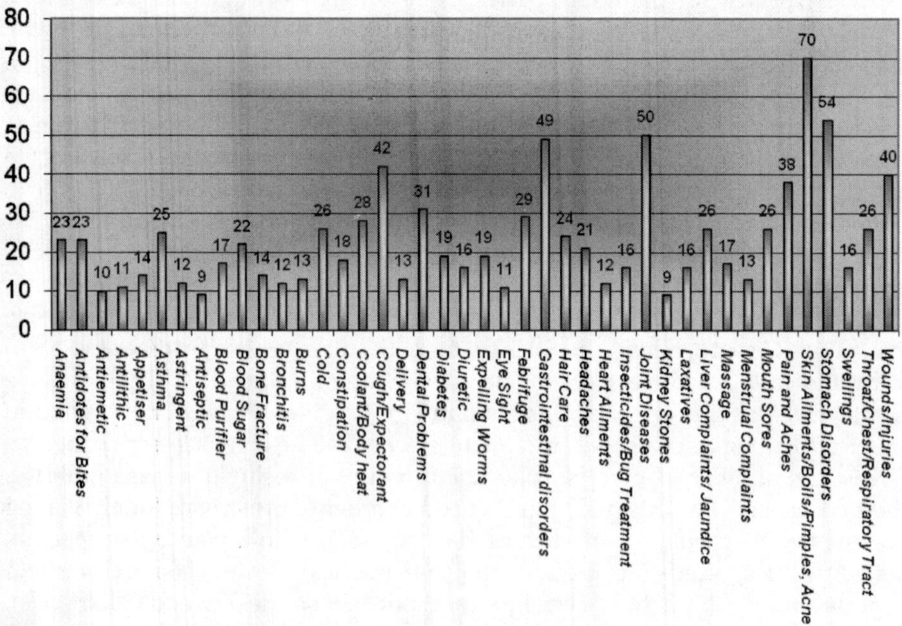

Fig. 3.22. Quantitative Uses of Plants for Different Ailments in the Study Area.

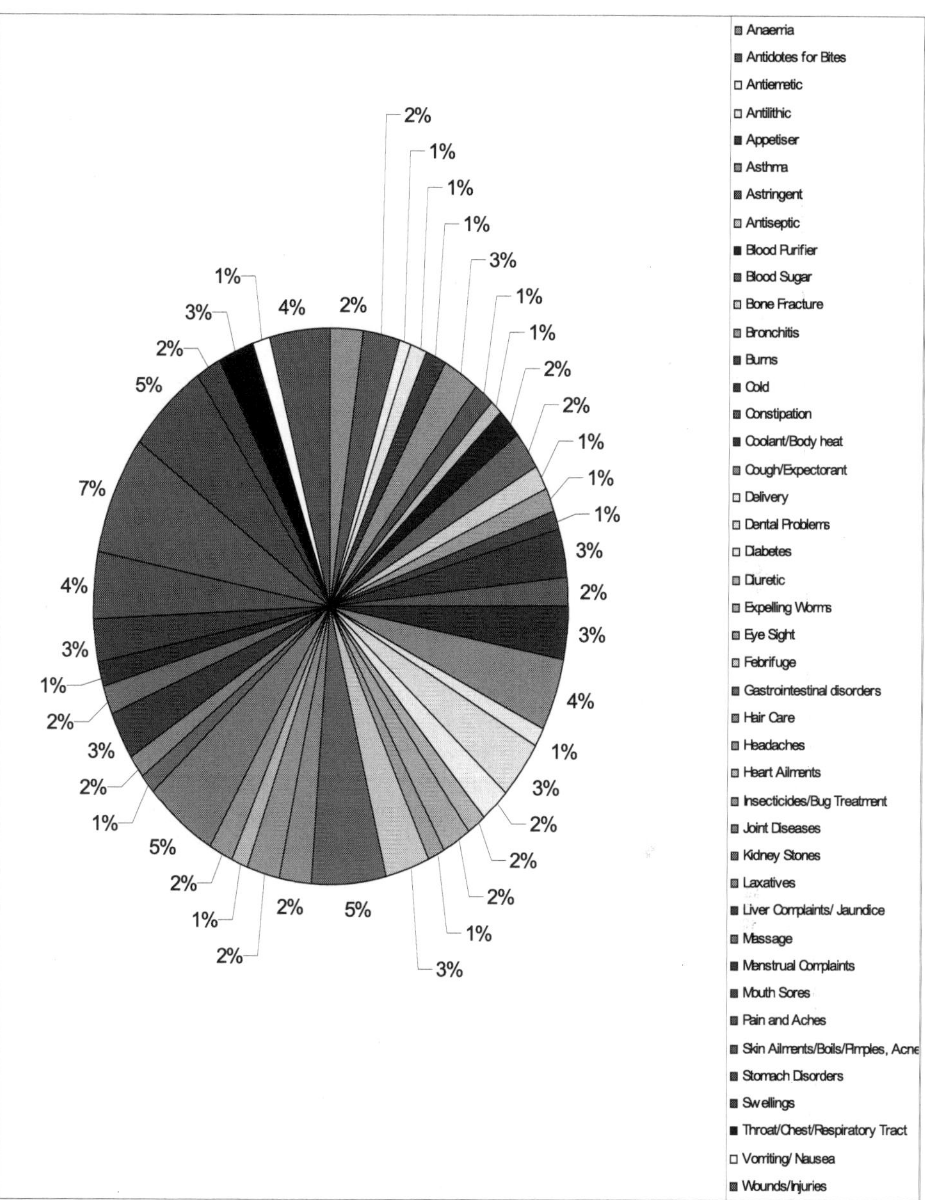

Fig. 3.23. Percentage Uses of Plants for Various Ailments and Disorders in the Study Area.

Table 3.17: Plants Used for Various Ailments in District Hamirpur

Ailment/s	Plant Species
Anaemia	*Actinidia chinensis*, Amaranthus paniculatus*, Amaranthus spinosus*, Amaranthus viridis*, Annona squamosa*, Brassica campestris, Brassica nigra, Bryophyllum calycinum, Carissa carandus*, Carissa spinarum*,Cinnamomum camphora, Daucus carota*, Eugenia cerasoides*, Eugenia jambolana, Fragaria chiloensis*, Lycopersicum esculentum*, Mangifera indica, Physalis peruviana, Psidium guajava*, Punica granatum*, Solanum melongena, Spinacia oleracea, Vitis vinifera*
Antidote for bites (Scorpion, Wasp Stings, Insect bite)	*Achyranthes aspera, Achyranthes bidentata, Aerva lanata, Ageratum conyzoides*, Boerhavia diffusa*, Canscora diffusa, Cannabis sativa*, Capsicum annuum, Catharanthus roseus, Cedrela toona, Eugenia jambolana, Litchi chinensis*, Martynia annua, Mesua ferrea*, Papaver rhoeus*, Pedilanthus tithymaloides*, Rauwolfia serpentina*, Roylea cinerea*, Santalum album, Sapindus mukorossi, Tagetes erecta*, Verbascum thapsus, Vitex negundo**
Antiemetic	*Allium cepa, Asparagus adscendens, Brassica juncea, Citrus medica, Eugenia jambolana*, Ocimum basilicum, Ocimum sanctum, Mentha piperita, Mentha spicata, Roylea cinerea*
Antilithic	*Aerva lanata, Artemisia vulgaris, Bryophyllum caylcinum*, Butea monosperma, Centella asiatica, Chenopodium album*, Chenopodium ambrosioides*, Cicer arietinum, Cissampelos pareira, Cucumis sativus, Hordeum vulgare*
Antiseptic	*Amorphophallus campanulatus, Curcuma domestica, Euphorbia pulcherrima, Euphorbia royleana*, Lannea coromandelica, Lavandula angustifolia, Nerium indicum, Ocimum sanctum, Randia dumetorum**
Appetiser	*Amaranthus paniculatus*, Amaranthus spinosus*, Amaranthus viridis*, Emblica officinalis, Eriobotrya japonica*, Ficus hispida, Ficus palmata*, Lycopersicum esculentum, Momordica charantia, Piper longum, Pyrus communis*, Ruta graveolens*, Terminalia bellirica*, Terminalia chebula*
Asthma	*Acacia concinna*, Achyranthes aspera, Adhatoda vasica, Aegle marmelos*, Amorphophallus campanulatus*, Centipeda minima, Cinnamomum tamala, Datura metel, Datura stramonium, Daucus carota, Elaeocarpus ganitrus, Helicteres isora, Lepidium sativum, Madhuca indica, Mangifera indica, Ocimum sanctum, Piper longum, Pistacia integerrima, Solanum viarum, Tylophora indica, Urtica dioica*, Vernonia anthelmintica*, Viola canescens, Withania somnifera*, Zingiber officinale*
Astringent	*Acacia nilotica, Alstonia scholaris, Centella asiatica*, Centipeda minima, Cucumis sativus, Curcuma domestica, Eugenia cerasoides, Eugenia jambolana, Helicteres isora, Madhuca indica, Momordica charantia, Rumex hastatus**
Bandage	*Lavandula angustifolia, Ricinus communis*

Contd...

Table 3.17: *contd...*

Ailment/s	Plant Species
Blood purifier	*Azadirachta indica, Berberis asiatica, Boerhavia diffusa*, Brassica nigra*, Carissa carandas*, Carissa spinarum, Catharanthus roseus*, Centella asiatica, Lawsonia alba*, Mallotus philippinensis, Mentha spicata*, Momordica charantia, Murraya koenigii*, Rauwolfia serpentina, Sesamum indicum, Thevetia neriifolia*, Tinospora cordifolia**
Blood sugar (diabetes)	*Aegle marmelos, Aloe barbedensis*, Argemone maxicana*, Azadirachta indica, Berberis asiatica*, Brassica nigra, Cassia occidentalis, Catharanthus roseus, Carica papaya, Eugenia jambolana, Ficus palmata, Luffa aegyptiaca, Madhuca indica, Momordica charantia, Murraya koenigii, Nasturtium officinale, Pongamia pinnata, Rauwolfia serpentina, Solanum tuberosum, Terminalia chebula, Tinospora cordifolia, Trigonella foenum-graecum**
Bone fracture, sprains	*Abrus precatorius, Artemisia japonica, Bambusa arundinacea*, Bambusa nutans*, Cuscuta reflexa, Dalbergia sissoo*, Dendrocalamus hamiltonii*, Dodonaea viscosa, Equisetum debile*, Oroxylum indicum, Rhynchostylis retusa, Ricinus communis, Soymida febrifuga, Vanda roxburghii*
Bronchitis	*Abelmoschus moschatus, Achyranthes bidentata, Acorus calamus, Adhatoda vasica, Carum copticum, Celastrus paniculatus*, Cinnamomum tamala, Elettaria cardamomum, Glycyrrhiza glabra, Lepidium sativum, Terminalia bellirica, Tylophora indica*
Burning sensation (soles)	*Annona squamosa, Holarrhena pubescens, Lawsonia alba, Lagenaria siceraria*, Luffa aegyptica*, Solanum melongena*
Burns	*Aloe vera, Aloe barbedensis, Annona squamosa, Artemisia vulgaris*, Cucumis sativus*, Curcuma domestica*, Hypericum cernum, Impatiens balsamina, Linum usitatissimum, Madhuca indica*, Martynia annua*, Prunus amygdalus, Prunus persica**
Clotting	*Ageratum conyzoides, Amaranthus viridis, Cassia occidentalis, Cynodon dactylon, Cuscuta reflexa, Lantana camara, Lawsonia alba, Mimosa pudica, Reinwardtia indica*
Cold	*Abelmoschus moschatus*, Abrus precatorius, Andrographis paniculata*, Centipeda minima, Cicer arietinum, Cinnamomum tamala, Cymbopogon citratus, Datura metel*, Elettaria cardamomum, Eucalyptus umbellata, Foeniculum vulgare, Glycyrrhiza glabra, Helicteres isora, Mesua ferrea, Nasturtium officinale, Nerium indicum*, Ocimum sanctum, Piper longum, Pongamia pinnata, Prunus persica*, Terminalia bellirica, Terminalia chebula, Urtica dioica*, Viola canescens, Zingiber officinale, Zizyphus jujuba**
Communicable diseases	*Azadirachta indica, Vitex negundo*

Contd...

Table 3.17: *contd...*

Ailment/s	Plant Species
Constipation	*Abelmoschus moschatus*, Aegle marmelos*, Cassia fistula*, Carica papaya*, Chenopodium ambrosioides*, Cissampelos pareira, Emblica officinalis, Mallotus philippinensis, Oroxylum indicum, Prunus amygdalus*, Psidium guajava, Randia dumetorum*, Raphanus sativus*, Ricinus communis, Rosa damascena*, Verbascum thapsus, Vitis vinifera, Woodfordia floribunda**
Coolant (refrigerant)/Body heat	*Abelmoschus esculentus, Abelmoschus moschatus, Aegle marmelos, Aloe vera, Asparagus adscendens, Centella asiatica, Coriandrum sativum*, Emblica officinalis*, Foeniculum vulgare, Holarrhena pubescens*, Lawsonia alba, Litchi chinensis*, Mallotus philippinensis, Mangifera indica, Mentha spicata, Mesua ferrea, Mirabilis jalapa*, Morus alba, Ocimum basilicum*, Oryza sativa, Punica granatum, Pyrus communis, Rosa damascena*, Roylea cinerea*, Saccharum officinarum, Santalum album, Triticum aestivum, Vitis vinifera*
Cough, Expectorant	*Abelmoschus moschatus*, Abrus precatorius, Achyranthes bidentata, Acorus calamus*, Adhatoda vasica, Aerva lanata*, Andrographis paniculata*, Carum copticum, Celastrus paniculata*, Centipeda minima, Cicer arietinum, Cinnamomum tamala, Citrus aurantium, Cymbopogon citratus, Datura metel*, Elettaria cardamomum, Emblica officinalis, Eucalyptus umbellata, Foeniculum vulgare, Glycyrrhiza glabra, Helicteres isora, Lepidium sativum, Mesua ferrea, Moringa oleifera, Nasturtium officinale, Ocimum sanctum*, Opuntia dillenii, Oxalis corniculata, Papaver rhoeus, Piper longum, Pistacia integerrima, Pogostemon benghalensis*, Pongamia pinnata, Punica granatum, Saraca asoca, Terminalia bellirica*, Terminalia chebula*, Tylophora indica, Urtica dioica, Viola canescens, Withania somnifera*, Zingiber officinale*
Dark circles	*Cucumis sativus, Lens culinaris, Solanum tuberosum*
Delivery	*Achyranthes bidentata, Albizia lebbeck, Asparagus adscendens*, Bambusa arundinacea, Bambusa nutans, Cardiospermum halicacabum*, Dendrocalamus hamiltonii, Foeniculum vulgare, Linum usitatissimum, Moringa oleifera*, Pongamia pinnata, Sesamum indicum, Soymida febrifuga**
Dental problems (toothache, gum diseases, pyorrhoea, etc.)	*Accacia catechu, Achyranthes aspera, Argemone maxicana, Azadirachta indica*, Calotropis procera, Clematis grata*, Curcuma domestica, Duranta repens*, Ficus glomerata, Hypericum cernum, Jasminum humile, Jasminum multiflorum, Juglans regia, Jatropha curcas, Lawsonia alba*, Madhuca indica, Murraya koenigii, Murraya paniculata, Nicotiana tobaccum, Ocimum sanctum, Pistacia integerrima*, Plumbago zeylanica, Psidium guajava, Ricinus communis, Spilanthes acmella, Solanum surattense*, Sonchus asper*, Vitex negundo, Wendlandia heynei, Xanthium strumarium, Zanthoxylum armatum*

Contd...

Table 3.17: *contd...*

Aliment/s	Plant Species
Diabetes	*Aegle marmelos, Aloe barbedensis*, Argemone mexicana*, Azadirachta indica*, Berberis asiatica*, Carica papaya, Cassia occidentalis*, Catharanthus roseus, Eugenia cerasoides*, Eugenia jambolana, Luffa aegyptica*, Madhuca indica, Momordica charantia, Murraya koenigii, Nasturtium officinale, Pongamia pinnata*, Premna latifolia*, Terminalia chebula, Trigonella foenum-graecum*
Diuretic	*Abelmoschus esculentus, Achras zapota, Achyranthes bidentata, Aerva lanata, Allium cepa, Boerhavia diffusa, Eugenia cerasoides, Eugenia jambolana, Hordeum vulgare, Lepidium sativum, Murraya koenigii, Oryza sativa, Raphanus sativus, Rosmarinus officinalis, Viola canescens, Withania somnifera*
Ear complaints (earache)	*Allium cepa, Allium sativum, Brassica campestris*, Cardiospermum halicacabum*, Colebrookia oppositifolia*, Vitex negundo*
Eczema	*Aloe barbedensis, Aloe vera, Carica papaya, Holarrhena pubescens*
Expelling worms	*Allium cepa*, Allium sativum*, Brassica juncea*, Cassia occidentalis*, Chenopodium album*, Chenopodium ambrosioides, Cordia dichotoma, Cucurbita hispida, Lycopersicum esculentum*, Mallotus philippinensis, Momordica charantia, Nasturtium officinale*, Nicotiana tobaccum*, Prunus persica, Raphanus sativus, Rauwolfia serpentina, Ricinus communis, Tabernaemontana divaricata, Vernonia anthelmintica**
Eyesight	*Aerva lanata, Berberis asiatica, Cynodon dactylon*, Eclipta prostrata*, Eupatorium adenophorum*, Rauwolfia serpentina*, Rosa damascena*, Roylea cinerea, Terminalia bellirica, Terminalia chebula, Vitex negundo*
Febrifuge (fever)	*Achras zapota*, Achyranthes aspera, Achyranthes bidentata, Adhatoda vasica, Adiantum lunulatum, Alstonia scholaris, Andrographis paniculata, Azadirachta indica, Cedrela toona, Cinnamomum camphora, Cissampelos pareira, Datura metel, Datura stramonium, Foeniculum vulgare, Fumaria indica, Glycyrrhiza glabra, Helicteres isora, Jasminum multiflorum*, Lepidium sativum*, Ocimum sanctum, Oxalis corniculata, Piper longum, Pyrus communis, Solanum nigrum, Solanum surattense, Soymida febrifuga, Tinospora cordifolia, Viola canescens, Vitis vinifera*
Flow of saliva	*Glycyrrhiza glabra*, Spilanthes acmella**
Flatulence	*Acorus calamus, Cinnamomum tamala, Citrus limon, Citrus medica, Coriandrum sativum*, Dalbergia sissoo*, Elettaria cardamomum*, Ficus hispida*, Stephania glabra*, Zingiber officinale*
Gastrointestinal *disorders (loose motion dysentery, indigestion, diarrhoea)*	*Aegle marmelos, Aerva lanata, Allium cepa*, Allium sativum, Andrographis paniculata*, Anetham sowa*, Artocarpus integrifolia*, Asparagus racemosus, Bauhinia vahlii, Bauhinia variegata, Cannabis sativa, Canscora diffusa, Carum copticum, Cassia fistula, Cassia occidentalis, Chenopodium ambrosioides*

Contd...

Table 3.17: *contd...*

Ailment/s	Plant Species
	Cinnamomum camphora, Cissampelos pareira, Citrus medica, Cordia dichotoma, Cuscuta reflexa, Elaeocarpus ganitrus, Foeniculum vulgare, Elettaria cardamomum, Ficus palmata*, Glycyrrhiza glabra, Lagenaria siceraria, Mentha spicata, Mesua ferrea, Momordica charantia, Moringa oleifera*, Musa sapientum, Ocimum basilicum, Oryza sativa, Piper longum, Prunus domestica, Punica granatum, Pyrus communis*, Pyrus pashia, Raphanus sativus, Roylea cinerea, Ruta graveolens, Sechium edule*, Solanum melongena*, Solanum nigrum, Terminalia chebula, Terminalia bellirica, Trigonella foenum-graecum, Tylophora indica**
Hair-care (baldness, dandruff, hair-tonic)	*Abrus precatorius, Acacia concinna, Albizia stipulata*, Allium sativum*, Annona squamosa, Cardiospermum halicacabum, Datura stramonium, Datura metel, Dodonaea viscosa*, Eclipta prostrata*, Emblica officinalis, Glycyrrhiza glabra, Hypericum cernum, Impatiens balsamina*, Jacaranda acutifolia*, Lawsonia alba, Melia azedarach*, Pyrus pashia*, Ricinus communis*, Rosmarinus officinalis, Sapindus mukorossi, Trigonella foenum graecum, Withania somnifera, Zizyphus jujuba**
Headaches	*Aerva lanata, Artemisia vulgaris*, Asparagus adscendens, Cestrum nocturnum*, Cymbopogon citratus, Elaeocarpus ganitrus, Eucalyptus umbellata, Ipomoea carnea*, Jasminum sambac, Melissa officinalis, Mentha piperata*, Mentha spicata, Nerium indicum, Ocimum sanctum, Oxalis corniculata*, Pogostemon benghalensis, Santalum album*, Sonchus asper*, Spilanthes acmella*, Woodfordia floribunda, Xanthium strumarium*
Heart ailments	*Actinidia chinensis*, Aegle marmelos, Carica papaya, Emblica officinalis, Habenaria pectinata, Helicteres isora, Lagenaria siceraria*, Litchi chinensis*, Luffa aegyptica*, Opuntia dillenii*, Solanum melongena*, Terminalia arjuna*
Hernia	*Dendrocalamus hamiltonii*
Impotency	*Malvaviscus conzatii*
Insecticide/Bugs treatment	*Annona squamosa*, Azadirachta indica, Berberis asiatica*, Boerhavia diffusa*, Cannabis sativa*, Cedrela toona, Citrus limon, Duranta repens, Eucalyptus umbellata, Lantana camara, Lavandula angustifolia, Nerium indicum, Phoenix sylvestris, Roylea cinerea, Vitex negundo, Wendlandia heynei*
Joint diseases (arthritis, knee pain, rheumatism, inflammation of joints, rheumatic pains)	*Acorus calamus, Adhatoda vasica*, Adiantum capillus-veneris*, Allium sativum*, Aloe vera*, Amaranthus paniculatus*, Amaranthus spinosus*, Amaranthus viridis, Artemisia japonica*, Boerhavia diffusa, Bombax ceiba, Brassica campestris*, Calotropis procera, Cannabis sativa*, Carthamus oxycantha, Chenopodium album, Chenopodium ambrosioides*, Colocasia esculenta, Curcuma domestica, Cuscuta reflexa, Cynodon dactylon*, Datura metel, Datura stramonium, Dioscorea bulbifera*, Dodonaea viscosa, Duranta repens*, Elaeocarpus ganitrus, Eugenia jambolana*, Juglans regia*, Lannea coromandelica, Linum usitatissimum*, Mesua ferrea*, Moringa*

Contd...

Table 3.17: *contd...*

Ailment/s	Plant Species
	oleifera, Nicotiana tobaccum, Plumbago zeylanica, Pyrus pashia, Rhynchostylis retusa*, Ricinus communis, Rosmarinus officinalis*, Roylea cinerea*, Sesamum indicum*, Soymida febrifuga*, Tinospora cordifolia*, Trichodesma indicum, Trigonella foenum-graecum, Urtica dioica*, Vanda roxburghii, Verbascum thapsus*, Vitex negundo, Zingiber officinale*
Kidney stones	*Aerva lanata*, Bryophyllum calycinum*, Celosia argentea*, Cissampelos pareira*, Cucumis sativus*, Hordeum vulgare, Momordica charantia, Oryza sativa*, Raphanus sativus*
Laxative	*Amaranthus paniculatus, Amaranthus spinosus, Amaranthus viridis, Avena sativa*, Berberis asiatica, Brassica campestris, Carduus nutans, Cassia fistula, Ficus palmata, Jatropha curcas, Rubus ellipticus*, Solanum nigrum, Spinacia oleracea, Terminalia bellirica, Terminalia chebula, Vitis jacquemontii**
Leucorrhoea	*Cissampelos pareira*, Eclipta prostrata, Jacaranda acutifolia*, Ocimum basilicum*, Saraca asoca*
Liver complaints (congestion, hepatic complaints, jaundice)	*Aegle marmelos, Albizia stipulata*, Aloe vera, Andrographis paniculata, Azadirachta indica, Boerhavia diffusa*, Butea monosperma, Carica papaya, Citrus limon, Citrus medica*, Daucus carota*, Eclipta prostrata*, Emblica officinalis*, Hordeum vulgare*, Lagenaria siceraria, Lycopersicum esculentum*, Mallotus philippinensis, Prunus domestica*, Punica granatum, Raphanus sativus*, Saccharum officinarum, Terminalia bellirica, Terminalia chebula, Tinospora cordifolia, Vitis vinifera, Woodfordia floribunda*
Massage	*Acorus calamus*, Adhatoda vasica, Adiantum capillus-veneris, Adiantum lunulatum, Aloe vera, Arachis hypogaea, Brassica campestris, Brassica juncea, Celastrus paniculatus, Cuscuta reflexa, Eclipta prostrata, Linum usitatissimum, Mesua ferrea, Prunus amygdalus, Ricinus communis, Sesamum indicum, Trigonella foenum-graecum*
Menstrual complaints (excessive bleeding, pains, amenorrhoea, menorrhagia, irregular cycle)	*Abroma augusta, Aloe barbedensis*, Amaranthus spinosus*, Amaranthus viridis*, Asparagus adscendens, Bauhinia variegata*, Butea monosperma, Cedrela toona*, Cinnamomum tamala*, Dodonaea viscosa, Holarrhena pubescens, Ruta graveolens, Saraca asoca*
Mouth sores, Ulcer	*Abelmoschus moschatus*, Acacia catechu, Acacia nilotica, Carissa spinarum*, Cassia occidentalis*, Cestrum nocturnum*, Cinnamomum tamala*, Colebrookia oppositifolia*, Cordia dichotoma*, Curcuma domestica, Cynodon dactylon, Elettaria cardamomum, Eugenia jambolana, Flacourtia ramontchii*, Jatropha curcas, Ocimum sanctum, Piper longum*, Psidium guajava*, Pyrus pashia*, Roylea cinerea, Rumex hastatus*, Solanum melongena*, Spilanthes acmella*, Stephania glabra*, Vitex negundo, Zanthoxylum armatum**

Contd...

Table 3.17: *contd...*

Ailment/s	Plant Species
Neuralagic disorders	*Abelmoschus moschatus, Bacopa monnieri, Canscora diffusa, Centella asiatica, Elaeocarpus ganitrus, Habenaria pectinata, Litchi chinensis, Musa sapientum**, Prunus amygdalus, Rauwolfia serpentina, Terminalia bellirica, Triticum aestivum, Withania somnifera**
Nose bleeding	*Cynodon dactylon*
Pains and aches (backache, bodyache, muscular pains)	*Acorus calamus, Adhatoda vasica, Adiantum capillus-veneris, Allium sativum, Aloe vera, Boerhavia diffusa, Bombax ceiba*, Butea monosperma, Carissa carandas, Carissa spinarum, Cuscuta reflexa, Cymbopogon citratus, Dalbergia sissoo, Datura metel, Duranta repens, Eruca sativa, Eucalyptus umbellata, Lagerstroemia indica*, Lantana camara, Linum usitatissimum*, Moringa oleifera, Murraya koenigii, Parthenium hysterophorus, Nicotiana tabaccum, Nerium indicum, Oxalis corniculata, Pogostemon benghalensis, Santalum album, Sesamum indicum, Sonchus asper, Soymida febrifuga, Spilanthes acmella, Tagetes erecta, Tinospora cordifolia, Trigonella foenum-graecum*, Verbascum thapsus, Vitex negundo Woodfordia floribunda*
Piles	*Allium cepa, Aloe vera*, Amorphophallus campanulatus, Calotropis procera, Emblica officinalis*, Lepidium sativum, Murraya koenigii*, Raphanus sativus, Sesamum indicum, Solanum melongena*, Terminalia bellirica, Viola canescens*
Psychosis/ Fear	*Adiantum lunulatum*, Aerva lanata, Cucurbita hispida, Datura metel, Dioscorea bulbifera, Paspalum scrobiculatum, Sapindus mukorossi**
Rabies	*Capsicum annuum, Sapindus mukorossi*
Sex disease	*Bombax ceiba, Bauhinia variegata, Butea monosperma, Cissampelos pareira, Eclipta prostata, Ficus hispida*, Ficus palmata*, Jacaranda acutifolia*, Prunus amygdalus, Saraca asoca*
Skin ailments (acne, boils, cracked skin, pimples, skin eruptions, skin rashes, dermatitis, emollients, itch)	*Abroma augusta, Abrus precatorius, Acacia nilotica*, Adhatoda vasica*, Allium cepa, Aloe barbedensis, Aloe vera, Artemisia japonica, Artocarpus lakoocha, Asparagus racemosus, Azadirachta indica, Berberis asiatica, Bombax ceiba, Brassica campestris, Bryophyllum calycinum, Butea monosperma, Calotropis procera, Capsicum annuum*, Carica papaya, Carthamus oxycantha, Catharanthus roseus, Cinnamomum camphora*, Cinnamomum tamala, Cassia occidentalis, Celastrus paniculatus, Clematis grata*, Coriandrum sativum, Cucumis sativus, Curcuma domestica*, Cuscuta reflexa, Cynodon dactylon, Dalbergia sissoo, Euphorbia pulcherrima, Euphorbia royleana, Ficus benghalensis*, Ficus palmata*, Helianthus annuus*, Helicteres isora, Hibiscus rosa-sinensis, Holarrhena pubescens*, Ipomoea carnea, Jacaranda acutifolia, Jatropha curcas*, Premna latifolia, Madhuca indica, Mallotus philippinensis*, Melissa officinalis*, Mesua ferrea*, Mirabilis jalapa, Momordica charantia*, Murraya koenigii, Nicotiana tabaccum*, Ocimum sanctum, Phyllanthus urinaria, Physalis peruviana, Plumbago zeylanica, Pongamia pinnata, Premna*

Contd...

Table 3.17: *contd...*

Ailment/s	Plant Species
	latifolia, Prunus amygdalus, Randia dumetorum*, Ranunculus muricatus*, Reinwardtia indica, Ricinus communis, Rosa damascena, Rosmarinus officinalis*, Roylea cinerea*, Rumex hastatus*, Santalum album*, Solanum tuberosum, Terminalia arjuna*
Slipped neck	*Parthenium hysterophorus**
Stomach disorders (abdominal pains, acidity, dyspepsia, flatulence, stomachache)	*Acorus calamus*, Aegle marmelos, Aerva lanata, Allium sativum*, Aloe vera*, Alstonia scholaris, Andrographis paniculata*, Anethum sowa, Asparagus adscendens, Avena sativa, Bombax ceiba, Canna indica, Canscora diffusa*, Carum copticum, Cassia fistula, Catharanthus roseus*, Cinnamomum tamala, Citrus medica, Coriandrum sativum, Cymbopogon citratus*, Elettaria cardamomum, Foeniculum vulgare, Jatropha curcus, Lagenaria siceraria*, Mallotus philippinensis*, Melissa officinalis, Mentha piperata, Mentha spicata*, Mesua ferrea, Momordica charantia*, Murraya koenigii, Musa sapientum*, Phyllanthus urinaria*, Physalis peruviana, Piper longum*, Pogostemon benghalensis*, Pongamia pinnata*, Psidium guajava, Punica granatum*, Pyrus communis, Randia dumetorum*, Raphanus sativus, Rosmarinus officinalis*, Saccharum officinarum, Stephania glabra, Taraxacum officinale*,Terminalia bellirica, Terminalia chebula, Tinospora cordifolia, Trichodesma indicum, Vernonia anthelmintica, Vitex negundo, Woodfordia floribunda*, Zingiber officinale*
Sweating	*Emblica officinalis, Mesua ferrea, Mirabilis jalapa, Luffa aegyptiaca, Morus alba, Solanum melongena**
Swelling (inflammation)	*Artemisia japonica*, Canna indica, Cannabis sativus*, Cinnamomum tamala, Cassia occidentalis*, Cuscuta reflexa, Cynodon dactylon, Santalum album, Sesamum indicum*, Solanum nigrum*, Soymida febrifuga, Stephania glabra, Tagetes erecta, Trichodesma indicum*, Vitex negundo, Withania somnifera*
Throat, chest and Respiratory tract infection (tonsils, sore throat)	*Abrus precatorius, Acacia catechu, Achyranthes aspera*, Achyranthes bidentata*, Adhatoda vasica, Aerva lanata*, Allium sativum, Andrographis paniculata, Carum copticum, Centipeda minima, Elettaria cardamomum*, Ficus palmata, Foeniculum vulgare, Glycyrrhiza glabra*, Helicteres isora, Martynia annua*, Melissa officinalis, Moringa oleifera*, Ocimum sanctum, Piper longum, Ranunculus muricatus*, Rosa damascena*, Rubus ellipticus*, Saraca asoca*, Spilanthes acmella, Vitis vinifera**
Tuberculosis	*Arachis hypogaea*, Cinnamomum tamala*, Ficus palmata, Pinus roxburghii*
Uterine disorder	*Abroma augusta, Holarrhena pubescens, Rauwolfia serpentina*, Ruta graveolens*, Soymida febrifuga*
Ulcers	*Carissa spinarum, Carissa carandas, Glycyrrhiza glabra, Jatropha curcas, Madhuca indica, Piper longum, Pongamia pinnata, Psidium guajava*

Contd...

Table 3.17: *contd...*

Ailment/s	Plant Species
Vomiting/ Nausea	*Allium cepa, Asparagus adscendens, Citrus grandis, Citrus limon, Citrus medica, Mentha piperata, Mentha spicata, Ocimum basilicum*, Ocimum sanctum, Oxalis corniculata, Tamarindus indica, Tylophora indica*
Wounds	*Abrus precatorius, Acorus calamus, Adhatoda vasica, Agave americana*, Ageratum conyzoides, Amorphophallus campanulatus, Artemisia japonica, Artocarpus lakoocha*, Bryophyllum calycinum, Carthamus oxycantha*, Celastrus paniculatus*, Colebrookia oppositifolia, Curcuma domestica, Cuscuta reflexa*, Cynodon dactylon, Ficus glomerata, Helianthus annuus, Jacaranda acutifolia, Jatropha curcas*, Lannea coromandelica, Lantana camara, Lawsonia alba, Linum usitatissimum*, Martynia annua*, Mesua ferrea*, Mimosa pudica, Mirabilis jalapa, Murraya koenigii*, Oroxylum indicum, Pedilanthus tithymaloides, Premna latifolia, Prunus persica, Reinwardtia indica, Rhynchostylis retusa*, Roylea cinerea, Solanum erianthum*, Terminalia arjuna, Trichodesma indicum*, Vanda roxburghii* Woodfordia floribunda*

The study also highlights the use of 42 tonics under 31 genera as iron source (*Amaranthus paniculatus*, Amaranthus spinosus*, Amaranthus viridis*, Brassica campestris*, Brassica juncea, Brassica nigra, Chenopodium álbum, Chenopodium ambrosioides, Cicer arietinum, Mentha piperata, Mentha spicata, Spinacia oleracea*), memory boosters (*Centella asiática, Emblica officinalis*, Habenaria pectinata),* refreshing drink (*Cucumis sativus, Daucas carota, Saccharum officinarum*), vit. C (*Citrus aurantium, Citrus medica, Emblica officinalis*, Oxalis corniculata, Physalis peruviana, Rumex hastatus*), cardiac tonic (*Carica papaya, Habenaria pectinata, Terminalia arjuna*), B-complex (*Brassica juncea, Brassica nigra*), nervous tonic (*Centella asiática, Habenaria pectinata*), liver tonic (*Saccharum officinarum, Terminalia bellirica, Terminalia chebula*), protein source (*Arachis hypogaea*, Glycine max*), and for strength (*Linum usitatissimum, Sesamum indicum*), blood formation and purification (*Eugenia cerasoides, Eugenia jambolana*) as well as for overcoming anaemic conditions (*Annona squamosa*, Carissa carandus, Carissa spinarum, Fragaria chiloensis, Lycopersicon esculentum, Morus alba*) and fatigue (*Punica granatum*) (Table 3.18). For a desired effect these are being given to the patients for a period ranging from several days to months. Of these, the use of members belonging to Amaranthaceae, Annonaceae, Brassicaceae, Euphorbiaceae, Fabaceae (families), viz., *Amaranthus paniculatus, A. spinosus, A. viridis, Annona squamosa, Arachis hypogaea, Brassica campestris and Emblica officinalis* as tonics have been recorded for the first time from this area.

Table 3.18: Tonic Plants of Hamirpur

Plant Species	Curative Property
*Amaranthus paniculatus**	Iron Source
*Amaranthus spinosus**	Iron Source
*Amaranthus viridis**	Iron Source
*Annona squamosa**	Anaemia
*Arachis hypogaea**	Protein Source
*Brassica campestris**	Iron Source
Brassica juncea	Iron Source
Brassica nigra	B-Complex, Iron Source
Carica papaya	Cardiac Troubles
Carissa carandus	Anaemia
Carissa spinarum	Anaemia
Centella asiatica	Memory
Chenopodium album	Iron Source
Chenopodium ambrosioides	Iron Source
Citrus aurantium	Vitamin C
Citrus medica	Vitamin C
Cicer arietinum	Iron Source
Cucumis sativus	Refreshing Drink
Daucas carot	Iron, Refreshing Drink
*Emblica officinalis**	Memory, Vitamin C
Eugenia cerasoides	Blood formation & Purification
Eugenia jambolana	Blood formation & Purification
Fragaria chiloensis	Anaemia
Glycine max	Proteins
Habenaria pectinata	Heart & Nervous Tonic
Lens culinaris	Protein Source
Linum usitatissimum	Strength
Lycopersicon esculentum	Anaemia
Mentha piperata	Iron Source
Mentha spicata	Iron Source
Morus alba	Anaemia
Oxalis corniculata	Vitamin C
Physalis peruviana	Vitamin C
Psidium guajava	Anaemia
Punica granatum	Anaemia, Fatigue
Rumex hastatus	Vitamin C
Saccharum officinarum	Refreshing drink
Sesamum indicum	Strength
Spinacia oleracea	Iron Source
Terminalia arjuna	Cardiac
Terminalia bellirica	Liver
Terminalia chebula	Liver, Stomach

Like other Indian traditional societies, the rural populace of the region is very much aware of the use of a wide range of plants species in their vicinity for their beauty, body tones, oral healthcare needs and perfumery (Tables 3.19, 3.20).

Table 3.19: Plants used as Body Tones, Shampoo Detergents, Chewing Gum, Facial Cosmetics, Scouring Teeth, Haircare, Dental Problems and Killing Lice

Use/s	Plant Species
Massage	*Acorus calamus, Adhatoda vasica, Aloe vera, Arachis hypogaea, Boerhavia diffusa, Brassica campestris, Brassica juncea, Brassica nigra, Cuscuta reflexa, Datura metel, Datura stramonium, Eclipta prostrata, Linum usitatissimum, Mesua ferrea, Prunus amaygdalus, Ricinus communis, Rosmarinus officinalis, Sesamum indicum, Trigonella foenum- graecum*
Facial cosmetics	*Aloe barbedensis, Aloe vera, Artocarpus lakoocha, Azadirachta indica, Berberis asiatica, Brassica campestris, Bryophyllum calycinum, Carica papaya, Celastrus paniculatus, Cicer arietinum, Citrus aurantium, Cucumis sativus, Curcuma domestica, Lavendula angustifolia, Lens culinaris, Melissa officinalis, Rosa damascena, Santalum album, Solanum tuberosum*
Dental problems (toothache, gum diseases, pyorrhoea, etc.)	*Accacia catechu, Achyranthes aspera, Argemone maxicana, Azadirachta indica*, Calotropis procera, Clematis grata*, Curcuma domestica, Duranta repens*, Ficus glomerata, Hypericum cernum, Jasminum humile, Jasminum multiflorum, Juglans regia, Jatropha curcas, Lawsonia alba*, Madhuca indica, Murraya koenigii, Murraya paniculata, Nicotiana tobaccum, Ocimum sanctum, Pistacia integerrima*, Plumbago zeylanica, Psidium guajava, Ricinus communis, Spilanthes acmella, Solanum surattense*, Sonchus asper*, Vitex negundo, Wendlandia heynei, Xanthium strumarium, Zanthoxylum armatum*
Hair-care (baldness, dandruff, hair-tonic)	*Abrus precatorius, Acacia concinna, Albizia stipulata*, Allium sativum*, Annona squamosa, Cardiospermum halicacabum, Datura stramonium, Datura metel, Dodonaea viscosa*, Eclipta prostrata*, Emblica officinalis, Glycyrrhiza glabra, Hypericum cernum, Impatiens balsamina*, Jacaranda acutifolia*, Lawsonia alba, Melia azedarach*, Pyrus pashia*, Ricinus communis*, Rosmarinus officinalis, Sapindus mukorossi, Trigonella foenum-graecum, Zizyphus jujuba**
Pimples/Acne	*Adhatoda vasica, Allium cepa, Artocarpus lakoocha, Azadirachta indica, Berberis asiatica, Bryophyllum calycinum, Curcuma domestica, Ipomoea carnea, Nicotiana tabaccum, Santalum album*
Skin ailments (acne, boils, cracked skin, pimples, skin eruptions, skin rashes, dermatitis, emollients, itch)	*Abroma augusta, Abrus precatorius, Acacia nilotica*, Adhatoda vasica*, Allium cepa, Aloe barbedensis, Aloe vera, Artemisia japonica, Artocarpus lakoocha, Asparagus racemosus, Azadirachta indica, Berberis asiatica, Bombax ceiba, Brassica campestris, Bryophyllum calycinum, Butea monosperma, Calotropis procera, Capsicum annuum*, Carica papaya, Carthamus oxycantha, Catharanthus roseus, Cinnamomum camphora*, Cinnamomum tamala, Cassia occidentalis, Celastrus paniculatus, Clematis grata*, Coriandrum sativum, Cucumis sativus, Curcuma domestica*, Cuscuta reflexa, Cynodon dactylon, Dalbergia sissoo, Euphorbia pulcherrima, Euphorbia royleana, Ficus benghalensis*, Ficus palmata*, Helianthus annuus*, Helicteres isora, Hibiscus rosa-sinensis, Holarrhena pubescens*, Ipomoea carnea, Jacaranda acutifolia, Jatropha curcas*, Premna latifolia, Madhuca indica, Mallotus philippinensis*, Melissa officinalis*, Mesua ferrea*, Mirabilis jalapa, Momordica charantia*, Murraya koenigii, Nicotiana tabaccum*, Ocimum sanctum, Phyllanthus urinaria, Physalis peruviana, Plumbago zeylanica, Pongamia pinnata, Premna latifolia*, Prunus amygdalus, Randia dumetorum*, Ranunculus muricatus*, Reinwardtia indica, Ricinus communis, Rosa damascena, Rosmarinus officinalis*, Roylea cinerea*, Rumex hastatus*, Santalum album*, Solanum tuberosum, Terminalia arjuna*
Reddening lips	*Juglans regia, Vitex negundo*
Scrubber	*Luffa aegyptica*
Shampoo/ detergents	*Acacia concinna, Randia dumetorum, Sapindus mukorossi*
Tooth brush (scouring teeth)	*Acacia catechu, Acacia nilotica, Achyranthes aspera, Azadirachta indica, Cinnamomum, tamala, Citrus grandis*, Citrus limon*, Citrus jambhiri*, Citrus medica*, Hypericum cernum*, Jatropha curcas, Juglans regia, Murraya koenigii*, Murraya paniculata, Plumbago zeylanica*, Psidium guajava, Pyrus pashia, Ricinus communis*, Spilanthes acmella, Vitex negundo*, Wendlandia heynei*, Zanthoxylum armatum*

Table 3.20: Ethnobotanically Important Plants Used in Perfumery

Botanical Name	Family	*Part Used*
*Abelmoschus moschatus**	Malvaceae	Seed
Citrus medica	Rutaceae	Fruit, Leaves
Jasminum humile	Oleaceae	Flower
Jasminum multiflorum	Oleaceae	Flower
Jasminum sambac	Oleaceae	Flower
*Lavandula angustifolia**	Lamiaceae	Leaves
Lawsonia alba	Lythraceae	Leaves
*Madhuca indica**	Sapotaceae	Leaves, Flower
*Melissa officinalis**	Lamiaceae	Leaves
Mentha piperata	Lamiaceae	Leaves
Mentha spicata	Lamiaceae	Leaves
Ocimum basilicum	Lamiaceae	Leaves
Ocimum sanctum	Lamiaceae	Leaves
Pelargonium graveolens	Geraniaceae	Leaves
Rosa damascena	Rosaceae	Flowers
Rosmarinus officinalis	Lamiaceae	Leaves
*Roylea cinerea**	Lamiaceae	Leaves
Santalum album	Santalaceae	Wood, Oil

Amongst these two species (*Juglans regia, Vitex negundo*) are used as substitute for lip-stick; 3 species as shampoo/ detergents (*Acacia concinna, Randia dumetorum, Sapindus mukorossi*); 1 species, *Luffa aegyptica*, as body scrubber; 3 species for dark circles (*Cucumis sativus, Lens culinaris, Solanum tuberosum*); 11 species as chewing gum (*Aegle marmelos, Albizia lebbeck, Bambusa arundinacea, Bambusa nutans, Dendrocalamus hamiltonii, Emblica officinalis, Holarrhena pubescens, Musa sapientum, Ocimum sanctum, Saccharum spontaneum, Santalum album*); 9 species for killing lice (*Annona squamosa, Eruca sativa, Eucalyptus umbellata, Lantana camara, Mentha spicata, Millettia extensa, Nicotiana tabacum, Roylea cinerea, Wendlandia heynei*); 10 species for pimples (*Adhatoda vasica, Allium cepa, Artocarpus lakoocha, Azadirachta indica, Berberis asiatica, Bryophyllum calycinum, Curcuma domestica, Ipomoea carnea, Nicotiana tabaccum, Santalum album*); 19 species for facial cosmetics (*Aloe barbedensis, Aloe vera, Artocarpus lakoocha, Azadirachta indica, Berberis asiatica, Brassica campestris, Bryophyllum calycinum, Carica papaya, Celastrus paniculatus, Cicer arietinum, Citrus aurantium, Cucumis sativus, Curcuma domestica, Lavendula angustifolia, Lens culinaris, Melissa officinalis, Rosa damascena, Santalum album, Solanum tuberosum*); and 18 species under 14 genera belonging to 9 families with predominance of Lamiaceae (8 species) as a family, *Jasminum* as a genus and leaves (12 species) as a plant part employed for perfumery; 19 species under 16 genera for body massage; 22 species under 17 genera for scouring teeth; 23 species under 22 genera for hair care; 31 species under 29 genera for dental caries, gum diseases and pyorrhoea; and 70 species under 67 genera for skin ailments like cracked skin, skin eruption, skin rashes, dermatitis (Table 3.19). Usages of most of these plants have been recorded for the first time (Sood *et al.*, 2011a) and hold considerable scope for popularization to be developed as safe beauty aids for modern human societies.

Overall, a total of 133 (65 wild; 68 cultivated) species belonging to 108 genera under 54 families have been identified for edibility from the regíon (Tables 3.21-3.26; Fig. 3.24-3.29) thereby corresponding to about 1: 67% of diversity of wild edibles from India (Anonymous, 1994; Singh & Arora, 1978). Out of these, 52 species/ 39 genera (22 families) are used as vegetables/ pot herbs (Table 3.22); 24

species/ 20 genera (16 families) as pickles and for chutney preparations (Table 3.23; Fig. 3.25); 22 species/ 19 genera (12 families) for flavouring purposes (Table 3.24); and 26 species/ 25 genera (15 families) for culinary preparations (Table 3.25). Organwise analysis reveals that inhabitants of this part of Himalayas fancied fruits (75 species) more than other parts (Table 3.21; Fig. 3.24). Relatively, the most frequently used families for vegetables and pot herbs are represented by Moraceae (7 species), followed by Cucurbitaceae, Brassicaceae, Fabaceae (5 each), Amaranthaceae (4 species), Chenopodiaceae, Solanaceae, Poaceae (3 each), Apiaceae, Araceae, Malvaceae (2 each), and rest 11 families such as Asteraceae, Boraginaceae, Caricaceae, Cordiaceae, Dioscoriaceae, Moringaceae, Musaceae, Nyctaginaceae, Rosaceae, Urticaceae, Zingiberaceae represented by 1 species each. These findings for the región are contrary to observations of Sharma (2009) who recorded that the food of Himalayan people is largely derived from plants belonging to 3 families like Poaceae, Fabaceae and Cucurbitaceae. Importantly, 12 species (*Arachis hypogaea, Cassia fistula, Cicer arietinum, Glycine max, Lens culinaris, Linum usitatissimum, Pennisetum americanum, Pennisetum purpureum, Prunus amygdalus, Triticum aestivum, Zea mays, Vicia pallida*) are rich in proteins; 10 in fats (*Arachis hypogaea, Brassica campestris, Eruca sativa, Gossypium arboreum, Helianthus annuus, Jatropha curcus, Juglans regia, Linum usitatissimum, Ricinus communis, Sesamum indicum*) and 35 species (*Amaranthus paniculatus**, *Amaranthus spinosus**, *Amaranthus viridis**, *Annona squamosa, Brassica campestris, Brassica juncea, Brassica nigra, Carica papaya, Carissa carandas* var., *congesta, Carissa spinarum, Chenopodium album, Chenopodium ambrosioides, Cicer arietinum, Colocasia esculenta, Daucus carota, Eugenia cerasoides, Eugenia jambolana, Flacourtia ramontchi, Foeniculum vulgare, Fragaria chiloensis, Lycopersicon esculentum, Mentha piperata, Mentha spicata, Morus alba, Prunus domestica, Prunus persica, Psidium guajava, Punica granatum, Pyrus communis, Spinacia oleracea, Tinospora cordifolia, Trigonella foenum-graecum, Urtica dioica, Xanthium strumarium, Zizyphus jujuba*) in iron (Table 3.26). On comparing upon the present data with earlier Indian works, it is indicated that usage of edible species like *Bacopa monnieri, Citrus grandis**, *Citrus jambhiri**, *Cucurbita hispida**, *Madhuca indica**, *Murraya koenigii**, *Pennisetum americanum**, *Pinus roxburghii**, *Vitis jacquemontii** as fruits/ seeds/ leaves; *Abelmoschus moschatus**, *Amaranthus paniculatus**, *Amorphophallus campanulatus**, *Bambusa arundinacea**, *Bambusa nutans**, *Brassica juncea**, *Carduus nutans**, *Colocasia esculenta**, *Daucas carota**, *Ficus carica**, *Ficus palmata**, *Foeniculum vulgare**, *Lycopersicum esculentum**, *Moringa oleifera**, *Musa sapientum**, *Psidium guajava**, *Spinacia oleracea**, *Trigonella foenum-graecum** as vegetables and pot herbs; *Amorphophallus campanulatus**, *Bauhinia variegata**, *Bambusa arundinacea**, *Bambusa nutans**, *Capsicum annuum**, *Cassia occidentalis**, *Citrus medica**, *Daucas carota**, *Mangifera indica**, *Punica granatum**, *Zingiber officinale** for pickling and chutney preparations; *Cymbopogon citratus**, *Elettaria cardamomum**, *Mentha piperata**, *Pelargonium graveolens**, *Piper longum**, *Rosmarinus officinalis**, *Trigonella foenum-graecum**, *Zingiber officinale* * for flavouring; and *Cannabis sativa**, *Cicer arietinum**, *Colocasia esculenta**, *Curcuma domestica**, *Daucas carota**, *Luffa aegyptiaca**, *Lycopersicum esculentum**, *Rumex hastatus**, *Sesamum indicum** for culinary preparations are unique and have not been hitherto reported. Another unique feature of the present study is the use of seeds of *Achyranthes bidentata* for suppressing hunger. This information is of great significance in training forces and commandos especially in jungle warfare. Undeniably, the information presented extends the base of our food bowl and also at the same time enhances our chances of survival in the event of any crisis like famine and war.

Table 3.21: Edible Plants of Hamirpur

S. No.	Botanical Name	Part Used
1.	*Achras zapota*	Fruit
2.	*Actinidia chinensis*	Fruit
3.	*Aegle marmelos*	Fruit
4.	*Aerva lanata*	Leaves
5.	*Agave americana*	Tender Shoot
6.	*Andrographis paniculata*	Leaves, Shoot
7.	*Annona squamosa*	Fruit
8.	*Arachis hypogaea*	Seed
9.	*Artocarpus integrifolia*	Fruit
10.	*Avena sativa*	Fruit (Grains)
11.	*Azadirachta indica*	Leaves
12.	*Bacopa monnieri**	Leaves
13.	*Bryophyllum caylcinum*	Leaves
14.	*Carica papaya*	Fruit
15.	*Carissa carandus*	Fruit
16.	*Carissa spinarum*	Fruit
17.	*Centella asiatica*	Leaves
18.	*Cicer arietinum*	Seed
19.	*Citrus aurantium*	Fruit
20.	*Citrus grandis**	Fruit
21.	*Citrus jambhiri**	Fruit
22.	*Citrus limon*	Fruit
23.	*Citrus medica*	Fruit
24.	*Colocasia esculenta*	All Parts
25.	*Cucumis sativus*	Fruit
26.	*Cucurbita hispida**	Fruit
27.	*Cucurbita pepo*	Fruit
28.	*Daucas carota*	Root
29.	*Elettaria cardamomum*	Fruit/ Seed
30.	*Emblica officinalis*	Fruit
31.	*Eriobotrya japonica*	Fruit
32.	*Eugenia cerasoides*	Fruit
33.	*Eugenia jambolana*	Fruit
34.	*Ficus carica*	Fruit
35.	*Ficus glomerata*	Fruit
36.	*Ficus hispida*	Fruit
37.	*Ficus palmata*	Fruit
38.	*Ficus roxburghii*	Fruit
39.	*Flacourtia ramontchi*	Fruit
40.	*Foeniculum vulgare*	Fruit/ Seed

contd...

Table 3.21: *contd...*

S. No.	Botanical Name	Part Used
41.	*Fragaria chiloensis*	Fruit
42.	*Glycine max*	Seed
43.	*Glycyrrhiza glabra*	Root
44.	*Hordeum vulgare*	Grain/Seed
45.	*Juglans regia*	Fruit
46.	*Lens culinaris*	Seed
47.	*Linum usitatissimum*	Seed
48.	*Litchi chinensis*	Fruit
49.	*Luffa aegyptiaca*	Fruit
50.	*Lycopersicum esculentum*	Fruit
51.	*Madhuca indica**	Flower
52.	*Mangifera indica*	Fruit
53.	*Mentha piperata*	Leaves
54.	*Mentha spicata*	Leaves
55.	*Momordica charantia*	Fruit
56.	*Morus alba*	Fruit
57.	*Murraya koenigii**	Leaves/ Fruit
58.	*Murraya paniculata*	Fruit
59.	*Musa sapientum*	Fruit
60.	*Opuntia dillenii*	Fruit
61.	*Oryza sativa*	Seed/Grains
62.	*Pennisetum americanum**	Seed/Grains
63.	*Phoenix sylvestris*	Seed
64.	*Physalis peruviana*	Fruit
65.	*Pinus roxburghii**	Seed
66.	*Prunus amygdalus*	Seed
67.	*Prunus domestica*	Fruit
68.	*Prunus persica*	Fruit
69.	*Psidium guajava*	Fruit
70.	*Punica granatum*	Fruit
71.	*Pyrus communis*	Fruit
72.	*Pyrus pashia*	Fruit
73.	*Rubus ellipticus*	Fruit
74.	*Saccharum officinarum*	Stem
75.	*Sesamum indicum*	Seed
76.	*Spilanthes acmella*	Inflorescence
77.	*Tagetes erecta*	Receptacle
78.	*Terminalia arjuna*	Bark
79.	*Triticum aestivum*	Grains/Seed
80.	*Vitis jacquemontii**	Fruit
81.	*Vitis vinifera*	Fruit
82.	*Zea mays*	Grains/Seed
83.	*Zizyphus jujuba*	Fruit

Table 3.22: Important Vegetable & Pot Herbs of District Hamirpur

S. No.	Botanical Name	Family	Part Used
1.	*Abelmoschus esculentus*	Malvaceae	Fruit
2.	*Abelmoschus moschatus**	Malvaceae	Fruit
3.	*Amaranthus paniculatus**	Amaranthaceae	Leaves/Seed
4.	*Amaranthus spinarum*	Amaranthaceae	Leaves
5.	*Amaranthus viridis*	Amaranthaceae	Leaves
6.	*Aerva lanata*	Amaranthaceae	Leaves
7.	*Amorphophallus campanulatus**	Araceae	Rhizome
8.	*Artocarpus integrifolia*	Moraceae	Fruit
9.	*Artocarpus lakoocha*	Moraceae	Fruit
10.	*Bambusa arundinacea**	Poaceae	Young Shoot
11.	*Bambusa nutans**	Poaceae	Young Shoot
12.	*Bauhinia variegata*	Fabaceae	Bud, Flower
13.	*Boerhavia diffusa*	Nyctaginaceae	Leaves
14.	*Bombax ceiba*	Boraginaceae	Flower Bud
15.	*Brassica campestris*	Brassicaceae	Leaves
16.	*Brassica juncea**	Brassicaceae	Leaves
17.	*Brassica nigra**	Brassicaceae	Leaves
18.	*Carduus nutans**	Astraceae	Leaves
19.	*Carica papaya*	Caricaceae	Fruit
20.	*Chenopodium album*	Chenopodiaceae	Leaves
21.	*Chenopodium ambrosioides*	Chenopodiaceae	Leaves
22.	*Cicer arietinum*	Fabaceae	Seed
23.	*Colocasia esculenta**	Araceae	Rhizome, Leaves
24.	*Cordia dichotoma*	Cordiaceae	Fruit
25.	*Cucurbita pepo*	Cucurbitaceae	Fruit
26.	*Curcuma domestica*	Zingiberaceae	Rhizome
27.	*Daucas carota**	Apiaceae	Root
28.	*Dendrocalamus hamiltonii*	Poaceae	Young Shoot
29.	*Dioscorea bulbifera*	Dioscoriaceae	Rhizome
30.	*Ficus carica**	Moraceae	Fruit
31.	*Ficus glomerata*	Moraceae	Fruit
32.	*Ficus hispida*	Moraceae	Fruit
33.	*Ficus palmata**	Moraceae	Fruit
34.	*Ficus roxburghii*	Moraceae	Fruit
35.	*Foeniculum vulgare**	Apiaceae	Leaves/Fruit
36.	*Glycine max*	Fabaceae	Seed/ Pod
37.	*Lagenaria siceraria*	Cucurbitaceae	Fruit
38.	*Luffa aegyptica*	Cucurbitaceae	Fruit
39.	*Lycopersicum esculentum**	Solanaceae	Fruit
40.	*Momordica charantia*	Cucurbitaceae	Fruit
41.	*Moringa oleifera**	Moringaceae	Pod
42.	*Musa sapientum**	Musaceae	Fruit
43.	*Nasturtium officinale*	Brassicaceae	Leaves
44.	*Psidium guajava**	Rosaceae	Fruits
45.	*Raphanus sativus*	Brassicaceae	Root/Leaves
46.	*Sechium edule*	Cucurbitaceae	Fruit
47.	*Solanum melongena*	Solanaceae	Fruit
48.	*Solanum tuberosum*	Solanaceae	Tuber
49.	*Spinacia oleracea**	Chenopodiaceae	Leaves
50.	*Trigonella foenum-graecum**	Fabaceae	Leaves
51.	*Urtica dioica*	Urticaceae	Leaves
52.	*Vicia pallida*	Fabaceae	Pod

Table 3.23: Plants Used for Pickles & Chutney by the Rural Populace of District Hamirpur

S. No.	Botanical Name	Family	Part Used
1.	*Aloe vera*	Liliaceae	Young Shoot
2.	*Amorphophallus campanulatus**	Araceae	Rhizome
3.	*Artocarpus lakoocha*	Moraceae	Fruit
4.	*Artocarpus integrifolia*	Moraceae	Fruit
5.	*Bauhinia variegata**	Fabaceae	Bud, Flower
6.	*Bambusa arundinacea**	Poaceae	Young Shoot
7.	*Bambusa nutans**	Poaceae	Young Shoot
8.	*Capsicum annuum*	Solanaceae	Fruit
9.	*Carissa spinarum*	Apocynaceae	Fruit
10.	*Cassia occidentalis**	Fabaceae	Seed
11.	*Citrus jambhiri*	Rutaceae	Fruit
12.	*Citrus limon*	Rutaceae	Fruit
13.	*Citrus medica**	Rutaceae	Fruit
14.	*Cordia dichotoma*	Cordiaceae	Fruit
15.	*Daucas carota**	Apiaceae	Root
16.	*Dendrocalamus hamiltonii*	Poaceae	Tender Shoot
17.	*Emblica officinalis*	Euphorbiaceae	Fruit
18.	*Mangifera indica*	Anacardiaceae	Fruit
19.	*Raphanus sativus*	Brassicaceae	Root
20.	*Zingiber officinale**	Zingiberaceae	Rhizome
Chutney Plants			
1.	*Bauhinia variegata**	Fabaceae	Flower, Bud
2.	*Capsicum annuum**	Solanaceae	Fruit
3.	*Cassia occidentalis*	Fabaceae	Seed
4.	*Mangifera indica**	Anacardiaceae	Fruit
5.	*Mentha spicata*	Lamiaceae	Leaves
6.	*Murraya koenigii*	Rutaceae	Leaves
7.	*Punica granatum**	Punicaceae	Seed
8.	*Tamarindus indica*	Fabaceae	Pod

Table 3.24: Flavouring Plants of District Hamirpur

Plant	Family	Part Used
Allium cepa	Liliaceae	Bulb/Leaves
Allium sativum	Liliaceae	Bulbs/Leaves
Anethum sowa	Apiaceae	Leaves/Fruit
Brassica nigra	Brassicaceae	Leaves/Seed
Capsicum annuum	Solanaceae	Fruit
Carum copticum	Apiaceae	Seed
Cinnamomum tamala	Lauraceae	Leaves/Bark
Citrus limon	Rutaceae	Fruit/Leaves
Coriandrum sativum	Apiaceae	Leaves/Fruit/Stem
Cymbopogon citratus*	Poaceae	Leaves
Elettaria cardamomum*	Zingiberaceae	Fruit/Seed
Foeniculum vulgare	Apiaceae	Fruit/Seed
Mentha piperata*	Lamiaceae	Leaves
Mentha spicata	Lamiaceae	Leaves
Murraya koenigii	Rutaceae	Leaves
Ocimum basilicum	Lamiaceae	Leaves
Ocimum sanctum	Lamiaceae	Leaves
Pelargonium graveolens*	Geraniaceae	Leaves
Piper longum*	Piperaceae	Fruit
Rosmarinus officinalis*	Lamiaceae	Leaves
Trigonella foenum- graecum*	Fabaceae	Seed
Zingiber officinale *	Zingiberaceae	Rhizome

Table 3.25: Plants Used in Culinary Preparations

Arachis hypogaea	Seed
Bauhinia variegata	Bud, Flower
Capsicum annuum	Fruit
Cannabis sativa* Leaves	Leaves
Cassia occidentalis	Seed
Cicer arietinum* Seed	Seed
Colocasia esculenta*	Rhizome
Cucumis sativus Fruits, Seeds	Fruit
Cucurbita hispida Fruits	Fruit
Curcuma domestica*	Rhizome
Daucas carota*	Root
Juglans regia	Fruit
Lagenaria siceraria	Fruit
Luffa aegyptiaca*	Fruit
Lycopersicum esculentum*	Fruit
Mangifera indica Fruits	Fruit
Mentha piperata Leaves	Leaves
Mentha spicata	Leaves
Ocimum basilicum	Leaves
Oxalis corniculata	Leaves
Prunus persica Fruits	Fruit
Punica granatum	Seed
Rumex hastatus*	Leaves
Sesamum indicum*	Seed
Tamarindus indica	Pod
Zingiber officinale *	Rhizome

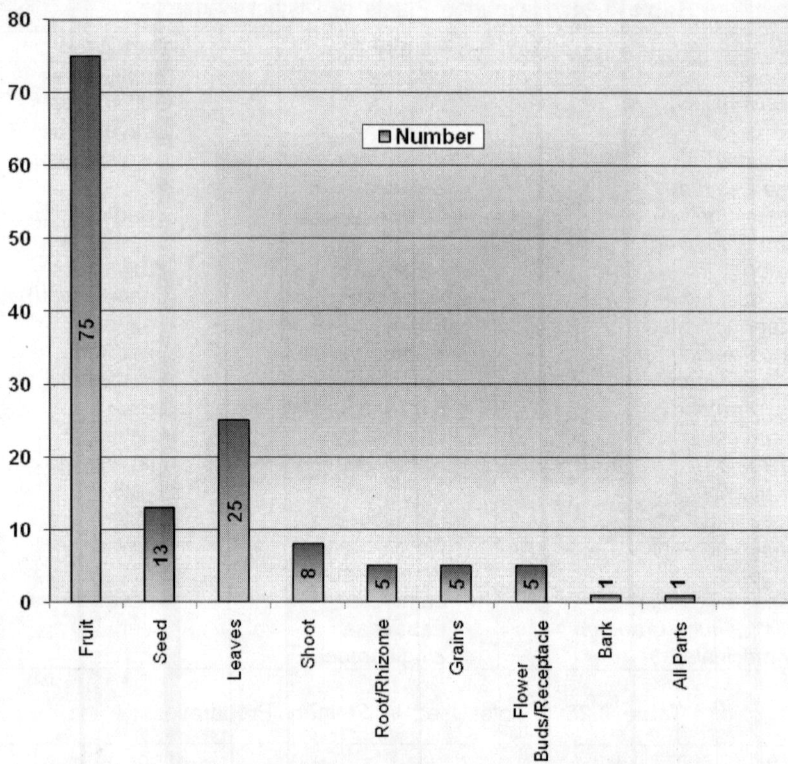

Fig. 3.24. Relative Disposition of Plant Parts Used for Edible Purposes in District Hamirpur.

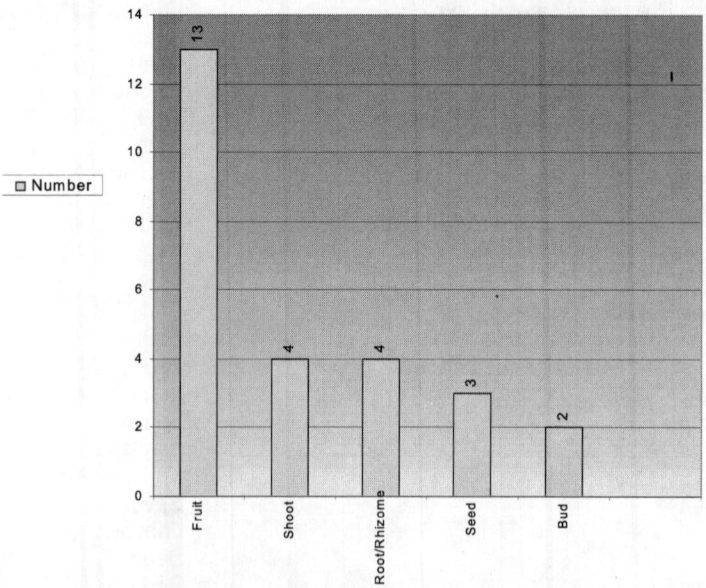

Fig. 3.25. Relative Number of Plant Parts Used for Pickling and Chutney Preparations in the Study Area.

Table 3.26: Nutritive Components Food Plants of District Hamirpur

Iron Rich Plants	Protein Rich Plants	*Fat Rich Plants*
*Amaranthus paniculatus**	*Arachis hypogaea*	*Arachis hypogaea*
*Amaranthus spinosus**	*Cassia fistula*	*Brassica campestris*
*Amaranthus viridis**	*Cicer arietinum*	*Eruca sativa*
Annona squamosa	*Glycine max*	*Gossypium arboreum*
Brassica campestris	*Lens culinaris*	*Helianthus annuus*
Brassica juncea	*Linum usitatissimum*	*Jatropha curcus*
Brassica nigra	*Pennisetum americanum*	*Juglans regia*
Carica papaya	*Pennisetum purpureum*	*Linum usitatissimum*
Carissa carandas var. congesta	*Prunus amygdalus*	*Ricinus communis*
Carissa spinarum	*Triticum aestivum*	*Sesamum indicum*
Chenopodium album	*Zea mays*	
Chenopodium ambrosioides	*Vicia pallida*	
Cicer arietinum		
Colocasia esculenta		
Daucus carota		
Eugenia cerasoides		
Eugenia jambolana		
Flacourtia ramontchi		
Foeniculum vulgare		
Fragaria chiloensis		
Lycopersicon esculentum		
Mentha piperata		
Mentha spicata		
Morus alba		
Prunus domestica		
Prunus persica		
Psidium guajava		
Punica granatum		
Pyrus communis		
Spinacia oleracea		
Tinospora cordifolia		
Trigonella foenum-graecum		
Urtica dioica		
Xanthium strumarium		
Zizyphus jujuba		

On comparing nutritional values of two of the presently investigated ethnobotanical taxa, the analysis reveals high contents of calcium (199.19 mg/100g), phosphorus (133.7 mg/100g), magnesium (84.69 mg/100g), carbohydrates (69.42g/100g) in fruits of *Zizyphus jujuba* as well as high contents of saponins (+++) in its leaves (Figs. 3.26, 3.27; Tables 3.27, 3.29); high values for calcium (128.56 ppm), magnesium (30.57 ppm) and iron (24.78%) in gum, 'Banslochan' obtained from *Bambusa nutans* (Figs. 3.28, 3.29; Table 3.28) and justifies the reason for it being given in the form of 'panjiri' to women after delivery to overcome weakness and fatigue due to deficiency of the above minerals. Importantly, nutritional comparision of data (present study) with the earlier published literature (Table 3.30) also point to their high neutraceutical potentiality for incorporation as a part of human diet in one form or other in this part of Himalayas and also at the same time offers a challenging task to the agronomists for improving upon their growth form for domestication by employing modern scientific tools.

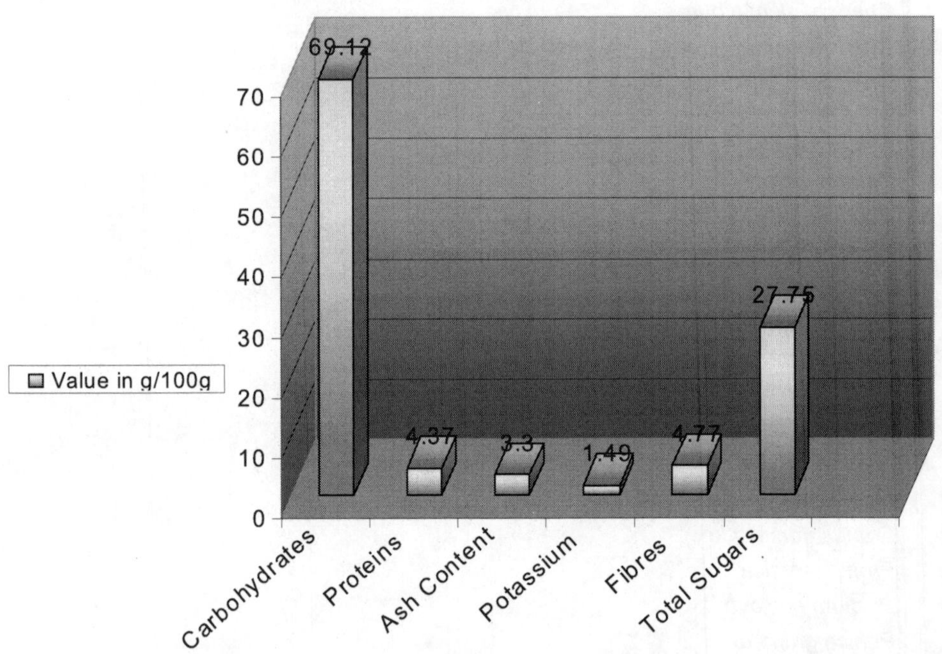

Fig. 3.26. Quantitative Reflection of Nutrients (g/100g) in Fruits of *Zizyphus jujuba*

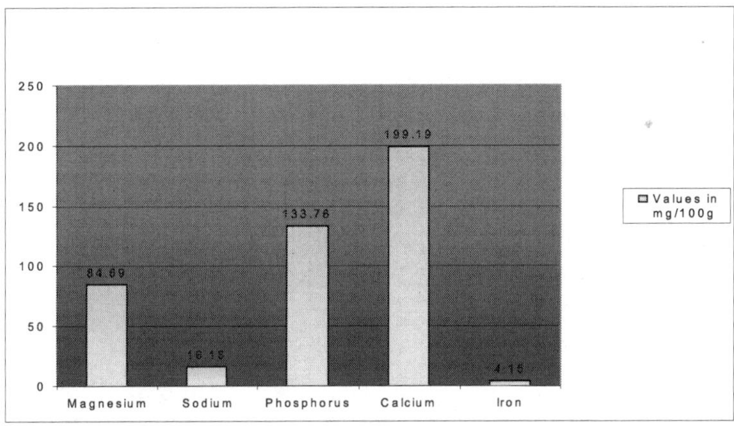

Fig. 3.27. Histogram Showing Quantitative Reflection of Minerals (mg/100g) in Fruits of *Zizyphus jujuba*.

Table 3.27: Nutritional Components of *Zizyphus jujuba* Fruits (On Dry Weight Basis)

Carbo-hydrates g/100g	Proteins g/100g	Total Sugars g/100g	Fibres g/100g	Ash content g/100g	Pota-ssium g/100g	Phos-phorus mg/100g	Sodium mg/100g	Cal-cium mg/100g	Iron mg/100g	Magne-sium mg/100g
69.12	4.37	27.75	4.77	3.30	1.49	133.76	16.18	199.19	4.15	84.69

Table 3.28: Nutritional Components of Gum 'Banslochan' (*Bambusa nutans*) (On Dry Weight Basis)

Carbo-hydrates (%)	Fats (%)	Proteins (%)	Iron (%)	Calcium ppm	Magnesium ppm	Phos-phorus (%)	Ash Content (%)	Vit. (C) (%)
1.70	Nil	Nil	24.78	128.56	30.57	Nil	90.80	Nil

Table 3.29: Phytochemical Analysis of Fruits of Mallotus philippinensis and Leaves of *Zizyphus jujuba*

Phytochemicals	*Mallotus philippinensis* (Unripe Seeds)	*Mallotus philippinensis* (Ripe Seeds)	*Zizyphus jujuba* Leaves (Dry powder)
Glycosides: Molisch's Test	---	---	---
Tannins: $FeCl_3$ Test	+	+	+
Saponins (a) LB Test, (b) Foam Test	+	+	+++
Terpenoids $CHCl_3$ + Conc. H_2SO_4 (Chloroform Test)	--	--	---
Alkaloids (a) Meyer's Test (b) Dragendroff Test	+	+	+
Flavonoids (a) Shinoda Test (b) Lead Acetate Test	+	+	+

Note---+ive (Present), +++ (present in excess) and ---(not detected).

Analysis carried out in Pharmacognosy Laboratory University Institute of Phamaceutical Sciences Punjab University Campus Chandigarh.

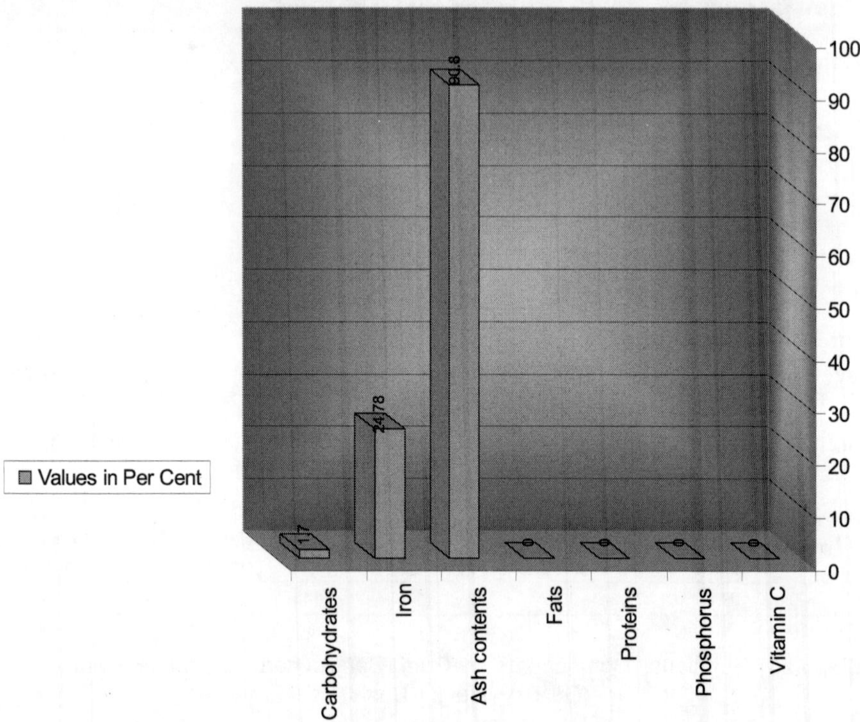

Fig. 3.28. Nutritional Components of Gum 'Banslochan' (%) on Dry Weight Basis.

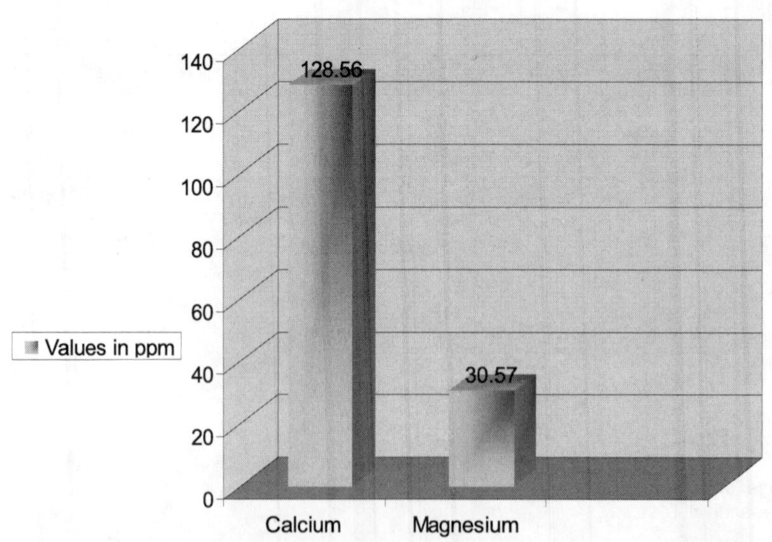

Fig. 3.29. Quantitative Reflection of Minerals (ppm) in 'Gum Banslochan'.

Table 3.30: Index to Nutritional Value of Ethnobotanically Important Plants of Hamirpur

Part used/Category	Moisture	Protein	Fat	Minerals	Fibres	Carbohydrates	Calories
Leaves							
Amaranthus spinosus	85.0	3.0	0.7	3.6	1.1	7.0	43
Amaranthus viridis	81.8	5.2	0.3	2.8	6.1	3.8	38
Bauhinia variegata	78.1	3.6	1.0	2.1	5.5	9.7	62
Chenopodium album	89.6	3.7	0.4	2.6	0.8	2.9	30
Colocasia esculenta	82.7	3.9	1.5	2.2	2.9	6.8	56
Fruits							
Aegle marmelos	61.5	1.8	0.3	1.7	2.9	31.8	137
Annona squamosa	69.3	1.7	0.6	1.1	3.7	23.6.	107
Artocarpus integrifolia	76.2	1.9	0.1	0.9	1.1	19.8	88
Artocarpus lakoocha	82.1	0.7	0.1	0.8	2.0	13.3	66
Cordia dichotoma	82.5	1.8	1.0	2.2	0.3	12.2	65
Eugenia jambolana	83.7	0.7	0.3	0.4	0.9	14.0	62
Flacourtia indica	67.8	1.7	1.8	1.3	4.7	22.7	114
Mangifera indica	86.1	0.6	0.1	0.3	1.1	11.8	51
Morus alba	87.5	1.5	0.4	0.9	8.3	1.4	
Phoenix sylvestris	59.2	1.2	0.4	1.7	3.7	33.8	144
Prunus persica	80.4	0.3	0.3	0.4	5.2	13.4	58
Tamarindus indica	20.9	3.1	0.1	2.9	5.6	67.4	283
Zizyphus jujuba	81.6	0.8	0.3	0.3	-	17.0	74
Underground Parts							
Allium cepa	86.8	1.2	-	0.4	0.6	11.0	-
Allium sativum	62	0.5	0.1	1.0	0.8	29	-
Amorphophallus campanulatus	73.7	1.2	0.1	0.8	0.8	18.4	79
Colocasia esculenta	73.1	3.0	0.1	1.7	1.0	21.1	97
Dioscorea bulbifera	67.6	4.5	0.8	1.3	1.1	24.7	124
Raphanus sativus	94.4	0.07	0.01	0.6	-	3.4	-
Spices							
	Moisture	**Protein**	**Fat**	**Min-erals**	**Fibres**	**Carbohydrates**	**Total Ash**
Capsicum annuum	82.60	2.90	0.60	-	6.80	6.10	1.0
Coriandrum sativum (leaves)	87.0	3.3	0.6	-	-	6.5	-
Coriandrum sativum (seed)	6.3	1.3	-	-	19.6	31.5	-
Curcuma domestica	5.8	8.6	8.9	-	6.9	63.0	-
Elettaria cardamomum	7-10	7.0-14.0	-	-	6.7-12.8	-	3.8-6.9
Foeniculm vulgare	6.30	9.5	10	-	18.5	42.3	-
Murraya paniculata	66.3	6.1	1.0	-	6.4	16.0	-
Ocimum basilicum	6.1	1.9	3.6	-	20.5	41.2	16.7
Punica granatum (Anardana)	78.0	1.6	0.1	-	5.1	14.5	-
Tamarindus indica	18.2	28.2		-	-	-	-
Trigonella foenum-graecum	6.3	9.5	10.0	-	18.5	42.3	-
Zingiber officinale	10.85	-	-	-	7.17	-	6.64

**Source:* Bennet et al., 1991; Dwivedi, 1993; Pruthi, 1976. (-) Values not available.

Regarding plants used for ornamental purposes, a wide range of plants numbering 67 (61 genera) belonging to 36 families are being used (Table 3.31). In order of usefulness, Fabaceae (7 species) is at the top, followed by Apocynaceae (5 species), Euphorbiaceae and Moraceae (4 species each), Bignoniaceae, Myrtaceae, Oleaceae (3 species), Amaranthaceae, Asteraceae, Lamiaceae, Liliaceae, Malvaceae, Meliaceae, Nyctaginaceae, Poaceae, Rosaceae (2 species each) and rest of families like Annonaceae, Arecaceae, Convolvulaceae, Crassulaceae, Cupressaceae, Cannaceae, Caricaceae, Elaeocarpaceae, Geraniaceae, Hypericaceae, Juglandaceae, Lythraceae, Orchidaceae, Papaveraceae, Proteaceae, Lauraceae, Solanaceae and Zingiberaceae are represented by single species each. Importantly, most frequently used genera are *Jasminum* (3 species), *Eugenia* (2 species), *Bauhinia* (2 species), *Ficus* (2 species) and *Tecoma* (2 species) etc. As a matter of fact use of species like *Tagetes erecta* for henna effect and that of *Butea monosperma**, *Callistemon lanceolatus**, *Catharanthus roseus**, *Celosia argentea**, *Cestrum nocturnum**, *Cycas revoluta**, *Cymbopogon citratus**, *Elaeocarpus ganitrus**, *Euphorbia pulcherrima**, *Hibiscus rosa-sinensis**, *Hypericum cernum**, *Ipomoea cairica**, *Jacaranda acutifolia**, *Jasminum humile**, *Juglans regia**, *Malvaviscus conzatii**, *Mimosa pudica**, *Papaver rhoeus**, *Pedilanthes tithymaloides**, *Polyalthia longifolia**, *Rosa damascena**, *Saraca asoca**, *Tagetes erecta**, *Tecoma grandiflora**, *Tecoma stans** and *Thuja orientalis** for ornamental purposes are being reported for the first time from the regiόn. By looking at the nativity, the study records 33 alien (11.14 %) ornamentals with large majority belonging to families like Bignoniaceae (3 species), Apocynaceae, Fabaceae, Lamiaceae, Liliaceae, Nyctaginaceae, Rosaceae (2 species each), Amaranthaceae, Caricaceae, Convolvulaceae, Crassulaceae, Euphorbiaceae, Geraniaceae, Hypericaceae, Malvaceae, Moraceae, Myrtaceae, Oleaceae, Proteaceae, Rutaceae, Solanaceae, etc. with 1 species each. Apparently, most of the alien/exotic species due to their usefulness have been introduced intentionally over a period of time in this part of Himalayas.

Table 3.31: Plants Used for Ornamental Purposes by Local Populace of Hamirpur

Use	Plant Species
Ornamentals	*Aloe vera, Alstonia scholaris, Amaranthus paniculatus, Artocarpus integrifolia, Azadirachta indica, Bauhinia vahlii, Bauhinia variegata, Bougainvillea spectabilis, Bryophyllum calycinum, Butea monosperma*, Callistemon lanceolatus*, Canna indica, Carica papaya, Cassia fistula, Catharanthus roseus*, Celosia argentea*, Cestrum nocturnum*, Cinnamomum camphora, Cycas revoluta*, Cymbopogon citratus*, Cynodon dactylon, Elaeocarpus ganitrus*, Elettaria cardamomum, Emblica officinalis, Eriobotrya japonica, Eugenia cerasoides, Eugenia jambolana, Euphorbia pulcherrima*, Ficus benghalensis, Ficus religiosa, Grevillea robusta, Helianthus annuus, Hibiscus rosa-sinensis*, Hypericum cernum*, Ipomoea cairica*, Jacaranda acutifolia*, Jasminum humile*, Jasminum multiflorum, Jasminum sambac, Juglans regia*, Lagerstroemia indica, Malvaviscus conzatii*, Melia azedarach, Mimosa pudica*, Mirabilis jalapa, Murraya paniculata, Nerium indicum, Papaver rhoeus*, Pedilanthes tithymaloides*, Pelargonium graveolens, Phoenix sylvestris, Polyalthia longifolia*, Pongamia pinnata, Populus nigra, Putranjiva roxburghii, Rhynchostylis retusa, Rosa damascena*, Rosmarinus officinalis, Salvia coccinea, Saraca asoca*, Tabernaemontana divaricata, Tagetes erecta*, Tecoma grandiflora*, Tecoma stans*, Thevetia neriifolia, Thuja orientalis*, Yucca aloifolia*

From analysis of data presented in Tables 3.32 and 3.33, it is indicated that 52 species under 42 genera 29 families with predominant families (Table 3.32), like Apocynaceae, Anacardiaceae, Asteraceae, Bignoniaceae, Brassicaceae, Cannaceae, Cucurbitaceae, Cupressaceae, Euphorbiaceae, Fabaceae, Lamiaceae, Lauraceae, Lythraceae, Liliaceae, Malvaceae, Meliaceae, Menispermaceae, Moraceae, Musaceae, Oleaceae, Papaveraceae, Pedaliaceae, Poaceae, Punicaceae, Rosaceae, Rhamnaceae, Rutaceae, Solanaceae, Zingiberaceae and with high used genera such as *Acacia, Achyranthes, Aegle, Aerva, Asclepias, Asparagus, Azadirachta, Bambusa, Butea, Calotropis, Canna, Cannabis, Capsicum, Cassia, Catharanthus, Cestrum, Cicer, Cinnamomum, Citrus, Coriandrum, Cucurbita, Curcuma, Cynodon, Datura, Dendrocalamus, Dioscorea, Elaeocarpus, Emblica, Eugenia, Ficus gossypium, Holarrhena, Jasminum, Lagerstroemia, Madhuca, Malvaviscus, Mangifera, Mimosa, Mirabilis, Musa, Ocimum, Salvia* are employed as sacred; 34 species (*Abrus precatorius, Achyranthes aspera, Achyranthes bidentata, Adiantum lunulatum, Aerva lanata**, *Aloe vera, Asparagus adscendens, Cannabis sativa**, *Capsicum annuum**, *Cassia fistula, Citrus limon**, *Citrus medica, Coriandrum sativum, Cucurbita hispida**, *Curcuma domestica, Cynodon dactylon, Datura metel**, *Datura stramonium**, *Dioscorea bulbifera**, *Elaeocarpus ganitrus**, *Ficus benghalensis**, *Ficus glomerata**, *Ficus religiosa, Hordeum vulgare, Mimosa pudica**, *Opuntia dillenii**, *Paspalum scrobiculatum, Randia dumetorum, Sapindus mukorossi**, *Solanum tuberosum, Stephania glabra**, *Urtica dioica**, *Vitex negundo, Zizyphus jujuba*) and 29 genera of 20 families for magico-religious purposes; and 11 species under 10 genera (*Aegle marmelos**, *Bambusa arundinacea**, *Bambusa nutans, Cynodon dactylon, Dendrocalamus hamiltonii**, *Emblica officinalis, Holarrhena pubescens, Musa sapientum**, *Ocimum sanctum, Saccharum spontaneum**, *Santalum album*) for performing rituals associated with departed souls (Table 3.33) thereby indicating that the local populace is deeply religious and superstitious. Moreover, some of the practices like the use of *Asparagus adscendens** and *Asparagus racemosus** for Maternal 'milini' (welcome to maternal uncles at the time of marriage in the family); *Asparagus adscendens** in 'hawan' (purification) ceremonies; *Abrus precatorius** for making rosaries for worshipping; *Cestrum nocturnum** as incense; *Bombax ceiba** for making 'bedi' (around which both bride and bridegroom circumbulate and take holy pledge) and 'toran dwar' (main entrance at the time of marriage); *Thuja orientalis** as 'vidya patri'; *Holarrhena pubescens** in 'sharradh' (ceremony for peace to the departed soul) rituals; *Dendrocalamus hamiltonii** for bier making; *Abelmoschus esculentus** for purifying jaggery; *Acacia catechu** for tooth brush, *Acacia concinna** for wine making; *Achyranthes aspera** for suppressing hunger; *Adhatoda vasica** for ripening of bananas; *Albizia lebbeck** for making body of trucks; *Aloe barbedensis** for making 'gulkand'; *Azadirachta indica** for packing woollen clothes and for grinding; *Butea monosperma** for shuttering frames; *Celastrus paniculatus** as horse feed; *Cucurbita pepo** as utensil; *Phragmites karka** for broom making, *Pinus roxburghii** female cones as torch light, *Pistacia integerrima** for smoking, *Punica granatum** for removing bitterness, *Putranjiva roxburghii** for construction, and *Pyrus pashia** for agricultural implements have not been described earlier and unique to the science of ethnobotany. It is a fact little realized that much of what remains of our rich biological heritage has been preserved through the ages because of a wealth of conservation-oriented cultural and religious traditions (Gadgil & Vartak, 1981).

Table 3.32: Sacred Families of District Hamirpur

Family	Plant Species
Poaceae	*Cynodon dactylon, Saccharum officinarum*, Saccharum spontaneum*, Triticum aestivum*, Bambusa nutans*
Apocynaceae	*Catharanthus roseus, Holarrhena pubescens, Nerium indicum, Tabernaemontana divaricata**
Fabaceae	*Acacia nilotica, Butea monosperma, Cicer arietinum*, Saraca asoca*
Malvaceae	*Hibiscus rosa-sinensis, Gossypium arboreum, Malvaviscus conzatii*
Moraceae	*Ficus benghalensis, Ficus glomerata, Ficus religiosa, Cannabis sativa**
Euphorbiaceae	*Emblica officinalis, Putranjiva roxburghii*
Lamiaceae	*Ocimum basilicum, Ocimum sanctum, Salvia coccinea*
Liliaceae	*Aloe vera*, Asparagus adscendens**
Oleaceae	*Jasminum humile*, Jasminum multiflorum*, Jasminum sambac*
Solanaceae	*Cestrum nocturnum*, Solanum tuberosum*
Anacardiaceae	*Mangifera indica*
Asteraceae	*Tagetes erecta**
Bignoniaceae	*Tecoma grandiflora**
Brassicaceae	*Raphanus sativus**
Cannaceae	*Canna indica*
Cucurbitaceae	*Cucurbita hispida, Cucurbita pepo*
Cupressaceae	*Thuja orientalis**
Lythraceae	*Lagerstroemia indica*
Lauraceae	*Cinnamomum camphora**
Meliaceae	*Mirabilis jalapa*
Menispermaceae	*Tinospora cordifolia*
Musaceae	*Musa sapientum**
Paparveraceae	*Papaver rhoeus*
Pedaliaceae	*Sesamum indicum*
Punicaceae	*Punica granatum*
Rosaceae	*Rosa damascena*
Rhamnaceae	*Zizyphus jujuba*
Rutaceae	*Aegle marmelos*
Zingiberaceae	*Curcuma domestica**

Table 3.33: Plants Used for Sacred, Magico-Religious and Death Ritual Purposes

Use/s	Plant Species
Sacred Plants	*Acacia catechu, Acacia nilotica, Achyranthes aspera, Aegle marmelos, Aerva lanata*, Asclepias curassavica*, Asparagus adscendens*, Asparagus racemosus*, Azadirachta indica, Bambusa nutans*, Butea monosperma, Calotropis procera*, Canna indica, Cannabis sativa, Capsicum annuum*, Cassia fistula, Catharanthus roseus, Cestrum nocturnum*, Cicer arietinum, Cinnamomum camphora*, Citrus jambhiri, Citrus medica, Coriandrum sativum*, Cucurbita hispida*, Cucurbita pepo*, Curcuma domestica*, Cynodon dactylon, Datura metel, Datura stramonium, Dendrocalamus hamiltonii, Dioscorea bulbifera, Elaeocarpus ganitrus, Emblica officinalis*, Eugenia jambolana, Ficus benghalensis, Ficus glomerata, Ficus religiosa, Gossypium arboreum, Holarrhena pubescens*, Jasminum humile*, Jasminum multiflorum*, Jasminum sambac, Lagerstroemia indica*, Madhuca indica, Malvaviscus conzatii*, Mangifera indica, Mimosa pudica, Mirabilis jalapa*, Musa sapientum*,* ***Ocimum basilicum, Ocimum sanctum, Salvia coccínea***
Magico-religious belief	*Abrus precatorius, Achyranthes aspera, Achyranthes bidentata, Adiantum lunulatum, Aerva lanata*, Aloe vera, Asparagus adscendens, Cannabis sativa*, Capsicum annuum*, Cassia fistula, Citrus limon*, Citrus medica, Coriandrum sativum, Cucurbita hispida*, Curcuma domestica, Cynodon dactylon, Datura metel*, Datura stramonium*, Dioscorea bulbifera*, Elaeocarpus ganitrus*, Ficus benghalensis*, Ficus glomerata*, Ficus religiosa, Hordeum vulgare, Mimosa pudica*, Opuntia dillenii*, Paspalum scrobiculatum, Randia dumetorum, Sapindus mukorossi*, Solanum tuberosum, Stephania glabra*, Urtica dioica*, Vitex negundo, Zizyphus jujuba*
Death rituals	*Aegle marmelos*, Bambusa arundinacea*, Bambusa nutans, Cynodon dactylon, Dendrocalamus hamiltonii*, Emblica officinalis, Holarrhena pubescens, Musa sapientum*, Ocimum sanctum, Saccharum spontaneum*, Santalum album*
Incense	*Cestrum nocturnum*

The findings indicated that rural inhabitants of the region are conscious for the health of their livestock as evident from the usage of as many as 123 plant species belonging to 105 genera under 49 families (Table 3.34). The predominant ailments prevailing in the region are: stomach disorders/dysentery/bloating/worms/constipation (28 species: *Aloe barbedensis, Aloe vera, Anetham sowa, Bauhinia vahlii, Berberis asiatica, Bombax ceiba, Brassica juncea, Carum copticum, Cassia fistula, Chenopodium ambrosioides, Cissampelos pareira, Citrus grandis, Cordia dichotoma, Dalbergia sissoo, Eupatorium adenophorum, Grewia oppositifolia, Lens culinaris, Mallotus philippinensis, Oroxylum indicum, Pogostemon benghalensis, Prunus persica, Psidium guajava, Rosa damascena, Saccharum officinarum, Terminalia bellirica, Trigonella foenum-graecum, Verbascum Thapsus, Zea mays*), increasing lactation (16 species: *Arachis hypogaea, Avena sativa, Bauhinia variegata, Brassica nigra, Carduus nutans, Eruca sativa, Ficus religiosa, Glycine max, Gossypium arboreum, Grewia oppositifolia, Hordeum vulgare, Leucaena leucocephala, Pennisetum americanum, Sesamum indicum, Tinospora cordifolia, Trifolium alexandrinum*), dhanku/cough/throat problems (9 species: *Adhatoda vasica, Adiantum capillus-veneris, Emblica officinalis, Euphorbia pulcherrima, Glycyrrhiza glabra, Ocimum basilicum, Ocimum sanctum, Terminalia bellirica, Terminalia chebula*), expulsión of placenta (8 species: *Achyranthes aspera, Achyranthes bidentata, Argemone mexicana, Bambusa arundinacea, Chenopodium album, Dendrocalamus hamiltonii, Mangifera indica Setaria viridis*), mastitis (8 species: *Chenopodium ambrosioides, Eruca sativa, Foeniculum vulgare, Lepidium sativum, Linum usitatissimum, Plumbago zeylanica, Vernonia anthelmintica, Vitex negundo*), wound healing (6 species: *Abrus precatorius, Agave americana, Brassica campestris, Curcuma domestica, Prunus persica, Woodfordia floribunda)*, killing exoparasites (6 species: *Annona squamosa, Citrus limon, Eucalyptus umbellata, Mentha spicata, Millettia extensa, Wendlandia heynei)*, checking abortion (5 species: *Brassica campestris, Cicer arietinum, Gossypium arboreum, Musa sapientum, Triticum aestivum*), prolapsed uterus (5 species: *Arisaema tortuosum, Centipeda minima, Colocasia esculenta, Holarrhena pubescens, Zingiber officinale)*, fractured bones (5 species: *Agave americana, Dodonaea viscosa, Rhynchostylis retusa, Soymida febrifuga, Vanda roxburghii)*, tympany (4 species: *Cardiospermum halicacabum, Cymbopogon citratus, Murraya paniculata, Zingiber officinale*), bites (3 species: *Cannabis sativa, Capsicum annuum, Sapindus mukorossi)*, magico-religious treatment (2 species: *Datura metel, Datura stramonium)*, fever (2 species: *Jasminum humile, Jasminum multiflorum)*, toe infection (2 species: *Randia dumetorum, Euphorbia royleana)*, and 'abhjana', diabetes, glossitis, foot sores with one species each thereby indicating less prevalence of these ailments in the region (Table 3.34). High used family for this purpose is Fabaceae (14 species), followed by Poaceae (13 species), Brassicaceae (6 species), Rutaceae (5 species), Lamiaceae (5 species), Asteraceae (5 species), Euphorbiaceae (4 species), Moraceae (4 species), Rosaceae (4 species), Solanaceae (4 species), Combretaceae (3 species), Menispermaceae (3 species), Araceae (3 species), Liliaceae (3 species), Sapindaceae (3 species), Orchidaceae (2 species), Linaceae (2 species), Rubiaceae, Zingiberaceae (2 species), Oleaceae (2 species), Amaranthaceae (2 species), Chenopodiaceae (2 species), Cucurbitaceae (2 species) and the remaining 26 families are represented by one species each. Interestingly, the projected analysis does not corroborate with the earlier findings of Jain & Srivastava (1999) who recorded the predominant use of Fabaceae, Asteraceae, Euphorbiaceae, Poaceae, Rubiaceae, Lamiaceae, Araceae, Cucurbitaceae, Solanaceae, Ranunculaceae, Verbenaceae, Apiaceae, Apocynaceae, Moraceae and Malvaceae etc. in their book entitled *'Dictionary of Ethnoveterinary Plants of India'*. In order of preference, the commonly used plant parts are: leaves (41%), flowers (6%), fruits (10%), seed/oil (21%), roots (6%), rhizomes/corms (5%), whole plant (5%), bark (3%), milky latex (2%), spine (1%) etc. (Fig. 3.30). Furthermore, a botanical analysis of ethnoveterinary recipes in India by Jain (2003) pointed out that 10 predominant genera of veterinary medicine in order of number of species are *Euphorbia, Ficus, Acacia, Cassia, Polygonum, Solanum, Arisaema, Crotolaria, Curcuma* and *Prunus*. Contrarily, these for the present study are viz., *Brassica* (3 species), *Terminalia* (3 species), *Achyranthes* (2 species), *Aloe* (2 species), *Bambusa* (2 species), *Bauhinia* (2 species), *Chenopodium* (2 species), *Citrus* (2 species), *Datura* (2 species), *Euphorbia* (2 species), *Ficus* (2 species), *Jasminum* (2 species), *Murraya* (2 species), *Ocimum* (2 species), *Pennisetum* (2 species), etc. Another unique feature of the present study is that usages of as many as 70 plants marked with asterisk in Table 3.34 are new to the science of ethnoveterinary medicine (Jain, 2003; Jain & Srivastava, 1999) and hold a great potential for drug discovery programme through clinical validation not only for ethnoveterinay purposes but for humans also.

Table 3.34: Ethnoveterinary Important Plants of District Hamirpur

S.No.	Botanical Name	Ailment	Part Used	Local Name	Family
1.	*Abrus precatorius* L.	Wound healing	Leaves	Ratti, Raktgunja	Fabaceae
2.	*Achyranthes aspera* L.	Expulsion of placenta	Root	Puthkanda	Amaranthaceae
3.	*Achyranthes bidentata* Blume	Expulsion of placenta	Root	Puthkanda	Amaranthaceae
4.	*Acorus calamus* L.	Joint pains	Rhizome	Baryan	Araceae
5.	*Adhatoda vasica* Nees	Loss of appetite, cough	Leaves	Basuti	Acanthaceae
6.	*Adiantum capillus-veneris* L.	Cough & Throat trouble	Leaves, Rhizome	Hansraj	Adiantaceae
7.	*Agave americana* L.	Wounds, Fractured bones	Mucilage, Leaves	Ramban	Agavaceae
8.	*Aloe barbedensis* Mill.	Dysentery	Leaves	Ghi-kawer	Liliaceae
9.	*Aloe vera* Webb. & Benth.	Stomach disorders	Leaves	Ghi-kawer	Liliaceae
10.	*Anetham sowa** Roxb. *ex* Flem.	Bay (Bloat)	Seeds	Kaudi saunf	Apiaceae
11.	*Annona squamosa* L.	Killing bugs	Leaves	Sitaphal	Annonaceae
12.	*Arachis hypogaea* L.	Lactation	Fruit, Oil cake	Moongphali	Fabaceae
13.	*Argemone mexicana* *L.	Placental evagination	Seed	Lea	Papaveraceae
14.	*Arisaema tortuosum** (Wall.) Schott	Prolapsed uterus	Rhizome	Kirre-Ri-Chhhali	Araceae
15.	*Artemisia vulgaris** L.	Burn injuries	Leaves	Charmad	Asteraceae
16.	*Artocarpus lakoocha** Roxb.	Increaseing Milk Content	Leaves	Dheun	Moraceae
17.	*Asparagus adscendens* Roxb.	Emetic, Coolant	Root	Sanspai	Liliaceae
18.	*Avena sativa** L.	Lactation	Seed	Joi, Javi	Poaceae
19.	*Bambusa arundinacea* (Retz.) Roxb.	Expulsion of placenta	Leaves	Bainjh	Poaceae
20.	*Bambusa nutans** Wall.	Abhjana (Anoestrus)	Leaves	Bans	Poaceae
21.	*Bauhinia vahlii** Wight & Arn.	Stomach disorders	Leaves	Tor	Fabaceae
22.	*Bauhinia variegata** L.	Lactation	Leaves	Kachnar, Karale	Fabaceae
23.	*Berberis asiatica** Roxb. *ex* DC.	Eye Wash, Dysentery	Leaves	Kasmal	Berberidaceae
24.	*Boerhavia diffusa** L.	Cudding	Seed, Plant	Itsat, Utsata	Nyctaginaceae
25.	*Bombax ceiba* L.	Stomach disorders	Leaves, Flower	Simbal, Simal	Boraginaceae

contd...

Table 3.34: *contd...*

S.No.	Botanical Name	Ailment	Part Used	Local Name	Family
26.	*Brassica campestris** L.	Checking abortions, Wounds	Leaves, Seed, oil cake	Sarson	Brassicaceae
27.	*Brassica juncea** (L.) Czern. & Coss.	Intestinal Worms	Oil cake	Rai	Brassicaceae
28.	*Brassica nigra* (L.) Koch	Lactation	Seed	Rai, Ohari	Brassicaceae
29.	*Cannabis sativa** L.	Skin eruptions, Insect bites	Leaves	Bhang	Moraceae
30.	*Capsicum annuum* L.	Rabied dog's Bite	Fruit	Mirch	Solanaceae
31.	*Cardiospermum halicacabum** L.	'Patrara', (acute tympany)	Seed, Fruit	Tikku-malu	Sapindaceae
32.	*Carduus nutans**L.	Lactation	Whole Plant	Lea	Asteraceae
33.	*Carum copticum** Hiern.	Abdominal pain, Bloating	Seed	Ajwain	Apiaceae
34.	*Cassia fistula** L.	Constipation, bloating	Fruit	Allia, Aahali, Kanyar	Fabaceae
35.	*Celastrus paniculatus** Willd.	Strength	Leaves, Seed	Sankhiryan, Sankheere	Celastraceae
36.	*Centipeda minima** (L.) A. Br. & Asch.	Dandu (Prolapsed Uterus)	Leaves	Nakchikdu	Asteraceae
37.	*Chenopodium album* L.	Expulsion of placenta	Stem, Leaves	Kunnah, Bathu	Chenopodiaceae
38.	*Chenopodium ambrosioides** L.	Constipation, Basaihar (mastitis)	Seed,	Kah-ajwain	Chenopodiaceae
39.	*Cicer arietinum** L.	Weakness, Frequent abortions	Seed, Leaves	Chollae	Fabaceae
40.	*Cissampelos pareira* L.	Dysentery	Whole Plant	Patindu, Bhatindu	Menispermaceae
41.	*Citrus grandis** (L.) Osbeck.	Removing indigestible matter	Fruit	Chakotra-Khatta	Rutaceae
42.	*Citrus limon* (L.) Burm. f.	Expelling ectoparasites	Fruit	Khatta, Galgal	Rutaceae
43.	*Colebrookia oppositifolia** Sm.	Burn injuries, Appetizer	Leaves, Stem	Dushane, Dus	Lamiaceae
44.	*Colocasia esculenta* *(L.) Schott	Dandu (prolapsed uterus)	Leaves	Kachalu, Geju	Araceae
45.	*Cordia dichotoma* Forst. f.	Expelling worms	Leaves, Fruit	Lasura	Cordiaceae
46.	*Cucumis sativus** L.	Burn injuries	Fruit	Kheera, Kakri	Cucurbitaceae
47.	*Cucurbita hispida** Thunb.	Pregnancy Loss	Fruit	Petha, Dhuda Petha	Cucurbitaceae
48.	*Curcuma domestica* (Medik.) Valeton	Wound healing	Rhizome	Haldi, Haldan	Zingiberaceae
49.	*Cymbopogon citratus** (DC.) Stapf	Patrara (acute Tympany)	Stem, Leaves	Lemongrass	Poaceae
50.	*Cynodon dactylon** (L.) Pers.	Burn injuries	Leaves	Doob, Dhruv, Doorva	Poaceae
51.	*Dalbergia sissoo** Roxb. *ex* DC.	Burn injuries, dysentery, abortions	Leaves	Tahali, Shisham	Fabaceae
52.	*Datura metel* auct. non L.	Magico-religious	Leaves, Flower	Datura	Solanaceae

contd...

Table 3.34: *contd...*

S.No.	Botanical Name	Ailment	Part Used	Local Name	Family
53.	*Datura stramonium* L.	Magico-religious treatment	Seed, Fruit	Datura, Dhatura	Solanaceae
54.	*Dendrocalamus hamiltonii** Nees & Arn. *ex* Munro	Expulsion of placenta	Leaves, Stem	Baingh, Maggar	Poaceae
55.	*Dodonaea viscosa* L. Jacq.	Fractured bones	Leaves	Mehandru	Sapindaceae
56.	*Emblica officinalis* Gaertn.	Dhaknu (Cough & cold)	Fruit	Ambla, Amla	Euphorbiaceae
57.	*Equisetum debile** Roxb. *ex* Vauch.	Evagination of uterus	Stem, Branches	Ghornali	Equisetaceae
58.	*Eruca sativa** Lam.	Enhancing lactation, Basaihar (mastitis	Seed	Taramira	Brassicaceae
59.	*Eucalyptus umbellata** Domin	Skin ectoparasites	Leaves	Safeda	Myrtaceae
60.	*Eupatorium adenophorum** Spreng.	Constipation	Leaves	Kali Basuti	Asteraceae
61.	*Euphorbia pulcherrima ** Willd.	Throat problems	Latex	Lalpatta	Euphorbiaceae
62.	*Euphorbia royleana** Boiss.	Stimulating ovulation, Tail infection	Latex	Chu, Chui	Euphorbiaceae
63.	*Ficus palmata ** Forsk.	Padjibhi (glossitis)	Leaves, Latex	Phalgu, Anjhir	Moraceae
64.	*Ficus religiosa** L.	Lactation	Leaves	Peepal	Moraceae
65.	*Foeniculum vulgare** Mill.	Mastitis	Seed	Saunf, Soay	Apiaceae
66.	*Fumaria indica** (Haussk.) Pugsley.	Pit (Jaundice)	Whole Plant	Pitpapara, Papara, Pitpapada	Fumariaceae
67.	*Glycine max* Merr.	Increasing Lactation	Plant, Seed	Soyabean	Fabaceae
68.	*Glycyrrhiza glabra* L.	Throat problems	Leaves, Roots	Malathi	Fabaceae
69.	*Gossypium arboreum** L.	Check abortions, Enhancing lactation	Oil cakes, Seed	Kapah, Kapas	Malvaceae
70.	*Grewia oppositifolia** Roxb.	Cud , Enhancing lactation	Leaves	Beul, Bhuel	Tiliaceae
71.	*Holarrhena pubescens ** Wall. *ex* G. Don	Prolapsed uterus	Bark	Kurchi, Keur, Inderjaun	Apocynaceae
72.	*Hordeum vulgare** L.	Lactation Enhancer	Stem, Seed	Jau, Satu	Poaceae
73.	*Jasminum humile* L.	Febrifuge	Twig, Root	Chameli	Oleaceae
74.	*Jasminum multiflorum* Roth.	Febrifuge	Twig, Root	Malti	Oleaceae
75.	*Lens culinaris ** Medic.	Bloating	Seed	Masar	Fabaceae
76.	*Lepidium sativum** L.	Basaihar (mastitis)	Seed	Hallon	Brassicaceae
77.	*Leucaena leucocephala* (Lam.) de Wit.	Enhancing, Lactation	Leaves	Alseenia	Fabaceae
78.	*Linum usitatissimum** L.	Basaihar (mastitis)	Seed	Alsi	Linaceae

contd...

Table 3.34: *contd...*

S.No.	Botanical Name	Ailment	Part Used	Local Name	Family
79.	*Mallotus philippinensis** (Lam.) Muell.-Arg.	Stomach pain	Leaves, Fruit, Seed	Kambhal, Kambal	Euphorbiaceae
80.	*Mangifera indica** L.	Placenta (jer girana)	Leaves	Amb, Aam	Anacardiaceae
81.	*Mentha spicata* L.	Exoparasite	Leaves, Stem	Pudina	Lamiaceae
82.	*Millettia extensa* Benth. *ex* Baker	Ectoparasite	Root	Salangaya	Fabaceae
83.	*Murraya koenigii* (L.) Spreng.	Sun stroke	Leaves	Gandhla, Curry patta	Rutaceae
84.	*Murraya paniculata* (L.) Jack.	Patrara (acute tympany)	Leaves	Gandhla	Rutaceae
85.	*Musa sapientum* L.	Pregnancy loss	Inflorescence	Kela	Musaceae
86.	*Nasturtium officinale** R. Br.	Diabetes	Stem, Leaves	Chuch	Brassicaceae
87.	*Ocimum basilicum* L.	Cough	Leaves	Bhabhri	Lamiaceae
88.	*Ocimum sanctum* L.	Cough	Leaves	Ram-Tulsi, Baramasi	Lamiaceae
89.	*Oroxylum indicum* Vent.	Constipation	Bark	Tat-palanga	Bignoniaceae
90.	*Pennisetum americanum* L.	Enhancing lactation	Whole Plant	Bajra, Barley	Poaceae
91.	*Pennisetum purpureum* Schum.	Milk consistency	Whole Plant	Chari	Poaceae
92.	*Plumbago zeylanica** L.	Basaihar (mastitis)	Root	Chitrak, Chitra	Plumbaginaceae
93.	*Pogostemon benghalensis* Burm. f.	Cudding	Leaves, Whole Plant	Kali Basuti, Dusyan	Lamiaceae
94.	*Prunus persica* Benth. & Hook. f.	Wounds, Intestinal worms	Leaves	Aru, Adu	Rosaceae
95.	*Psidium guajava** L.	Dysentery	Leaves	Amrood	Rosaceae
96.	*Pyrus pashia ** Ham.	Foot sore	Leaves, Fruit	Kainth, Kaenth	Rosaceae
97.	*Randia dumetorum** (Retz). Lam.	Toe infection, 'gahli' (Tail gangrene)	Trunk, Spines	Rada	Rubiaceae
98.	*Reinwardtia indica* Dum.	Boils	Leaves, Flower	Basant, Piyan	Linaceae
99.	*Rhynchostylis retusa* Blume	Fractured bones	Leaves, Flower	Bhangru	Orchidaceae
100.	*Rosa damascena** Mill.	Constipation	Flower	Gulab	Rosaceae
101.	*Saccharum officinarum** L.	Abdominal pain, Constipation	Stem, Leaves	Ganna	Poaceae
102.	*Sapindus mukorossi** Gaertn.	Dog bite	Leaves, Fruit	Reetha, Doda, Doddae	Sapindaceae

contd...

Table 3.34: *contd...*

S.No.	Botanical Name	Ailment	Part Used	Local Name	Family
103.	*Sesamum indicum* DC.	Enhancing milk production	Oil, Oil cake	Til	Pedaliaceae
104.	*Setaria viridis** Beauv.	Plancental evagination	Leaves	Hathi grass	Poaceae
105.	*Solanum tuberosum* L.	Burn injuries	Tuber	Aloo	Solanaceae
106.	*Soymida febrifuga* A. Juss.	Bone fractures, injuries	Stem Bark	Ruhan, Rehana	Meliaceae
107.	*Stephania glabra* *(Roxb.) Miers	Flatulence, Mouth ulceration, Poisonous teats	Fruit, Corm	Bis-Khappar	Menispermaceae
108.	*Terminalia arjuna* * (Roxb.) Wt. & Arn.	Burn injuries	Bark	Arjun Tree	Combretaceae
109.	*Terminalia bellirica** (Gaertn.) Roxb.	Cough*, Dysentery, Constipation	Seed husk, Fruit	Behra, Bahera	Combretaceae
110.	*Terminalia chebula* Retz.	Cough	Fruit	Harar	Combretaceae
111.	*Tinospora cordifolia** (Willd.) Miers.	Enhancing lactation	Stem, Leaves	Gulaja, Giloy, Galon	Menispermaceae
112.	*Trifolium alexandrinum** L.	Lactation	Whole Plant	Barseem	Fabaceae
113.	*Trigonella foenum-graecum** L.	Bloating	Seeds	Methe, Methi	Fabaceae
114.	*Triticum aestivum** L.	'Taruna' (abortion)	Grains/ Seed	Gehun	Poaceae
115.	*Vanda roxburghii** R. Br.	Fractured bones	Leaves, Flower	Bhangru	Orchidaceae
116.	*Verbascum thapsus** L.	Constipation, 'Ban' (Obstruction)	Aerial plant parts	Ban Tambakhoo	Scrophulariaceae
117.	*Vernonia anthelmintica** Willd.	'Basaihar' (Mastitis)	Seed, Leaves	Brahmjiri	Asteraceae
118.	*Vitex negundo* L.	Appetiser, Mastitis	Leaves, Twig	Banna	Verbenaceae
119.	*Wendlandia heynei* (Roem. & Schult.) Sant. & Merch.	Expelling leeches	Leaves	Panseera	Rubiaceae
120.	*Woodfordia floribunda* Salisb.	Wound healing	Leaves, Flower	Dhai, Dhavae, Dhaun	Lythraceae
121.	*Zanthoxylum armatum* * DC.	Indigestion	Seeds, Fruits	Tirmer	Rutaceae
122.	*Zea mays* L.	Constipation	Seed	Makki, Bhuta	Poaceae
123.	*Zingiber officinale* * Rosc.	Polapsed uterus	Rhizome	Adrak	Zingiberaceae

contd...

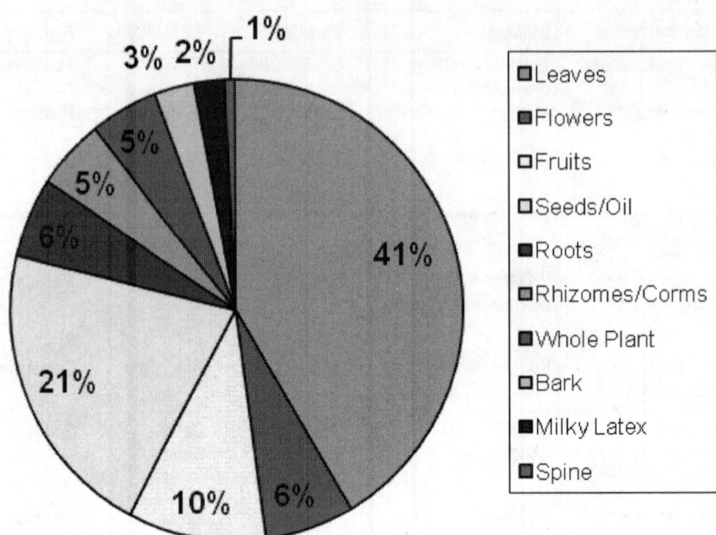

Fig. 3.30. Relative Percentage of Plant Parts Used for Treating Veterinary Ailments in District Hamirpur

Against various types of bites (scorpion, wasp, insect) the usage and acceptability of 22 plants under 21 genera like *Achyranthes, Aerva, Ageratum, Boerhavia, Cannabis, Capsicum, Catharanthus, Cedrela, Eugenia, Litchi, Martynia, Mesua, Papaver, Pedilanthus, Rauwolfia, Roylea, Santalum, Sapindus, Tagetes, Verbascum, Vitex* are quite high among the rural folks of the study area (Table 3.35). Their tropical application is believed to help in neutralizing some of the inflammatory agents present in toxins and checks pain and swelling. Of these, the use of *Ageratum conyzoides**, *Boerhavia diffusa**, *Cannabis sativa**, *Litchi chinensis**, *Mesua ferrea**, *Papaver rhoeus**, *Pedilanthus tithymaloides**, *Rauwolfia serpentina**, *Roylea cinerea**, *Tagetes erecta**, *Vitex negundo** against bites form the first records of its type for the rich medicinal herbal flora of India. Undeniably, clinical and experimental status of these claims are necessary for better utilization of the knowledge in future.

Table 3.35: Plants Used Against Bites

Use/s	Plant Species
Antidote for bites (Scorpion, Wasp Stings, Insect bite)	*Achyranthes aspera, Achyranthes bidentata, Aerva lanata, Ageratum conyzoides*, Boerhavia diffusa*, Cannabis sativa*, Capsicum annuum, Catharanthus roseus, Cedrela toona, Eugenia jambolana, Litchi chinensis*, Martynia annua, Mesua ferrea*, Papaver rhoeus*, Pedilanthus tithymaloides*, Rauwolfia serpentina*, Roylea cinerea*, Santalum album, Sapindus mukorossi, Tagetes erecta*, Verbascum thapsus, Vitex negundo**

For their fodder requirements, altogether 108 species under 92 genera and 46 families are held in high esteem by the local populace in this part of N.W. Himalayas (Table 3.36). Species richness in forages predominantly belong to Poaceae (17 species: *Avena sativa, Bambusa arundinacea, Bambusa nutans, Cynodon dactylon, Dendrocalamus hamiltonii, Hordeum vulgare, Oryza sativa, Paspalum scrobiculatum, Pennisetum americanum, Pennisetum purpureum, Phragmites karka, Saccharum arundinaceum, Saccharum officinarum, Saccharum spontaneum, Setaria viridis, Triticum aestivum, Zea mays*), Fabaceae (15 species: *Albizia lebbeck, Albizia stipulata, Arachis hypogaea, Bauhinia variegata, Bauhinia vahlii, Butea monosperma, Cassia fistula, Cicer arietinum, Dalbergia sissoo, Glycine max, Glycyrrhiza glabra, Lens culinaris, Millettia extensa, Trifolium alexandrinum, Vicia pallida*), Moraceae (6 species: *Ficus carica, Ficus glomerata, Ficus hispida, Ficus palmata, Ficus roxburghii, Morus alba*), Asteraceae (5 species: *Ageratum conyzoides, Carduus nutans, Eclipta prostrata, Taraxacum officinale, Vernonia anthelmintica*), Brassicaceae (4 species: *Brassica campestris, Brassica juncea, Brassica niger, Lepidium sativum*), Rosaceae (4 species: *Eriobotrya japonica, Prunus domestica, Pyrus pashia, Rubus elipticus*), Chenopodiaceae (3 species: *Chenopodium álbum, Chenopodium ambrosioides, Spinacia oleracea*), Euphorbiaceae (3 species: *Emblica officinalis, Mallotus philippinensis, Phyllanthus urinaria*), Lamiaceae (3 species: *Colebrookia oppositifolia, Pogostemon benghalensis, Salvia coccinea*), Amaranthaceae (2 species: *Amaranthus viridis, Amaranthus spinosus*), Anacardiaceae (2 species: *Lannea coromandelica, Pistacia integérrima*), Apocynaceae (2 species: *Carissa carandus, Carissa spinarum*), Meliaceae (2 species: *Cedrela toona, Melia azedarach*), Menispermaceae (2 species: *Cissampelos pareira, Tinospora cordifolia*), Myrtaceae (2 species: *Eugenia cerasoides, Eugenia jambolana*), Nyctaginaceae (2 species: *Boerhavia diffusa, Bougainvillea spectabilis*), Sapindaceae (2 species: *Cardiospermum halicacabum, Dodonaea viscosa*) and rest of the families such as Acanthaceae, Agavaceae, Apiaceae, Araceae, Bignoniaceae, Boraginaceae, Cannaceae, Caricaceae, Celastraceae, Convolvulaceae, Flacourtiaceae, Fumariaceae, Liliaceae, Linaceae, Lythraceae, Malvaceae, Orchidaceae, Oxalidaceae, Papaveraceae, Polygonaceae, Ranunculaceae, Rhamnaceae, Salicaceae, Sapotaceae, Sterculiaceae, Teliaceae, Urticaceae and Verbenaceae are poorly represented by one species each (Table 3.36). Notably high used genera in the region for forages are : *Ficus* (5 species), *Brassica, Carissa, Saccharum* (3 species each), *Albizia, Amaranthus, Bambusa, Chenopodium, Eugenia, Pennisetum* (2 species each), etc. As a matter of fact, species marked with asterisk in Table 3.36 are new records for Indian forages (Anonymous, 1994; Sood & Kaushal, 2008) and hold considerable scope for popularization.

With regard to data on nutritional analysis of three fodder plants presently investigated, leaves of *Mallotus philippinensis, Zizyphus jujuba,* and *Grewia oppositifolia* justify being given as forages in the region as these are rich in carbohydrates, proteins, fats, calcium and magnesium (Table 3.37, 3.38; Figs. 3.31, 3.32, 3.34) and considered essential for improving quality of milk from lactating animals. Besides, leaves of *Mallotus philippinensis* also possess 549.48 ppm iron, 20.93 mg/100g vit. A, 12.80 mg/100g vit. C (Table 3.37; Figs. 3.31, 3.33), and that of *Zizyphus jujuba* and *Grewia oppositifolia* in possessing starch, sodium, potassium and magnesium (Table 3.38). Unlike the earlier nutritional analytic work on forages (Table 3.39), the present study is unique as details of vitamins and minerals have also been worked out for the first time.

Table 3.36: Fodder Resources of District Hamirpur

Botanical Name	Local Name/s	Family
Adhatoda vasica	Basuti	Acanthaceae
Agave americana	Ramban	Agavaceae
*Ageratum conyzoides**	Neela Fulunu	Asteraceae
Albizia lebbeck	Sirin	Fabaceae
Albizia stipulata	Oyee	Fabaceae
Amaranthus viridis	Cholaee	Amaranthaceae
*Amaranthus spinosus**	Cholaee	Amaranthaceae
*Arachis hypogaea**	Moogphali	Fabaceae
Argemone maxicana	Lea	Papaveraceae
*Arisaema tortuosum**	Kirre-Ri-Chhhali	Araceae
Asparagus racemosus	Satavari	Liliaceae
Avena sativa	Joi, Javi	Poaceae
*Bambusa arundinacea**	Bainjh	Poaceae
*Bambusa nutans**	Bans	Poaceae
Bauhinia variegata	Kachnar, Karale	Fabaceae
*Bauhinia vahlii**	Tor	Fabaceae
Boerhavia diffusa	Itsat, Utsata	Nyctaginaceae
Bombax ceiba	Simbal	Boraginaceae
*Bougainvillea spectabilis**	Garden glory	Nyctaginaceae
*Brassica campestris**	Sarson	Brassicaceae
*Brassica juncea**	Rai	Brassicaceae
Brassica nigra	Rai, Ohari	Brassicaceae
Butea monosperma	Palah, Palash, Plah	Fabaceae
*Canna indica**	Sudarshan, Hardarshan	Cannaceae
Cardiospermum halicacabum	Tikku-Malu	Sapindaceae
Carduus nutans	Lea	Asteraceae
*Carica papaya**	Papita	Caricaceae
*Carissa carandus**	Karonda, Bada Garna	Apocynaceae
*Carissa spinarum**	Chhota Garna	Apocynaceae
Cassia fistula	Allia, Aahali, Kanyar	Fabaceae
Cedrela toona	Tooni	Meliaceae
*Celastrus paniculatus**	Sankhiryan, Sankheere	Celastraceae
Chenopodium album	Kunnah, Bathu	Chenopodiaceae
Chenopodium ambrosioides	Kah-ajwain	Chenopodiaceae
Cicer arietinum	Cholle	Fabaceae

contd...

Table 3.36: *contd...*

Botanical Name	Local Name/s	Family
Cissampelos pareira	Patindu, Bhatindu	Menispermaceae
Colebrookia oppositifolia	Dushane, Dus	Lamiaceae
Cynodon dactylon	Doob, Dhruv, Doorva	Poaceae
Dalbergia sissoo	Tahali, Shisham	Fabaceae
*Daucus carota**	Gajjar	Apiaceae
*Dendrocalamus hamiltonii**	Baingh, Maggar	Poaceae
Dodonaea viscosa	Mahendru	Sapindaceae
Eclipta prostrata	Bhringraj	Asteraceae
Emblica officinalis	Aamla	Euphorbiaceae
*Eriobotrya japonica**	Lukat, Laukhat	Rosaceae
*Eruca sativa**	Taramira	Brassicaceae
Eugenia cerasoides	Kathman	Myrtaceae
Eugenia jambolana	Jamun	Myrtaceae
Ficus carica	Anjir	Moraceae
Ficus glomerata	Umreyan, Umara	Moraceae
Ficus hispida	Devryan	Moraceae
Ficus palmata	Phalgu, Anjhir.	Moraceae
Ficus roxburghii	Triamble, Trayambalu.	Moraceae
Flacourtia ramontchi	Kangu	Flacourtiaceae
*Fumaria indica**	Pitpapara	Fumariaceae
*Glycine max**	Soybean	Fabaceae
*Glycyrrhiza glabra**	Malathi	Fabaceae
Gossypium arboreum	Kapas	Malvaceae
Grewia oppositifolia	Beul	Tiliaceae
*Habenaria pectinata**	Orchid	Orchidaceae
*Helicteres isora**	Marorphali	Sterculiaceae
*Hordeum vulgare**	Jau, Satu	Poaceae
Ipomoea carnea	Badi Basuti	Convolvulaceae
*Lannea coromandelica**	Kehmble	Anacardiaceae
*Lens culinaris**	Masar	Fabaceae
*Lepidium sativum**	Hallon	Brassicaceae
Leucaena leucocephala	Alseenia	Fabaceae
*Linum usitatissimum**	Alsi	Linaceae
*Madhuca indica**	Mahua	Sapotaceae
Mallotus philippinensis	Kambhal, Kambal	Euphorbiaceae
Melia azedarach	Draek	Meliaceae

contd...

Table 3.36: *contd...*

Botanical Name	Local Name/s	Family
*Millettia extensa**	Salangaya	Fabaceae
Morus alba	Toot	Moraceae
Oroxylum indicum	Tatpalanga	Bignoniaceae
Oryza sativa	Dhan, Jari	Poaceae
Oxalis corniculata	Maroli	Oxalidaceae
*Paspalum scrobiculatum**	Kodri grass	Poaceae
*Pennisetum americanum**	Bajra, Barley	Poaceae
Pennisetum purpureum	Chari	Poaceae
*Phragmites karka**	Elephant grass	Poaceae
Phyllanthus urinaria	Bhumi- Amla	Euphorbiaceae
*Pistacia integerrima**	Kakarsinghi	Anacardiaceae
*Pogostemon benghalensis**	Kali Basuti, Dusyan	Lamiaceae
Populus nigra	Poplar	Salicaceae
Premna latifolia	Bhankar	Verbenaceae
*Prunus domestica**	Plum, Aloobukhara	Rosaceae
Pyrus pashia	Kainth	Rosaceae
*Ranunculus muricatus**	Buttercup	Ranunculaceae
Raphanus sativus	Mooli	Brassicaceae
Rubus ellipticus	Akhe	Rosaceae
Rumex hastatus	Maroli	Polygonaceae
Saccharum arundinaceum	Sarphar	Poaceae
Saccharum officinarum	Ganna	Poaceae
Saccharum spontaneum	Kans	Poaceae
*Salvia coccinea**	Phultulsi	Lamiaceae
Setaria viridis	Hathi grass	Poaceae
Spinacia oleracea	Palak	Chenopodiaceae
*Taraxacum officinale**	Dudhli, Dulal	Asteraceae
Tinospora cordifolia	Gulaja, Giloy, Galon	Menispermaceae
Trifolium alexandrinum	Barseem	Fabaceae
Triticum aestivum	Gehun, Gandam	Poaceae
Urtica dioica	Bichhu-buti	Urticaceae
Vernonia anthelmintica	Brahmjiri	Asteraceae
Vicia pallida	Badakaer, Roothi	Fabaceae
Wendlandia heynei	Panseera	Rubiaceae
Woodfordia floribunda	Dhai, Dhavae, Dhaun	Lythraceae
Zea mays	Makki	Poaceae
*Zizyphus jujuba**	Ber	Rhamnaceae

Table 3.37: Nutritional Analysis of Fodder Leaves of *Mallotus philippinensis* (On Dry Weight Basis) (Present Study)

Name of Plant	Carbo-hydrates (%)	Pro-teins (%)	Fats (%)	Ash Content	Phos-phorus	Cal-cium (%)	Magn-esium (%)	Vita-min E) Mg/ 100g	Vita-min (A) Mg/ 100g	Iron ppm	Vita-min (C) Mg/ 100g
Mallotus Philippi-nensis	45.03	15.67	3.05	9.22	0.19	1.53	0.28	12.80	20.93	549.48	0.41

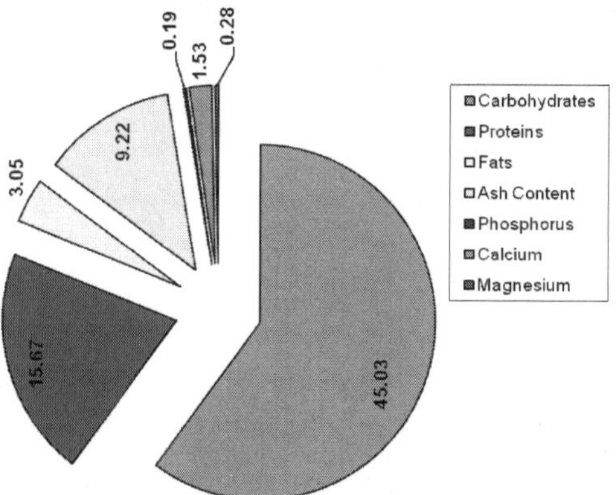

Fig. 3.31. Nutritional Analysis (%) of Fodder Plant *Mallotus philippinensis* (on dry weight basis).

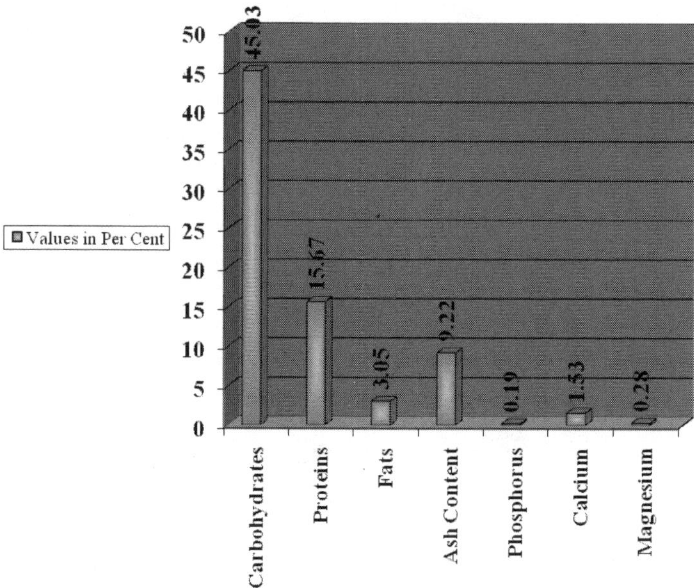

Fig. 3.32. Nutritional Components of Leaves of *Mallotus philippinensis*.

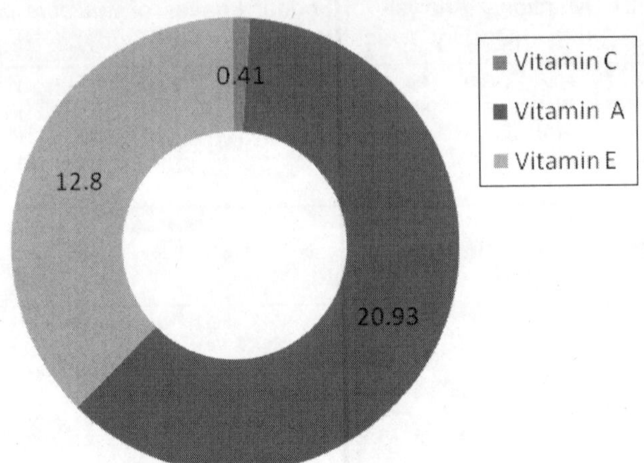

Fig. 3.33. Doughnut Showing Respective Value of Vitamins in Leaves of *Mallotus philippinensis.*

Table 3.38: Nutritional Analysis of Two Other Important Fodder Plants of District Hamirpur

Name of Plant	Carbo-hydrates (%)	Proteins (%)	Fats (%)	Starch (%)	Sodium ppm	Calcium (%)	Magne-sium (%)	Potassi-um (%)
Zizyphus jujuba	49.83	16.66	2.54	19.44	254.88	2.06	0.20	1.68
Grewia oppositifolia	41.21	20.20	3.71	19.91	68.42	2.36	0.26	1.76

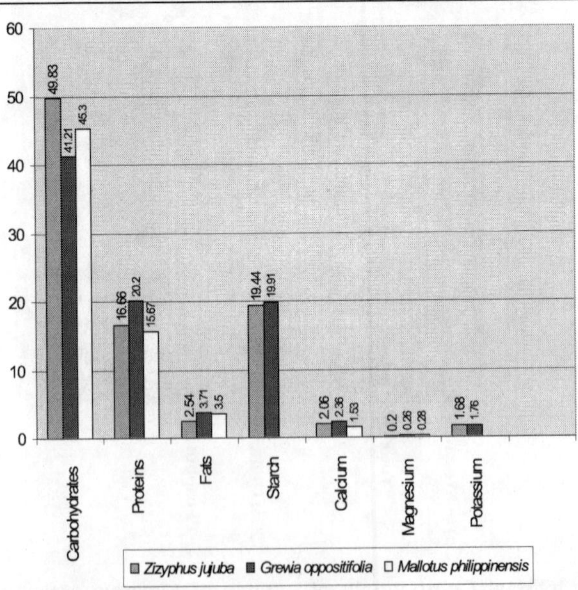

Fig. 3.34. Nutritional Components of Three Main Fodder Plants of Hamirpur.

Table 3.39: Index to Nutritional Value of Ethnobotanically Important Fodder Plants of Hamirpur

Fodder	Crude protein (%)	Crude fibre	Nitrogen free extract	Ether extract	Total ash	Calcium	Phosphorus
Acacia catechu	13.03	21.88-22.55	46.9-50.56	3.05-4.55	9.66-9.80	2.46-2.74	0.14-0.17
Acacia nilotica	18.72-19.75	6.49-9.81	-	2.16-4.20	4.79-11.23	1.11-2.80	0.25-0.52
Bauhinia vahlii	2.94	-	-	14.55	4.30	-	-
Bauhinia variegata	10.73-15.91	25.28-32.97	40.87-51.83	1.33-3.93	6.27-12.37	1.76-4.13	0.20-0.38
Cedrela toona	-	-	-	-	9.45-10.86	2.63-4.23	Mg-0.60 N-2.24-2.49
Cynodon dactylon	10.47	28.17	47.81	1.80	11.75		
Dalbergia sissoo	15.00	34.10 neutral detergent fibre	-	-	-	2.00	0.17
Ficus benghalensis	7.74-11.17	13.91-29.95	37.61-51.59	2.46-5.08	9.30-16.92	2.53-4.12	0.08-0.42
Ficus glomerata	11.06-15.20	12.27-16.50	55.52-59.00	2.23-2.87	11.88-18.39	1.73-2.96	0.16-0.45
Ficus roxburghii	12.28-13.35	7.71-17.79	50.02-64.87	4.49-4.65	9.42-15.42	1.31-2.19	0.17-0.22
Flacourtia indica	-	-	-	-	6.60	2.21	-
Grewia oppositifolia	15.6-20.0	18.08-22.5	41.7-49.0	2.1-6.7	2.5-4.0	0.15-0.30	-
Lannea coromandelica	11.44-16.17	16.12-19.76	53.10-59.05	2.47-4.20	8.50-9.19	1.75-2.41	1.75-2.41
Leucaena leucocephala	22.22	46.30 neutral deterg-ent fibre	-	3.56	-	-	-
Mallotus philippinensis	13.37	29.05	44.64	3.65.	8.69	1.34	0.24
Melia azedarach	-	13.30	-	-	-	2.55	0.23
Morus alba	15.0-27.64	9.07-15.27	47.98-49.70	2.30-8.04	14.32-22.87	2.42-4.71	0.23-0.97
Pennisetum purpureum	1.0	7.4	10.2	-	-	0.12	0.07
Zizyphus jujuba	12.86-16.93	13.47-17.07	55.33-57.90	1.52-3.23	10.24-11.72	1.42-3.59	0.21-0.33

* *Source:* Bennet et al., 1991; Dwivedi, 1993; Pruthi, 1976. (-) Values not available.

For rural-based economy like elsewhere the study highlights the use of a total of 296 species under 243 genera belonging to 86 families to be associated with material culture of the region (Tables 3.40-3.44). For example, under fuel wood there are 45 (15.20%) species, followed by agricultural tools and implements with 32 (10.81%) species, fibre yielding with 29 (9.79%) species, fencing purposes with 29 (9.79%) species, timber with 28 (9.45%) species, furniture with 27 (9.12%) species, insect repellent with 25 (8.44%) species, dye with 17 (5.74%) species, gum with 9 (3.04%) species, bedding for animals with 7 (2.36%) species, souring with 6 (2.02%) species, grafting with 3 (1.01%) species, rearing of silk-worm with 1

(.33%) species, and for various other purposes like building works, house-hold utensils, house-building, ladder, mats, measure of weight, musical instruments, roofing, plastering walls, shuttering, thatching, pollution absorbent, torch light, water container, writing pen with 73 (24.66%) species, (Tables 3.40-3.44). Of these, use of species with asterisk against their names have been hitherto unrecorded and unique to the study area. It is evident from the above forgoing information that the region is very rich in plants of great economic utilities and the resultant information is of great significance for achieving the goal of economic benefits for the local communities by setting up several cottage and rural industries in the region (Table 3.45). At the same time, it is a big challenge to agronomists for more agronomically and economically desirable traits to meet the needs and aspirations of future generations who may require altogether new sources of genes to thwart unforeseen adverse changes in global climate, severe abiotic stresses, unpredictable climate and pests (Chang, 1985; Paroda & Arora, 1991).

Table 3.40: Fuel Wood Plants of District Hamirpur

S.No.	Botanical Name	Local Name/s	Family
1.	*Acacia nilotica*	Kikker, Shami	Fabaceae
2.	*Albizia lebbeck*	Sirin	Fabaceae
3.	*Albizia stipulata*	Oayee	Fabaceae
4.	*Arachis hypogaea**	Moongphali	Fabaceae
5.	*Bambusa arundinacea*	Bainjh (Magar)	Poaceae
6.	*Bambusa nutans**	Bans	Poaceae
7.	*Bauhinia vahlii**	Tor	Fabaceae
8.	*Berberis asiatica*	Kasmal	Berberidaceae
9.	*Carissa carandus**	Karonda, Bada Garna	Apocynaceae
10.	*Carissa spinarum**	Chhota Garna	Apocynaceae
11.	*Cassia fistula*	Allia, Aahali, Kanyar	Fabaceae
12.	*Colebrookia oppositifolia**	Dushane, Dus	Lamiaceae
13.	*Delbergia sissoo*	Tahali, Shisham	Fabaceae
14.	*Dendrocalamus hamiltonii**	Baingh, Maggar	Poaceae
15.	*Dodonaea viscosa*	Mehandru	Sapindaceae
16.	*Emblica officinalis*	Ambla, Amla	Euphorbiaceae
17.	*Eucalyptus umbellata*	Safeda	Myrtaceae
18.	*Ficus carica*	Anjir	Moraceae
19.	*Ficus hispida**	Devryan	Moraceae
20.	*Ficus palmata*	Phalgu, Anjhir	Moraceae
21.	*Ficus roxburghii**	Triamble, Trayambalu	Moraceae
22.	*Flacourtia ramontchi**	Kangu	Flacoutiaceae
23.	*Grewia oppositifolia*	Beul, Bhuel	Tiliaceae

contd...

Table 3.40: *contd...*

S.No.	Botanical Name	Local Name/s	Family
24.	*Lagerstroemia indica**	Har-shingar	Lythraceae
25.	*Lannea coromandelica**	Kehmble	Anacardiaceae
26.	*Lantana camara**	Phulnu	Verbenaceae
27.	*Mallotus philippinensis*	Kambhal	Euphorbiaceae
28.	*Mangifera indica*	Amb, Aam	Anacardiaceae
29.	*Parthenium hysterophorus*	Congress grass	Asteraceae
30.	*Pinus roxburghii**	Chil,Chir	Pinaceae
31.	*Pistacia integerrima**	Kakarsinghi, Kakad-singhi	Anacardiaceae
32.	*Populus nigra**	Poplar	Moraceae
33.	*Premna latifolia**	Bhankar	Verbenaceae
34.	*Prunus amygdalus**	Badam	Rosaceae
35.	*Prunus domestica**	Plum, Aloo Bukhara	Rosaceae
36.	*Prunus persica*	Aaru	Rosaceae
37.	*Pyrus pashia*	Kainth, Kaenth	Rosaceae
38.	*Rubus ellipticus**	Akhe	Rosaceae
39.	*S a ccharum arundinaceum**	Sarphar	Poaceae
40.	*Saccharum officinarum*	Ganna	Poaceae
41.	*Terminalia arjuna*	Arjun	Combretaceae
42.	*Wendlandia heynei*	Panseera	Rubiaceae
43.	*Woodfordia floribunda**	Dhai, Dhavae, Dhaun	Lythraceae
44.	*Zea mays**	Makki, Bhuta	Poaceae
45.	*Zizyphus jujuba**	Ber	Rhamnaceae

Table 3.41: Fibre Resources of District Hamirpur

Botanical Name	Family	Part Used
Abelmoschus esculentum	Malvaceae	Stem
Abelmoschus moschatus	Malvaceae	Stem
*Abroma augusta**	Sterculiaceae	Stem
Agave Americana	Agavaceae	Leaves
*Aloe barbedensis**	Liliaceae	Leaves
*Aloe vera**	Liliaceae	Leaves
Artocarpus lakoocha	Moraceae	Branches
*Asclepias curassavica**	Asclepiadaceae	Stem
Bambusa arundinacea	Poaceae	Stem
Bambusa nutans	Poaceae	Stem
Bauhinia vahlii	Fabaceae	Stem
Bombax ceiba	Boraginaceae	Stem/Branches
Calotropis procera,	Asclepiadaceae	Stem

contd...

Table 3.41: *contd...*

Botanical Name	Family	Part Used
*Cannabis sativa**	Moraceae	Stem
*Clematis grata**	Ranunculaceae	Stem
Dendrocalamus hamiltonii	Poaceae	Stem
Gossypium arboreaum	Malvaceae	Seeds
Grewia oppositifolia	Tiliaceae	Stem/Branches
*Helicteres isora**	Sterculiaceae	Stem
Hordeum vulgare	Poaceae	Stem
Ipomoea cairica	Convolvulaceae	Stem
Linum usitatissimum	Linaceae	Stem
*Millettia extensa**	Fabaceae	
*Morus alba**	Moraceae	Stem/Branches
Phragmites karka	Poaceae	Stem
Saccharum arundinaceum	Poaceae	Inflorescence
Tylophora indica	Asclepiadaceae	Stem
Urtica dioica	Urticaceae	Stem
*Yucca aloifolia**	Liliaceae	Leaves

Table 3.42: Important Timber Resources of District Hamirpur

Botanical Name	Family	Local Names
*Acacia catechu**	Fabaceae	Khair
*Acacia concinna**	Fabaceae	Shika- kai
*Achras zapota**	Sapotaceae	Chikku
Albizia lebbeck	Fabaceae	Sirin
*Albizia stipulata**	Fabaceae	Oayee
*Alstonia scholaris**	Apocynaceae	Shaitan Ka Per
Artocarpus integrifolia	Moraceae	Kathal
Butea monosperma	Fabaceae	Palah, Palash, Plah
Cassia fistula	Fabaceae	Allia, Aahali, Kanyar
Cedrela toona	Meliaceae	Tooni
Dalbergia sissoo	Fabaceae	Tahali, Shisham
*Eriobotrya japonica**	Rosaceae	Lukat, Laukhat
Eucalyptus umbellata	Myrtaceae	Safeda
Eugenia cerasoides	Myrtaceae	Kathman
Eugenia jambolana	Myrtaceae	Jamun
*Ficus glomerata**	Moraceae	Umreyan, Umara
Ficus hispida *	Moraceae	Devryan
*Grewia oppositifolia**	Tiliaceae	Beul, Bhuel
*Jacaranda acutifolia**	Bignoniaceae	Nili-Gulmohar
*Juglans regia**	Juglandaceae	Akharot
Lannea coromandelica	Anacardiaceae	Kehmble
*Leucaena leucocephala**	Fabaceae	Alseenia
*Madhuca indica**	Sapotaceae	Mahua
Mangifera indica	Anacardiaceae	Amb, Aam
*Pinus roxburghii**	Pinaceae	Chil
*Populus nigra**	Moraceae	Poplar
*Putranjiva roxburghii**	Euphorbiaceae	Putranjan
Terminalia arjuna	Combretaceae	Arjun Tree

Table 3.43: Dye Yielding Plants of District Hamirpur

Plant Spp.	Family	Parts Used
Acacia concinna	Fabaceae	Pod
*Bauhinia variegate**	Fabaceae	Bark
*Carthamus oxycantha**	Asteraceae	Flower
Cedrela toona	Fabaceae	Flower
*Cucurbita hispida**	Cucurbitaceae	Fruit
Curcuma domestica	Zingiberaceae	Rhizome
Emblica officinalis	Euphorbiaceae	Fruit
Juglans regia	Juglandaceae	Fruit Rind
Impatiens balsamina	Balsaminaceae	Leaves
Lawsonia alba	Lythraceae	Leaves
Mallotus philippinensis	Euphorbiaceae	Fruit
*Opuntia dillenii**	Cactaceae	Fruit
Pinus roxburghii	Pinaceae	Pollen grains/Male cones
Pyrus pashia	Rosaceae	Leaves
Soymida febrifuga	Meliaceae	Bark
Tagetes erecta	Asteraceae	Inflorescence
Terminalia chebula	Combretaceae	Fruit

Table 3.44: Plant Resources of District Hamirpur for Miscellaneous Purposes

Use/s	Plant Species
Agricultural implements	*Acacia catechu, Acacia nilotica, Albizia lebbeck*, Albizia stipulata*, Bombax ceiba, Cassia fistula, Cedrela toona, Dalbergia sissoo, Dendrocalamus hamiltonii, Eriobotrya japonica*, Eucalyptus umbellata, Eugenia cerasoides*, Eugenia jambolana, Ficus glomerata, Ficus hispida*, Ficus roxburghii*, Flacourtia ramontchi, Grewia oppositifolia, Jacaranda acutifolia, Juglans regia, Lannea coromandelica*, Leucaena leucocephala, Mallotus philippinensis*, Mangifera indica, Melia azedarach*, Morus alba, Pistacia integerrima, Populus nigra, Putranjiva roxburghii, Pyrus pashia*, Saraca asoca, Terminalia arjuna*
Bedding for animals (Pattri)	*Cedrela toona*, Eucalyptus umbellata, Eugenia cerasoides*, Eugenia jambolana, Mallotus philippinensis*, Pinus roxburghii*, Triticum aestivum*
Fencing	*Agave americana, Argemone mexicana*, Asclepias curassavica*, Bambusa arundinacea, Bambusa nutans, Bougainvillea spectabilis*, Canna indica, Carduus nutans*, Carissa carandas, Carissa spinarum*, Dendrocalamus hamiltonii, Duranta repens, Euphorbia pulcherrima*, Euphorbia royleana*, Flacourtia*

contd...

Table 3.44: *contd...*

Use/s	Plant Species
	*ramontchi**, *Ipomoea cairica**, *Ipomoea carnea**, *Jasminum humile, Jasminum multiforum, Jasminum sambac, Lantana camara, Nerium indicum, Opuntia dillenii, Phoenix sylvestris, Saccharum arundinaceum, Sacchrum officinarum, Saccharum spontaneum, Yucca aloifolia**, *Zizyphus jujuba*
Furniture	*Acacia catechu**, *Acacia nilotica, Albizia lebbeck, Albizia stipulata, Bambusa arundinacea**, *Bambusa nutans**, *Bombax ceiba, Cedrela toona, Dalbergia sissoo, Dendrocalamus hamiltonii, Eriobotrya japonica, Eucalyptus umbellata**, *Eugenia cerasoides**, *Eugenia jambolana, Ficus hispida, Grewia oppositifolia, Jacaranda acutifolia**, *Juglans regia, Lannea coromandelica, Mallotus philippinensis**, *Mangifera indica**, *Melia azedarach, Morus alba, Pinus roxburghii**, *Pistacia integerrima, Populus nigra, Putranjiva roxburghii**
Grafting	*Citrus jambhiri**, *Citrus lemon**, *Pyrus pashia**
Gum (adhesive)	*Aegle marmelos, Albizia lebbeck**, *Bambusa arundinacea, Bambusa nutans, Cordia dichotoma, Dendrocalamus hamiltonii, Moringa oleifera, Prunus persica**, *Soymida febrifuga*
Insect repellent	*Abrus precatorius, Ageratum conyzoides, Allium cepa**, *Allium sativum, Annona squamosa**, *Azadirachta indica, Berberis asiatica**, *Cannabis sativa**, *Cedrela toona, Cestrum nocturnum, Chenopodium album, Cordia dichotoma, Datura metel, Datura stramonium, Duranta repens, Eruca sativa, Eucalyptus umbellata**, *Lavendula angustifolia**, *Millettia extensa, Nasturtium officinale, Nerium indicum, Phoenix sylvestris, Roylea cinerea, Vitex negundo, Wendlandia heynei*
Other Purposes (bowls, building works, games, house building, ladder, mats, measure of weight, musical instruments, necklaces, plates, roofing, plastering walls, pollution absorbent, purifying jaggery, shuttering purposes, thatching, torch light, water container, writing pen etc.)	*Abelmoschus esculentus**, *Abrus precatorius**, *Acacia catechu**, *Acacia concinna**, *Acacia nilotica, Abrus precatorius, Achras zapota, Achyranthes aspera, Achyranthes bidentata, Adhatoda vasica**, *Agave americana, Albizia lebbeck, Albizia stipulata, Aloe barbedensis, Alstonia scholaris, Amaranthus viridis, Arachis hypogaea, Asparagus adscendens, Asparagus racemosus, Azadirachta indica, Bambusa arundinacea, Bambusa nutans, Bauhinia vahlii**, *Bombax ceiba, Butea monosperma**, *Carissa spinarum, Cassia fistula, Cedrela toona, Celastrus paniculatus, Cucurbita pepo, Dalbergia sissoo, Dendrocalamus hamiltonii, Eriobotrya japonica, Eucalyptus umbellata, Eugenia cerasoides, Eugenia jambalana, Ficus glomerata, Ficus hispida, Ficus roxburghii, Flacourtia ramontchi, Glycine max, Grewia oppositifolia, Holarrhena pubescens, Jacaranda acutifolia, Juglans regia, Lannea coromandelica**, *Lantana camara**, *Leucaena leucocephala, Mangifera indica, Mimosa pudica, Nerium indicum**, *Parthenium hysterophorus, Phragmites karka**, *Phoenix sylvestris**, *Pinus roxburghii**, *Pistacia integerrima, Polyalthia longifolia, Populus nigra, Prunus domestica, Punica granatum**, *Putranjiva roxburghii, Pyrus pashia, Saccharum arundinaceum**, *Saccharum officinarum, Sapindus mukorossi**, *Sesamum indicum, Tagetes erecta, Tamarindus indica, Terminalia arjuna, Thuja orientalis, Woodfordia floribunda, Zanthoxylum armatum, Zizyphus jujuba*
Rearing silkworms	*Morus alba*
Souring	*Citrus jambhiri**, *Citrus grandis**, *Citrus limon**, *Citrus medica, Mangifera indica, Tamarindus indica*

Table 3.45: Ethnobotanical Plants of District Hamirpur with Additional Uses (Present Study)

S. No.	Botanical Names	*Uses*
1.	*Abelmoschus esculentus**	Purifying Jaggery*
2.	*Abrus precatorius**	Weights, Necklace*, Bracelets, Rosaries*, Decorating baskets
3.	*Acacia catechu**	Tooth brush, Construction of poles, Furniture
4.	*Acacia concinna**	Wine (sur) making, Detergent soaps
5.	*Acacia nilotica**	Tooth brushes, Tent pegs
6.	*Achras zapota*	Building works
7.	*Actinidia chinensis*	Jams and Jellies,
8.	*Achyranthes aspera**	Tooth brush, Suppressing hunger
9.	*Achyranthes bidentata*	Tooth brush, Suppressing hunger, making wine (Sur)
10.	*Adhatoda vasica**	Packing unripe fruits
11.	*Albizia lebbeck**	Agricultural implements, Body of trucks
12.	*Albizia stipulate**	Wine (sur) making, Agricultural implements
13.	*Aloe barbedensis**	Gulkand making
14.	*Alstonia scholaris**	Making slates for writing
15.	*Amaranthus viridis**	Appetiser
16.	*Asparagus adscendens**	Maternal 'Milini' in marriages
17.	*Asparagus racemosus**	'Hawan', marriages
18.	*Arachis hypogaea**	Adulterant in ghee, Snacks
19.	*Azadirachta indica**	Packing woollen clothes, Danda for grinding,
20.	*Bambusa arundinacea**	Winnowing treys, Tool handles, Mats, Paper, Baskets, Stools, Chhaj
21.	*Bambusa nutans**	Winnowing treys, Tool handles, Mats, Paper, Baskets, Stools, Chhaj
22.	*Bauhinia vahlii**	Pattal (plates), Duna (bowl)
23.	*Boerhavia diffusa**	Cudding
24.	*Bombax ceiba**	'Bedi' in marriages and 'toran' dawar
25.	*Butea monosperma**	Shuttering, Frames
26.	*Carissa spinarum*	Comb making
27.	*Cassia fistula*	Agricultural implements, Doors, Roofs
28.	*Cedrela toona**	Agricultural implements, Construction
29.	*Celastrus paniculatus**	Added to horse feed
30.	*Cucurbita pepo**	Utensil making, Confectionary
31.	*Dalbergia sissoo*	Agricultural implements, Furniture, building works
32.	*Dendrocalamus hamiltonii*	Basket and Bier making
33.	*Eriobotrya japonica*	Agricultural implements, Furniture
34.	*Elettaria cardamomum*	Flavouring tea and coffee
35.	*Eucalyptus umbellata*	Kills pests, Insect repellent
36.	*Eugenia cerasoides*	Bedding for animals, Agricultural implements
37.	*Eugenia jambolana*	Bedding for animals, Agricultural implements
38.	*Ficus glomerata*	House construction
39.	*Ficus hispida*	Agricultural implements, Construction
40.	*Ficus roxburghii*	Meal plates (pattals), Construction and agricultural implements

Table 3.45: *contd...*

S. No.	Botanical Names	Uses
41.	*Flacourtia ramontchi*	Agricultural implements, fencing
42.	*Glycine max*	Milk and Cheese formation
43.	*Grewia oppositifolia*	Cot frames, Tool handles, Agricultural implements, Furniture
44.	*Holarrhena pubescens*	'Sharradh' rituals
45.	*Ipomoea cairica*	Antifire property
46.	*Ipomoea carnea*	Packing unripe bananas
47.	*Jacaranda acutifolia*	Furniture and construction works
48.	*Juglans regia*	Furniture and construction works, reddening lips
49.	*Lannea coromandelica*	Agricultural implements, construction
50.	*Lantana camara*	Killing lice and bed bugs
51.	*Leucaena leucocephala*	Construction purposes
52.	*Mallotus philippinensis*	Bedding for cattle
53.	*Mangifera indica*	Furniture, Contruction of houses
54.	*Mimosa pudica*	Magico-religious purposes
55.	*Morus alba*	Walking sticks, Hockey sticks, Baskets, Sticks for beating
56.	*Nerium indicum*	Insect and worm repellent
57.	*Nicotiana tabacum*	Flavouring coffee
58.	*Oroxylum indicum*	Decorating caps, Bedding for cattle
59.	*Parthenium hysterophorus*	Plastering walls
60.	*Paspalum scrobiculatum*	Phychosis treatment
61.	*Phoenix sylvestris*	Brooms, Mats, baskets, Soil conservation
62.	*Phragmites karka**	Brooms,
63.	*Pinus roxburghii**	Torch light, Turpentine, Varnish, Paints, Filling Material
64.	*Pistacia integerrima**	Furniture making, Smoking
65.	*Polyalthia longifolia**	Festoons of leaves strung across doors.
66.	*Populus nigra*	Plywood, building works
67.	*Prunus domestica*	Jams and jellies
68.	*Punica granatum**	Removing bitterness
69.	*Putranjiva roxburghii**	Costruction and furniture
70.	*Pyrus pashia**	Agricultural implements
71.	*Randia dumetorum*	Detergent for washing woollen clothes, Soaps
72.	*Saccharum arundinaceum*	Pen for writing, Brooms,
73.	*Saccharum officinarum*	Wine making, Jaggery, Thatching roof
74.	*Saccharum spontaneum*	Winnowing trays, 'kalam'(pen) making
75.	*Santalum album*	'Tilak' making
76.	*Sapindus mukorossi*	Necklace, Detergent, Soap
77.	*Solanum surattense*	Smoke inhaled
78.	*Tagetes erecta*	Henna effects
79.	*Tamarindus indica*	Seeds for playing
80.	*Terminalia arjuna*	Agricultural implements, building works
81.	*Thuja orientalis*	'Vidya patri'
82.	*Woodfordia floribunda*	Flowers sprinkled over wounds
83.	*Zanthoxylum armatum*	Tooth brushes
84.	*Zizyphus jujuba*	Fencing

Observed Density and Availability of Underutilized Plant Resources

As far as the observed density and availability of ethnobotanically important plant resources is concerned, 128 species have found to be abundant, 93 as very abundant and 75 species rare (*Abelmoschus moschatus, Abroma augusta, Acacia concinna, Achyranthes aspera, Acorus calamus, Acorus calamus, Actinidia chinensis* var. *deliciosa, Adiantum capillus-veneris, Aerva lanata, Aloe barbedensis, Aloe vera, Alstonia scholaris, Andrographis paniculata, Annona squamosa, Artocarpus integrifolia, Artocarpus lakoocha, Asclepias curassavica, Asparagus racemosus, Bryophyllum calycinum, Butea monosperma, Callistemon lanceolatus, Calotropis procera, Carum copticum, Celosia argentea, Centella asiatica, Cinnamomum camphora, Cinnamomum tamala, Cucurbita hispida, Cycas revoluta, Cymbopogon citratus, Eclipta prostrata, Elaeocarpus ganitrus, Elettaria cardamomum, Fragaria chiloensis, Grevillea robusta, Habenaria pectinata, Holarrhena pubescens, Hypericum cernuum, Jasminum sambac, Juglans regia, Lagerstroemia indica, Lavandula angustifolia, Lawsonia alba, Lens culinaris, Martynia annua, Melissa officinalis, Mentha piperata, Mesua ferrea, Mimosa pudica, Moringa oleífera, Nicotiana tabacum, Pedilanthus tithymaloides, Pelargonium graveolens, Pennisetum americanum, Plumbago zeylanica, Polyalthia longifolia, Pongamia pinnata, Putranjiva roxburghii, Rauwolfia serpentina, Reinwardtia indica, Rhynchostylis retusa, Rosmarinus officinalis, Ruta graveolens, Santalum album, Saraca asoca, Tamarindus indica, Thuja orientalis, Trichodesma indicum, Tylophora indica, Vanda roxburghii, Vicia pallida, Withania somnifera, Yucca aloifolia*) in distribution (Table 3.46). Of the taxa reported presently, species like *Ipomoea carica* has been listed as rare and endangered flowering plant species in India (Rao, 2006); *Datura stramonium* as threatened medicinal plant resources in W. Himalayas (Kaul, 1996); *Andrographis paniculata, Bacopa monnieri, Centella asiatica, Plumbago zeylanica* as endangered medicinal plants due to unsustainable harvesting (Rao & Rajsekharon, 2002); *Rauwolfia serpentina* as endangered medicinal plant species in H.P. with status 'A' and Ranking I and *Santalum album* a vulnerable (Nayar & Shastri, 1987-1988); *Acorus calamus* as IUCN status endangered medicinal plant having threat level of 2, 1, 5; *Abrus precatorius, Achyranthes aspera, Ageratum conyzoides, Chenopodium ambrosoides, Cissampelos pareira, Fumaria indica, Oxalis corniculata, Solanum nigrum, Taraxacum officinale, Xanthium strumarium* as species which may be exported subject to licensing and ceiling limit (Subramani *et al.*, 2005) whereas *Acorus calamus, Rauwolfia serpentina* and *Withania somnifera* are listed in negative list of export by Government of India (Ahuja, 2001). For warding off the danger of extinction of these listed species, there is a dire need for habitat conservation of the wild resources of these medicinal plants as well as their conservation and cultivation be encouraged and strictly enforced with participation of local people village Panchayats, NGOs and forest department. In addition, biotechnological methods can also provide new means for conservation and propagation of these invaluable plants.

Table 3.46: Observed Density and Availability (ODA) of Ethnobotanically Important Plants of District Hamirpur

S.No.	Botanical Name	Local Name	Observed Density
1.	*Abelmoschus esculentus*	Bhindi	+++
2.	*Abelmoschus moschatus*	Kasturi Bhindi	+
3.	*Abroma augusta*	Ulat Kamal	+
4.	*Abrus precatorius*	Ratti, Raktgunja	++
5.	*Acacia catechu*	Khair	+++
6.	*Acacia concinna*	Shika- kai	+
7.	*Acacia nilotica*	Kikker, Shami	++
8.	*Achras zapota*	Chikku	++
9.	*Achyranthes aspera*	Puthkanda	+
10.	*Achyranthes bidentata*	Puthkanda	++
11.	*Acorus calamus*	Baryan	+
12.	*Actinidia chinensis* var. *deliciosa*	Kiwi	+
13.	*Adhatoda vasica*	Basuti	+++
14.	*Adiantum capillus-veneris*	Hansraj	+
15.	*Adiantum lunulatum*	Hansraj	++
16.	*Aegle marmelos*	Bil	++
17.	*Aerva lanata*	Bhadra	+
18.	*Agave americana*	Ramban	++
19.	*Ageratum conyzoides*	Neela Fulunu	+++
20.	*Albizia lebbeck*	Sirin	+++
21.	*Albizia stipulata*	Oayee	+++
22.	*Allium cepa*	Piaz	++
23.	*Allium sativum*	Lahsun	++
24.	*Aloe barbedensis*	Ghi-kawer	+
25.	*Aloe vera*	Ghi-kawer	+
26.	*Alstonia scholaris*	Shaitan-Ka-Per	+
27.	*Amaranthus paniculatus*	Cholai	++
28.	*Amaranthus spinosus*	Cholaee	+++
29.	*Amaranthus viridis*	Cholaee	+++
30.	*Amorphophallus campanulatus*	Zimmikand	++
31.	*Andrographis paniculata*	Kalmegh	+
32.	*Anetham sowa*	Kaudi saunf	++
33.	*Annona squamosa*	Sitaphal	+
34.	*Arachis hypogaea*	Moongphali	++
35.	*Argemone mexicana*	Lea	+++
36.	*Arisaema tortuosum*	Kirre-Ri-Chhhali	++
37.	*Artemisia japonica*	Chamber	+++
38.	*Artemisia vulgaris*	Charmad	+++
39.	*Artocarpus integrifolia*	Kathal	+
40.	*Artocarpus lakoocha*	Dheun	+

Contd...

Table 3.46: *contd...*

S.No.	Botanical Name	Local Name	Observed Density
41.	*Asclepias curassavica*	Ghanera	+
42.	*Asparagus adscendens*	Sanspai	++
43.	*Asparagus racemosus*	Satavari	+
44.	*Avena sativa*	Joi, Javi	+++
45.	*Azadirachta indica*	Neem	++
46.	*Bacopa monnieri*	Jal Brahmi	++
47.	*Bambusa arundinaceae*	Bainjh	+++
48.	*Bambusa nutans*	Bans	+++
49.	*Bauhinia vahlii*	Tor	++
50.	*Bauhinia variegata*	Kachnar, Karale, Karayalae	++
51.	*Berberis asiatica*	Kasmal	++
52.	*Boerhavia diffusa*	Itsat, Utsata	++
53.	*Bombax ceiba*	Simbal, Simal	++
54.	*Bougainvillea spectabilis*	Garden glory	++
55.	*Brassica campestris*	Sarson	+++
56.	*Brassica juncea*	Rai	++
57.	*Brassica nigra*	Rai, Ohari	++
58.	*Bryophyllum calycinum*	Patharchat, Gillar pattar	+
59.	*Butea monosperma*	Palah, Palash, Plah	+
60.	*Callistemon lanceolatus*	Bottle brush	+
61.	*Calotropis procera*	Ak, Madar	+
62.	*Canna indica*	Sudarshan, Hardarshan	++
63.	*Cannabis sativa*	Bhang	+++
64.	*Canscora diffusa*	Shankhpushpi	++
65.	*Capsicum annuum*	Mirch	+++
66.	*Cardiospermum halicacabum*	Tikku-malu	++
67.	*Carduus nutans*	Lea	++
68.	*Carica papaya*	Papita	++
69.	*Carissa carandas* var. *congesta*	Karonda, Bada Garna	++
70.	*Carissa spinarum*	Chhota Garna	+++
71.	*Carthamus oxycantha*	Pohli	++
72.	*Carum copticum*	Ajwain	+
73.	*Cassia fistula*	Allia, Aahali, Kanyar	++
74.	*Cassia occidentalis*	Bada-relu, Alliuna	++
75.	*Catharanthus roseus*	Sadabahar	+++
76.	*Cedrela toona*	Tooni	+++
77.	*Celastrus paniculatus*	Sankhiryan, Sankheere	++
78.	*Celosia argentea*	Kukkar-phul	+
79.	*Centella asiatica*	Brahmi	+
80.	*Centipeda minima*	Nakchikdu	++

Contd...

Table 3.46: *contd...*

S.No.	Botanical Name	Local Name	Observed Density
81.	*Cestrum nocturnum*	Rat-ki-Rani	++
82.	*Chenopodium album*	Kunnah, Bathu	+++
83.	*Chenopodium ambrosioides*	Kah-ajwain	+++
84.	*Cicer arietinum*	Chollae	++
85.	*Cinnamomum camphora*	Kapur, Kapoor	+
86.	*Cinnamomum tamala*	Tejpatra	+
87.	*Cissampelos pareira*	Patindu, Bhatindu	+++
88.	*Citrus aurantium*	Santara, Santra	++
89.	*Citrus grandis*	Chakotra-Khatta	++
90.	*Citrus jambhiri*	Jhambiri, Jambirdi, Kimb	+++
91.	*Citrus limon*	Khatta, Galgal	+++
92.	*Citrus medica*	Nimbu	++
93.	*Clematis grata*	Dhagali Bel	++
94.	*Colebrookia oppositifolia*	Dushane, Dus	+++
95.	*Colocasia esculenta*	Kachalu, Geju	+++
96.	*Cordia dichotoma*	Lasura	++
97.	*Coriandrum sativum*	Dhania	+++
98.	*Cucumis sativus*	Kheera, Kakri	+++
99.	*Cucurbita hispida*	Petha, Dhuda Petha	+
100.	*Cucurbita pepo*	Kaddu	+++
101.	*Curcuma domestica*	Haldi, Haldan	++
102.	*Cuscuta reflexa*	Akashbel, Amarvela, Halwan	+++
103.	*Cycas revoluta*	Sago Palm	+
104.	*Cymbopogon citratus*	Lemongrass	+
105.	*Cynodon dactylon*	Doob, Dhruv, Doorva	+++
106.	*Dalbergia sissoo*	Tahali, Shisham	+++
107.	*Datura metel*	Datura	++
108.	*Datura stramonium*	Datura, Dhatura	+++
109.	*Daucus carota*	Gajjar	++
110.	*Dendrocalamus hamiltonii*	Baingh, Maggar	+++
111.	*Dioscorea bulbifera*	Tardi	+++
112.	*Dodonaea viscosa*	Mehandru	++
113.	*Duranta repens*	Neelkanthi, Neel Kanta	++
114.	*Eclipta prostrata*	Bhringraj	+
115.	*Elaeocarpus ganitrus*	Rudraksha	+
116.	*Elettaria cardamomum*	Ellachi, Elaichi	+
117.	*Emblica officinalis*	Ambla, Amla	+++
118.	*Equisetum debile*	Ghornali	++
119.	*Eriobotrya japonica*	Lukat, Laukhat	++
120.	*Eruca sativa*	Taramira	++
121.	*Eucalyptus umbellata*	Safeda	+++
122.	*Eugenia cerasoides*	Kathman	++

Contd...

Table 3.46: *contd...*

S.No.	Botanical Name	Local Name	Observed Density
123.	*Eugenia jambolana*	Jamun	++
124.	*Eupatorium adenophorum*	Kali Basuti	++
125.	*Euphorbia pulcherrima*	Lalpatta	++
126.	*Euphorbia royleana*	Chu, Chui	+++
127.	*Ficus benghalensis*	Bar, Bad, Barh	+++
128.	*Ficus carica*	Anjir	+++
129.	*Ficus glomerata*	Umreyan, Umara	++
130.	*Ficus hispida*	Devryan	+++
131.	*Ficus palmata*	Phalgu, Anjhir	+++
132.	*Ficus religiosa*	Peepal	+++
133.	*Ficus roxburghii*	Triamble, Trayambalu	+++
134.	*Flacourtia ramontchi*	Kangu	+++
135.	*Foeniculum vulgare*	Saunf, Soay	++
136.	*Fragaria chiloensis*	Strawberry	+
137.	*Fumaria indica*	Pitpapara, Papara, Pitpapada	++
138.	*Glycine max*	Soyabean	++
139.	*Glycyrrhiza glabra*	Mallathi	++
140.	*Gossypium arboreum*	Kapah, Kapas	++
141.	*Grevillea robusta*	Silver oak	+
142.	*Grewia oppositifolia*	Beul, Bhuel	+++
143.	*Habenaria pectinata*	Orchid	+
144.	*Helianthus annuus*	Sooraj-Mukhi	++
145.	*Helicteres isora*	Marorphali	++
146.	*Hibiscus rosa-sinensis*	China-Rose, Shoe Flower	+++
147.	*Holarrhena pubescens*	Kurchi, Keur, Inderjaun	+
148.	*Hordeum vulgare*	Jau, Satu	++
149.	*Hypericum cernuum*	Pinli	+
150.	*Impatiens balsamina*	Teur, Balsam	+++
151.	*Ipomoea cairica*	Railway Creeper	++
152.	*Ipomoea carnea*	Badi Basuti	+++
153.	*Jacaranda acutifolia*	Nili-Gulmohar	++
154.	*Jasminum humile*	Chameli	++
155.	*Jasminum multiflorum*	Malti	++
156.	*Jasminum sambac*	Motia	+
157.	*Jatropha curcus*	Rattanjot, Jamnotta	++
158.	*Juglans regia*	Akharot	+
159.	*Lagenaria siceraria*	Bitter Gourd	++
160.	*Lagerstroemia indica*	Har-shingar	+
161.	*Lannea coromandelica*	Kehmble	+++
162.	*Lantana camara*	Phulnu	+++
163.	*Lavandula angustifolia*	Lavender	+

Contd...

Table 3.46: *contd...*

S.No.	Botanical Name	Local Name	Observed Density
164.	*Lawsonia alba*	Mehendi	+
165.	*Lens culinaris*	Masar	+
166.	*Lepidium sativum*	Hallon	++
167.	*Leucaena leucocephala*	Alseenia	++
168.	*Linum usitatissimum*	Alsi	++
169.	*Litchi chinensis*	Lichi	++
170.	*Luffa aegyptiaca*	Ghiya-Tori	+++
171.	*Lycopersicum esculentum*	Tamatar	++
172.	*Madhuca indica*	Mahua	++
173.	*Mallotus philippinensis*	Kambhal, Kambal	+++
174.	*Malvaviscus conzatii*	Gudhal	++
175.	*Mangifera indica*	Amb, Aam	+++
176.	*Martynia annua*	Kaun	+
177.	*Melia azedarach*	Draek	+++
178.	*Melissa officinalis*	Balm	+
179.	*Mentha piperata*	Pipermint	+
180.	*Mentha spicata*	Pudina	+++
181.	*Mesua ferrea*	Nagkesar	+
182.	*Millettia extensa*	Salangaya	++
183.	*Mimosa pudica*	Chhui Muui, Lajwanti	+
184.	*Mirabilis jalapa*	Gulal, 4 o'clock	+++
185.	*Momordica charantia*	Karela	++
186.	*Moringa oleifera*	Suhanjan, Sungna	+
187.	*Morus alba*	Toot	++
188.	*Murraya koenigii*	Gandhla, Curry patta	+++
189.	*Murraya paniculata*	Gandhla	++
190.	*Musa sapientum*	Kela	+++
191.	*Nasturtium officinale*	Chuch	++
192.	*Nerium indicum*	Kaner	++
193.	*Nicotiana tabacum*	Tambaku, Tammaku	+
194.	*Ocimum basilicum*	Bhabhri	++
195.	*Ocimum sanctum*	Ram-Tulsi, Baramasi	+++
196.	*Opuntia dillenii*	Chu, Thor	++
197.	*Oroxylum indicum*	Tat-palanga	+++
198.	*Oryza sativa*	Dhan, Jari	++
199.	*Oxalis corniculata*	Maroli	+++
200.	*Papaver rhoeus*	Gulla	++
201.	*Parthenium hysterophorus*	Congress grass	+++
202.	*Paspalum scrobiculatum*	Kodro grass	++
203.	*Pedilanthus tithymaloides*	Nagbel	+
204.	*Pelargonium graveolens*	Geranium	+
205.	*Pennisetum americanum*	Bajra, Barley	+
206.	*Pennisetum purpureum*	Chari	++
207.	*Phoenix sylvestris*	Jangali Khajoor	++
208.	*Phragmites karka*	Elephant grass	++
209.	*Phyllanthus urinaria*	Bhumi- Amla	++

Contd...

Table 3.46: *contd...*

S.No.	Botanical Name	Local Name	Observed Density
210.	*Physalis peruviana*	Rasbhari	++
211.	*Pinus roxburghii*	Chil	+++
212.	*Piper longum*	Mugh	++
213.	*Pistacia integerrima*	Kakarsinghi, Kakad-singhi	+++
214.	*Plumbago zeylanica*	Chitrak, Chitra	+
215.	*Pogostemon benghalensis*	Kali Basuti, Dusyan	++
216.	*Polyalthia longifolia*	Ashok	+
217.	*Pongamia pinnata*	Sukh- Chain	+
218.	*Populus nigra*	Poplar	+++
219.	*Premna latifolia*	Bhankar	++
220.	*Prunus amygdalus*	Badam	++
221.	*Prunus domestica*	Aloo Bukhara, Plum	++
222.	*Prunus persica*	Aru, Adu	++
223.	*Psidium guajava*	Amrood	+++
224.	*Punica granatum*	Daaru, Daadu	++
225.	*Putranjiva roxburghii*	Putranjan	+
226.	*Pyrus communis*	Nakh, Nashpati	++
227.	*Pyrus pashia*	Kainth, Kaenth	+++
228.	*Randia dumetorum*	Rada	++
229.	*Ranunculus muricatus*	Buttercup	++
230.	*Raphanus sativus*	Mooli	+++
231.	*Rauwolfia serpentina*	Sarpgandha	+
232.	*Reinwardtia indica*	Basant, Piyan	+
233.	*Rhynchostylis retusa*	Bhangru	+
234.	*Ricinus communis*	Airand, Arind, Erand	+++
235.	*Rosa damascena*	Gulab	++
236.	*Rosmarinus officinalis*	Rosemary	+
237.	*Roylea cinerea*	Korru	++
238.	*Rubus ellipticus*	Akhe	+++
239.	*Rumex hastatus*	Maroli	+++
240.	*Ruta graveolens*	Sudab	+
241.	*Saccharum arundinaceum*	Sarphar	+++
242.	*Saccharum officinarum*	Ganna	++
243.	*Saccharum spontaneum*	Kans	+++
244.	*Salvia coccinea*	Phultulsi	+++
245.	*Santalum album*	Chandan	+
246.	*Sapindus mukorossi*	Reetha, Doda, Doddae	++
247.	*Saraca asoca*	Ashoka	+
248.	*Sechium edule*	Lonku	+++
249.	*Sesamum indicum*	Til	++
250.	*Setaria viridis*	Hathi grass	+++
251.	*Solanum erianthum*	Rada	++
252.	*Solanum melongena*	Baingan	++
253.	*Solanum nigrum*	Makko	+++
254.	*Solanum surattense*	Choti-Kandayali	++

Contd...

Table 3.46: *contd...*

S.No.	Botanical Name	Local Name	Observed Density
255.	*Solanum tuberosum*	Aloo	++
256.	*Sonchus asper*	Bhursalae	+++
257.	*Soymida febrifuga*	Ruhan, Rehana	++
258.	*Spilanthes acmella*	Karkara	++
259.	*Spinacia oleracea*	Palak, sag	++
260.	*Stephania glabra*	Bis-Khappar	++
261.	*Tabernaemontana divaricata*	Tagar	++
262.	*Tagetes erecta*	Ghatak	+++
263.	*Tamarindus indica*	Imali	+
264.	*Taraxacum officinale*	Dudhli, Dulal	+++
265.	*Tecoma grandiflora*	Shehnai Phool	++
266.	*Tecoma stans*	Tecoma, Shehnai Phool	++
267.	*Terminalia arjuna*	Arjun Tree	++
268.	*Terminalia bellirica*	Behra, Bahera	++
269.	*Terminalia chebula*	Harar	+++
270.	*Thevetia neriifolia*	Kaner	++
271.	*Thuja orientalis*	Saru , Morphankh	+
272.	*Tinospora cordifolia*	Gulaja, Giloy, Galon	+++
273.	*Trichodesma indicum*	Chhota Kalpa, Rukhabi	+
274.	*Tridax procumbens*	Lamb	+++
275.	*Trifolium alexandrinum*	Barseem	+++
276.	*Trigonella foenum- graecum*	Methe, Methi	+++
277.	*Triticum aestivum*	Gehun	+++
278.	*Tylophora indica*	Damabuti	+
279.	*Urtica dioica*	Bichhubuti, Choti kogsi	+++
280.	*Vanda roxburghii*	Bhangru	+
281.	*Verbascum thapsus*	Ban Tambakhoo	++
282.	*Vernonia anthelmintica*	Brahmjiri	++
283.	*Vicia pallida*	BadaKaer, Roothi	+
284.	*Viola canescens*	Banfsha, Banaksha	++
285.	*Vitex negundo*	Banna	+++
286.	*Vitis jacquemontii*	Geedar Dhakh	++
287.	*Vitis vinifera*	Angoor	++
288.	*Wendlandia heynei*	Panseera	+++
289.	*Withania somnifera*	Asgandh, Ashwagandha	+
290.	*Woodfordia floribunda*	Dhai, Dhavae, Dhaun	+++
291.	*Xanthium strumarium*	Bhakhadda	+
292.	*Yucca aloifolia*	Ramban	+
293.	*Zanthoxylum armatum*	Tirmer	++
294.	*Zea mays*	Makki, Bhuta	+++
295.	*Zingiber officinale*	Adrak	++
296.	*Zizyphus jujuba*	Ber	+++

Economic Valuation of the Source Flora

The results of economic valuation of all the 296 ethnobotanical species presented in Table 3.47 indicate that *Grewia oppositifolia* ('beul') tops the list in possessing highest total importance value/ potentiality (70%), followed by *Dendrocalamus hamiltonii, Linum usitatissimum, Sesamum indicum* (65%) each, *Arachis hypogaea, Emblica officinalis, Juglans regia* (60%) each, *Brassica campestris, Dalbergia sissoo, Mangifera indica, Pyrus pashia, Saccharum officinarum, Triticum aestivum, Zea mays* (55%) each, *Aegle marmelos, Bambusa arundinacea, Bambusa nutans, Carissa carandas* var. *congesta, Cedrela toona, Cicer arietinum, Eruca sativa, Eugenia cerasoides, Eugenia jambolana, Mallotus philippinensis, Pinus roxburghii, Prunus persica, Spinacia oleracea, Zizyphus jujuba* (50%) each, etc. Undoubtedly, the resulted information can be judiciously exploited by using modern technologies and also at the same time are the fundamental requisite for any developmental planning concerned with the welfare of the rural inhabitants of this region of N. W. Himalayas.

Table 3.47: Total Importance Values (TIV) of Ethnobotanically Important Plants of Hamirpur, Himachal Pradesh

Sr. No.	Botanical Name	Dye	Fibre	Fodder	Fuel	Medi-cine	Oil	Tannin/ Gum	Timber	Others	Edible	TIV
1.	*Grewia oppositifolia*		+++	+++	++	+			++	+++		70
2.	*Dendro-calamus hamiltonii*			++	+++	++		++		+++	+	65
3.	*Linum usitatissimum*		++	++		+++	++			++	++	65
4.	*Sesamum indicum*			++		+++	+++			++	+++	65
5.	*Arachis hypogaea*			+++	+	++	++			+	+++	60
6.	*Emblica officinalis*	++		+		+++				+++	+++	60
7.	*Juglans regia*	++				++			+++	++	+++	60
8.	*Brassica campestris*			++		+++	+++				+++	55
9.	*Dalbergia sissoo*			+++	++	++			+++	+		55
10.	*Mangifera indica*				+	++			++	+++	+++	55
11.	*Pyrus pashia*	+		++	++	++				++	++	55
12.	*Saccharum officinarum*			+++	++	+				++	+++	55
13.	*Triticum aestivum*			+++		++				+++	+++	55
14.	*Zea mays*			+++	+++					++	+++	55
15.	*Aegle marmelos*					+++		++		+++	++	50
16.	*Bambusa arundinacea*			++	++	+		++		++	+	50
17.	*Bambusa nutans*			++	++	+		++		++	+	50
18.	*Carissa carandas* var. *congesta*			++	++	++				+	+++	50
19.	*Cedrela toona*	+		++		++			+++	++		50
20.	*Cicer arietinum*			+++		+				+++	+++	50
21.	*Eruca sativa*			+++		+++	++			++		50
22.	*Eugenia cerasoides*			++	+	+++				++	++	50
23.	*Eugenia jambolana*			++	+	+++				++	++	50
24.	*Mallotus philippinensis*	++		++	++	++				++		50
25.	*Pinus roxburghii*				+++	+			++	+++	+	50
26.	*Prunus persica*				++	++		++		++	++	50
27.	*Spinacia oleracea*			++		+++				++	+++	50
28.	*Zizyphus jujuba*			+++	+++	+				+	++	50
29.	*Bauhinia vahlii*		++	++	+	+				+++		45
30.	*Bauhinia variegata*			+++		+++					+++	45

contd...

Table 3.47: *contd...*

Sr. No.	Botanical Name	Dye	Fibre	Fodder	Fuel	Medi-cine	Oil	Tannin/ Gum	Timber	Others	Edible	TIV
31.	*Carica papaya*			++		+++				+	+++	45
32.	*Carissa spinarum*			++	++	++				+	++	45
33.	*Chenopodium album*			++		+++				++	++	45
34.	*Daucus carota*			++		+++				+	+++	45
35.	*Ficus carica*			++	++	+				++	++	45
36.	*Ficus roxburghii*			++	++					++	+++	45
37.	*Glycine max*			+++		++				+	+++	45
38.	*Helicteres isora*		++	++		+++				++		45
39.	*Madhuca indica*			++		+++	++			++		45
40.	*Moringa oleifera*					+++		++		++	++	45
41.	*Pistacia integerrima*			++	++	+++			++			45
42.	*Prunus amygdalus*				++	++	++				+++	45
43.	*Prunus domestica*			++	++	++					+++	45
44.	*Punica granatum*					+++				+++	+++	45
45.	*Raphanus sativus*			++		+++				++	++	45
46.	*Solanum tuberosum*			+		++				+++	+++	45
47.	*Terminalia arjuna*			++		+++				++	++	45
48.	*Tinospora cordifolia*			++		+++				++	++	45
49.	*Trigonella foenum-graecum*					+++				+++	+++	45
50.	*Vitis vinifera*					+++				+++	+++	45
51.	*Abelmoschus esculentus*		+	+		++				+	+++	40
52.	*Abelmoschus moschatus*			+	+	+++				++	+	40
53.	*Acacia nilotica*			+	++	++			+	++		40
54.	*Allium cepa*					+++				++	+++	40
55.	*Allium sativum*					+++				++	+++	40
56.	*Amaranthus spinosus*			++		++		+		+	++	40
57.	*Amaranthus viridis*			++		++		+		+	++	40
58.	*Azadirachta indica*					+++				+++	++	40
59.	*Bacopa monnieri*			++		+++				+	++	40
60.	*Capsicum annuum*					++				+++	+++	40
61.	*Catharanthus roseus*					+++				+++	++	40
62.	*Centella asiatica*			+		+++				++	++	40
63.	*Citrus aurantium*					+++				++	+++	40
64.	*Citrus limon*					+++				++	+++	40
65.	*Citrus medica*					+++				++	+++	40
66.	*Elettaria cardamomum*					+++				++	+++	40
67.	*Eucalyptus umbellata*				++	++	++			++		40
68.	*Ficus glomerata*			++		+				++	+++	40
69.	*Foeniculum vulgare*					+++				++	+++	40
70.	*Lycopersicon esculentum*					++				+++	+++	40
71.	*Melia azedarach*			++		++			++	++		40
72.	*Momordica charantia*					+++				++	+++	40
73.	*Musa sapientum*					++				+++	+++	40
74.	*Ocimum sanctum*					+++				+++	++	40
75.	*Oryza sativa*			+++		++					+++	40
76.	*Pongamia pinnata*			+++		+++				++		40
77.	*Populus nigra*			++	++				++	++		40
78.	*Rubus elipticus*			++	+	++				+	++	40
79.	*Saccharum arundinaceum*			+++	++					+++		40
80.	*Solanum melongena*					++				+++	+++	40

contd...

Table 3.47: *contd...*

Sr. No.	Botanical Name	Dye	Fibre	Fodder	Fuel	Medi-cine	Oil	Tannin/ Gum	Timber	Others	Edible	TIV
81.	*Soymida febrifuga*					+++		+++		++		40
82.	*Terminalia bellirica*					+++				++	+++	40
83.	*Terminalia chebula*	++				+++					+++	40
84.	*Acacia catechu*			++		++			+	++		35
85.	*Acacia concinna*	++		+		++			+	+		35
86.	*Adhatoda vasica*			++		+++				++		35
87.	*Albizia lebbeck*			++	+	+			+	++		35
88.	*Albizia stipulata*			++	++	+			++			35
89.	*Aloe vera*					+++				+++	+	35
90.	*Berberis asiatica*			+	+	++				++	+	35
91.	*Bombax ceiba*		++			++				+++		35
92.	*Cassia fistula*			+	+	+++				+	+	35
93.	*Cissampelos pareira*			++		+++				++		35
94.	*Citrus grandis*					++				++	+++	35
95.	*Colebrookia oppositifolia*				++	+++				++		35
96.	*Coriandrum sativum*					++				++	+++	35
97.	*Cucurbita hispida*	+				+				++	+++	35
98.	*Curcuma domestica*					+++				++	+++	35
99.	*Dodonaea viscosa*			++	++	+++						35
100.	*Eriobotrya japonica*			++						++	+++	35
101.	*Ficus hispida*				++	+				++	++	35
102.	*Ficus palmata*				+	++				++	++	35
103.	*Flacourtia ramontchi*			++		+				++	++	35
104.	*Glycyrrhiza glabra*			++		+++					++	35
105.	*Gossypium arboreum*			++			++				+++	35
106.	*Lannea coromandelica*			++	+	++			++			35
107.	*Luffa aegyptiaca*					++				++	+++	35
108.	*Millettia extensa*		++	+++		++						35
109.	*Morus alba*			++						+++	++	35
110.	*Ocimum basilicum*					+++				++	++	35
111.	*Opuntia dillenii*	++				++				++	+	35
112.	*Phoenix sylvestris*			++	+					+++	+	35
113.	*Piper longum*					+++				++	++	35
114.	*Pyrus communis*					++				++	+++	35
115.	*Ricinus communis*					+++	++			++		35
116.	*Santalum album*			++		++				+++		35
117.	*Setaria viridis*			+++		++				++		35
118.	*Tylophora indica*		++			+++				++		35
119.	*Urtica dioica*		+	++		++				++		35
120.	*Wendlandia heynei*			++	++	+				++		35
121.	*Abrus precatorius*					+++				+++		30
122.	*Aloe barbedensis*					+++				+++		30
123.	*Amorpho-phallus campanulatus*					++				++	++	30
124.	*Anethum sowa*					++				++	++	30
125.	*Annona squamosa*					++				++	++	30
126.	*Artocarpus integrifolia*			+		+				+	+++	30
127.	*Artocarpus lakoocha*			+		+				+	+++	30
128.	*Brassica juncea*			++		++					++	30
129.	*Brassica nigra*			++		++					++	30
130.	*Butea monosperma*			++		++				++		30
131.	*Cannabis sativa*		++			++				++		30

contd...

Table 3.47: *contd...*

Sr. No.	Botanical Name	Dye	Fibre	Fodder	Fuel	Medi-cine	Oil	Tannin/ Gum	Timber	Others	Edible	TIV
132.	*Chenopodium ambrosioides*					++				++	++	30
133.	*Cinnamomum tamala*					+++				++	+	30
134.	*Citrus jambhiri*					++				+	+++	30
135.	*Colocasia esculenta*					+				++	+++	30
136.	*Cordia dichotoma*					++		++			++	30
137.	*Cucumis sativus*					++				++	++	30
138.	*Cucurbita pepo*				+					++	+++	30
139.	*Cymopogon citratus*			++		+				++	+	30
140.	*Helianthus annuus*					++	++			++		30
141.	*Hordeum vulgare*			++		++					++	30
142.	*Jacaranda acutifolia*					++			++	++		30
143.	*Jatropha curcas*					++	++			++		30
144.	*Lantana camara*				++	++				++		30
145.	*Lavandula angustifolia*					++	++			++		30
146.	*Lens culinaris*			++		++					++	30
147.	*Melissa officinalis*					++	++			++		30
148.	*Murraya koenigii*					+++				++	+	30
149.	*Oxalis corniculata*			++		++				++		30
150.	*Pelargonium graveolens*					++	++			++		30
151.	*Pennisetum americanum*			+++						+	++	30
152.	*Phragmites karka*			+++						+++		30
153.	*Pogostemon benghalensis*			++		+++				+		30
154.	*Psidium guajava*					++				+	+++	30
155.	*Randia dumetorum*					++				++	++	30
156.	*Rumex hastatus*			++		+				+	++	30
157.	*Sechium edule*					+				++	+++	30
158.	*Tagetes erecta*					++				+++	+	30
159.	*Tamarindus indica*					+				++	+++	30
160.	*Vitex negundo*					+++				+++		30
161.	*Woodfordia floribunda*			++	++	++						30
162.	*Zingiber officinale*					+++				+++		30
163.	*Achras zapota*					+			+		+++	25
164.	*Actinidia chinensis* var. *deliciosa*					++					+++	25
165.	*Agave americana*		++	+		+					+	25
166.	*Amaranthus paniculatus*					++				+	++	25
167.	*Argemone mexicana*			++		+				++		25
168.	*Asparagus adscendens*					+++				++		25
169.	*Asparagus racemosus*			+		++				++		25
170.	*Boerhavia diffusa*			+		+++				+		25
171.	*Calotropis procera*		+			++				++		25
172.	*Cardiospermum halicacabum*					+++				++		25
173.	*Carum copticum*					+++					++	25
174.	*Celastrus paniculatus*			++		+++						25
175.	*Cinnamomum camphora*					++				+++		25
176.	*Cynodon dactylon*			+++		++						25

contd...

Table 3.47: *contd...*

Sr. No.	Botanical Name	Dye	Fibre	Fodder	Fuel	Medi-cine	Oil	Tannin/ Gum	Timber	Others	Edible	TIV
177.	*Datura metel*					+++				++		25
178.	*Datura stramonium*					+++				++		25
179.	*Dioscorea bulbifera*					+				++	++	25
180.	*Duranta repens*					++				+++		25
181.	*Eclipta prostrata*			++		+++						25
182.	*Elaeocarpus ganitrus*					++					+++	25
183.	*Ficus benghalensis*					++					+++	25
184.	*Ficus religiosa*			++						+++		25
185.	*Habenaria pectinata*			++		++				+		25
186.	*Impatiens balsamina*	++				+				++		25
187.	*Lagerstroemia indica*				++	+				++		25
188.	*Lawsonia alba*	++				++				+		25
189.	*Leucaena leucocephala*			+++						++		25
190.	*Litchi chinensis*					++					+++	25
191.	*Mentha piperata*					+++				++		25
192.	*Mentha spicata*					+++				++		25
193.	*Mimosa pudica*			+		++				++		25
194.	*Nerium indicum*					++				+++		25
195.	*Nicotiana tabacum*					+++				++		25
196.	*Oroxylum indicum*			++		+++						25
197.	*Parthenium hysterophorus*			+	+	++				+		25
198.	*Pennisetum purpureum*			+++	+					+		25
199.	*Plumbago zeylanica*					+++				++		25
200.	*Putranjiva roxburghii*								++	+++		25
201.	*Rhynchostylis retusa*					+++				++		25
202.	*Rosa damascena*					+++				++		25
203.	*Rosmarinus officinalis*					+++	++					25
204.	*Roylea cinerea*					+++				++		25
205.	*Saccharum spontaneum*			++						+++		25
206.	*Sapindus mukorossi*					++				+++		25
207.	*Spilanthes acmella*					+++				++		25
208.	*Vanda roxburghii*					+++				++		25
209.	*Vernonia anthelmintica*			++		+++						25
210.	*Vicia pallida*			++						+	++	25
211.	*Viola canescens*			++		+++						25
212.	*Withania somnifera*					+++				++		25
213.	*Yucca aloifolia*		++							+++		25
214.	*Zanthoxylum armatum*					+++				++		25
215.	*Abroma augusta*		+			+++						20
216.	*Achyranthes aspera*					++				++		20
217.	*Achyranthes bidentata*					++				++		20
218.	*Acorus calamus*					+++				+		20

contd...

Table 3.47: *contd...*

Sr. No.	Botanical Name	Dye	Fibre	Fodder	Fuel	Medi-cine	Oil	Tannin/ Gum	Timber	Others	Edible	TIV
219.	*Aerva lanata*					+++					+	20
220.	*Ageratum conyzoides*			+		++				+		20
221.	*Alstonia scholaris*					++				++		20
222.	*Andrographis paniculata*					+++					+	20
223.	*Arisaema tortuosum*			++		++						20
224.	*Avena sativa*			++		+					+	20
225.	*Canna indica*			+		+				++		20
226.	*Carduus nutans*			++						+	+	20
227.	*Cassia occidentalis*					++				+	+	20
228.	*Celosia argentea*					++				++		20
229.	*Cestrum nocturnum*					++				++		20
230.	*Clematis grata*			+		++				+		20
231.	*Cuscuta reflexa*					+++				++		20
232.	*Cycas revoluta*									++	++	20
233.	*Equisetum debile*			+		+++						20
234.	*Euphorbia pulcherrima*					++					++	20
235.	*Euphorbia royleana*					++					++	20
236.	*Fragaria chiloensis*			+		+					++	20
237.	*Fumaria indica*			++		++						20
238.	*Holarrhena pubescens*					++				++		20
239.	*Ipomoea cairica*			++						++		20
240.	*Ipomoea carnea*					++				++		20
241.	*Jasminum humile*					++				++		20
242.	*Jasminum multiflorum*					++				++		20
243.	*Jasminum sambac*					++				++		20
244.	*Lagenaria siceraria*					++					++	20
245.	*Lepidium sativum*			++		++						20
246.	*Malvaviscus conzatii*					+				+++		20
247.	*Mirabilis jalapa*					++				++		20
248.	*Nasturtium officinale*					++					++	20
249.	*Papaver rhoeus*					++				++		20
250.	*Paspalum scrobiculatum*			++		+				+		20
251.	*Pedialanthes tithymaloides*					++				++		20
252.	*Physalis peruviana*					++					++	20
253.	*Premna latifolia*			++		++						20
254.	*Reinwardtia indica*					++				++		20
255.	*Saraca asoca*					++				++		20
256.	*Solanum surattense*			++		++						20
257.	*Tabernae-montana divaricata*					++				++		20
258.	*Thevetia neriifolia*					++				++		20
259.	*Tridax procumbens*			++		++						20
260.	*Adiantum capillus-veneris*			+		++						15
261.	*Adiantum lunulatum*			+		++						15
262.	*Bougainvillea spectabilis*			+						++		15

contd...

Table 3.47: *contd...*

Sr. No.	Botanical Name	Dye	Fibre	Fodder	Fuel	Medi-cine	Oil	Tannin/ Gum	Timber	Others	Edible	TIV
263.	*Bryophyllum calycinum*					++					+	15
264.	*Carthamus oxycantha*	+				++						15
265.	*Centipeda minima*					+++						15
266.	*Grevillea robusta*									+++		15
267.	*Hibiscus rosa-sinensis*					+				++		15
268.	*Mesua ferrea*					+++						15
269.	*Murraya paniculata*									++	+	15
270.	*Phyllanthus urinaria*			++		+						15
271.	*Ranunculus muricatus*			++		+						15
272.	*Rauwolfia serpentina*					+++						15
273.	*Ruta graveolens*					+++						15
274.	*Salvia coccinea*			++						+		15
275.	*Stephania glabra*					+++						15
276.	*Taraxacum officinale*			++		+						15
277.	*Tecoma grandiflora*									+++		15
278.	*Tecoma stans*									+++		15
279.	*Trichodesma indicum*					+++						15
280.	*Trifolium alexandrinum*			+++								15
281.	*Vitis jacquemontii*					+				+	+	15
282.	*Artemisia japonica*					++						10
283.	*Artemisia vulgaris*					++						10
284.	*Asclepias curassavica*		+							+		10
285.	*Callistemon lanceolatus*									++		10
286.	*Canscora diffusa*					++						10
287.	*Eupatorium adenophorum*					++						10
288.	*Hypericum cernum*									++		10
289.	*Martynia annua*					++						10
290.	*Polyalthia longifolia*									++		10
291.	*Solanum erianthum*					++						10
292.	*Solanum nigrum*					++						10
293.	*Sonchus asper*					++						10
294.	*Thuja orientalis*									++		10
295.	*Verbascum thapsus*					++						10
296.	*Xanthium strumarium*					+++						10

Others*: House building, musical instruments, plates, combs, bowls, roofing, torch light, plastering walls, thatching, water container, ladder, mats, measure of weight, shuttering purposes, writing pen, ornamental, fencing, bites (scorpion, wasp, insects), death rituals, fibre (cordage, chicks, basketry, stuffing), facial cosmetics, massage, bedding for animals ('pattri'), insect repellent, magico-religious beliefs, chewing gum, expelling leeches, games, grafting purposes, incense, rearing silkworm, scrubber, shampoo and detergent, souring teeth, suppressing hunger, etc.
Edible: Eaten as such, pickle, jam, jellies, sauces, chutney, fruits, dry fruit, grains, vegetable, pot-herb, seed, endosperm, murraba, gums, paneer, etc.

Native Status of the Ethnobotanical Flora

Present inventorization reveals that total alien ethnobotanical flora of the region is represented by 136 species belonging to 121 genera under 51 families (Tables 3.50-3.55). Relatively, the exotics with 45.94% of the overall representation are less utilized for ethnobotanical purposes than the native species (54.05%) (Tables 350, 3.53; Figs. 3.35, 36). On comparison with the estimated values of aliens for Indian flora (18%; Nayar, 1977), it is seen that this region has significantly higher percentage. Habit wise analysis indicates that the flora abounds in 68 herbs, 32 shrubs, 4 undershrubs, 21 trees, 7 climbers and 4 lianas (Table 3.49). Amongst these, the herbaceous elements predominate the alien flora (Table 3.49), and 117 species of dicotyledonous taxa grouped under 104 genera of 43 families constitute maximum (85.27%) proportion of utilization in the study area (Tables 3.50, 3.53) followed by monocots (15species, 13 genera, 4 families); Poaceae (*Avena sativa, Dendrocalamus hamiltonii, Paspalum scrobiculatum, Pennisetum purpureum, Setaria viridis, Triticum aestivum, Zea mays*), Liliaceae (*Allium cepa, Allium sativum, Aloe barbadensis, Aloe vera, Yucca aloifolia*), Agavaceae (*Agave americana*), Araceae (*Amorphophallus campanulatus*) (Table 3.52), Gymnosperms (*Cycas revoluta* and *Thuja orientalis*) which include families Cycadaceae and Cupressaceae, and Pteridophytes (*Adiantum capillus-veneris* and *Equisetum debile*) comprising of families Adiantaceae and Equisetaceae (Table 3.50). Importantly, families with largest number of alien representatives are Asteraceae (13 species), Fabaceae (11 species), Solanaceae (8 species), Brassicaceae, Poaceae (7 species each), Lamiaceae (6 species), Rosaceae, Liliaceae (5 species each), Amaranthaceae, Apiaceae, Rutaceae, Euphorbiaceae (4 species each), Apocynaceae, Bignoniaceae, Cucurbitaceae, Chenopodiaceae, Moraceae, (3 species each), Asclepiadaceae, Myrtaceae, Nyctaginaceae, Ranunculaceae, Sapindaceae, Verbenaceae, Convolvulaceae, Vitaceae (2 species each), etc. Families represented by single plant species in the total alien flora are Actinidiaceae, Agavaceae, Annonaceae, Balsaminaceae, Berberidaceae, Cactaceae, Caricaceae, Crassulaceae, Cuscutaceae, Geraniaceae, Hypericaceae, Malvaceae, Martyniaceae, Meliaceae, Oleaceae, Oxalidaceae, Proteaceae, Rhamnaceae, Sapotaceae and Violaceae (Tables 3.51, 3.52, 3.54). Evidently, the top 20 families contribute 94 species with proportion of 72.86%. As a matter of fact, the predominant genera with highest number of aliens are *Amaranthus* (3 species), *Brassica* (3 species), *Artemisia* (2 species), *Prunus* (2 species), *Solanum* (2 species), *Mentha* (2 species), *Allium* (2 species), *Aloe* (2 species), *Euphorbia* (2 species), *Tecoma* (2 species), *Chenopodium* (2 species), *Ipomoea* (2 species), *Vitis* (2 species) (Table 3.57). Of the crops, 22 species are aliens (Table 3.58). It appears that a large majority of the aliens were intentionally introduced as edible. Such a baseline information providing authentic identification, documentation, characterization and utilization of aliens and native elements separately are of paramount importance for formulating strategies for their effective management and predictive systems.

Table 3.48: Native Status of Aliens of District Hamirpur, H.P.

S.No.	Continent/Country	Species	Genera	Families
1.	Asia (Afghanistan, Bangladesh, Bhutan, India, Nepal, Pakistan, Srilanka, Thailand, etc.)	17	17	13
2.	Central, North and South America (Brazil, Caribbean Territories, Chile, Mexico, Peru, Venezuela, United States, etc.)	38	36	23
3.	African Continent (Algeria, Egypt, Ethiopia, Ghana, Kenya, South Africa, Tunisia, Zimbabwe, etc.)	09	08	07
4.	Europe (Bulgaria, France, Germany, Greece, Italy, Spain, U.K., etc.)	09	04	04
5.	Australia (New South Wales, Queensland, Victoria, etc.)	06	06	05
6.	Common origin (Countries of Asia, North, Central & South America, Australia, Europe, Africa)	57	54	27

Table 3.49: Percentage of Aliens Belonging to Different Categories

Categories	Sub Categories	Percentage of spp.
Origin	Asian	12.5
	American (Central, South & North)	27.94
	African	6.61
	Australian	4.41
	European	3.67
	Common origin	41.91
Habit	Herbs	68
	Shrubs	32
	Under Shrubs	4
	Trees	21
	Climbers	7
	Lianas	4

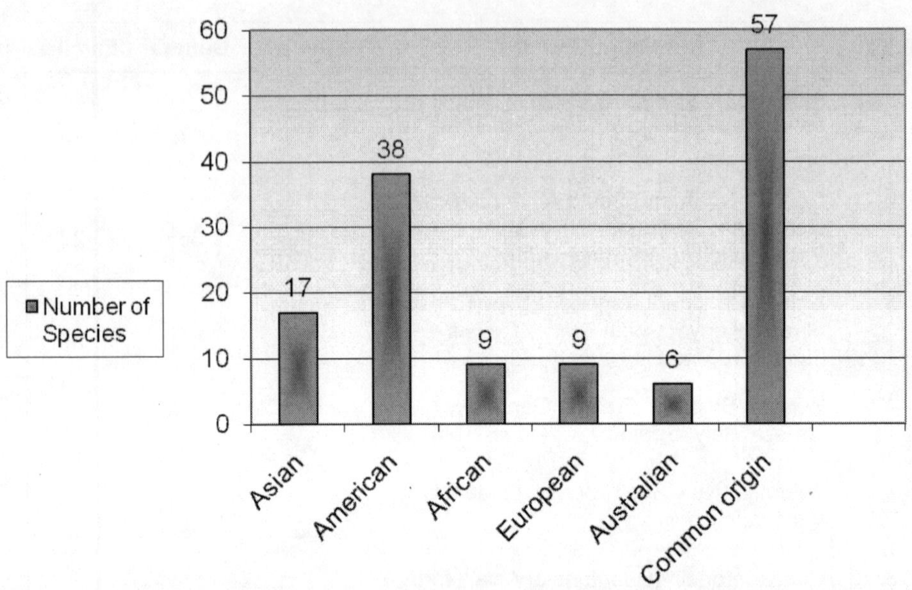

Fig. 3.35. Histogram Showing Native Status of Ethnobotanically Important Plants of District Hamirpur

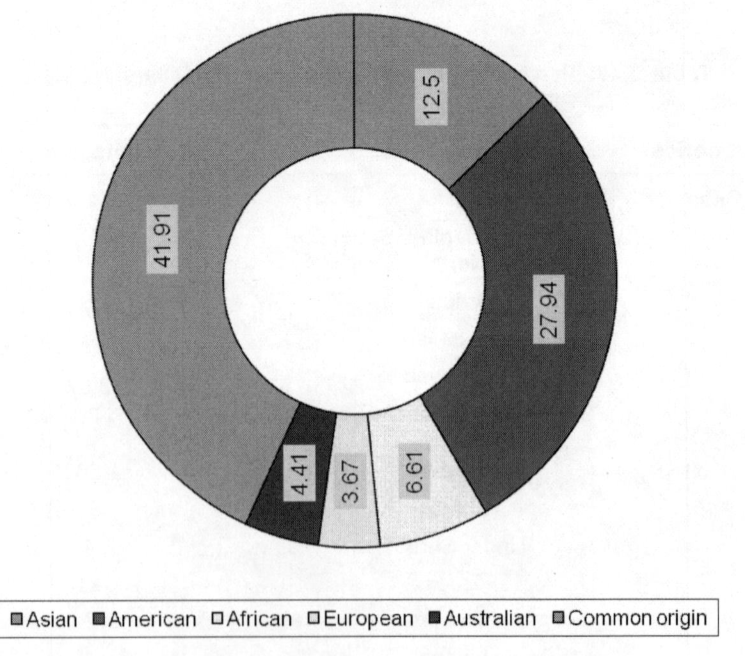

Fig. 3.36. Doughtnut Showing Native Status of Ethnobotanically Important Aliens of Hamirpur.

Of the estimated 136 species of invasive alien flora of India (Reddy, 2008), 17 species namely *Ageratum conyzoides* (Asteraceae), *Asclepias curassavica* (Asclepiadaceae), *Calotropis procera* (Asclepiadaceae), *Cassia occidentalis* (Fabaceae), *Celosia argentea* (Amaranthaceae), *Cuscuta reflexa* (Cuscutaceae), *Datura metel* (Solanaceae), *Impatiens balsamina* (Balsaminaceae), *Ipomoea carnea* (Convolvulaceae), *Lantana camara* (Verbenaceae), *Martynia annua* (Martyniaceae), *Mimosa pudica* (Fabaceae), *Mirabilis jalapa* (Nyctaginaceae), *Oxalis corniculata* (Oxalidaceae), *Sonchus asper*, *Tridax procumbens* and *Xanthium strumarium* (Asteraceae) are well represented in the study area.

With regard to the contribution of source floras to the aliens of district Hamirpur, present study revealed that common origin elements (countries of Asia, North, Central and South America, Australia, Europe, Africa) contribute maximum percentage of species (41.91), followed by American (27.94), African (6.61), Asian (12.5), Australian (4.41) and European (3.67) (Tables 3.48, 3.49). Factors like increasing global trade and transport have been attributed (Drake *et al.*, 1989; Kowarik, 2003; Perrings *et al.*, 2005) for the translocation of species either deliberately or by accident and ultimately in the homogenisation of earth's biota; and are a serious but underestimated problem in biodiversity conservation (Cronk & Fuller, 1995).

Table 3.50: Native Status of Ethnobotanically Important Plants of District Hamirpur

S.No.	Name of the Plant	Family	Origin
1.	*Abelmoschus esculentus* (L.) Moench.	Malvaceae	India
2.	*Abelmoschus moschatus* Medic.	Malvaceae	India, Tropical regions
3.	*Abroma augusta* Jacq.	Sterculiaceae	Hotter parts of India
4.	*Abrus precatorius* L.	Fabaceae	Indonesia, Tropical Africa, Sudan, Kenya, Tanzania, Uganda, Rwanda, Nigeria, Togo, Namibia, Madagascar, India
5.	*Acacia catechu* (L. f.) Willd.	Fabaceae	Asia, China, India, Indian Ocean
6.	*Acacia concinna* (Willd.) DC.	Fabaceae	Asia, Central and South India warmer parts.
7.	*Acacia nilotica* (L.) Delile	Fabaceae	Egypt, Maghreb and Sahel, Mozambiue and Natal (South Africa) Indian subcontinent, Australia.
8.	*Achras zapota* L.	Sapotaceae	South Mexico, Central America, Coastal Yucatan, Tropical America
9.	*Achyranthes aspera* L.	Amaranthaceae	India, Pakistan, Asia
10.	*Achyranthes bidentata* Blume	Amaranthaceae	East Asia, China and Java
11.	*Acorus calamus* L.	Araceae	Europe, India, Myanmar, Sri-Lanka, Arabia, North America, Northern and Eastern Asia.

contd...

Table 3.50: *contd...*

S.No.	Name of the Plant	Family	Origin
12.	*Actinidia chinensis* var. *deliciosa* (A. Chev.) A. Chev.	Actinidiaceae	China, Asia, New-Zealand
13.	*Adhatoda vasica* Nees	Acanthaceae	Asia, Sri-Lanka, Nepal, India, Pakistan
14.	*Adiantum capillus-veneris* L.	Adiantaceae	Central and South America, Eurasia, Tropical origin, North America.
15.	*Adiantum lunulatum* Burm.	Adiantaceae	Goa, India, West coast
16.	*Aegle marmelos* (L.) Correa *ex* Roxb.	Rutaceae	India, Central Mexico
17.	*Aerva lanata* (L.) Juss. *ex* Schult.	Amaranthaceae	Sierra Leone, S. Nigeria, E. Africa, Asia and China, Ceylon, Malesia, Asia to India, North East Tropical- Africa, Ethiopia, Somalia. East-Tropical- Africa, Kenya, Tanzania, Uganda. West-Central Tropical-African Cameroon, Rwanda, Zaire. West-Tropical- Africa. South-Tropical- Africa, Malawi, Mozambique, Zimbabwe, Southern Africa, South Africa.
18.	*Agave americana* L.	Agavaceae	Mexico, United States, Central and tropical South America.
19.	*Ageratum conyzoides* L.	Asteraceae	Tropical America; India, South America, Africa, Australia, Southeast Asia and U.S.A.
20.	*Albizia lebbeck* (L.) Willd.	Fabaceae	Pakistan, India, Bangladesh, the Andaman Islands, Myanmar, S.China, N E Thailand and Malaysia, Sri Lanka, Eastern Islands of Indonesia, Africa and N. Australia.
21.	*Albizia stipulata* Boiv.	Fabaceae	Cambodia, China, India, Indonesia, Laos, Myanmar, Thailand, Vietnam
22.	*Allium cepa* L.	Liliaceae	West or Central Asia. In Europe bronze age Iliad and Odyssey
23.	*Allium sativum* L.	Liliaceae	Central Asia, (Africa, Europe, Ancient Egyptian) Mediterranean region.
24.	*Aloe barbedensis* Mill.	Liliaceae	Somalia, Sudan, Africa
25.	*Aloe vera* Tourn. *ex* L.	Liliaceae	Originated in the warm dry, climate of Africa. Tropical Nothern Africa, Greek, Egyptians, Roman Eras
26.	*Alstonia scholaris* R.Br.	Apocynaceae	Indian subcontinent and Southeast Asia
27.	*Amaranthus paniculatus* L.	Amaranthaceae	Eastern America
28.	*Amaranthus spinosus* L.	Amaranthaceae	Tropical Americas

contd...

Table 3.50: *contd...*

S.No.	Name of the Plant	Family	Origin
29.	*Amaranthus viridis* L.	Amaranthaceae	South America, Tropical and Sub tropical region of world, Asian origin
30.	*Amorphophallus campanulatus* (Roxb.) Blume *ex* Decne.	Araceae	Sumatra
31.	*Andrographis paniculata* Wall. *ex* Nees	Acanthaceae	East Indies, China, India, Sri-Lanka, South Asia
32.	*Anetham sowa* Roxb. *ex* Flem.	Apiaceae	Meditteranean and West Asia and South of Russia
33.	*Annona squamosa* L.	Annonaceae	Tropical Americas, Colombia, Caribbean, Central America, N. South America
34.	*Arachis hypogaea* L.	Fabaceae	Peru, South America, Introduced in Mexico, West Indies
35.	*Argemone mexicana* L.	Papaveraceae	Mexico, West Indies, United states, India, Ethiopia, Western U.S.
36.	*Arisaema tortuosum* (Wall.) Schott.	Araceae	Eastern and Central Africa, Asia and Eastern North America, North India
37.	*Artemisia japonica* Thunb.	Asteraceae	North and South Hemisphere
38.	*Artemisia vulgaris* L.	Asteraceae	Temperate Europe, Asia, Northern Africa, Alaska
39.	*Artocarpus integrifolia* L. f.	Moraceae	Rainforest of Western Ghats, India, Florida
40.	*Artocarpus lakoocha* Roxb.	Moraceae	Indo-Pakistan subcontinent, Myanmar, Malaysia, India
41.	*Asclepias curassavica* L.	Asclepiadaceae	Florida, Mexico, Eastern and Western U.S.
42.	*Asparagus adscendens* Roxb.	Liliaceae	Moluccas, Western Himalayas, Himachal Predesh, Kumaon
43.	*Asparagus racemosus* Willd.	Liliaceae	Upper Gangetic plains, Tropical and Subtropical parts of India including the Himalayas and Andamans.
44.	*Avena sativa* L.	Poaceae	Europe, Asia, North west Africa
45.	*Azadirachta indica* A. Juss.	Meliaceae	Asia, Northern Mayanmar, Assam in India, Myanmar, Sri-Lanka, Tropical and Subtropical Zones, S.E.Asia, Pacific Islands
46.	*Bacopa monnieri* (L.) Penn.	Scrophulariaceae	Southern Asia, India, Hawaii, Puerto Rico, Southern coastal U.S.
47.	*Bambusa arundinacea* (Retz.) Roxb.	Poaceae	South China, Forests of India, Thailand
48.	*Bambusa nutans* Wall.	Poaceae	India, Myanmar, Bangladesh and Thailand
49.	*Bauhinia vahlii* Wight & Arn.	Fabaceae	North Africa, Canary Islands and Spain

Contd...

Table 3.50: *contd...*

S.No.	Name of the Plant	Family	Origin
50.	*Bauhinia variegata* L.	Fabaceae	India and S.E. Asia
51.	*Berberis asiatica* Roxb. *ex* DC.	Berberidaceae	Temperate and subtropical regions of Europe, Asia, Africa, North America and South America
52.	*Boerhavia diffusa* L.	Nyctaginaceae	India, Brazil, Old World tropics, Tropical Africa, Throughout Old World
53.	*Bombax ceiba* L.	Boraginaceae	Western Africa, Indian Subcontinent, South east Asia and northern Australia
54.	*Bougainvillea spectabilis* Willd.	Nyctaginaceae	South America, Brazil, Peru southern Argentina, South America
55.	*Brassica campestris* L.	Brassicaceae	Wild Western Europe, The Mediterranean, Temperate Asia, North America, South America, Australia
56.	*Brassica juncea* (L.) Czern. & Coss.	Brassicaceae	Australia, Mediterranean, Queensland, New southwales, Victoria
57.	*Brassica nigra* (L.) Koch.	Brassicaceae	Southern Mediterranean region of Europe
58.	*Bryophyllum calycinum* Salisb.	Crassulaceae	South Africa, Temperate Asia, Australia, New-Zealand, West India, Macaronesia, Mascarenes, Galapagos, Melanesia, Polynesia, Hawaii
59.	*Butea monosperma* (Lam.) Taub.	Fabaceae	Tropical and Subtropical parts of Indian subcontinent and south Asia, Ranging from India, Bangladesh, Nepal, Pakistan, Sri-Lanka, Mayanmar, Thailand, Laos, Cambodia, Vietnam, Malaysia, Western Indonesia
60.	*Callistemon lanceolatus* DC.	Myrtaceae	States of Queens land, New South Wales and Victoria in Australia
61.	*Calotropis procera* (Ait.) R.Br.	Asclepiadaceae	Israel, North Africa, Tropical Africa, Western Asia, South Asia and Indo-China
62.	*Canna indica* L.	Cannaceae	South America, Caribbean, North and Central America
63.	*Cannabis sativa* L.	Moraceae	Central Asia and Himalayas, Caucasus, China, Iran, India, Middle east
64.	*Canscora diffusa* (Vahl.) R.Br.	Gentianaceae	India, Forests of Karnataka
65.	*Capsicum annuum* L.	Solanaceae	Southern North America and Northern south America, Brazil, Colombia
66.	*Cardiospermum halicacabum* L.	Sapindaceae	East Indies, India, Nepal, Africa
67.	*Carduus nutans* L.	Asteraceae	Europe and asia except for far north

Contd...

Table 3.50: *contd...*

S.No.	Name of the Plant	Family	Origin
68.	*Carica papaya* L.	Caricaceae	Tropics of America, southern Mexico, Central America, West Indies and Bahamas, Bermuda
69.	*Carissa carandas* L. var. *congesta* (Wt.) Bedd.	Apocynaceae	Tropical and subtropical regions of Africa, Australia, Asia, Indian Subcontinent, Myanmar, Malacca, Sri-Lanka
70.	*Carissa spinarum* auct. non L.	Apocynaceae	Australia, Central Australia
71.	*Carthamus oxycantha* Bieb.	Asteraceae	India, Afghanistan, Ethiopia, Turkey, Syria, Lebnan, Iraq
72.	*Carum copticum* Benth.	Apiaceae	Egypt, Eastern Mediterranean, Persia, India, Myanmar, Sri Lanka, Thailand
73.	*Cassia fistula* L.	Fabaceae	Southern Asia, Pakistan, India to Myanmar, Sri Lanka
74.	*Cassia occidentalis* L.	Fabaceae	Central and South America
75.	*Catharanthus roseus* (Linn.) G. Don	Apocynaceae	Madagascar
76.	*Cedrela toona* Roxb. *ex* Rottl.	Meliaceae	Australia, New South wales and Queensland
77.	*Celastrus paniculatus* Willd.	Celastraceae	U.S.
78.	*Celosia argentea* L.	Amaranthaceae	Tropical Origin, Tropical Africa, Benin, Nigeria, Asia, Togo, Ghana, Cameroon, Tropical and Subtropical Asia and America
79.	*Centella asiatica* (L.) Urban.	Apiaceae	Throughout South East Asia China to Madagascar, Asia, India
80.	*Centipeda minima* (L.) A. Br. & Aschers.	Asteraceae	Eastern Asia to China, Malaysia, Tropical Australia
81.	*Cestrum nocturnum* L.	Solanaceae	West Indies
82.	*Chenopodium album* L.	Chenopodiaceae	America, Eastern Asia, Europe
83.	*Chenopodium ambrosioides* L.	Chenopodiaceae	Central and Southern Mexico
84.	*Cicer arietinum* L.	Fabaceae	Central and Southern Mexico
85.	*Cinnamomum camphora* (L.) Nees & Eberm.	Lauraceae	Taiwan, Southern Japan, Southeast China and Indo-china
86.	*Cinnamomum tamala* Nees & Eberm.	Lauraceae	South slopes of Himalayas
87.	*Cissampelos pareira* L.	Menispermaceae	India
88.	*Citrus aurantium* L.	Rutaceae	Southeast Asia, Northeastern India, Myanmar and Yunan, Province of China
89.	*Citrus grandis* (L.) Osbeck.	Rutaceae	Latin America, Tropical and Subtropical parts of India, China, Northern Australia and New Caledonia

Contd...

Table 3.50: *contd...*

S.No.	Name of the Plant	Family	Origin
90.	*Citrus jambhiri* Lush.	Rutaceae	South East Asia, Northestern India, Myanmar, Yunan and China
91.	*Citrus limon* (L.) Burm. f.	Rutaceae	Asia, India, Myanmar, China, South and Southeast Asia
92.	*Citrus medica* L.	Rutaceae	Persia and the land of Medes
93.	*Clematis grata* Wall.	Ranunculaceae	China, Japan and Europe
94.	*Colebrookia oppositifolia* Smith	Lamiaceae	India, Nepal, Kumaon, East Indies
95.	*Colocasia esculenta* (L.) Schott.	Araceae	Wetland of Indo-Malayan region, Tropical Asia, Malyasia
96.	*Cordia dichotoma* Forst. f.	Cordiaceae	India, Myanmar, Nepal, IndoMalaya ecozone, Northern Australia, Western Melanesia
97.	*Coriandrum sativum* L.	Apiaceae	Eastern Mediterranean region, Southern Europe, North Africa, South-western Asia
98.	*Cucumis sativus* L.	Cucurbitaceae	India, Western Asia, Europe, Romans (France), England, North America
99.	*Cucurbita hispida* Wall.	Cucurbitaceae	Asia, China, Southeast Asia
100.	*Cucurbita pepo* L.	Cucurbitaceae	Central America, Mexico, Europe, Asia
101.	*Curcuma domestica* (Medik.) Valeton.	Zingiberaceae	India, Southern Asia, North eastern, Nizamabad, South India
102.	*Cuscuta reflexa* Roxb.	Cuscutaceae	Northern Europe
103.	*Cycas revoluta* L.	Cycadaceae	Far East, Southern Japan
104.	*Cymbopogon citratus* (DC.) Stapf.	Poaceae	India, Sri-Lanka, Malaysia, South Asia, Australia, Myanmar, Thailand
105.	*Cynodon dactylon* (L.) Pers.	Poaceae	Africa, Eurasia, Tropical America, India, Australia, New Southwales, New-Zealand
106.	*Dalbergia sissoo* Roxb. *ex* DC.	Fabaceae	North America, Indian subcontinent, Southern Iran, Persian Myanmar (Kenya, Tanzania)
107.	*Datura metel* auct. non L.	Solanaceae	Central America, Southwestern region of North America
108.	*Datura stramonium* L.	Solanaceae	Caspian region of Europe, North America and Indian subcontinent
109.	*Daucus carota* L.	Apiaceae	Temperate region of Europe, Southwest Asia, Northeast North, America and Australia
110.	*Dendrocalamus hamiltonii* Nees & Arn.	Poaceae	Southern Myanmar
111.	*Dioscorea bulbifera* L.	Dioscoreaceae	S.W. Asia, India, Africa, West Africa, Tropical Asia, East Africa
112.	*Dodonaea viscosa* L. Jacq.	Sapindaceae	Southwest US, Northern Mexico, Hawaii

Contd...

Table 3.50: *contd...*

S.No.	Name of the Plant	Family	Origin
113.	*Duranta repens* L.	Verbenaceae	Mexico, South America, Caribbean
114.	*Eclipta prostrata* Roxb.	Asteraceae	India, China, Thailand and Brazil
115.	*Elaeocarpus ganitrus* Roxb. *ex* G. Don	Elaeocarpaceae	India, Nepal, Indonesia, Queensland, Australia
116.	*Elettaria cardamomum* (L.) Maton.	Zingiberaceae	Southern India, Sri-Lanka
117.	*Emblica officinalis* Gaertn.	Euphorbiaceae	India
118.	*Equisetum debile* Roxb. *ex* Vauch.	Equisetaceae	Europe, North Africa, Southeast China
119.	*Eriobotrya japonica* Lindl.	Rosaceae	Southeastern China
120.	*Eruca sativa* Mill.	Brassicaceae	Central and Southern Europe, Nothern America, Mediterranean region, Morocco and Portugal, Lebnon, Turkey
121.	*Eucalyptus umbellata* Domin.	Myrtaceae	Eastern Australia
122.	*Eugenia cerasoides* Roxb.	Myrtaceae	Western Ghats India
123.	*Eugenia jambolana* Lam.	Myrtaceae	India, Nepal, Pakistan and Indonesia
124.	*Eupatorium adenophorum* Spreng.	Asteraceae	Central America
125.	*Euphorbia pulcherrima* Willd.	Euphorbiaceae	Mexico, Central America, Guatemala
126.	*Euphorbia royleana* Boiss.	Euphorbiaceae	Ethiopian
127.	*Ficus benghalensis* L.	Moraceae	Indian Subcontinent
128.	*Ficus carica* L.	Moraceae	Western Asia, Mediterranean Countries of Europe, North Africa, England, Mexico
129.	*Ficus glomerata* Roxb.	Moraceae	Australasia, South-east Asia, Indian Subcontinent
130.	*Ficus hispida* L.f.	Moraceae	India, Asia
131.	*Ficus palmata* Forsk.	Moraceae	South East Asia
132.	*Ficus religiosa* L.	Moraceae	India and Asia, Tropica Indonesia, India, China
133.	*Ficus roxburghii* Wall. *ex* Miq.	Moraceae	China, Bhutan, India, Nepal, Malaysia, Myanmar; Pakistan, Thailand, Vietnam
134.	*Flacourtia ramontchi* L' Herit.	Flacourtiaceae	Africa, Tropical and Temperate parts of Asia
135.	*Foeniculum vulgare* Mill.	Apiaceae	Mediterranean and North Monastries

Contd...

Table 3.50: *contd...*

S.No.	Name of the Plant	Family	Origin
136.	*Fragaria chiloensis* (L.) Duchesne	Rosaceae	India, Coastal California, Montara, Pacific Ocean, North and South America, Hawaii, Chile, Argentina
137.	*Fumaria indica* (Haussk.) Pugsley	Fumariaceae	Europe, Temperate- Asia, Afganistan, Persia, India
138.	*Glycine max* Merrill.	Fabaceae	North America, South East Asia, North and Central China, India, Korea and Taiwan
139.	*Glycyrrhiza glabra* L.	Fabaceae	Spain, Germany, France, Italy, Russia
140.	*Gossypium arboreum* L.	Malvaceae	India, Pakistan, East Africa
141.	*Grevillea robusta* A. Cunn. *ex* R. Br.	Proteaceae	Eastern Coastal Australia
142.	*Grewia oppositifolia* Roxb.	Tiliaceae	India, East Asia
143.	*Habenaria pectinata* Don	Orchidaceae	Temperate and Tropical Grasslands, North America, North Mexico
144.	*Helianthus annuus* L.	Asteraceae	Masoamerica, Mexico
145.	*Helicteres isora* L.	Sterculiaceae	Nepal, Bihar, Bengal, central, Western and Southern India, Andaman Islands
146.	*Hibiscus rosa-sinensis* L.	Malvaceae	Southeastern Asia, Tropical and Subtropical parts, China
147.	*Holarrhena pubescens* (Buch.-Ham.) Wall. *ex* G. Don	Apocynaceae	Indian Subcontinent
148.	*Hordeum vulgare* L.	Poaceae	Israel-Jordan, Near East to South Asia, Egypt, Ethopia, Tibet
149.	*Hypericum cernum* Roxb.	Hypericaceae	Europe and Asia
150.	*Impatiens balsamina* L.	Balsaminaceae	India, Myanmar, China
151.	*Ipomoea cairica* (L.) Sweet	Convolvulaceae	Africa
152.	*Ipomoea carnea* Jacq.	Convolvulaceae	South America
153.	*Jacaranda acutifolia* auct. non- Humb. & Bonpl.	Bignoniaceae	South America
154.	*Jasminum humile* L.	Oleaceae	Switzerland, Central Asia
155.	*Jasminum multiflorum* (Burm. f.) Andr.	Oleaceae	India, Penninsular part of Florida
156.	*Jasminum sambac* (L.) Ait	Oleaceae	South East Asia, South China, India
157.	*Jatropha curcus* L.	Euphorbiaceae	India

Contd...

Table 3.50: *contd...*

S.No.	Name of the Plant	Family	Origin
158.	*Juglans regia* L.	Juglandaceae	Asia, Temperate Afghanistan, Erbaijan, Iran, Iraq, Turkey, Tajikistan, Tropical Asia, Pakistan, Europe, Bulgaria, Yugoslavia, Greece
159.	*Lagenaria siceraria* (Molina) Standl.	Cucurbitaceae	Mexico, Peru, New Mexico, Zimbabwe, Arizona, Colorado, Pennysylvania
160.	*Lagerstroemia indica* L.	Lythraceae	Indian subcontinent, Southern Asia, Northern Australia, Oceania, China
161.	*Lannea coromandelica* (Houtt.) Merrill.	Anacardiaceae	Bangladesh; Bhutan; Combodia; China; India-Andaman and Nicobar, Arunachal Pradesh, Assam, Bihar, Gujarat, Himachal Pradesh, Jammu and Kashmir, Karnatka, Kerala, Madhya Pradesh, Maharashtra, Meghalaya, Orissa, Rajasthan, Sikkim, Tamil Nadu, Tripura, Uttar Pradesh, West Benga; Laos; Mayanmar; Nepal; Pakistan; Sri- Lanka; Thailand; Vietnam
162.	*Lantana camara* L.	Verbenaceae	Tropical regions of America & Africa, Australian Pacific region, Jamaica
163.	*Lavandula angustifolia* Mill.	Lamiaceae	Cape Verde & Canary Islands, Europe, Northern & Eastern Africa, Mediterranean, South-West Asia, South-East India
164.	*Lawsonia alba* Lam.	Lythraceae	Tropical & Subtropical regions of Africa, Southern Asia, Northern Australasia, Tropical Savannah
165.	*Lens culinaris* Medic.	Fabaceae	Near East Asia
166.	*Lepidium sativum* L.	Brassicaceae	Iran, Western or Central Asia, Southern America, Peru, Europe, Spanish
167.	*Leucaena leucocephala* (Lam.) de Wit.	Fabaceae	Southern Mexico, Northern Central America (Belize & Guatemala)
168.	*Linum usitatissimum* L.	Linaceae	Mediterranean to India, Ethiopia, Ancient Egypt, Republic of Georgia
169.	*Litchi chinensis* (Gaertn.) Sonn.	Sapindaceae	Provinces of Kwangtung and Fukien in Southern China, Southeastern Asia
170.	*Luffa aegyptiaca* Mill.	Cucurbitaceae	Africa, Asia, Tropical & Subtropical parts of the World, West Africa, India
171.	*Lycopersicon esculentum* Mill.	Solanaceae	South America, Andes region of Peru, Northern Chile
172.	*Madhuca indica* J. F. Gmel.	Sapotaceae	Tropical India, Central & Northern plains and forests of India

Contd...

Table 3.50: *contd...*

S.No.	Name of the Plant	Family	Origin
173.	*Mallotus philippensis* (Lam.) Muell.-Arg.	Euphorbiaceae	Asia Afghanistan, Australia, Mount Keira, Sydney, Philippines
174.	*Malvaviscus conzatii* Greenm.	Malvaceae	Mexico
175.	*Mangifera indica* L.	Anacardiaceae	East Asia, Philippines, Africa, Brazil, Portuguese, Southern Asia, Myanmar, Andaman Islands, India
176.	*Martynia annua* L.	Martyniaceae	North America, Mexico
177.	*Melia azedarach* L.	Meliaceae	China, Baluchistan, Kashmir, Western Himalayas, Pakistan, India, Indochina, Southeast Asia
178.	*Melissa officinalis* L.	Lamiaceae	Europe, North America, Iran
179.	*Mentha piperata* L.emend. Huds.	Lamiaceae	Europe
180.	*Mentha spicata* L. emend. Nathh.	Lamiaceae	Eurasia, Britain
181.	*Mesua ferrea* L.	Clusiaceae	Tropical Sri-Lanka, Southern Nepal, Indochina, Malay Peninsula
182.	*Millettia extensa* Benth. *ex* Baker	Fabaceae	India
183.	*Mimosa pudica* L.	Fabaceae	South America, Central America, Tropical America
184.	*Mirabilis jalapa* L.	Nyctaginaceae	Mexico, Peruvian Andes
185.	*Momordica charantia* L.	Cucurbitaceae	Tropical & Subtropical Asia, Africa, Caribbean
186.	*Moringa oleifera* Lam.	Moringaceae	Tropical & Subtropical World, Southern Africa, Cambodia, Foothills of Himalayas, Nepal, Indonesia, Malaysia, Mexico, Central & South America, Sri-Lanka
187.	*Morus alba* L.	Moraceae	Northern China
188.	*Murraya koenigii* (L.) Spreng.	Rutaceae	India Ceylon
189.	*Murraya paniculata* (L.) Jack.	Rutaceae	Asia (Southern China to Northeast Pakistan, Sri-Lanka, Malaysia, Malesia, Northern Australia
190.	*Musa sapientum* L.	Musaceae	Tropical Southeast Asia, Philippines, Papua New guinea, South Asia
191.	*Nasturtium officinale* R. Br.	Brassicaceae	Europe, Central Asia
192.	*Nerium indicum* Mill.	Apocynaceae	AFRICA-Algeria, Ethiopia, Libya, Morocco, Tunisia; ASIA-TEMPERATE-Afganistan, China, Cyprus, Iran, Iraq, Israel, Jordan, Lebnon, Oman, United Arab Emirates, Syria, Turkey; ASIA-TROPICAL- India-Jammu and Kashmir, Punjab, Nepal, Pakistan; EUROPE- Albania, Former Yugoslavia, Greece, Italy, Malta; France; Portugal, Spain

Contd...

Table 3.50: *contd...*

S.No.	Name of the Plant	Family	Origin
193.	*Nicotiana tabacum* L.	Solanaceae	Tropical America, Hispaniola, Africa
194.	*Ocimum basilicum* L.	Lamiaceae	Tropical Asia, Africa, India, France
195.	*Ocimum sanctum* L.	Lamiaceae	South, Western Asia, Australia
196.	*Opuntia dillenii* (Ker-Gawl.) Haw.	Cactaceae	Mexico
197.	*Oroxylum indicum* Vent.	Bignoniaceae	India, Bhutan, Southern China, Indo China, Malesia
198.	*Oryza sativa* L.	Poaceae	Asia, Myanmar, Indonesia, Southeast Asia
199.	*Oxalis corniculata* L.	Oxalidaceae	Cosmopolitan
200.	*Papaver rhoeas* L.	Papaveraceae	Europe, Mediterranean, Northern Africa, Western Asia, Indian subcontinent-Pakistan
201.	*Parthenium hysterophorus* L.	Asteraceae	TROPICAL and SUBTROPICAL America, India
202.	*Paspalum scrobiculatum* L.	Poaceae	Hawaii, Lower states of America, Canada, Greenland, France, North America
203.	*Pedilanthus tithymaloides* (L.) Poir.	Euphorbiaceae	Tropical & Subtropical North America, Central America, Florida, Mexico, Caribbean
204.	*Pelargonium graveolens* L'Hert.	Geraniaceae	South Africa
205.	*Pennisetum americanum* L.	Poaceae	Africa, West Africa, Indian Subcontinent-Pakistan, India
206.	*Pennisetum purpureum* Schum.	Poaceae	Tropical Africa, Kenya, Tanzania, Uganda, Ethiopia, Angola, Malawi, Mozambique, Zambia, Zimbabwe, Ghana, Guinea, Liberia, Nigeria, Sierra Leone, Togo, Cameroon
207.	*Phoenix sylvestris* Roxb.	Arecaceae	India, Nepal, Pakistan
208.	*Phragmites karka* Trin. *ex* Steud.	Poaceae	South Asia, North America, India Australia
209.	*Phyllanthus urinaria* L.	Euphorbiaceae	India, Tropical east Asia
210.	*Physalis peruviana* L.	Solanaceae	South America, South Africa, Tropical Peru, Colombia, Ecuador
211.	*Pinus roxburghii* Sarg.	Pinaceae	Himalayas, Pakistan (North-West Frontier Province, Azad Kashmir, Margala Hills, Islamabad, North India, Nepal and Bhutan
212.	*Piper longum* L.	Piperaceae	North-eastern, Southern India, Sri-Lanka, Bengal
213.	*Pistacia integerrima* Stewart *ex* Brandis.	Anacardiaceae	Mediterranean region to East Asia, Himalayan ranges on the North-West
214.	*Plumbago zeylanica* L.	Plumbaginaceae	Southern India, Malaysia, Tropical Old World

Contd...

Table 3.50: *contd...*

S.No.	Name of the Plant	Family	Origin
215.	*Pogostemon benghalensis* Kuntze.	Lamiaceae	Bangladesh, China, India, Myanmar, Nepal, Sri-Lanka, Thailand, Asia, Philippines, Vietnam, West Africa
216.	*Polyalthia longifolia* (Sonn.) Thwaites	Annonaceae	India, Sri-Lanka
217.	*Pongamia pinnata* Pierre.	Fabaceae	Tropical & Temperate Asia, India, China, Japan, Malesia, Australia, Pacific islands
218.	*Populus nigra* L.	Moraceae	Europe, Southwest Central Asia, Northwest Africa
219.	*Premna latifolia* Roxb.	Verbenaceae	East Africa, India, Bangladesh, Japan, China, Taiwan, Guinea, Australia, Polynesia, Melanesia
220.	*Prunus amygdalus* Batsch.	Rosaceae	Middle East, South Asia, Mediterranean, Northern Africa, Southern Europe, California, West Asia
221.	*Prunus domestica* L.	Rosaceae	Northern temperate region of globe, North America
222.	*Prunus persica* Batsch.	Rosaceae	China, Persia
223.	*Psidium guajava* L.	Rosaceae	Mexico, Central America, Northern South America, Africa, Southeast Asia, Caribbean, North America, Australia
224.	*Punica granatum* L.	Punicaceae	Iran, Saudi Arabia. Afghanistan, Baluchistan, India
225.	*Putranjiva roxburghii* Wall.	Euphorbiaceae	Indo-China
226.	*Pyrus communis* L.	Rosaceae	Central to Eastern Europe, South-West Asia
227.	*Pyrus pashia* Ham.	Rosaceae	Southern Asia, China, India, Iran, Afghanistan
228.	*Randia dumetorum* Poir.	Rubiaceae	Ceylon & India, America, Texas, Florida, China
229.	*Ranunculus muricatus* L.	Ranunculaceae	Europe, Africa, Australia, Western & Eastern America, California, Oregon, Washington
230.	*Raphanus sativus* L.	Brassicaceae	Europe, Roman, Asia
231.	*Rauwolfia serpentina* Benth. *ex* Kurz.	Apocynaceae	South-east Asia, (India & China, Indonesia)
232.	*Reinwardtia indica* Dum.	Linaceae	India, China
233.	*Rhynchostylis retusa* Blume	Orchidaceae	Myanmar, India, Malaya, Singapore, Thailand, Sri-Lanka
234.	*Ricinus communis* L.	Euphorbiaceae	Africa, India, Mediterranean Basin
235.	*Rosa damascena* Mill.	Rosaceae	Middle East, Syria, Europe, East Asia, Western or Central Asia, North Africa
236.	*Rosmarinus officinalis* L.	Lamiaceae	Mediterranean region, Asia

Contd...

Table 3.50: *contd...*

S.No.	Name of the Plant	Family	Origin
237.	*Roylea cinerea* (D. Don) Baillon.	Lamiaceae	N.W.India, Nepal
238.	*Rubus ellipticus* Sm.	Rosaceae	S.E.Asia, India, Himalayas
239.	*Rumex hastatus* D. Don	Polygoniaceae	Asia, India, Afghanistan
240.	*Ruta graveolens* L.	Rutaceae	Mediterranean, Macaronesia, Southwest Asia
241.	*Saccharum arundinaceum* Reitz.	Poaceae	China, India
242.	*Saccharum officinarum* L.	Poaceae	South Asia, Hawaii, South pacific Islands, New Guinea, Florida to Texas
243.	*Saccharum spontaneum* L.	Poaceae	Tropical Asia, Europe, Turkmenistan, Iran, Southeast Asia, New Guinea, India, Eastern Africa,
244.	*Salvia coccinea* L.	Lamiaceae	America-Florida, Mexico, South western U.S.A., Mexico, Central America, Caribbean, South America, Colombia, Peru, Brazil
245.	*Santalum album* L.	Santalaceae	India, Bangladesh, Sri-Lanka, Australia, Indonesia, Asia, Sri-Lanka, Malaysia, Indonesia, Taiwan
246.	*Sapindus mukorossi* Gaertn.	Sapindaceae	Temperate & Tropical regions, Old World, China, Japan
247.	*Saraca asoca* (Roxb.) De Wilde.	Fabaceae	Indian Subcontinent, Deccan plateau, Western ghats in India, Myanmar, Malaysia
248.	*Sechium edule* Sw.	Cucurbitaceae	Mexico, Costa Rica, Tropical & Central America, Masoamerica
249.	*Sesamum indicum* L.	Pedaliaceae	Africa, India, Egypt, Babylon, Assyria, Turkey
250.	*Setaria viridis* Beauv.	Poaceae	Euasia, China, Europe
251.	*Solanum erianthum* D. Don	Solanaceae	Southern North America, Northern South America, Galapagos Islands
252.	*Solanum melongena* L.	Solanaceae	Indian subcontinent, Southern-eastern Asia
253.	*Solanum nigrum* L.	Solanaceae	India, Asia, Africa
254.	*Solanum surattense* Burm. f.	Solanaceae	India
255.	*Solanum tuberosum* L.	Solanaceae	Southern Peru, N W Bolivia, Chile, Europe, South America, Europe, Africa, Asia, North America
256.	*Sonchus asper* Hill.	Asteraceae	India, Europe
257.	*Soymida febrifuga* A. Juss.	Meliaceae	Indo Malaysia
258.	*Spilanthes acmella* auct. non (L.) Murr.	Asteraceae	South Asia, India
259.	*Spinacia oleracea* L.	Chenopodiaceae	Central, South-Western Asia, Ancient Persia, Ancient China

Table 3.50: *contd...*

S.No.	Name of the Plant	Family	Origin
260.	*Stephania glabra* (Roxb.) Miers.	Menispermaceae	Eastern & Southern Asia, Australasia
261.	*Tabernaemontana* divaricata (L.) R. Br.	Apocynaceae	North India, Tropics & Southeast Asia & India, Thailand
262.	*Tagetes erecta* L.	Asteraceae	North America, South America, South Europe
263.	*Tamarindus indica* L.	Fabaceae	Tropical Africa, Sudan, Cameroon, Nigeria, Tanzania, Arabia, Oman, Dhofar, Africa, South Asia, Australia, South East Asia, Taiwan, China
264.	*Taraxacum officinale* Weber *ex* Wigg.	Asteraceae	Eurasia
265.	*Tecoma grandiflora* Loisel.	Bignoniaceae	China, Japan
266.	*Tecoma stans*(L.) H.B. &K.	Bignoniaceae	South-western America, Florida, Mexico, Caribbean, Bahamas, Central America, South America, Argentina
267.	*Terminalia arjuna* (Roxb.) Wt. & Arn.	Combretaceae	India
268.	*Terminalia bellirica* (Gaertn.) Roxb.	Combretaceae	South-east Asia, Bangladesh, Cambodia, Bhutan
269.	*Terminalia chebula* Retz.	Combretaceae	India, South-east Asia, China, Nepal, Sri-Lanka, Malaysia, Vietnam
270.	*Thevetia neriifolia* Juss. *ex* Steud.	Apocynaceae	West Tropical Africa
271.	*Thuja orientalis* L.	Cupressaceae	North America
272.	*Tinospora cordifolia* (Willd.) Miers ex Hook. f. &Thoms.	Menispermaceae	India, Myanmar, Sri-Lanka
273.	*Trichodesma indicum* R.Br.	Boraginaceae	India, Mayanmar, Philippines, Mauritius
274.	*Tridax procumbens* L.	Asteraceae	Tropical America, India
275.	*Trifolium alexandrinum* L.	Fabaceae	England, Europe, Mediterranean region, Central and Northern Asia
276.	*Trigonella foenum-graecum* L.	Fabaceae	Near-East, Eastern Mediterranean, Greece
277.	*Triticum aestivum* L.	Poaceae	North Africa, Europe, East Asia, North America, Near East, Ethiopian Highlands
278.	*Tylophora indica* (Burm.f.) Merrill.	Asclepiadaceae	India
279.	*Urtica dioica* L.	Urticaceae	India, Europe, Asia, North America, North Africa
280.	*Vanda roxburghii* R.Br.	Orchidaceae	North-east India, Asia, Myanmar, Thailand, Indo China, Malaya
281.	*Verbascum thapsus* L.	Scrophulariaceae	Europe, Northern Africa, Asia, Temperate & Tropical Asia

Contd...

Table 3.50: *contd...*

S.No.	Name of the Plant	Family	Origin
282.	*Vernonia anthelmintica* Willd.	Asteraceae	Africa
283.	*Vicia pallida* Facquem.	Fabaceae	Africa, South-West Asia, Egypt, Ethiopia, South Africa, Sudan, Tunisia, Uganda, Zaire, Zimbabwe, Keniya, Algeria
284.	*Viola canescens* Wall.	Violaceae	Europe, Britain, Asia
285.	*Vitex negundo* L.	Verbenaceae	Eastern & Southern Africa, Asia, India
286.	*Vitis jacquemontii* Parker	Vitaceae	Northern Hemisphere, North America, Asia, Southern Europe, South-Western Asia
287.	*Vitis vinifera* L.	Vitaceae	Mediterranean region, Central Europe, South-Western Asia, Moroco, Portugal, Southern Germany, East to Northern Iran
288.	*Wendlandia heynei* (Roem. & Schult.) Sant. & Merch.	Rubiaceae	India, Asia tropical
289.	*Withania somnifera* (L.) Dunal	Solanaceae	AFRICA- Algeria, Angola, Cape Verde, Egypt, Liberia, Morocco, Nigeria, Somalia, south Africa, Sudan, Tanzania, Tunisia, Uganda, Zambia, Zimbabwe, ASIA TEMPERATE- Afganistan, Arabia, Iran, Iraq, Israel, Jorden, Lebnon, Syria, Turkey; ASIA-TROPICAL- India, Pakistan, Sri Lanka; Spain EUROPE- Greece, Italy
290.	*Woodfordia floribunda* Salisb.	Lythraceae	Africa, South East Asia
291.	*Xanthium strumarium* L.	Asteraceae	America, India, Sri Lanka
292.	*Yucca aloifolia* L.	Liliaceae	North America, South America, Florida
293.	*Zanthoxylum armatum* DC.	Rutaceae	Europe, North America, South America
294.	*Zea mays* L.	Poaceae	Mesoamerica, Southern Mexico, Cuba, South-East Europe, Northern Africa, Weastern China, Philippines & East Indies, France, Italy
295.	*Zingiber officinale* Rosc.	Zingiberaceae	South Asia, East Africa, Caribbean
296.	*Zizyphus jujuba* (Lam.) Gaertn. non-Mill.	Rhamnaceae	China, Far East, Middle East, Africa

Table 3.51: Exotic Dicot Families of District Hamirpur, H.P.

Family	Botanical Names
Actinidiaceae	*Actinidia chinensis* var. *deliciosa*
Amaranthaceae	*Amaranthus paniculatus, Amaranthus spinosus, Amaranthus viridis, Celosia argentea*
Annonaceae	*Annona squamosa*
Apiaceae	*Anetham sowa, Coriandrum sativum, Daucus carota, Foeniculum vulgare*
Apocynaceae	*Carissa spinarum, Catharanthus roseus, Thevetia neriifolia*
Asclepiadaceae	*Asclepias curassavica, Calotropis procera*
Asteraceae	*Ageratum conyzoides, Artemisia japonica, Artemisia vulgaris, Carduus nutans, Centipeda minima, Eupatorium adenophorum, Helianthus annuus, Sonchus asper, Tagetes erecta, Taraxacum officinale, Tridax procumbens, Vernonia anthelmintica, Xanthium strumarium*
Balsaminaceae	*Impatiens balsamina*
Berberidaceae	*Berberis asiatica*
Bignoniaceae	*Jacaranda acutifolia, Tecoma grandiflora, Tecoma stans*
Brassicaceae	*Brassica campestris, Brassica juncea, Brassica nigra, Eruca sativa, Lepidium sativum, Nasturtium officinale, Raphanus sativus*
Cactaceae	*Opuntia dillenii*
Caricaceae	*Carica papaya*
Celastraceae	*Celastrus paniculatus*
Chenopodiaceae	*Chenopodium album, Chenopodium ambrosioides, Spinacia oleracea*
Convolvulaceae	*Ipomoea cairica, Ipomoea carnea*
Crassulaceae	*Bryophyllum calycinum*
Cucurbitaceae	*Cucurbita pepo, Lagenaria siceraria, Sechium edule*
Cuscutaceae	*Cuscuta reflexa*
Euphorbiaceae	*Euphorbia pulcherrima, Euphorbia royleana, Mallotus philippinensis, Pedilanthus tithymaloides*
Fabaceae	*Arachis hypogaea, Bauhinia vahlii, Cassia occidentalis, Cicer arietinum, Glycyrrhiza glabra, Lens culinaris, Leucaena leucocephala, Mimosa pudica, Trifolium alexandrinum, Trigonella foenum-graecum, Vicia pallida*
Geraniaceae	*Pelargonium graveolens*
Hypericaceae	*Hypericum cernum*
Lamiaceae	*Melissa officinalis, Mentha piperata, Mentha spicata, Ocimum sanctum, Rosmarinus officinalis, Salvia coccinea*
Malvaceae	*Malvaviscus conzatii*
Martyniaceae	*Martynia annua*

Contd...

Table 3.51: *contd...*

Family	Botanical Names
Meliaceae	*Cedrela toona,*
Moraceae	*Ficus carica, Morus alba, Populus nigra*
Myrtaceae	*Callistemon lanceolatus, Eucalyptus umbellata*
Nyctaginaceae	*Bougainvillea spectabilis, Mirabilis jalapa*
Oleaceae	*Jasminum humile*
Oxalidaceae	*Oxalis corniculata*
Proteaceae	*Grevillea robusta*
Ranunculaceae	*Clematis grata, Ranunculus muricatus*
Rhamnaceae	*Zizyphus jujuba*
Rosaceae	*Eriobotrya japonica, Prunus domestica, Prunus persica, Pyrus communis, Rosa damascena*
Rutaceae	*Citrus medica, Murraya paniculata, Ruta graveolens, Zanthoxylum armatum*
Sapindaceae	*Dodonaea viscosa, Sapindus mukorossi*
Sapotaceae	*Achras zapota*
Solanaceae	*Capsicum annuum, Cestrum nocturnum, Datura metel, Lycopersicon esculentum, Nicotiana tabacum, Physalis peruviana, Solanum erianthum, Solanum tuberosum*
Verbenaceae	*Duranta repens, Lantana camara*
Violaceae	*Viola canescens*
Vitaceae	*Vitis vinifera, Vitis jacquemontii*

Table 3.52: Exotic Monocot Families of District Hamirpur

Family	Botanical Names
Agavaceae	*Agave americana*
Araceae	*Amorphophallus campanulatus*
Liliaceae	*Allium cepa, Allium sativum, Aloe barbedensis, Aloe vera, Yucca aloifolia*
Poaceae	*Avena sativa, Dendrocalamus hamiltonii, Oryza sativa, Paspalum scrobiculatum, Pennisetum purpureum, Setaria viridis, Triticum aestivum, Zea mays*

Table 3.53: Ethnobotanically Important Indigenous Plants of District Hamirpur

S.No.	Name of the Plant	Family	Origin
1.	*Abelmoschus esculentus* (L.) Moench.	Malvaceae	India
2.	*Abelmoschus moschatus* Medic.	Malvaceae	India, Tropical regions
3.	*Abroma augusta* Jacq.	Sterculiaceae	Hotter parts of India
4.	*Abrus precatorius* L.	Fabaceae	Indonesia, Tropical Africa, Sudan, Kenya, Tanzania, Uganda, Rwanda, Nigeria, Togo, Namibia, Madagascar, India
5.	*Acacia catechu* (L. f.) Willd.	Fabaceae	Asia, China, India, Indian Ocean
6.	*Acacia concinna* (Willd.) DC.	Fabaceae	Asia, Central and South India warmer parts.
7.	*Acacia nilotica* (L.) Delile	Fabaceae	Egypt, Maghreb and Sahel, Mozambiue and Natal (South Africa) Indian subcontinent, Australia.
8.	*Achyranthes aspera* L.	Amaranthaceae	India, Pakistan, Asia
9.	*Achyranthes bidentata* Blume	Amaranthaceae	East Asia, China and Java, India
10.	*Acorus calamus* L.	Araceae	Europe, India, Myanmar, Sri-Lanka, Arabia, Native to North America, Nothern and Eastern Asia.
11.	*Adhatoda vasica* Nees	Acanthaceae	Asia, Sri Lanka, Nepal, India, Pakistan
12.	*Adiantum lunulatum* Burm.	Adiantaceae	Goa, India, West coast
13.	*Aegle marmelos* (L.) Correa *ex* Roxb.	Rutaceae	India, Central Mexico
14.	*Aerva lanata* (L.) Juss. *ex* Schult.	Amaranthaceae	Sierra Leone, S. Nigeria, E. Africa, Asia and China, Sri Lanka, Malesia, Asia to India, North East Tropical-Africa, Ethiopia, Somalia. East-Tropical-Africa, Kenya, Tanzania, Uganda. West-Central Tropical-African Cameroon, Rwanda, Zaire. West-Tropical- Africa. South-Tropical- Africa, Malawi, Mozambique, Zimbabwe, Southern Africa, South Africa.
15.	*Albizia lebbeck* (L.) Willd.	Fabaceae	Pakistan, India, Bangladesh, the Andaman Islands, Myanmar, S.China, N E Thailand and Malaysia, Sri Lanka, Eastern Islands of Indonesia, Africa and N. Australia.
16.	*Albizia stipulata* Boiv.	Fabaceae	Cambodia, China, India, Indonesia, Laos, Myanmar, Thailand, Vietnam
17.	*Alstonia scholaris* R.Br.	Apocynaceae	Indian subcontinent and Southeast Asia

Table 3.53: Ethnobotanically Important Indigenous Plants of District Hamirpur

S.No.	Name of the Plant	Family	Origin
18.	*Andrographis paniculata* Wall. *ex* Nees	Acanthaceae	East Indies; China, India, Sri-Lanka, South Asia
19.	*Argemone mexicana* L.	Papaveraceae	Mexico, West Indies, United states, India, Ethiopia, Western U.S.
20.	*Arisaema tortuosum* (Wall.) Schott.	Araceae	Eastern and Central Africa, Asia and Eastern North America, North India
21.	*Artocarpus integrifolia* L. f.	Moraceae	Rainforest of Western Ghats, India, Florida
22.	*Artocarpus lakoocha* Roxb.	Moraceae	Indo-Pakistan subcontinent, Myanmar, Malaysia, India
23.	*Asparagus adscendens* Roxb.	Liliaceae	Moluccas, Western Himalayas, Hamachal Predesh, Kumaon
24.	*Asparagus racemosus* Willd.	Liliaceae	Upper Gangetic plains, Tropical and Sub-Tropical parts of India including the Himalayas and Andamans.
25.	*Azadirachta indica* A. Juss.	Meliaceae	Asia, Northern Mayanmar, Assam in India, Myanmar, Sri-Lanka, Tropical and subtropical Zones, S.E.Asia, Pacific Islands
26.	*Bacopa monnieri* (L.) Penn.	Scrophulariaceae	Southern Asia, India, Hawaii, Puerto Rico, Southern coastal U.S.
27.	*Bambusa arundinaceae* (Retz.) Roxb.	Poaceae	South China, Forests of India, Thailand
28.	*Bambusa nutans* Wall.	Poaceae	India, Myanmar, Bangladesh, Myanmar and Thailand
29.	*Bauhinia variegata* L.	Fabaceae	India and S.E. Asia
30.	*Boerhavia diffusa* L.	Nyctaginaceae	India, Brazil, Old World tropics, Tropical Africa, Throughout Old World
31.	*Bombax ceiba* L.	Boraginaceae	Western Africa, Indian Subcontinent, South east Asia and northern Australia
32.	*Butea monosperma* (Lam.) Taub.	Fabaceae	Tropical and Subtropical parts of Indian subcontinent and south Asia, Ranging from India, Bangladesh, Nepal, Pakistan, Sri-Lanka, Mayanmar, Thailand, Laos, Cambodia, Vietnam, Malaysia, Western Indonesia
33.	*Canna indica* L.	Cannaceae	South America, Caribbean, North and Central America, India
34.	*Cannabis sativa* L.	Moraceae	Central Asia and Himalayas, Caucasus, China, Iran, India, Middle east
35.	*Canscora diffusa* (Vahl.) R.Br.	Gentianaceae	India, Forests of Karnataka
36.	*Cardiospermum halicacabum* L.	Sapindaceae	East Indies, India, Nepal, Africa

Contd...

Table 3.53: *contd...*

S.No.	Name of the Plant	Family	Origin
37.	*Carissa carandas* L. var. *congesta* (Wt.) Bedd.	Apocynaceae	Tropical and subtropical regions of Africa, Australia, Asia, Indian Subcontinent, Myanmar, Malacca, Sri-Lanka
38.	*Carthamus oxycantha* Bieb.	Asteraceae	India, Afghanistan, Ethiopia, Turkey, Syria, Lebnan, Iraq
39.	*Carum copticum* Benth.	Apiaceae	Egypt, Eastern Mediterranean, Persia, India, Myanmar, Sri Lanka, Thailand
40.	*Cassia fistula* L.	Fabaceae	Southern Asia, Pakistan, India to Myanmar, Sri Lanka
41.	*Centella asiatica* (L.) Urban.	Apiaceae	Native to Throughout South East Asia China to Madagascar , Asia, India
42.	*Cinnamomum camphora* (L.) Nees & Eberm.	Lauraceae	Taiwan, southern Japan, Southeast China and Indo-china
43.	*Cinnamomum tamala* Nees & Eberm.	Lauraceae	South slopes of Himalayas
44.	*Cissampelos pareira* L.	Menispermaceae	India
45.	*Citrus aurantium* L.	Rutaceae	Southeast Asia, Northeastern India, Myanmar and Yunan, Province of China
46.	*Citrus grandis* (L.) Osbeck.	Rutaceae	Latin America, Tropical and Subtropical parts of India, China, Northern Australia and New Caledonia
47.	*Citrus jambhiri* Lush.	Rutaceae	South East Asia, Northestern India, Myanmar, Yunan and China
48.	*Citrus limon* (L.) Burm. f.	Rutaceae	Asia, India, Myanmar, China, South and Southeast Asia
49.	*Colebrookia oppositifolia* Smith	Lamiaceae	India, Nepal, Kumaon, East Indies
50.	*Colocasia esculenta* (L.) Schott.	Araceae	Wetland of Indo-Malayan region, Tropical Asia, Malyasia
51.	*Cordia dichotoma* Forst. f.	Cordiaceae	India, Myanmar, Nepal, IndoMalaya ecozone, Northern Australia, Western Melanesia
52.	*Cucumis sativus* L.	Cucurbitaceae	India, Western Asia, Europe, Romans (France), England, North America
53.	*Cucurbita hispida* Wall.	Cucurbitaceae	Asia, China, Southeast Asia
54.	*Curcuma domestica* (Medik.) Valeton.	Zingiberaceae	India, Southern Asia, North eastern, Nizamabad, South India
55.	*Cymbopogon citratus* (DC.) Stapf.	Poaceae	India, Sri-Lanka, Malaysia, South Asia, Australia, Myanmar, Thailand
56.	*Cynodon dactylon* (L.) Pers.	Poaceae	Africa, Eurasia, Tropical America, India, Australia, New Southwales, New-Zealand

Contd...

Table 3.53: *contd...*

S.No.	Name of the Plant	Family	Origin
57.	*Dalbergia sissoo* Roxb. *ex* DC.	Fabaceae	North America, Indian subcontinent, Southern Iran, Persian Myanmar (Kenya, Tanzania)
58.	*Datura stramonium* L.	Solanaceae	Caspian region of Europe, North America and Indian subcontinent
59.	*Dioscorea bulbifera* L.	Dioscoreaceae	S.W. Asia, India, Africa, West Africa, Tropical Asia, East Africa
60.	*Eclipta prostrata* Roxb.	Asteraceae	India, China, Thailand and Brazil
61.	*Elaeocarpus ganitrus* Roxb. *ex* G. Don	Elaeocarpaceae	India, Nepal, Indonesia, Queensland, Australia
62.	*Elettaria cardamomum* (L.) Maton.	Zingiberaceae	Southern India, Sri-Lanka
63.	*Emblica officinalis* Gaertn.	Euphorbiaceae	India
64.	*Eugenia cerasoides* Roxb.	Myrtaceae	Western Ghats India
65.	*Eugenia jambolana* Lam.	Myrtaceae	India, Nepal, Pakistan and Indonesia
66.	*Ficus benghalensis* L.	Moraceae	Indian Subcontinent
67.	*Ficus glomerata* Roxb.	Moraceae	Australasia, South-east Asia, Indian Subcontinent
68.	*Ficus hispida* L.f.	Moraceae	India, Asia
69.	*Ficus pamata* Forsk.	Moraceae	South East Asia
70.	*Ficus religiosa* L.	Moraceae	India and Asia, Tropica Indonesia, China
71.	*Ficus roxburghii* Wall. *ex* Miq.	Moraceae	China, Bhutan, India, Nepal, Malaysia, Myanmar, Pakistan, Thailand, Vietnam
72.	*Flacourtia ramontchi* L' Herit.	Flacourtiaceae.	Africa, Tropical and Temperate parts of Asia
73.	*Fragaria chiloensis* (L.) Duchesne	Rosaceae	India, Coastal California, Montara, Pacific Ocean, North and SouthAmerica, Hawaii, Chile, Argentina
74.	*Fumaria indica* (Haussk.) Pugsley	Fumariaceae	Europe, Temperate Asia, Afganistan, Persia, India
75.	*Glycine max* Merrill.	Fabaceae	North America, South East Asia, North and Central China, India, Korea and Taiwan
76.	*Gossypium arboreum* L.	Malvaceae	India and Pakistan, East Africa
77.	*Grewia oppositifolia* Roxb.	Tiliaceae	India, East Asia
78.	*Helicteres isora* L.	Sterculiaceae	Nepal, Bihar, Bengal, Central, Western and Southern India, Andaman Islands
79.	*Hibiscus rosa-sinensis* L.	Malvaceae	Southeastern Asia, Tropical and Subtropical parts, China

Contd...

Table 3.53: *contd...*

S.No.	Name of the Plant	Family	Origin
80.	*Holarrhena pubescens* (Buch.-Ham.) Wall. *ex* G. Don	Apocynaceae	Indian Subcontinent
81.	*Hordeum vulgare* L.	Poaceae	Israel-Jordan, Near East to South Asia, Egypt, Ethopia, Tibet
82.	*Jasminum multiflorum* (Burm. f.) Andr.	Oleaceae	India, Penninsular part of Florida
83.	*Jasminum sambac* (L.) Ait	Oleaceae	South East Asia, South China, India
84.	*Jatropha curcus* L.	Euphorbiaceae	India
85.	*Juglans regia* L.	Juglandaceae	Asia, Temperate Afghanistan, Erbaijan, Iran, Iraq, Turkey, Tajikistan, Tropical Asia, Pakistan, Europe, Bulgaria, Yugoslavia, Greece
86.	*Lagerstroemia indica* L.	Lythraceae	Indian subcontinent, Southern Asia, Northern Australia, Oceania, China
87.	*Lannea coromandelica* (Houtt.) Merrill.	Anacardiaceae	Bangladesh, Bhutan, Combodia, China, India-Andaman and Nicobar, Arunachal Pradesh, Assam, Bihar, Gujarat, Himachal Pradesh, Jammu and Kashmir, Karnatka, Kerala, Madhya Pradesh, Maharashtra, Meghalaya, Orissa, Rajasthan, Sikkim, Tamil Nadu, Tripura, Uttar Pradesh, West Bengal, Laos, Mayanmar, Nepal, Pakistan, Sri lanka, Thailand, Vietnam
88.	*Lavandula angustifolia* Mill.	Lamiaceae	Cape Verde & Canary Islands, Europe, Northern & Eastern Africa, Mediterranean, South-West Asia, South-East India
89.	*Lawsonia alba* Lam.	Lythraceae	Tropical & Subtropical regions of Africa, Southern Asia, Northern Australasia, Tropical Savannah
90.	*Linum usitatissimum* L.	Linaceae	Mediterranean to India, Ethiopia, Ancient Egypt, Republic of Georgia
91.	*Litchi chinensis* (Gaertn.) Sonn.	Sapindaceae	Provinces of Kwangtung and Fukien in Southern China, Southeastern Asia
92.	*Luffa aegyptiaca* Mill.	Cucurbitaceae	Africa, Asia, Tropical & Subtropical parts of the World, West Africa, India
93.	*Madhuca indica* J. F. Gmel.	Sapotaceae	Tropical India, Central & Northern plains and forests of India
94.	*Mangifera indica* L.	Anacardiaceae	East Asia, Philippines, Africa, Brazil, Portuguese, Southern Asia, Myanmar, Andaman Islands, India
95.	*Melia azedarach* L.	Meliaceae	China, Baluchistan, Kashmir, Western Himalayas, Pakistan, India, Indochina, Southeast Asia

Contd...

Table 3.53: *contd...*

S.No.	Name of the Plant	Family	Origin
96.	*Mesua ferrea* L.	Clusiaceae	Tropical Sri-Lanka, Southern Nepal, Indo-china, Malay Peninsula
97.	*Millettia extensa* Benth. *ex* Baker	Fabaceae	India
98.	*Mirabilis jalapa* L.	Nyctaginaceae	Mexico, Peruvian Andes
99.	*Momordica charantia* L.	Cucurbitaceae	Tropical & Subtropical Asia, Africa, Caribbean
100.	*Moringa oleifera* Lam.	Moringaceae	Tropical & Subtropical World, Southern Africa, Cambodia, Foothills of Himalayas, Nepal, Indonesia, Malaysia, Mexico, Central & South America, Sri-Lanka
101.	*Murraya koenigii* (L.) Spreng.	Rutaceae	India Sri-Lanka
102.	*Musa sapientum* L.	Musaceae	Tropical Southeast Asia, Philippines, Papua New guinea, South Asia
103.	*Nerium indicum* Mill.	Apocynaceae	AFRICA-Algeria, Ethiopia, Libya, Morocco, Tunisia; ASIA-TEMPERATE-Afganistan, China, Cyprus, Iran, Iraq, Israel, Jordan, Lebnon, Oman, United Arab Emirates, Syria, Turkey; ASIA-TROPICAL- India-Jammu and Kashmir, Punjab, Nepal, Pakistan; EUROPE- Albania, Former Yugoslavia, Greece, Italy, Malta; France; Portugal, Spain
104.	*Ocimum basilicum* L.	Lamiaceae	Tropical Asia, Africa, India, France
105.	*Oroxylum indicum* Vent.	Bignoniaceae	India, Bhutan, Southern China, Indo China, Malesia
106.	*Papaver rhoeas* L.	Papaveraceae	Europe, Mediterranean, Northern Africa, Western Asia, Indian subcontinent-Pakistan
107.	*Parthenium hysterophorus* L.	Asteraceae	TROPICAL and SUBTROPICAL America, India
108.	*Pennisetum americanum* L.	Poaceae	Africa, West Africa, Indian Subcontinent-Pakistan, India
109.	*Phoenix sylvestris* Roxb.	Arecaceae	India, Nepal, Pakistan
110.	*Phragmites karka* Trin. *ex* Steud.	Poaceae	South Asia, North America, India Australia
111.	*Phyllanthus urinaria* L.	Euphorbiaceae	India, Tropical East Asia
112.	*Pinus roxburghii* Sarg.	Pinaceae	Himalayas, Pakistan (North-West Frontier Province, Azad Kashmir, Margala Hills, Islamabad, North India, Nepal and Bhutan
113.	*Piper longum* L.	Piperaceae	North-eastern, Southern India, Sri-Lanka, Bengal
114.	*Pistacia integerrima* Stewart *ex* Brandis.	Anacardiaceae	Mediterranean region to East Asia, Himalayan ranges on the North-West
115.	*Plumbago zeylanica* L.	Plumbaginaceae	Southern India, Malaysia, Tropical Old World

Contd...

Table 3.53: *contd...*

S.No.	Name of the Plant	Family	Origin
116.	*Pogostemon benghalensis* Kuntze.	Lamiaceae	Bangladesh, China, India, Myanmar, Nepal, Sri-Lanka, Thailand, Asia, Philippines, Vietnam, West Africa
117.	*Polyalthia longifolia* (Sonn.) Thwaites	Annonaceae	India, Sri-Lanka
118.	*Pongamia pinnata* Pierre.	Fabaceae	Tropical & Temperate Asia, India, China, Japan, Malesia, Australia, Pacific islands
119.	*Premna latifolia* Roxb.	Verbenaceae	East Africa, India, Bangladesh, Japan, China, Taiwan, Guinea, Australia, Polynesia, Melanesia
120.	*Prunus amygdalus* Batsch.	Rosaceae	Middle East, South Asia, Mediterranean, Northern Africa, Southern Europe, California, West Asia
121.	*Psidium guajava* L.	Rosaceae	Mexico, Central America, Northern South America, Africa, Southeast Asia, Caribbean, North America, Australia
122.	*Punica granatum* L.	Punicaceae	Iran, Saudi Arabia. Afghanistan, Baluchistan, India
123.	*Putranjiva roxburghii* Wall.	Euphorbiaceae	Indo-China
124.	*Pyrus pashia* Ham.	Rosaceae	Southern Asia, China, India, Iran, Afghanistan
125.	*Randia dumetorum* Poir.	Rubiaceae	Sri-Lanka & India, America, Texas, Florida, China
126.	*Rauwolfia serpentina* Benth. *ex* Kurz.	Apocynaceae	South-East Asia, (India & China, Indonesia)
127.	*Reinwardtia indica* Dum.	Linaceae	India, China
128.	*Rhynchostylis retusa* Blume	Orchidaceae	Myanmar, India, Malaya, Singapore, Thailand, Sri-Lanka
129.	*Ricinus communis* L.	Euphorbiaceae	Africa, India, Mediterranean Basin
130.	*Roylea cinerea* (D. Don) Baillon.	Lamiaceae	N.W.India, Nepal
131.	*Rubus ellipticus* Sm.	Rosaceae	S.E.Asia, India, Himalaya
132.	*Rumex hastatus* D. Don	Polygoniaceae	Asia, India, Afghanistan
133.	*Saccharum arundinaceum* Reitz.	Poaceae	China, India
134.	*Saccharum officinarum* L.	Poaceae	South Asia, Hawaii, South pacific Islands, New Guinea, Florida to Texas
135.	*Saccharum spontaneum* L.	Poaceae	TROPICAL Asia, Europe, Turkmenistan, Iran, Southeast Asia, New Guinea, India, Eastern Africa,
136.	*Santalum album* L.	Santalaceae	India, Bangladesh, Sri-Lanka, Australia, Indonesia, Asia, Sri-Lanka, Malaysia, Indonesia, Taiwan
137.	*Saraca asoca* (Roxb.) De Wilde.	Fabaceae	Indian Subcontinent, Deccan plateau, Western Ghats in India, Myanmar, Malaysia
138.	*Sesamum indicum* L.	Pedaliaceae	Africa, India, Egypt, Babylon, Assyria, Turkey

Contd...

Table 3.53: *contd...*

S.No.	Name of the Plant	Family	Origin
139.	*Solanum melongena* L.	Solanaceae	Indian subcontinent, Southern-eastern Asia
140.	*Solanum nigrum* L.	Solanaceae	India, Asia, Africa
141.	*Solanum surattense* Burm. f.	Solanaceae	India
142.	*Soymida febrifuga* A. Juss.	Meliaceae	Indo Malaysia
143.	*Spilanthes acmella* auct. non (L.) Murr.	Asteraceae	South Asia, India
144.	*Stephania glabra* (Roxb.) Miers.	Menispermaceae	Eastern & Southern Asia, Australasia
145.	*Tabernaemontana* divaricata (L.) R. Br.	Apocynaceae	North India, Tropics & Southeast Asia & India, Thailand
146.	*Tamarindus indica* L.	Fabaceae	Tropical Africa, Sudan, Cameroon, Nigeria, Tanzania, Arabia, Oman, Dhofar, Africa, South Asia, Australia, South East Asia, Taiwan, China
147.	*Terminalia arjuna* (Roxb.) Wt. & Arn.	Combretaceae	India
148.	*Terminalia bellirica* (Gaertn.) Roxb.	Combretaceae	South-East Asia, Bangladesh, Cambodia, Bhutan
149.	*Terminalia chebula* Retz.	Combretaceae	India, South-east Asia, China, Nepal, Sri-Lanka, Malaysia, Vietnam
150.	*Tinospora cordifolia* (Willd.) Miers ex Hook. f. &Thoms.	Menispermaceae	India, Myanmar, Sri-Lanka
151.	*Trichodesma indicum* R.Br.	Boraginaceae	India, Mayanmar, Philippines, Mauritius
152.	*Tylophora indica* (Burm.f.) Merrill.	Asclepiadaceae	India
153.	*Urtica dioica* L.	Urticaceae	India, Europe, Asia, North America, North Africa
154.	*Vanda roxburghii* R.Br.	Orchidaceae	North-east India, Asia, Myanmar, Thailand, Indo China, Malaya
155.	*Verbascum thapsus* L.	Scrophulariaceae	Europe, Northern Africa, Asia, Temperate & Tropical Asia
156.	*Vitex negundo* L.	Verbenaceae	Eastern & Southern Africa, Asia, India
157.	*Wendlandia heynei* (Roem. & Schult.) Sant. & Merch.	Rubiaceae	India, Asia tropical
158.	*Withania somnifera* (L.) Dunal	Solanaceae	AFRICA- Algeria, Angola, Cape Verde, Egypt, Liberia, Morocco, Nigeria, Somalia, south Africa, Sudan, Tanzania, Tunisia, Uganda, Zambia, Zimbabwe, ASIA TEMPERATE- Afganistan, Arabia, Iran, Iraq, Israel, Jorden, Lebnon, Syria, Turkey; ASIA-TROPICAL-India, Pakistan, Sri Lanka; Spain EUROPE- Greece, Italy
159.	*Woodfordia floribunda* Salisb.	Lythraceae	Africa, South East Asia
160.	*Zingiber officinale* Rosc.	Zingiberaceae	South Asia, East Africa, Caribbean

Table 3.54: Taxonomic Components of the Ethnobotanical Aliens of District Hamirpur

S.No.	Plant Group	No. of Families	No. of Genera	No. of Species
1.	Dicots	43	104	117
2.	Monocots	4	13	15
3.	Gymnosperms	2	2	2
4.	Pteridophytes	2	2	2

Table 3.55: Exotic Alien Flora of District Hamirpur

Category	Number	Percentage
Species	136	45.94
Genera	121	49.79
Number of Families	51	59.30
Herbs	68	58.62
Shrubs	32	54.23
Under-Shrubs	4	40
Trees	21	23.86
Climbers	7	41.17
Lianas	4	66.66

Table 3.56: Predominant Ethnobotanical Alien Families of District Hamirpur

S.No.	Family	Total Number of Species
1.	Asteraceae	13
2.	Fabaceae	11
3.	Solanaceae	8
4.	Brassicaceae	7
5.	Poaceae	7
6.	Lamiaceae	6
7.	Rosaceae	5
8.	Liliaceae	5
9.	Amaranthaceae	4
10.	Apiaceae	4
11.	Rutaceae	4
12.	Euphorbiaceae	4
13.	Apocynaceae	3
14.	Bignoniaceae	3
15.	Cucurbitaceae	3
16.	Chenopodiaceae	3
17.	Moraceae	3
18.	Asclepiadaceae	2
19.	Myrtaceae	2
20.	Nyctaginaceae	2
21.	Ranunculaceae	2
22.	Sapindaceae	2
23.	Verbenaceae	2
24.	Convolvulaceae	2
25.	Vitaceae	2

Table 3.57 : Predominant Alien Genera of District Hamirpur

S.No.	Genus	Number of Species
1	*Amaranthus*	3
2	*Brassica*	3
3	*Artemisia*	2
4	*Prunus*	2
5	*Solanum*	2
6	*Mentha*	2
7	*Allium*	2
8	*Aloe*	2
9	*Euphorbia*	2
10	*Tecoma*	2
11	*Chenopodium*	2
12	*Ipomoea*	2
13	*Vitis*	2
14	*Anethum*	1
15	*Coriandrum*	1
16	*Carissa*	1
17	*Daucas*	1
18	*Foeniculum*	1
19	*Asclepias*	1
20	*Calotropis*	1
21	*Carduus*	1
22	*Centipeda*	1
23	*Helianthus*	1
24	*Sonchus*	1
25	*Taraxacum*	1
26	*Tridax*	1

Table 3.58 : Predominant Alien Crops of District Hamirpur

S.No.	Name	Family
1	*Allium cepa*	Liliaceae
2	*Allium sativum*	Liliaceae
3	*Aloe vera*	Liliaceae
4	*Arachis hypogaea*	Fabaceae
5	*Avena sativa*	Poaceae
6	*Brassica campestris*	Brassicaceae
7	*Brassica juncea*	Brassicaceae
8	*Brassica nigra*	Brassicaceae
9	*Capsicum annuum*	Solanaceae
10	*Citrus medica*	Rutaceae
11	*Lycopersicon esculentum*	Solanaceae
12	*Nicotiana tabacum*	Solanaceae
13	*Oryza sativa*	Poaceae
14	*Pennisetum purpureum*	Poaceae
15	*Raphanus sativus*	Brassicaceae
16	*Solanum tuberosum*	Solanaceae
17	*Trigonella foenum-graecum*	Fabaceae
18	*Triticum aestivum*	Poaceae
19	*Vitis vinifera*	Vitaceae
20	*Zea mays*	Poaceae
21	*Cucurbita pepo*	Cucurbitaceae
22	*Sechium edule*	Cucurbitaceae

Conclusion

It is obvious from the foregoing discussion that the local communities of the region are basically wise, eco-friendly and have a self-sufficient and self-reliant subsistence system as reflected by their dependence on a great diversity of plant species to ensure a year-round supply. At the same time, the practice reflects their conservational wisdom as it exerts less pressure on the available natural resources. Hopefully, the generated data will serve as a very valuable document for the planners, policy-makers, foresters, academicians and the scientific organizations on the basis of which better developmental schemes can be formulated for socio-economic uplifting of the rural inhabitants of the study area.

Bibliography

Abraham, Z. 1981. Ethnobotany of the Todas, the Kotas and the Irulas of the Nilgiris, 308-320. *In*: Jain S.K. (*ed.*) *Glimpses of Indian Ethnobotany*. Oxford & IBH, New Delhi.

Agarwal, A.K. 1997. Therapeutic efficacy of herbal gel for skin affections in dogs. *The Indian Vet. J.* **74**: 417-419.

Agarwal, V. S. 1986. *Economic Plants of India*. BSMPS, Dehradun & Kailash Prakashan, Calcutta.

Agarwal, V.S. 2003. *Directory of Indian Economic Plants*. BSMPS, Dehradun.

Aggarwal, A., Prakash, V. & Sharma, S. 2004. Ethnobotanical studies on some wild plants of lower hills. *Appl. Bot. Abs.* **24**(2): 136.

Agrawal, S.R. 1981. Trees, flowers and fruits in Indian folk songs, folk proverbs and folk tales, 3-12. *In:* Jain S.K. (*ed.*) *Glimpses of Indian Ethnobotany*. Oxford & IBH, New Delhi.

Ahluwalia, K.S. 1952. Medicinal plants of Kangra valley. *Indian For.* **78**(4): 181-194.

Ahuja, B.S. 1965. *Medicinal Plants of Saharanpur*. Survey of Medicinal Plants, CCAR Gurukul Kangri Vishwavidalya, Hardwar.

Ahuja, P.S. 2001. Current status of propagation of medicinal plants in Indian Himalaya: 207-230. *In:* Samant, S.S., Dhar, U. & Palni, L.M.S. (*eds*) *Himalayan Medicinal Plants-Potential and Prospects*. Gynodaya Prakash, Nainital.

Alcorn, J.B. 1984. *Huastee Mayan Ethnobotany*. Univ. of Texas press, Austin.

Ambasta, S.P. (*ed.*) 1986. *The Useful Plants of India*. CSIR, New Delhi.

Aminuddin & Girach, R.D. 1991. Ethnobotanical studies on Bondo tribe of district Karaput (Orissa), India. *Ethnobotany* **3**: 15-19.

Aminuddin & Girach, R.D. 1993. Observations on ethnobotany of the Bhujia-a tribe of Sonabera plateau. *Ethnobotany* **5**: 83-86.

Aminuddin & Girach, R.D. 1996. Native phytotheraphy among the Paudi Bhuinya of Bonai Hills. *Ethnobotany* **8**: 66-70.

Anonymous, 1948-1976. *The Wealth of India -Raw Materials,* Vols. 1-11. CSIR, Delhi.

Anonymous, 1966. *The Indian Pharmacopoeia,* 2nd ed., Govt. of India Publ., Delhi.

Anonymous, 1985. *The Wealth of India, Raw Materials.* Revised & enlarged Vol. IA., P.I.D., C. S.I.R., New Delhi.

Anonymous, 1990. *Phytochemical Investigation of Certain Medicinal Plants Used in Ayurveda.* CCARS, New Delhi.

Anonymous, 1994. *Ethnobiology in India-A Status Report.* Ministry of Environment and Forest, Govt. of India.

Anonymous, 1998a. *All India Coordinated Project on Ethnobotany.* Final Technical Report 1992-1998. Ministry of Environment & Forests, Govt. of India.

Anonymous, 1998b. *The Wealth of India. A Dictionary of Indian Raw Material and Industrial Products.* Vols. I-XIV. CSIR, New Delhi.

Anonymous, 2008. *Agro Techniques of Selected Medicinal Plants.* Vol. 1, National Medicinal Plants Board. Department of AYUSH, ministry of health and family welfare, (TERI press The Energy and resources Institute. New Delhi).

Ansari, A.A. 1991. Ethnobotanical notes on some plants of Khirsu, Pauri Garhwal, U.P. *Ethnobotany* **3**: 105-107.

Arizpe, L. 1996. Culture and environment. *Nature & Res.* **32**: 1-30.

Arora, R.K. 1995. Ethnobotanical studies on plant genetic resources-national efforts and concern. *Ethnobotany* **1**: 125-136.

Arora, R.K., Maheswari, M.L., Chandel, K.P.S. & Gupta, R. 1980. Mano (*Inula racemosa*): little known aromatic plants of Lahaul valley, India. *Econ. Bot.* **34** (2): 175-180.

Arora, R.K. & Pandey, A. 1996. *Wild Edible Plants of India -Diversity, Conservation and Use.* NBPGR, New Delhi.

Arulpriya, P., Lalitha, P. & Hemlata, S. 2011. Competence of different solvent extraction methods in the isolation of bioactive principles from *Samanea saman* (Jacq.) Merr. *In: Pharma Sci. Monitor an Int. J. Pharm. Sci.* **2** (2) suppl.-1.

Asolkar, L.V., Kakkar, K.K. & Chakre, O.J. 1992. *Second Supplement to Glossary of Indian Medicinal Plants with Active Principles.* Part -1 (A -K). CSIR, New Delhi.

Aswal, B.S. 1996. Conservation of ethno-medicinal plants diversity of Garhwal Himalaya in India, 133-135. *In:* Jain, S.K. (*ed.*) *Ethnobiology in Human Welfare.* Deep Publ., New Delhi.

Aswal, B.S. & Mehrotra, B.N. 1994. *Flora of Lahaul Spiti (A Cold Desert in North West Himalaya).* BSMPS, Dehradun.

Atkinson, E.T. 1882. *Economic Botany of the Himalayan Region.* Cosmo Publ., New Delhi.

Augustine, Jomy & Sivadasan, M. 2004. Ethnomedicinal plants of Periyar Tiger Reserve, Kerela India. *Ethnobotany* **16**: 44-49.

Badola, H.K. 2001. Medicinal plant diversity in Himachal Pradesh, 407-430. *In*: Samant, S.S., Dhar, U. & Palani, L.M.S. (*eds.*) *Himalayan Medicinal Plants: Potential and Prospects.* Himvikas Occasional Publication, No. 14, Gyanodya Prakashan.

Badola, H.K. & Pal, M. 2002. Endangered medicinal plant species in Himachal Pradesh. *Curr. Sci.* **83:** 797-798.

Bajpayee, K. K. & Dixit, G. 1996. Ethnobotanical studies on food stuffs of tribals of Tarai-Region, Uttar Pradesh. *J. Econ. Taxon. Bot.* Add. Series **12**: 128-132.

Balasubramaniam, P. & Prasad, Narendra 1996. Medicinal plants among the Irulars of Attapady and Bolumampatti forests in the Nilgiri Biosphere Reserve. *J. Econ. Taxon. Bot.* Add. Series **12**: 253-259.

Banerjee, Archana 1996. Medicinal uses of some flowers by Santhals of West Bengal. *J. Econ. Taxon. Bot.* Add. Series **12**: 314-317.

Banerjee, D.K. & Pal, D.C. 1996. Plant used by the tribals of North Indian plains for hair and scalp protection, 187-188. *In*: Jain, S.K. (*ed.*) *Ethnobiology in Human Welfare*. Deep Publ., New Delhi.

Baruah, P. & Sarma, G.C. 1984. Studies on the medicinal uses of plants by the Bodo tribals of Assam II. *J. Econ. Taxon. Bot.* **5**: 599-604.

Baruah, P. & Sarma, G.C. 1987. Studies on the medicinal uses of plants by the north east tribes. Pt. 3. *J. Econ. Taxon. Bot.* **11**: 71-76.

Belal, Ahmed Esmat & Springuel, Irina 1996. Economic value of plant diversity in arid environments. *Nature & Res.* **32** (1): 33-39.

Bennet, S.S.R. 1985. Ethnobotanical studies in West Sikkim. *J. Econ. Taxon. Bot.* **7:** 317-321.

Bennet, S.S.R. 1986. *Name Changes in Flowering Plants of India and Adjacent Regions.* Triseas Publishers, Dehra Dun-248001, India.

Bennet, S.S.R., Biswas, S., Chandra, V., Singh, P. J., Chandra, S. & Dayal, R. 1991. *Food from Forest.* FRI, ICFRE, Dehradun.

Bhalla, N.P., Sahu, T.R., Mishra, G.P. & Dakwale, R. N. 1982. Traditional plant medicines of Sagar Distt. Madhya Pradesh, India. *J. Econ. Taxon. Bot.* **3**: 23-32.

Bhalla, S. & Bhalla, T.C. 2006. Traditional food and beverages of Himachal Pradesh. *Indian J. Trad. Knowledge* **6**(**1**): 17-24.

Bhandary, M.J., Chandrashekar, K.R. & Kaveriappa, K.M. 1996. Ethnobotany of Gowlis of Uttar Kannada District, Karnataka. *J. Econ. Taxon. Bot.* Add. Series **12:** 244-249.

Bhardwaj, S. & Gakhar, S.K. 2003. Ethnomedicinal plants used by the tribals of Mizoram to cure dysentery. *Ethnobotany* **15**: 51-54.

Bhargava, K.S. 1959. Unusual and supplementary food plants of Kumaon. *J. Bombay Nat. Hist. Soc.* **56**: 26-31.

Bhargava, N. 1981. Plants in folk life and folk-lore in Andaman & Nicobar Islands, 329-344. *In:* Jain, S.K. (*ed.*) *Glimpses of Indian Ethnobotany*. Oxford & IBH, New Delhi.

Bhatt, D.C., Mitaliya, K.D. & Mehta S.K. 2001. Some sacred plants of Gujarat and their medicinal uses. *Ethnobotany* **13**: 146-149.

Bhatt, D.C., Mitaliya, K.D, Mehta, S.K. & Joshi, P.N. 2002. Notes on some ethnomedicinal plants of Pachchham Hills of Kachchh Distt. Gujarat. *Ethnobotany* **14**: 34-35.

Bhatt, K.C. 1990. *Ethnobotanical studies of some remote areas of Pithoragarh and Chamoli Districts.* Ph.D. thesis, HNB Univ. Garhwal.

Bhatt, R.P. & Sabnis, S.D. 1987. Contribution to the ethnobotany of Khedbrahma region of north Gujarat. *J. Econ. Taxon. Bot.* **9**: 139-145.

Bhattacharjee, S., Tiwari, K.C., Majumdar, R. & Misra, A.K. 1980. Folk-lore medicine from district Kamrup (Assam). *Bull. Medico-Ethno-bot. Res.* **1**: 441-460.

Bhattacharjee, S. K. 1998. *Handbook of Medicinal Plants.* (3rd Revised Edition) Pointer Publ., Jodhpur (India).

Bhattacharyya, G. 1996. Medico-ethno-botanical value of Saurashtra weeds. *J.Econ.Taxon. Bot. Add. Series* **12**: 166-168.

Bhattarai, N.K. 1992. Folk uses of plants in veterinary medicine in central Nepal. *Fitoterapia* **63**(6): 497-505.

Binoj, K. M.S. & Balakrishnan, N.P. 1996. Ethnobotanical studies of one genus *Euphorbia* (Euphorbiaceae). *J. Econ. Taxon. Bol.* Add. Series **12**: 46-49.

Bondya, S.L. & Khanna, K.K. & Singh, K.P. 2006. Ethnomedicinal used of leafy vegetable from the tribals folk lore of Achanakmar-Amarkantak Biosphere Reserve (M.P. & Chattishgarh). *Ethnobotany* **18**: 145-148.

Bondya, S. L. & Sharma, H.P. 2003. Ethnobotanical studies on plants used in diabetes (Madumeha) under the Bharagora Block of Jarkhand. *Ethnobotany* **16**: 139-140.

Borthakur, S.K. 1980. *Medical flora of Karbi-Anglong (Mikir hills, Assam).* Ph.D. thesis, Guahati Univ.

Borthakur, S.K. 1993. Native phytotherapy for child and women diseases from Assam. *Ethnobotany* **5**(87): 87-91.

Borthakur, S.K. 1996a. Postnatal care of women in traditional system in Assam. *Ethnobotany* **8**: 51-55.

Borthakur, S.K. 1996b. Wild edible plants in markets of Assam, India-an ethnobotanical investigation, 31-34. *In*: Jain, S.K. (*ed.*) *Ethnobiology in Human Welfare*. Deep. Publ., New Delhi.

Borthakur, S.K., Chaudhary, B.T. & Gogoi, R. 2004. Folklore hepato-protective herbal recipes from Assam in Northeast India. *Ethnobotany* **16**: 76-82.

Borthakur, S.K., & Gogoi, P. 1994. Indigenous technology of making writing materials among the Tai Khamtis. *Ethnobotany* **6**: 5-8.

Borthakur, S.K. & Sarma, U.K. 1996. Ethnoveterinary medicine with special reference to cattle prevalent among the Nepalis of Assam, India, 209-210. *In*: Jain, S.K. (*ed.*) *Ethnobiology in Human Welfare.* Deep Publ., New Delhi.

Brahmam, M., Dhal, N.K. & Saxena, H.O. 1996. Ethnobotanical studies among the Tanla of Malayagiri Hills in Dhenkanal District, Orissa, India, 393-396. *In*: Jain, S.K. (ed.) *Ethnobiology in Human Welfare.* Deep Publ., New Delhi.

Brahmam, M. & Saxena, H.O. 1990 Ethnobotany of Gardamardan Hills–some noteworthy folk-medicinal uses. *Ethnobotany* **2**: 71-79.

Brown, D. 1995. *Encyclopaedia of Herbs and their Uses.* Dorling Kindersley, London.

Burkill, J. H. 1966. *A Dictionary of Economic Products of the Malay Peninsula.* 2 vols. Art Printing works, Kuala Lumur.

Caius J. F. 1935. The medicinal and poisonous ferns of India, *J Bombay Natl. Hist. Soc.,* **38**: 341-361.

Caius, J. F. 2003. *The Medicinal and Poisonous Plants of India.* Sci. Publ., Jodhpur (India).

Calabrese, C., Berman, S.H., Babish, J., Ma, X., Shinto, L. & Cynthia, A. 2000. A phase 1 trial of Andrographolide in HIV positive patients and normal volunteers. *Phytother. Res.* **14** (5): 333-338.

Chak, I.M. & Sharma, J.N. 1965. Effect of asarone on experimentally induced conflict neurosin in rats. *Indian J. Exp. Biol.* **3**: 252.

Chandra Sekar, K. & Srivastava, S.K. 2003. Ethnomedicinal studies in Pin valley National Park, Lahaul-Spiti, Himachal Pradesh. *Ethnobotany* **15**: 44-47.

Chandra Sekar, K. & Srivastava, S.K. 2005. New reports on aphrodisiac plants from Pin Valley National Park, Himachal Pradesh. *Ethnobotany* **17**: 189-190.

Chang, T.T. 1985. Crop history and genetic conservation in rice-a case study. *Iowa State J. Res.* **59:** 405-455.

Chatterjee, A. & Pakrashi, S.C. (*eds.*) 1997. *The Treatise on Indian Medicinal Plants.* Vols. 1 -5. CSIR, New Delhi.

Chauhan, N.S. 1974. Medicinal plants of Una forest division, Una district, H.P. lying in Shivalik range. *Nagarjun* **27**: 31-39.

Chauhan, N.S. 1983. Some rare and interesting medicinal plants of H.P. All India Summer Inst. on Agroforestry in the Western Himalaya, July 22 – August 7, 1983.

Chauhan, N.S. 1984. Medicinal Wealth of Pabbar valley in Himachal Pradesh, M.Sc. Thesis, HPKVV, Palampur.

Chauhan, N.S. 1988a, Exploitation and utilization of herbal resources of H.P., 23-32. Invited lecture delivered 'Workshop on Oppurtunities Identification in Agro-based and Food-Products', pp STEP, Holiday Home, Shimla.

Chauhan, N.S. 1988b. Ethnobotanical study of medicinal plants of Himachal Pradesh. 187-198. *In*: Kaushik, P. (*ed.*) *Indigenous Medicinal Plants.* Today and Tomorrow Printers & Publ., New Delhi.

Chauhan, N.S. 1989a. Medicinal and aromatic plants of Himachal Pradesh and their role in social Forestry programme. IIIrd Forestery Conference (proceeding) Vol. II: (89): 1-9, FRI Dehradun.

Chauhan, N.S. 1989b. Potential of aromatic plants flora in Himachal Pradesh. *Indian Perfum.* **33**(2): 118-122.

Chauhan, N.S. 1989c. Endangered ayurvedic pharmacopoeial plant resources of H.P., 199-205. *In:* Kaushik P. (*ed.*) *Indiginous Medicinal Plants. Including Microbes and Fungi.* M/S Today & Tomorrow's Printers and Pub., New Delhi.

Chauhan, N.S. 1990. Medicinal orchids of Himachal Pradesh. *J. Orchid Soc. India* **4**(2): 99-105.

Chauhan, N.S. 1995. *Plant Resources of Economic Use in Himachal Pradesh.* Dept. of Forest Products, Dr. Y.S. Parmar Univ. Hort. Forestry, Solan (H.P.), India.

Chauhan, N.S. 1996a. Ethno-medicobotany of Pabbar valley in District Shimla, Himachal Pradesh (India), 30-. *In*: *UHF-IUFRO International Workshop in Prospects of Medicinal Plants.*

Chauhan, N.S. 1996b. *Plant Resources of Economic Use in Himachal Pradesh.* DEE – UH& F, Solan, 77pp.

Chauhan, N.S. 1999. *Medicinal and Aromatic Plants of Himachal Pradesh.* Indus Publ. Co., New Delhi.

Chauhan, N.S. 2003. Important medicinal and aromatic plants of Himachal Pradesh. *Indian For.* **129**(8): 979-998.

Chauhan, N.S. & Lakhanpal, T.N. 2000. Ecological and ethnobotanical studies on the flora of Spiti, Himachal Pradesh-a cold desert in western Himalaya, 293-300. *In*: Kohli, R.K., Singh, H.P., Vij, S.P. Dhir, K.K., Batish, D.R. & Khurana, D.R. (*eds.*) *Man and Forests.* DNAES, ISTS, Bot. Dept., & C.N.S., Panjab Univ., Chandigarh.

Chaurasia, O.P., Ahmed, Z. & Ballabh, B. 2007. *Ethnobotany and Plants of Trans-Himalaya.* Satish Serial Publ. House, New Delhi.

Chaurasia, O. P., Singh, B. & Ballabh, B. 2001. Potential aromatic floras of Himalayan cold district: Ladakh & Lahul-Spiti. *J. Econ. Taxon. Bot.* **25** (1): 91-97.

Chhetri, D., Parajuli, P. & Subha, G. 2005. Antidiabetic plants used by Sikkim and Darjeeling Himalayan tribes, India. *J. Ethnopharmacol.* **99**: 199-202.

Chhetri, D. R. 2005. Ethnomedicinal plants of Khangchendzonga national park, Sikkim, India. *Ethnobotany* **17**: 176-178.

Chhetri, R.B. 1994. Further observations on ethnomedicobotany of Khasi Hills in Meghalaya, India. *Ethnobotany* **6**: 33-36.

Chinthu, T.U., Narainswami, B. K. & Ravi, S. 1998. Ethnoveterinary medicine for dairy cow, 13-14. *In*: Mathias, E., Rangnekar, D.V. & Mc Corkle, C.M. *(eds.) Ethnoveterinary Medicine-Alternatives for Livestock Development.* BAIF development Research Foundation, Pune, India.

Choondal, C. 1988. Some medicine for livestock in Kerala. *Folklore* **29**(9): 193-201.

Chopra, I.C., Abrolo, B.K. & Handa, K.L. 1983. *Medicinal Plants of the Arid Zones.* Today & Tomorrow's Printers & Publ., New Delhi.

Chopra, R.N., Badhwar, R.L. & Gosh, S. 1965. *Poisonous Plants of India.* 2 Vols. ICAR, New Delhi.

Chopra, R.N., Chopra, I.C. & Verma, B.S. 1969. *Supplement to Glossary of Indian Medicinal Plants.* C.S.I.R., New Delhi.

Chopra, R.N., Nayar, S.L. & Chopra, I.C. 1956. *Glossary of Indian Medicinal Plants.* CSIR, New Delhi.

Chopra, R.N., Nayar, S.L. & Chopra, I.C. 1986. *Glossary of Indian Medicinal Plants.* (*Including the Supplement*) CSIR, New Delhi.

Chowdhery, H.J. 1996. Arunachal Pradesh, its people, ethnobotanical diversity and conservation, 323-328. *In*: Jain, S.K. (*ed.*) *Ethnobiology in Human Welfare.* Deep Publ., New Delhi.

Chowdhery, H.J & Wadhwa, B.M. 1984. *Flora of Himachal Pradesh.* Vol. 1-3. BSI, Calcutta.

Chowdhury, J. U., Alam, M. K., & Hassan, M.A. 1996. Some traditional folk formularies against dysentery and diarrhoea in Bangladesh. *J. Econ. Taxon. Bot.* Addl. Series **12**: 20-23.

Chunekar, K.C., & Hota, N.P. 1998. *Plants of Bhava Prakash.* Nat. Acad. of Ayurveda.

Chunekar, K.C., & Pondel, K. 1999. *Plants of Sharangadhara Samhita.* Nat. Acad. of Ayurveda.

Collett, H. 1902. *Flora Simlensis.* Thacker Spink and Co., Calcutta and Shimla, Reprinted 1971, BSMPS, Dehradun.

Coon, J.T. & Ernst E. 2004. *Andrographis paniculata* in the treatment of upper respiratory tract infections: a systematic review of safety and efficacy, *Planta Med.* **70**(4): 293-298.

Cronk, Q.C.B. & Fuller, J.L. 1995. *Plant Invasion: The Threat to Natural ecosystem.* World Wide Fund for Nature. Chapman & Hall, London.

Dagar, H.S. 2003. Ethnobotanical evaluation of some endemic plants in Andaman and Nicobar Islands, 134-147. *In*: *Vistas in Ethnobotany*, Vol.-I, *Indian J. Pure & Appl. Biol.*

Dagar, H.S. & Dagar, J.C. 1996. Ethnobotanical studies of the Nicobarese of Chowra Island of Nicobar group of Islands. *J. Econ. Taxon. Bot.* Addl. Series **12**: 381-388.

Dagar, J.C. & Dagar, H.S. 1999. *Ethnobotany of Aborigines of Andaman-Nicobar Islands.* Surv. of India, Port Blair, Andaman & Nicobar Islands, India.

Dandiya, P.C. & Sharma, J. D. 1961. Pharmacological effects of α-asarone and β-asarone. *Nature* **192**: 1299.

Dandiya, P.C. & Sharma, J.D. 1962. Studies on *Acorus calamus* V. Pharmacological action of α-asarone and β-asarone in C.N.S. *Indian J. Med. Res.* **50**: 46.

Das, D. & Agarwal, V.S. 1991. *Fruits Drug Plants of India.* Kalyani Publ. New Delhi.

Das, H.S., Panda, P.C. & Patnaik, S.N. 1996. Traditional use of wetland plants of Eastern Orissa. *J. Econ. Taxon. Addl. Series* **12:** 306-313. Sci. Pub., Jodhpur (India).

Dash, S.S. & Mishra, M.K. 2000. Taxonomic survey and systematic census of economic plants of Narayana Patna Hills of Koraput District, Orissa: 473-498. *In*: Maheshwari, J.K. (*ed.*) *Ethnobotany and Medicinal Plants of Indian Subcontinent.* Sci. Publ., Jodhpur (India).

Dastur, J.F. 1964. *Useful Plants of India and Pakistan.* Taraporevala Sons & Co. Pvt. Ltd. Dadabhai Naro Ji Road, Bombay.

Dastur, J.F. 1970. *Medicinal Plants of India and Pakistan.* D.B. Taraporevala Sons & Co., Pvt. Ltd. 2nd ed. Bombay.

Datt, Bhaskar 1996. Ethnobotanical resources of Chhatarpur district (Madhya Pradesh), 400-402. *In*: Jain, S.K. (*ed.*) *Ethnobiology in Human Welfare.* Deep Publ., New Delhi.

Deeba, F., Muhammad, G., Iqbal, Z. & Hussain, I. 2009. Survey of ethno-veterinary practices used for different ailments in dairy animals in peri-urban areas of Faisalabad (Pakistan). *Int. J. Agric. & Biol.* **11**: 535-541.

Dey, A.C. 1980. *Indian Medicinal Plants Used in Ayurvedic Preparations.* BSMPS, Dehradun.

Dey, K.L. 1973. *The Indigenous Drugs of India.* The Chronica Bot. New Delhi.

Dey, K.L., Bahadur, R. & C.I.E. 1973. *The Indigenous Drugs of India: Short Descriptive Notices of the Principal Medicinal Products Met with in British India.* Pama Primlane. The Chronica Bot., New Delhi, India.

Dhaliwal, D.S. & Sharma, M. 1999. *Flora of Kullu District (Himachal Pradesh).* BSMPS, Dehradun.

Dhiman, A.K. 2003. *Sacred Plants and Their Medicinal Uses.* Daya Publ. House. Delhi.

Dhiman, D.R. 1976. *Himachal Pradesh-ki-Vanoshdiya Sampada.* Imperial Printing Press, Dharmshala, H.P.

Diwaker, P.G. & Ansari, A.A. 2003. Economic Importance of the plant resources of Buldhana district of Maharashtra, 592-600. *In*: Singh, V. & Jain, A.P. (*ed.*) *Ethnobotany and Medicinal Plants of India and Nepal.* Vol. 2. Sci. Publ., Jodhpur (India).

Dixit, R.D.,1974. Fern-A much neglected group of medicinal plants, 1. *J Res. Indian Med.* **9**: 74-90.

Dixit, R.D., & Pal, D.C. 1996. Role of household remedies in pry. health-care, 108-109. *In*: Jain, S.K. (*ed.*) *Ethnobiology in Human Welfare.* Deep Publ., New Delhi.

Dixit, S.K. 1987. *Studies on Pathogenesis, Diagnosis and Treatment of Experimentally Induced Gastric Ulcers in Swine,* M.V.Sc. Thesis, IVRI Deemed Univ., Izatnagar.

Dobriyal, R.M., Singh, G.S., Rao, K.S. & Saxena, K.G. 1997. Medicinal plant resources in Chhakinal watershed in north-western Himalaya. *J. Herbs Spices & Med. Plants* **5**: 15-27.

Drake, J.A., Mooney, H.A., Castri, F., Groves, R.H. , Kruger, F.J., Rejmanek, M. & Williamson, M. (*eds.*) 1989. *Biological Invasions: A Global Perspective.* John Wiley & Sons, New York.

Drury, H. 1978. *Useful Plants of India.* Periodical Experts Book Agency, Delhi.

Duhoon, S.S., Kopper, M.N., & Chandra, U. 1996. Seabuckthorn (*Hippophae spp*)-a less known wonder plant of ethno-medico-botanical importance in cold desert of India. *J. Econ. Taxon. Bot.* Addl. Series **12**:43-45.

Duke, J. A. & Wain, K.K.1981. *Medicinal Plants of the World.* Computer Index with more than 85,000 enteries. 3 vols. Longman group U.K. Ltd.

Duthie, J.E. 1888. *Fodder Grasses of Northern India.* Avishkar Publ., Dist. Jaipur (Raj.) India.

Dwarakan, P. & Ansari, A.A. 1996. Less known uses of plants of Kollimalai (Salem Dt., Tamil Nadu) in South India. *J. Econ. Taxon. Bot.* Addl. Series **12**: 284-286.

Dwivedi, A. P. 1993. *Forest -The Non Wood Resources.* Intl. Book Distri. Dehradun (India).

Dwivedi, S.W., Dwivedi, Sangetta, & Patel, P.C. 2005. Herbal remedies used by rural people in treatment of joint diseases. *Ethnobotany* **17**: 193-196.

Dymock, W., Warden, C.J. & Hooper, D. 1889-1891. *Pharmacographica Indica*. Thacker Sprink & Co., Calcutta.

Etkin, N. 1993. Anthropological methods in ethnophramacology. *J. Ethnopharmacol.* **38**: 93-104.

Facciola, S. 1990. *Cornucopia-A Source Book of Edible Plants.* Kampong Publ. Kampong.

Freedman, R. 1998. *Famine Foods-Amaranthaceae*. Purdue Univ.

Gadgil, M. & Vartak, V.D. 1981. Sacred groves of Maharashtra; an inventory, 279-294. *In:* Jain, S.K. (*ed.*) *Glimpses of Ethnobotany*, Oxford Univ. Press, Bombay.

Gangwar, A.K. & Ramkrishnan, P.S. 1990. Ethnobiological notes on some tribes of Arunachal Pradesh, Northeastern India. *Econ. Bot.* **44**: 94-105.

Gaur, R.D., Bhatt, K.C. & Tiwari, J.K. 1992. An ethnobotanical studies of Uttar Pradesh Himalaya in relation to veterinary medicine. *J. Econ. Taxon. Bot.* **5**: 795-799.

Gaur, R.D. & Singh, P.B. 1993. Ethno-medicinal plants of Mandi district, Himachal Pradesh. *Bull. Medico-Ethnobot. Res.* **14**: 1-11.

Geetha, S., Lakshmi, G. & Ranjithakani, P. 1996a. Ethnobotanical review: wild fibre yielding plants of Kolli Hills, Tamil Nadu. *J. Econ. Taxon. Bot.* Addl. Series. **12**: 250-252.

Geetha, S., Lakshmi, G. & Ranjithkani, P. 1996b. Ethnoveterinary medicinal plants of Kolli Hills, Tamil Nadu. *J. Econ. Taxon. Bot.* Addl. Series. **12**: 289-191.

Ghani, A. 1998. *Medicinal plants of Bangladesh Chemical Constituents and Uses*. Asiatic Soc. Bangladesh, Bangladesh.

Ghosh, A., Maity, S. & Maity, M. 1996. Ethnomedicine in Bankura district, West Bengal. *J. Econ. Taxon. Bot.* Addl. Series **12**: 318-320.

Gill, L.S., Idu, M. & Ogbor, D.N. 1997. Folk medicinal plants; practices and beliefs of the Bini People in Nigeria. *Ethnobotany* **9**: 1-5.

Gill, L.S. & Nyawuame, H.G.K., 1994. Leguminosae in ethnomedical practices in Nigeria. *Ethnobotany* **6**: 51-64.

Gill, L.S., Nyawuame, H.G.K., Esezobor, E.I., & Osagie, I.S. 1993. Nigerian folk medicine : practices and beliefs of Esan people. *Ethnobotany* **5**: 129-142.

Girach, R.D. 1992. Medicinal plants used by Kondh tribe of district Phulbani (Orissa) in Eastern India. *Ethnobotany* **4**: 53-66.

Girach, R.D. & Aminuddin 1995. Ethnomedicinal uses of plants among the tribals of Singhbhum district, Bihar, India. *Ethnobotany* **7**: 103-107.

Girach, R.D., Aminuddin, A. M., Brahmam, M. & Mishra, M.K. 1996. Native phytotherapy among rural population of district Bhadrak, Orissa, 162-164. *In:* Jain, S.K. (*ed.*) *Ethnobiology in Human Welfare.* Deep Publ., New Delhi.

Girach, R.D., Aminudhin, A., Brahmam, M. & Mishra, M.K. 1997. Observations on ethnomedicinal plants of Bhadrak district, Orissa, India. *Ethnobotany* **9**: 44-46.

Girach, R.D., Brahmam, M. & Mishra, M.K. 1998. Folk veterinary herbal medicine of Bhadrak, Orissa India. *Ethnobotany* **10**: 85-88.

Goel, A.K., Sahoo A.K. & Mudgal, V. 1984. *A Contribution to the Ethnobotany of Santal Pargana, Bihar*. BSI, Howrah.

Goel, A.K. & Tripathy, S. 2009. Ethnobotanical spectrum of Indian flora: an overview during the past 20 years in the journal Ethnobotany. *Ethnobotany* **21**: 131-134

Gogoi, J.C., Sharma, D.K. & Borthakur, S.K. 2003. Folk-lore hepatoprotective herbal recipes for Assam in Northeastern India, 1-21. *In*: *Vistas in Ethnobotany* Vol.-1. *Indian J. Applied & Pure Biol.*

Gogoi, P. & Borthakur, S.K. 1991. Plants in religio-cultural beliefs of Tai Khamtis of Assam (India). *Ethnobotany* **3**: 89-95.

Gogoi, R. & Borthakur, S.K. 2001. Notes on herbal recipes of Bodo tribe in Kamrup Distt. Assam. *Ethnobotany* **13**: 15-23.

Goud, P. Saiprasad & Pullaiah, T. 1996a. Ethnobotany of Kurnool district– some wild plants used as a food. *J. Econ. Taxon. Bot.* Addl. Series **12**: 224-227.

Goud, P. Saiprasad & Pullaiah, T. 1996b. Folk veterinary of Kurnool district, Andhra Pradesh. *Ethnobotany* **8**: 71-74.

Govaerts, R. 2001. How many species of seed plants are there? *Tax.* **50**: 1085-1090.

Grieve, M. 1984. *A Modern Herbal. Penguin*. Vol. 1, Harcourt, Brace & Com. Publ. U.S.A.

Gupta, B., Nayak, S. & Solanki, S. 2011. *Abroma augusta* Linn. f: a review. *Der Pharmacia Sinica* **2**(4): 253-261.

Gupta, M.P. 1988. Present status and future prospects for the cultivation and collection of Medicinal plants of Himachal Pradesh. *Indian For.* **114**(1): 19-25.

Gupta, R. 1961. Flora of Lam Dal. *Indian. For.* **87**(5): 316-324.

Gupta, R. 1964. Survey record of medicinal and aromatic plants of Chamba forest division, Himachal Pradesh. *Indian For.* **90**: 454-468.

Gupta, R. 1971. Medicinal and aromatic plants of Bhandal ranges and Churah forest division, Chamba district, Himachal Pradesh. *J. Bomb. Nat. Hist. Soc.* **68** (3): 791-803.

Gupta, R. 1981. Plants in folk medicine of the Himalaya, 83-90. *In*: Jain, S.K. (ed.) *Glimpses of Indian Ethnobotany*. Oxford & IBH Publ. Co., New Delhi.

Hajra, P.K. 1987. Plants in magico-religious beliefs and in Sanskrit literature, 98-103. *In:* Jain S.K. (ed.) *A Manual of Ethobotany.* Sci. Publ., Jodhpur.

Hajra, P.K. & Baishya, A.K. 1981. Ethnobotanical notes on the Miris (Mishings) of Assam plains, 161-169. *In:* Jain S.K.. (ed.) *Glimpses of Indian Ethnobotany.* Oxford & IBH, New Delhi.

Hajra, P.K. & Das, B.K. 1982. Vegetation of Gangtok with special reference to exotic plants. *Indian For.* **107**: 554-566.

Halim, E. 2002. Hypoglycaemic, hypolipidemic and antioxidant properties of combination of *Curcumin* from *Curcuma longa* L. and partially purified product from *Abroma augusta* L. in streptozotocin induced diabetes. *Indian J Clin. Biochem.* **17** (2): 33-43.

Harborne, J.B. 1973. *Phytochemical Methods, A Guide to Modern Techniques of Plant*

Analysis. Chapman & Hall Ltd, London.

Harshberger, J.W. 1895. Some new ideas: The plants cultivated by aboriginal people and how used in primitive commerce. *The Evening Telegraph (daily)*, Philadelphia **63**(134): 2.

Hedberg, I. 1993. Botanical methods in ethnopharmacology and the need for conservation of medicinal plants. *J. Ethnopharmacol.* **38**: 121-128.

Heiser, C.B. 1995. Edgar Anderson botanist and curator of useful plants. *Ann. Miss. Bot. Gard.* **82** (1): 54-60.

Hembrom, P.P. 1991. Tribal medicine in Chotanagpur and Santhap Parganas of Bihar, India. *Ethnobotany* **3**: 97-99.

Hembrom, P. P. 1996. Contact therapy practised by Mundas of Chotanagpur (Bihar). *Ethnobotany* **8**: 36-39.

Hemborm, P. P. & Goel, A. K. 2005. Horopathy: ethnomedicine of Mundas. *Ethnobotany* **17**: 89-95.

Henry, A.N., Hosagoudar, V.B. & Ravi Kumar, K. 1996. Ethno-medico-botany of the Southern Western Ghats of India, 173-180. *In*: Jain, S.K. (*ed.*) *Ethnobiology in Human Welfare*. Deep Publ., New Delhi.

Hooker, J.D. 1872-1897. *The Flora of British India,* Vols. I-VIII. Lalit Mohan Basu, Allahabad.

Hosagoudar, V.B. & Henry, A.N. 1996a. Ethnobotany of Soligas in Biligiri Rangana Betta, Karnataka, Southern India. *J. Econ. Taxon. Bot.* Addl. Series **12**: 228-243.

Hosagoudar, V.B. & Henry, A.N. 1996b. Ethnobotany of Kadars, Malasars and Kuthuvans of the Anamalais in Coimbatore District, Tamil Nadu, India. *J. Econ. Taxon. Bot.* Addl. Series **12**: 260-267.

Hosagoudar, V.B. & Henry, A.N. 1996c. Ethnobotany of Tribes Irular, Kurumban and Paniyan of Nilgiris in Tamil Nadu, Southern India. *J.Econ. Taxon. Bot.* Addl. Series **12**: 364-366.

Huidrom, B. K. S.1996. Plants used in medico-sexual purposes by Meitein community in Manipur state, India. *J. Econ. Taxon. Bot.* Addl. Series **12**: 364-366.

Hussain A., Virmani, O.P., Popli, S.R., Mishra, L.N., Gupta, M.N., Srivastava, G.N., Abraham, Z. & Singh, A.K. 1992. *Dictionary of Indian Medicinal Plants.* CIMAP, Lucknow.

Iqbal, Z., Jabbar, A., Muhammad, S.A., Muhammad, G. & Muhammad, L. 2005. Possible role of ethnoveterinary medicine in poverty reduction in Pakistan: Use of botanical anthelmintics as an example. *J. Agric. & Soc. Sci.* **2**-187-195.

Islam, M. 2002. Wetland biodiversity with reference to economic plants of Assam and its neighbouring areas, 169-204. *In:* Trivedi, P.C.(*ed.*) *Ethnobotany.* Avishkar Publ., Distri. Jodhpur (Raj.), India.

Jadhav, D. 2006. Ethnomedicinal survey of Maalgamdi in Ujjain distt. Madhya Pradesh, India. *Ethnobotany* **18**: 157-159.

Jain, A. K. & Patole, S. N. 2001. Less known medicinal uses of plants among some tribal and rural communities of Panchmarhi Forest (M.P.). *Ethnobotany* **13**: 96-100.

Jain, A.K. & Sharma, H.O. 1996. Ethnobiological studies of Shariya tribe of Central India, 397-399. *In*: Jain, S.K. (*ed.*) *Ethnobiology in Human Welfare*. Deep Publ., New Delhi.

Jain, S.K. 1965. Medicinal plantlore of the tribal of Bastar. *Econ. Bot.* **19**: 236-250.

Jain, S.K. 1967. Ethnobotany: Its scope and study. *Indian Museum Bulletin* **2:** 39-43.

Jain, S.K. 1968. *Medicinal Plants.* National Book Trust, New Delhi.

Jain, S.K. (*ed.*) 1981. *Glimpses of Indian Ethnobotany*. Oxford & IBH, Publ. Co., New Delhi.

Jain, S.K. 1986. Ethnobotany. *Interdiscipl. Sci. Rev.* **2**: 285-292.

Jain, S.K. 1987. Ethnobotany its concepts and relevance. pp. 1-12. *Press. Add. X Bot. Conf.* Patna.

Jain, S.K. 1989. Medicinal Plants and Ethnobotany, *In:* Johri, B.M. (*ed.*) *Progress of Botany* (1910-1985). *Retrospect and Prospect*. Indian Sci. Con. Assoc. Calcutta.

Jain, S.K. 1991. *Dictionary of Indian Folk Medicine and Ethnobotany*. Deep Publ., New Delhi.

Jain, S.K. 1995. Ethnobotanical diversity in Zingibers of India. *Ethnobotany* **1**: 83-88.

Jain, S.K. (*ed.*) 1996. *Ethnobiology in Human Welfare*. Deep Publ., New Delhi.

Jain, S. K. 2002. *Bibliography of Indian Ethnobotany*. Sci. Publ., Jodhpur.

Jain, S. K. 2003. Indian plants in Tibetan medicine. *Ethnobotany* **15**: 236.

Jain, S.K. 2004. Objective ethnobotany: Traditional knowledge and modern approach. *Ethnobotany* **16**: 1-9.

Jain, S. K. 2006. Ethnobotany in the new millennium-some thoughts on future direction in Indian ethnobotany. *Ethnobotany* **18:** 1-3.

Jain, S.K. & Borthakur, S.K. 1980. Ethnobotany of the Mikiris of India. *Econ. Bot.* **34:** 264-272.

Jain, S. K. & De Filipps, R. A. 1991. *Medicinal Plants of India.* (2 vols.) Ref. Publ., Algonac.

Jain S. K. & Goel, A. K. 2005. Some Indian plants in Tibetan traditional medicine-1. *Ethnobotany* **17**: 127-136.

Jain, S. K., Mudgal, V., Banerjee, D. K., Guha, A., Pal, D. C. & Das, D. 1984. *Bibliography of Ethnobotany*. BSI, Calcutta.

Jain, S.K., Ranjan, Vinay, Sikarwar, R.K.S., & Saklani, A. 1994. Botanical distribution of psychoactive plants of India. *Ethnobotany* **6**: 65-75.

Jain, S.K. & Rao, R.R. (*eds.*) 1977. *A Handbook of Field and Herbarium Methods.* Today's & Tomorrow's Printers & Publ., New Delhi.

Jain, S.K. & Saklani, Arvind 1992. Cross-cultural ethnobotanical studies in North East India. *Ethnobotany* **4**: 25-38.

Jain, S.K., Sikarwar, R.L.S. & Pathak, V. 1997. Ethnobotanical aspects of some plants in Latin America. *Ethnobotany* **9**: 16-23.

Jain, S.K., Sinha, B.K. & Gupta, R.C. (*eds.*) 1991. *Notable Plants in Ethnomedicine of India*. Deep Publ., New Delhi.

Jain, S. K. & Srivastava, S. 1999. *Directory of Ethnoveterinary Plants of India.* Deep Publ., New Delhi.

Jain, S. K. & Srivastava, S. 2001. Indian ethnobotanical literature in last two decades -a graphic review and future directions. *Ethnobotany* 13: 1-8.

Jain, S.K. & Tarafder, C.R. 1970. Medicinal plant lore of the Santals. A revival of P.O. Boddings's work. *Econ. Bot.* **24**: 241-278.

Jain, S.P. 1984. Ethnobotany of Morni and Kalesar (Ambala, Harayana). *J. Econ. Taxon. Bot.* **5**: 809-813.

Jain, S.P. & Puri, H.S. 1994. An ethno-medico-botanical survey of Parvati valley in Himachal Pradesh (India). *J. Econ. Taxon. Bot.* **10**: 231-222.

Janaki Ammal, E.K. 1955. *An Introduction to the Subsistence Economy of India. Background Paper No. 10.* Winner Gref Foundation Intl. Symp. 'Man's Role in Changing the Face of the Earth'. Princeton Inn. Princeton, N.J. June 16-22.

Jha, R.R. & Varma, S.K. 1996. Ethnobotany of Sauria Paharias of Santhal pargana, Bihar: I medicinal plants. *Ethnobotany* **8**: 31-35.

Jha, V. & Basak, P. K. 1994. Ethnobotanical aspects of Mithila Paintings. *Ethnobotany* **6**: 9-18.

Jha, V., Choudhary, U.N. & Saraswati, K.C. 1991. Botanical aspects of an ethnoveterinary prescription in Mithila, North Bihar (India). *Ethnobotany* **3**: 101-104.

Jha, V., Mishra, S., Kargupta, A.N. & Jha, A. 1996. Leaves and flowers utilised as supplementary vegetables in Darbhanga (North Bihar) and their ethnobotanical significace. *J. Econ. Taxon. Bot.* Addl. series **12**: 395-402.

Joseph, J. & Kharkonger, P. 1981. A preliminary ethnobotanical survey in the Khasi and Jaintia hills, Meghalaya, 115-123. *In:* Jain, S.K. (ed.) *Glimpses of Indian Ethnobotany*. Oxford & IBH, New Delhi.

Joshi, A.R. 1993. *EIA of Kali gandaki Hydropower Project (Biodiversity Report),* East Consult, Kathmandu, Nepal.

Joshi, P. 1995. *Ethnobotany of the Primitive Tribes in Rajasthan*. Rupa Books Pvt. Ltd., Jaipur.

Kadel, C. & Jain, A. K. 2006. Plant used in ethnoveterinary practices in Jhabua distt. Madhya Pradesh. *Ethnobotany* **18**: 149-152.

Kala, C.P. 2003. *Medicinal Plants of Indian Trans-Himalaya (Focus on Tibetan Use of Medicinal Resources)*. BSMPS, Dehradun.

Kala, C.P. & Manjrekar, N. 2000. Ethno-medicobotany of India Trans-Himalaya: a case study from Spiti. *J. Econ. Taxon. Bot.* **23** (1): 177-183.

Kalpesh, I., Rao, V. R., Parabia, F.M., Kothari I.L. & Parabia M.H. 2004. Ethnobotany of Little Rann of Kachh, Gujarat. *Ethnobotany* **16**: 40-43.

Kapur, S.K. 1993. Ethnomedico plants of Kangra valley (Himachal Pradesh). *J. Econ. Taxon. Bot.* **17**(2): 395-408.

Kapur, S.K. 1996. Traditionally important medicinal plants of Bhaderwah hills - Jammu Province III. *J. Econ. Taxon. Bot.*, Addl. Series **12**: 62-69.

Kapur, S.K. & Nanda, S. 1996. Traditionally important medicinal plants of Bhaderwah Hills (Jammu Province) Part–II. *J. Econ. Taxon. Bot.* Addl. Series **12**: 56-61.

Kapur, S.K., Nanda, S. & Srivastava, T.N. 1996. Ethnobotanical uses of RRL-Herbarium – III. *J. Econ. Taxon. Bot.* Addl. Series **12**: 50-55.

Kapur, S.K. & Singh, P. 1996. Traditionally important medicinal plants of Udhampur district (Jammu Province) Part -I. *J. Econ. Taxon. Bot.*, Addl. Series **12**: 82-88.

Karnick, C. R. 1994. *Pharmacopoieal Standards of Herbal Plants*. Vol. 1. Sri Satguru Publ.

Karnick, C.R., Tiwari, R., Majumdar, R. & Bhattacharjee, S. 1981. Newer ethnobotanical and folklore studies of some medicinal plants of Gauhati and surrounding areas. *Nagarjun* **24**: 240-245.

Katewa, S.S. & Arora, Asha 1997. Some plants in folk medicine of Udapur district (Rajasthan). *Ethnobotany* **9**: 48-51.

Katewa, S.S., Choudhary, B.L., Jain, Anita & Takhar H.K. 2001. Some plants in folk medicine of Rajsamand distt. (Rajasthan). *Ethnobotany* **13**: 129-134.

Katewa, S.S. & Sharma, Rajesh 1998. Ethnomedicinal observation from certain watershed Areas of Rajasthan. *Ethnobotany* **10**: 46-49.

Kaul, M.K. 1996. Strategies for sustainable use of threatened medicinal plant resources in Western Himalayas, 284-286. *In*: Jain, S.K. (*ed.*) *Ethnobiology in Human Welfare*. Deep Publ. New Delhi.

Kaul, M.K. 1997. *Medicinal Plants of Kashmir and Ladakh-Temperate and Cold Arid Himalaya.* Indus Publ. Co., New Delhi.

Kaur, H. & Sharma, M. 2004. *Flora of Sirmaur (Himachal Pradesh)*. BSMPS, Dehradun.

Kaushik, P. 1988. *Indigenous Medicinal Plants* (Including Microbes and Fungi). Today and Tomorrow's Printers & Publ., New Delhi.

Kempraja, Narainswami, B.K. & Kulkari, V. 1998. Effective ethnoveterinary practices for treating affections of skin and limbs in large animals, 33-34. *In:* Mathias, E, Rangnekar, D.V. & Mc Corkle, C.M. *(eds.) Ethnoveterinary Medicine-Alternatives for Livestock Development.* BAIF development Research Foundation, Pune, India.

Khan, A., Rahman, M. & Islam, S. 2007. Antibacterial, Antifungal and cytotoxic activities of tuberous roots of *Amorphophallus campanulatus*. *Turk. J. Biol.* **31**: 167-172.

Khan, A.A. 2001. Need for an index to compute the relative reliability of ethnobotanical claims. *Ethnobotany* **13**: 84-86.

Khan, A. A. 2005. Ethnobotany in twenty first Century in India. *Ethnobotany* **17**: 71-78.

Khan, I.A. & Khanum, A. 2005. *Role of Biotechnology in Medicinal and Aromatic Plants.* Ukaaz Publ., Hyderabad. Andhra Pradesh.

Khan, T., Ahmad, W., Bashir, S., Iqbal, Z. & Manzoor, A. 2003. Biological and Pharmacological Properties of *Abroma augusta* L. Seed Oil. *Pakistan J. Biol. Sci.*

6 (13): 1142-1144.

Khanna, K.K. 2002. Unreported ethnomedicinal uses of plants from the tribal and rural folklores of Gonda district, Uttar Pradesh. *Ethnobotany* **14**: 52-56.

Khanna, K. K. & Kumar, R. 2000. Ethnomedicinal plants used by the Gujjar tribe of Saharnpur district, Uttar Pradesh. *Ethnobotany* **12**: 17-22.

Khanna, K.K., Shukla, G. & Mudgal, V. 1996a. New traditional medicinal uses of plants from Jalaun district, Uttar Pradesh. *J. Econ. Taxon. Bot.* Addl. Series **12**: 108-111.

Khanna, K.K., Srivastava, P.K. & Mudgal, V. 1996b. Noteworthy medicinal plants uses from rural folklore of Raebareli district, Uttar Pradesh. *J. Econ. Taxon. Bot.* Addl. Series **12**: 118-122.

Khare, P.C. (*ed.*) 2007. *Indian Medicinal Plants. An Illustrated Dictionary*. Springer-India Pvt. Ltd. Akash Deep Building Barakhamba Road, New Delhi.

Kharkonger, P. & Joseph, J. 1981. Folk-lore medicobotany of rural Khasi and Jaintia tribes in Meghalaya, 124-136. *In*: Jain, S. K.(*ed.*) *Glimpses in Indian Ethnobotany*. Oxford & IBH Publ., New Delhi.

Kirtikar, K.R. & Basu, B.D. 1994. *Indian Medicinal Plants*. Vol. 1-4, 2nd *ed.*, BSMPS, Dehradun.

Kiruba, S., Jeeva, S. & Dhas, S.S.M. 2006. Enumeration of ethnoveterinary plants of Cape Comorin, Tamil Nadu. *Indian J. Trad. Knowledge* **5**: 576-578.

Koelz, W. N. 1979. Notes on the ethnobotany of Lahul, a province of the Punjab. *Quar. J. Crude Drug Res.* **17**: 1-56.

Kokate, C.K. 1994. *Practical Pharmacognosy, ed.* 4, Vallabh Prakashan, New Delhi.

Kothari, M.J. & Moorthy, S. 1996. Ethnobotany in human welfare of Raigad district–in Maharastra State, India, 403-407. *In*: Jain, S.K. (*ed.*) *Ethnobiology in Human Welfare*. Deep Publ., New Delhi.

Kothari, M.J. & Rao, K.M. 2000. Ethnobotanical studies of Thane district Maharashtra: 265-272. In Maheshwari, J.K. (*ed.*) *Ethnobotany and Medicinal Plants of Indian Subcontinent*, Sci. Publ., Jodhpur (India).

Kowarik, I. 2003. Human agency in biological invasions: secondary releases foster naturalization and population expansion of alien plant species. *Biol. Invasions* **5**: 293-312.

Krenzelok, E. P., Jacobsen, T. D. & Aronis, J. M.1996. *Poinsettia* exposures have good outcomes -Just as we thought in. *Am. J. Emerg. Med.* **14** (7): 671-674.

Krishna, B. & Singh, S. 1987. Ethnobotanical observations in Sikkim. *J. Econ. Taxon. Bot.* **9**: 1-7.

Kshirsagar, R.D. & Singh, N.P. 2000. Less known ethnomedicinal use of plants in Coorg Distt. of Karnataka State, Southern India. *Ethnobotany* **12**: 12-16.

Kshirsagar, S. R., Parobia, M.H. & Reddy M.N. 2003. Ethnobotany of coastal areas in Gujarat. *Ethnobotany* **15**: 60-63.

Kulkarni, D.K. & Kumbhojkar, M.S. 1993. Kitchen garden plants of Mahadeokali tribe in Maharastra. *Ethnobotany* **5**: 119-127.

Kulkarni, D.K. & Kumbhojkar, M.S. 1996. Pest control in tribal areas of western

Maharastra -an ethnobotanical approach. *Ethnobotany* **8**: 56-59.

Kumar, A. 1996. Some ethnomedicinal plants of the Murias of the Indravati Tiger reserve, Bastas (Madhya Pradesh). *J. Econ. Taxon. Bot.* Addl. Series **12**: 201-205.

Kumar, K. & Goel, A.K. 1998. Little known ethnomedical plants of Santhal and Paharia tribes in Santhal Paragana, Bihar India. *Ethnobotany* **10**: 66-67.

Kumar, M., Ramesh, M. & Sequiera, S. 2003. Medicinal pteridophytes of Kerala, South India, *Indian Fern J.* **20**: 1-28.

Kumar, R. V. & Pullaiah, T. 1998. Medicinal plants used by tribals of Prakasan Distt. Andhra Pradesh. *Ethnobotany* **10**: 97-102.

Kumar, R. V. & Pullaiah, T. 2003. A survey of wild edible plants of Chenchu tribes of Prakasam District (Andhra Pradesh), 22-31. *In*: *Vistas in Ethnobotany* Vol.–1. *Indian J. Applied & Pure Biology*.

Kumar, S. & Chauhan, A.K.S. 2005. Medicinal plant used by local inhabitants in Bhartapur Distt. Rajasthan. *Ethnobotany* **17**: 179-183.

Kumar, S. & Chauhan, A.K.S. 2006. Less known medicinal used of plant species in Koeladeo national Park, Bharatpur Rajasthan. *Ethnobotany* **18**: 153-156.

Kumar, S., Gozal, S. & Farzana, P. 2004. Ethnoveterinary plants in Indian arid zone. *Ethnobotany* **16**: 91-95.

Kumar, V. & Jain, S.K. 1998. A contribution to ethnobotany of Surguga Distt. in Madhya Pradesh India. *Ethnobotany* **10**: 86-89.

Kurian, J.C. (*ed.*) 1995. *Plants That Heals.* Oriental Watchman Publ. House, Pune, India.

Lal, B. & Dube, V.P. 1984. Ethnobotanical studies on *Grewia optiva* Drum. *ex* Berret among the Jaunsaris and the people of the adjoining areas of Tehri Garhwal, U.P. *Himal. Res. Dev.* **3**: 6-8.

Lal, B., Vats, S.K., Singh, R.D. & Gupta, A.K. 1996. Plants used as ethnomedicine and supplementary food by Gaddis of Himachal Pradesh, 384-387. *In:* Jain, S.K. (*ed.*) *Ethnobiology in Human Welfare*. Deep Publ., New Delhi.

Lalramnghinglova, J.H. 1996. Ethnobotany of Mizoram-A preliminary survey. *J. Econ. Taxon. Bot.* Addl. Series **12**: 439-459.

Lata, Sneh 1996. *Grewia optiva* Drumm. 'Bhimal' : A boon for rural people in Kumaun Hills, India, 471-471. *In*: Jain, S.K. (*ed.*) *Ethnobiology in Human Welfare,* Deep Publ., New Delhi.

Lipp, F. J. 1989. Methods for ethnophramacology field work. *J. Ethnopharmacol.* **25**: 139-150.

Mahato, A. Kumar & Mahato, P. 1996. Ethnobotanical wealth of Chhotanagpur Plateau–IV. Some medicinal plants used against intestinal worms. *J. Econ. Taxon. Bot.* Addl. Series **12**: 389-391.

Maheshwari, J.K. 1987a. Interdisciplinary approaches in ethnobotany, 19-34. *In*: Jain, S.K. (*ed.*) *A Manual of Ethnobotany*, Sci. Publ., Jodhpur, India.

Maheshwari, J.K. 1987b. Ethnobotany in development of conservation of resources, 135-143. *In:* Jain, S.K. (*ed.) A Manual of Ethnobotany*. Sci. Publ., Jodhpur, India.

Maheshwari, J.K. (*ed.*) 1992. *Ethnobotany in India*. Sci. Publ., Jodhpur.

Maheshwari, J.K. (*ed.*) 1996. *Ethnobotany in South Asia*. Sci. Publ., Jodhpur.

Maheshwari, J.K. & Paul, S.R. 1975. The exotic flora of Ranchi. *J. Bombay Nat. Hist. Soc.* **72**(1); 158-188.

Maheshwari, J.K. & Singh, J.P. 1984. Contribution to the ethnobotany of Bhoxa tribe of Bijnor and Pauri Garhwal districts, U.P. *J. Econ. Taxon. Bot.* **5**: 251-259.

Maheshwari, J.K. & Singh, J.P. 1987. Contribution to the ethnobotany of Bhoxa tribe of Bijnor and Pauri Gahwal District U.P. *J.Econ Taxon. Bot.* **5**(2): 251-259.

Maheshwari, J.K., Singh, K.K. & Saha, S. 1980. The ethnomedicinal uses of plants of Tharus in Kheri district, Uttar Pradesh. *Bull. Medico-Ethnobot. Res.* **1**: 318-337.

Maheshwari, J.K., Singh, K.K. & Saha, S. 1981. *The Ethnobotany of the Tharus of Kheri District, Uttar Pradesh*. NBRI, Lucknow.

Maheshwari, J.K., Singh, K.K. & Saha, S. 1986. *Ethnobotany of Tribals of Mirzapur District, U.P.* NBRI, Lucknow.

Maji, S. & Sikdar, J.K. 1982. A taxonomic survey and systematic census on the edible wild plants of Midnapore district, West Bengal. *J. Econ. Taxon. Bot.* **3**: 717-737.

Malik, S. 1996. Belief and health care: A study among the Saoras of Koraput distt. Orissa, 151-152. *In*: Jain S.K. (*ed.*) *Ethnobiology in Human Welfare*. Deep Publ., New Delhi.

Malik, V., Mohammad, I. & Bharagava, A.K. 2009. Enumeration of ethnoveterinary plants of Saharanpur and adjacent Siwalik hills. *Indian Bot. Soc.* **88**: 77-81.

Maliya, S.D. 2004. Some new or less known folk medicines of district Bahraich U. P. India. *Ethnobotany* **16**: 113-115.

Maliya, S.D. & Singh, K.K. 2003. Some new or less known folk medicines of Badraich distt., Uttar Pradesh. *Ethnobotany* **15**: 132-135.

Manandhar, N.P. 1980. *Medicinal Plants of Nepal Himalaya*. Ratna Pustak Bhandar, Kathmandu, Nepal.

Manandhar, N.P. 1985. Ethnobotanical notes on certain medicinal plants used by Tharus of Dang-deokhuri district, Nepal. *Intl. J. Crude Drug Res.* **25**(4): 153-159

Manandhar, N.P. 1987a. Traditional medicinal plants used by tribals of Lamjung district, Nepal. *Intl. J. Crude Drug Res.* **25**(4): 236-240.

Manandhar, N.P. 1987b. An ethnobotanical profile of Manang valley, Nepal. *J.Econ. Taxon. Bot.* **10**(1): 207-213.

Manandhar, N.P. 1988. Ethnoveterinary medicines drugs of central development region of Nepal. *Bull. Medico-ethnobot. Res.* **10**(3-4): 93-99.

Manandhar, N.P. 1989a. Medicinal plants used by Chepang tribes of Makawanpur district, Nepal. *Fitotreapia* **60** (1): 61-68.

Manandhar, N.P. 1989b. *Useful Wild Plants of Nepal*. Franz Steiner Verlag Wiesbaden GMBH, Stuttgart, W. Germany.

Manandhar, N.P. 1990. Folk-lore medicine of Chitwan district, Nepal. *Ethnobotany* **2**: 31-38.

Manandhar, N. P. 1996. Traditional practice for oral healthcare in Nepal. *J. Econ. Taxon. Bot.* Addl. Series **12**: 408-413.

Manandhar, N.P. 2000. Botany and ethnobotany of *Urtica dioica* L. (Stinging nettle), 38-40. *In*: Maheshwari, J.K. *(ed.) Ethnobotany and Medicinal Plants of Indian Subcontinent*. Sci. Publ., Jodhpur.

Manandhar, N.P. 2002. *Plants and People of Nepal*. Timber Press. Oregon.

Mandal, S.K. & Basu, S.K. 1996. Ethnobotanical studies among some tribals of Niligiri district, Tamil Nadu. *J. Econ. Taxon. Bot.* Addl. Series **12**: 268-271.

Manilal, K. S. 1989. Linkages of ethnobotany with other sciences and discipline *Ethnobotany* **1**: 15-24.

Mao, A.A. 1993. A preliminary report on the folklore botany of Mao Nagas of Manipur (India). *Ethnobotany* **5**: 143-147.

Masih, S.K. 2003, Diversity of ethnomedicinal wealth in Amar Kantak plateau region (India), 204-214. *In*: *Vistas in Ethnobotany*, Vol.-I, *Indian J. Applied & Pure Biology*.

Mathews, K.M. 1969. Exotic flora of Kodaikanal and Palni hills. *Rec. Bot. Surv. India*. **20**(1): 1-241.

Maya, S., Kumari, S. P. & Menon, V. S. 2003. Ethnobotanical notes on the flora of sacred tanks of Kerala. *Ethnobotany* **15**: 55-59.

Megoneitso, K. &. Rao, R.R. 1983. Ethnobotanical studies in Nagaland-4. Sixty two medicinal plants used by the Angami Nagas. *J. Econ. Taxon. Bot.* **4**: 167-172.

Mehrotra, B.N. 1990. Quality control requirement of medicinal plants used in traditional medicines. *Ethnobotany* **2**: 19-24.

Mehrotra, S. & Mehrotra, B.N. 2005. Role of traditional and folklore herbals in the development of new drugs. *Ethnobotany* **17**: 104-111.

Mehrotra, S., Rawat, A.K.S., Singh, H.K. & Shome, U. 1995. Standardization of popular Ayurvedic adaptogenic preparation 'Chayavanprash' and ethnobotany of its ingredients. *Ethnobotany* **7**: 1-15.

Mishra, D. K., Samanta, G. & Mishra, T. K. 1996a. Ethnomedicine of the tribe Kharia of Midnapore District, West Bengal. *J. Econ. Taxon. Bot.* Addl. Series **12**: 329-331.

Mishra, M.K. & Gokhale, S.S. 2003. *Baseline Information on Medicinal Plants Conservation and Sustainable Utilization (Himachal Pradesh)*. A project sponsored by UNDP/GEF, MOEF, Govt. of India.

Mishra, S., Jha, V., & Jha, S. 1996b. Plants in ethno-veternary practices in Darbhanga (North Bihar), *In*: Jain, S.K. (*ed.*) *Ethnobiology in Human Welfare*, Deep Publ., New Delhi.

Mishra, T.K., Bhunia, A., Sahoo, T.K. & Roy, S.B. 1997. Ethnobotany vis-à-vis joint forest management movement of Midnapore (West Bengal), India. *Ethnobotany* **9**: 52-55.

Mittre, V. 1981. Wild plants in Indian folk life -A historical perspective, 37-58. *In:* Jain, S.K. (*ed.*) *Glimpses of Indian Ethnobotany*. Oxford & IBH, New Delhi.

Mittre, V. 1997. Wild plants in Indian folk life-a historical perspective, 35-53. In Jain, S.K.(*ed.*) *Contribution to Indian Ethnobotany*. Sci. Publ., Jodhpur (India).

Mohanty, R.B. & Padhy, S.N. 1996. Traditional phytotherapy for diarrhoeal diseases in Ganjam and Phulbani districts of South Orissa, India. *Ethnobotany* **8**: 60-65.

Mohanty, S.K., Dash, S.K. & Padhy, S.N. 1998. Traditional phytotherapy for diarrhoea disease in India-A review. *Ethnobotany* **10**: 103-111.

Mokat, D.K. & Deokule, S.S. 2004. Plants used as veterinary medicine in Ratnagiri district of Maharashtra. *Ethnobotany* **16**: 131-135.

Molla, H.A. & Roy, B. 1996. Some ethnobotanical claims from the Jalpaiguri district of West Bengal. *J. Econ. Taxon. Bot.* Addl. Series. **12**: 322-324.

Mudgal, V. 1995. Recent ethnobotanical works on different states/ tribes of India, 48-57. *In*: Jain, S.K. *(ed.) A Manual of Ethnobotany.* Sci. Publ., Jodhpur, India.

Mudgal, V., Pal, D.C., Kayal, R.N. & Saha, S. 1999. *Ethnobotany of Totopora.* BSMPS, Dehradun.

Mukherjee, A., Mukherjee, S. & Mukherjee, M. 1986. Less known medicinal plants of Rubiaceae in Darjeeling Hills, West Bengal (India). *Indian J. Appl. & Pure Bot.* **1**: 86-91.

Mukherjee, A. & Namhata, D. 1990. Medicinal plants of tribals of Sundargarh district, Orissa. *Ethnobotany* **2**: 57-60.

Murthy, K.S., Sharma, P.C. & Kishore, P. 1986. Tribal remedies for snake-bite from Orissa. *Anc. Sci. Life* **6**: 122-123.

Murthy, S.R., Rani, S.S. & Pullaiah, T. 2003. Wild edible plants of Andhra Pradesh, India, 613-630. *In:* Singh, V. & Jain, A.P. *(eds) Ethnobotany and Medicinal Plants of India and Nepal.* Sci. Publ., Jodhpur.

Myers, N. 1990. The biodiversity challenge: expanded hot-spots analysis. *The Environmentalist* **10**: 242-256.

Nahar, L., Farhana, A. & Islam, K.M.S. 2010. Comparative study of antidiabetic effect of *Abroma Augusta* and *Syzygium cumini* on alloxan induced diabetic rat. *Agric. Biol. J. N. Am.* **1**(6), 1268-1272.

Nair, N.C. 1964. On a botanical tour to Lahaul and Spiti (Punjab Himalaya). *Bull. Bot. Surv. India.* **6**: 219-235.

Nair, R. C. 1977. *Flora of Bashahr Himalaya.* Intl. Biosci. Publ., Hissar.

Nath, S.C. & Borodoloi, D.N. 1989. Ethnobotanical observation on some medicinal folklore of Tirap district, Arunachal Pradesh. *J. Econ. Taxon. Bot.* **13**: 321-325.

Nautiyal, S. 1981. Some medicinal plants of Garhwal Hills -a traditional use. *J. Sci. Res. Pl. Med.* **2**: 12-17.

Nayar, M. P. 1977. Changing patterns of the Indian flora. *Bull. Bot. Surv. Ind.* **19**: 145-154.

Nayar, M. P., Ramamurthy, K. & Agarwal, V. S. 1989. *Economic Plants of India.* Vol. I & II. BSI, Calcutta.

Nayar, M. P. & Sastry, A.R. K. 1987-1988. *Red Data Book on Indian Plants* (Vol. 1-11). BSI, Calcutta.

Nazarudeen, A., Seeni, S., Koshy, K.C. & Pushangadan, P. 1996. Folk plants of food, medicine, adornment and repellent used by the Paniyar community in North Kerela. *J. Econ. Taxon. Bot.* Addl. Series **12**: 299-305.

Negi, K.S. & Gaur, R.D. 1991. A contribution to the edible wild fruits of Uttar Pradesh hills. *Bull. Bot. Surv. India* **43**: 233-266.

Negi, K.S. & Pant, K.C. 1990. Notes on ethnobotany of the Gangwal -a tribe of Garhwal Himalaya. *Ethnobotany* **2**: 81-85.

Negi, K.S., Tiwari, J.K. & Gaur, R.D. 1985. Economic importance of some common trees in Garhwal Himalaya: an ethnobotanical study. *Indian J. Forestr.* **8**: 276-289.

Negi, K.S., Tiwari, J.K., Gaur, R.D. & Pant, K.C. 1993. Notes on ethnobotany of five districts of Garhwal Himalaya, Uttar Pradesh, India. *Ethnobotany* **5**: 73-81.

Negi, P. S. & Hajra, P.K.2007. Alien flora of Doon valley, North West Himalaya. *Curr. Sci.* **92** (7): 968-978.

Negi, P. S. & Subramani, S. P. 2002. Ethnobotanical study in the village Chhitkul of Sangla Valley, district Kinnaur, Himachal Pradesh. *J. Non-Timber Forest Prod.* **9** (3-4): 113-120.

Negi, P.S. & Subramani, S.P. 2006. Ethnobotanical study in the Naggar Area of Kullu Valley, Himachal Pradesh. *J. Econ. Taxon. Bot*. **30** (2): 349-358.

Negi, Y.S. & Bhalla, P. 2002. Collection and marketing of important medicinal and aromatic plants in tribal areas of Himachal Pradesh. *Indian For.* **128**(6): 641-649.

Nellis, 1997a. *Poisonous Plants and Animals of Florida and the Caribbean*. Pineapple Press, Inc. 1st *(ed.)* Florida U.S.A. 182-183.

Nellis, 1997b. *Poisonous Plants and Animals of Florida and the Caribbean*. Pineapple Press, Inc. 1st *(ed.)* Florida U.S.A. 352.

Oommachan, M.1977. *The Flora of Bhopal (Angiosperms)* Anand Kumar Jain, J. K. Brothers. Bhopal (India).

Painuli, R.M. & Maheshwari, J.K. 1996. Some interesting ethnomedicinal plants used by Sahariya tribe of Madhya Pradesh. *J. Econ. Taxon. Bot.* Addl. Series. **12**: 179-185.

Pal, D.C. 1980. Observations on folk-lore about plants used in veterinary medicine in Bengal, Orissa and Bihar, 245-257. *In:* Jain, S.K. (*ed.*) *Glimpses of India Ethnobotany*. Oxford and IBH Publ., New Delhi.

Pal, D.C. 1981. Plants used in treatment of cattle and birds among tribals of Eastern India, 245-257. *In:* Jain, S.K. *(ed.) Glimpses of Indian Ethnobotany*. Oxford & IBH, New Delhi.

Pal, D.C. & Jain, S.K. 1998. *Tribal Medicine.* Nayaprakash, Calcutta.

Panda, S. 1996. Plant use–a recent perspective from Eastern-Himalayas. *J. Econ. Taxon. Bot*. Addl. Series **12**: 332-337.

Pandey, A. K., Bora, H. R. & Deka, S.C. 1996. An ethno-medico botanical study of Golghat district, Assam : native plant remedies for jaundice. *J. Econ. Taxon. Bot.* Addl. Series. **12**: 344-349.

Pandey, A. K. & Rout, S. D. 2006. Ethnobotanical uses of plants of Similipal Biosphere Reserve (Orissa). *Ethnobotany*: **18**: 102-106.

Pandey, D.N. 1998. *Ethno Forestry (Local Knowledge for Sustainable Forestry and Livelihood Security).* Himanshu Publ., Udaipur, New Delhi.

Pandey, P. H. & Verma, B.K. 2002. Plants in oral healthcare among the aboriginals of Gonda and Balrampur regions, U.P., India. *Ethnobotany* **14**: 81-86.

Pandey, P. H. & Verma, B.K. 2003. Magico-therapeutic plants of Uttar Pradesh, India, 178-187. *In*: *Vistas in Ethnobotany*,Vol.-I, *Indian J. Pure & Applied Biology*.

Pandey, R.P. & Parmar, P.J. 1994. The exotic flora of Rajasthan. *J. Econ. Taxon. Bot.* **18**(1): 105-121.

Parihar P. & Parihar L., 2006. Some pteridophytes of medicinal importance from Rajasthan. *Nat. Product Rad.* **5** (4), 297-301.

Paroda, R.S. & Arora, K.R. 1991. Plant genetic resources: General perspective, 1-24. *In*: Paroda, R.S. & Arora, R.K. (*eds*) *Plant Genetic Resources-Conservation and Management*. IBPGR, New Delhi.

Prakash, V., Aggarwal, A., Sharma, A. & Mehrotra, R.S. 2003. Ethnobotanical studies on some wild plants of district Hamirpur in Himachal Pradesh, 374. *In:* Prakash, P. *(ed) Proc. Nat. Sem. Biochem. Sci.: Health and Environment Aspects*. Allied Publ., Pvt. Ltd. New Delhi.

Parker, R.N. 1993. *A Forest Flora for the Punjab with Hazara and Delhi,* BSMPS & Periodical Experts, Delhi.

Parmar, C. & Kaushal, M. K. 1982. *Wild Fruits of the Sub-Himalayan Region*. Kalyani Publ. New Delhi-Ludhiana.

Parrotta, J. A. 2001. *Healing Plants of Pennisular India*. CAB Intl., U.K.

Patel, N.B., Sidaha, B.S. & Jain, B.K. 2003. Tribal artifacts of Dholwani forest of Sabarkantha (Gujrat). *Ethnobotany* **15**: 40-43.

Patil, D.A. 2000. Sanskrit plant names in an ethnobotanical perspectives. *Ethnobotany* **12:** 60-64.

Patil, D.A. 2004. Origin of medicines vis-a-vis doctrine of signatures. *Ethnobotany* **16**: 52-58.

Patil, S.H. & Merat, M. M. 2003. Ethnoveterinary practices of Satpudas of Nandurbar Distt. of Maharashtra. *Ethnobotany* **15**: 103-106.

Patole, S. N. & Jain, A. K. 2002. Some wild edible plants of Pachmarhi Biosphere Reserve (M.P.). *Ethnobotany* **14**: 48-51.

Pawar, S. & Patil, D.A. 2004. Herbal folk remedies for urinogenital complaints in Jalgaon distt. of Maharashtra. *Ethnobotany* **16**: 86-90.

Pearson, D. 1987. *Pearson's Chemical Analysis of Foods*. Longman Sci. & Tech., Harlone.

Pei, S.J. 2001. Ethnobotanical approaches of traditional medicine studies: Some experiences from Asia. *Pharmaceutical Biol.* **39:** 74-79.

Pei, S. J. & Long, C. L. 1998. *The Applied Ethnobotany*. Yunnan Ethnology Press, Kunmung.

Perrings, C., Dehnen-Schmutz, K., Touza, J. & Williamson, M. 2005. How to manage biological invasions under globalization. *Trends Ecol. Evol.* **20:** 212-215.

Polunin, O. & Stainton, A. 1984. *Flowers of the Himalaya*. Oxford Univ. Press, Delhi.

Polunin, O. & Stainton, A. 1987. *Concise Flowers of Himalaya. Oxford Univ. Press,* Delhi.

Prajapati, N.D., Purohit, S.S., Sharma, S.S. & Kumar, T. 2003 (*ed.*). *A Hand Book of Medicinal Plants, A Complete Source Book,* (Agrobios (India) Publ., New Delhi.

Prasad, P., Nagrajan, R. Singh, A.J.A., Narayanan, L.M. & Natarajan, C.R. 1996. Ethnobotany of the Kanikkars of South Tamil Nadu-1. *J. Econ. Taxon. Bot*. Addl. Series **12**: 292-298.

Prasad, V.K., Rajagopal, T. & Badarinath, K.V.S. 2002. Ethnobotany of Kondareddis of Rampa Agency, Eastern Ghats, Andhra Pradesh, 205-220. *In:* Trivedi, P.C. (*ed.*) *Ethnobotany.* Avishkar Publ., Distributors, Jodhpur (Raj.) India.

Pruthi, J.S. 1976. *Spices and Condiments.* National Book Trust, New Delhi.

Pullaiah, T. & Kumar, T. Dharma Chandra 1996. Herbal plants in Mannanur forest Mahaboobnagar district, Andhra Pradesh. *J. Econ. Taxon. Bot*. Addl. Series **12**: 218-220.

Punjani, B. K. 1998. Role of plants in field fencing in tribal area of Distt.-Sabarkantha (North Gujarat). *Ethnobotany* **10**: 56-60.

Purohit, K. & Samant, S.S. 1995. *Fodder Trees and Shrubs of Central Himalaya.* Gyanodaya Prakashan, Nainital (U.P.), India.

Pushpangadan, P. 1990. Health Status of Tribals of India. Key note addess, 67-71. Nat. Sem. Health of Tribal people. *Indian Society for Health Administrators* (ISHA), May 21-26, 1990.

Pushpangadan, P. 1994. *Ethnobiology in India, A Status Report*. Ministry of Environment and Forest, GOI, New Delhi.

Pushpangadan, P. & Kumar, B. 2005. Ethnobotany, CBD, WTO and The Biodiversity act of India. *Ethnobotany* **17**: 2-12.

Raaman, N. 2008. *Phytochemical Techniques.* New India Publ. Agency.

Radhakrishnan, K., Pandurangan, A. G., & Pushpangandan P. 1996. Ethnobotany of wild edible plants of Kerala, India, 48-51. *In* : Jain, S.K. (*ed.*) *Ethnobotany in Human Welfare*. Deep Publ., New Delhi.

Ragupathy, S. & Mahadevan, A. 1991. Ethnobotany of Kodiakkarai reserve forest, Tamil Nadu, South India. *Ethnobotany* **3**: 79-82.

Rahamtullah, M., Nasrin, D. & Jahan, I. 2010. Survey of medicinal plants used by folk medicinal practioners in Daudhandi sub-district of Comilla district, Bangladesh. *Advances in Natl. & Applied Sci.* **4**(2): 99-104.

Rajasekaran, A. & Singh, J. 2009. Ethnobotany of Indian Horse Chestnut (*Aesculus indica*) in Mandi district, Himachal Pradesh. *Indian J. Trad. Knowledge* **8**(2): 285-286.

Rajendran, A., Rama Rao, N. & Henry, A.N. 1996. Hepatic stimulant plants of Andhra Pradesh, India. *J. Econ. Taxon. Bot*. Addl. Series **12**: 221-223.

Raju, R.A. 1998. *Wild Plants of Indian Subcontinent and Their Economic Use.* CBS Publishers & Distributors, New Delhi.

Rajwar, G.S. 1983. Low altitude medicinal plants of south Garhwal. *Bull. Medico-Ethnobot. Res.* **4**:14-28.

Ram, A., Jeewan, R., K., Reddy, K. E. & Venkata, R. R. 2002. Medicinal plant-lore of Sugails of Gooty forests Andhra Pradesh. *Ethnobotany* **14**: 37-42.

Rama Rao, N. & Henry, A. N. (*ed.*) 1996. *Ethnobotany of Eastern Ghats in Andhra Pradesh, India*. Bot. Surv. of India, Calcutta.

Ramachandran, V.S. & Nair, N.C. 1981. Ethnobotanical conservations of Irulars of Tamil Nadu (India). *J. Econ. Taxon. Bot.* **2**: 183-190.

Ramdas, S.R., Ghotge, N.S., Ashalata, S., Mathur, N.P., Brome, V.G. & Rao, S. 2000. Ethnoveterinary remedies used in common surgical conditions in some districts of Andhra Pradesh and Maharashtra, India. *Ethnobotany* **12**: 100-112.

Rana, J. C. 2004. *Lesser Known Food Plants of Himachal Pradesh*. NBPGR, Regional Station, Phagli, Shimla.

Rana, T.S., Dutt, B. & Rao, R.R. 1996. Strategies of sustainable utilization of plant resources by the tribals of the Tons valley, Western Himalaya. *Ethnobotany* **8**: 96-104.

Ranade, V.V. 1998. Current techniques for pharmacological evaluation of indigenous drugs for veterinary use, 34-37. *In*: Diwedi, S.K. (*ed.) Techniques for Scientific Validation and Evaluation of Ethnoveterinary Practices.* IVRI, Izatnagar.

Ranjan, S. & Sethuraman, M. 1997. Traditional veterinary practices in rural areas of Dindigul district, Tamil Nadu, India. *Ind. Knowled. & Develop. Monitor.* **5**(3): 709.

Ranjan, V. 1996. Some ethnomedicinal plants of Lalitpur district, Uttar Pradesh, India, 149-150. *In*: Jain, S. K. (*ed.*) *Ethnobotany in Human Welfare*. Deep Publ. New Delhi.

Rao, B. U.V.U. 2006. Collection and documentation of ethno-medicinal and non-timber forest products (N.T.F.P.). Information from Royalseema region of Andhra Pradesh, India. *Ethnobot. Leaf.* **10:** 294-304.

Rao, D. M. & Pullaiah, T. 2001. Ethno-medico-botanical studies in Guntur distt. of Andhra Pradesh, India. *Ethnobotany* **13**: 40-44.

Rao, N.S. B., Rajasekhar, D., Raju, D. C. & Nagaraju, J. 1996. Ethnomedicinal notes on some plants of Tirumala Hills for dental disorders. *Ethnobotany* **8**: 88-91.

Rao, R.R. & Jamir, N.S. 1982. Ethnobotanical studies in Nagaland 1; Medicinal plants. *Econ. Bot.* 36: 176-181.

Rao, R. R. 1996. Traditional knowledge and sustainable development: key role of ethnobiologists. *Ethnobotany* **8**: 14-24.

Rao, V.V. & Rajasekharan, P.E. 2002. Threatened medicinal plant resources and conservation needs. *The Botanica* **52**: 53-63.

Rastogi, M.A. 1960. Medicines from the wild. A case study of the Great Himalayan Park. *Indian Magz. of Her People & Culture*: 74-75.

Rastogi, R.P. & Mehrotra, B.N. 1990. *Compendium of Indian Medicinal Plants.* Vol.1. (1960-1969). CDRI & CSIR.

Rastogi, R.P. & Mehrotra, B.N. 1991. *Compendium of Indian Medicinal Plants.* Vol.2. (1970-1979). CDRI & CSIR.

Rastogi, R.P. & Mehrotra, B.N. 1993. *Compendium of Indian Medicinal Plants.* Vol.3. (1980-1984). CDRI & CSIR.

Rastogi, R.P. & Mehrotra, B.N. 1995a. *Compendium of Indian Medicinal Plants*. Vol.4. (1985-1989). CDRI & CSIR.

Rastogi, R.P. & Mehrotra, B.N. 1995b. *Compendium of Indian Medicinal Plants*. Vol.5. (1990-1994). CDRI & CSIR.

Rau, M.A. 1961. A recent survey of medicinal plant resources of Lahaul valley (Panjab India). *Pakistan J. Soc. & Ind. Res. Spl. Nr. Int. Symp.* **4**: 201-205.

Ravi, R. 1994. *Ethnobotanical Studies of Badagas in Nilgiris*. Ph.D. Thesis Bharathiar Univ., Coimbatore.

Ravishankar, T. & Henry, A.N. 1992. Ethnobotany of Aditlabad District, Andhra Pradesh, India. *Ethnobotany* **4**: 45-52.

Rawat, D.S. & Kharwal, A.D. 2010. Indigenous traditional method for making liquid detergent 'chhoa' from *Zea mays* Linn. in Shivalik Hills (Himachal Pradesh), India. *Ethnobot. Leaf.* **14:** 5-7.

Rawat, D.S., Kharwal, A.D. & Rawat, S. 2009. Ethnobotanical studies on dental hygiene in District Hamirpur, Himachal Pradesh (H.P.), India. *Ethnobot. Leaf.* **13:** 1434-42.

Reddy, C.S. 2008. Catalogue of invasive alien flora of India. *Life Sci. Jour.* **5**(2): 84-89.

Reddy, C.S. & Raju, V.S. 2000. Folklore biomedicines for common veterinary diseases in Nalgonda district, Andhra Pradesh, India. *Ethnobotany* **12**: 113-117.

Reddy, C.S. & Reddy, K.N. 2004. *Cassia rotundifolia* Pers. (Ceasalpiniaceae): a new record for India. *J. Econ. Taxon. Bot.* **28**: 73-74.

Reddy, H. H., Vijayalakshmi, K. & Venkata Raju, R.R. 1996. Native phytotherapy for snake bite in Nallamalais Eastern Ghats, India. *J. Econ. Taxon. Bot.* Addl. Series **12**: 214-217.

Reddy, K.N., Bhanja, R., & Raju, V. S. 1998. Plant used in ethnoveterinary practices in Warangal Distt. Andhra Pradesh, India. *Ethnobotany* **10**: 75-84.

Reddy, K.N., Bhanja, R., & Raju, V.S. 2000. *Cassia uniflora* Miller: a new record for Andhra Pradesh, India. *Indian J. For.* **23**(3): 324-325.

Rekha, S., Jagdish, T., Mattri, S. & Narainswami, B.K. 1998. Ethnoveterinary medicine for ruminants, 53-54. *In:* Mathias, E., Rangnekar, D.V. & Mc Corkle, C.M.*(eds.) Ethnoveterinary Medicine-Alternatives for Livestock Development*. BAIF development Research Foundation, Pune, India.

Retnam, K. R. & Martin, P. 2006. *Ethnomedicinal Plants*. Agrobios, Jodhpur.

Ricarda, S., 2006. Testimony to Agriculture and Forestry. Committee of Senate of Canada, Canada.

Rodgers, W.A. & Panwar, H.S. 1990. *A Biogeographical Classification for Conservation Planning in India*. Wildlife Institute of India, Dehradun, India.

Sagwal, S.S. & Gupta, N.K. 1988. Role of forestry extension in exploitation on medicinal and aromatic plants of Himachal Pradesh, 50-54. *In*: Khosla, P.K. & Sehgal, R.N. *(eds) Trends in Tree Science*. ISTS Publ., Solan.

Sahoo, A.K. & Mudgal, V. 1995. Less known ethnobotanical uses of plants of Phulbani district, Orissa, India. *Ethbobotany* **7**: 63-67.

Sahni, K.C. 1998. *The Book of Indian Trees*. Bombay Nat. Hist. Soc., Oxford Univ. Press.

Sahu, T.R. 1996. Life support promising food plants among aboriginals of Bastar (M.P.), India, 26-30. *In*: Jain, S.K. (*ed.*) *Ethnobiology in Human Welfare*. Deep Publ. New Delhi.

Saini, D.C. 1996a. Ethnobotany of Tharus of Basti District, Uttar Pradesh. *J. Econ. Taxon. Bot.* Addl. Series **12**: 138-153.

Saini, V.K. 1996b. Plants in the welfare of tribal women and children in certain areas of Central India, 140-144. *In*: Jain, S.K. (*ed.*) *Ethnobiology in Human Welfare*. Deep. Publ., New Delhi.

Saklani, A. & Jain, S.K. 1994. *Cross-Cultural Ethnobotany of North East India*. Deep Publ. New Delhi.

Samati, H. 2004. Kitchen garden plants of Pnar tribe in Jaintia Hills Distt. Meghalaya. *Ethnobotany* **16**: 125-130.

Samwatsar, S. & Diwanji, V.B. 1996a. Plants used for skin diseases, cuts, wounds, and bruishes by the tribals of Western M.P. *J. Econ. Taxon. Bot.* Addl. Series. **12**: 192-195.

Samwatsar, S. & Diwanji, V.B. 1996b. Plants used in snake, scorpion and insect bites/stings by Adibasis of Jhabua (MP) India. *J. Econ. Taxon. Bot.* Addl. Series **12**: 199-200.

Sarangi, N. & Sahu R.K. 2004. Ethnomedicinal plants used in venereal and gynaecological disorders in Kalahandi Orissa. *Ethnobotany* **16**: 16-20.

Sarin, Y.K. 1967. A survey of vegetable raw resources of Lahaul. *Indian For.* **93**(7): 489-499.

Sarin, Y.K. & Chopra, C.L. 1985. Medicinal and aromatic plant resources of Himachal Pradesh, present status and future prospects. *Workshop on Sci. and Tech., Shimla*, Oct. 29-30, 1985.

Sarin, Y.K., Kapoor, L.D. & Chopra, I.C. 1963. *Physochlaina praelta* (Don) Miers: a hyoscyamine yielding plant important in Lahaul's economy. *Indian For.* **89**: 610-611.

Satapathy, K.B. & Brahmam, M. 1996. Some medicinal plants used by tribals of Sundargarh district, Orissa, India, 153-155. *In*: Jain, S.K. (*ed.*) *Ethnobiology in Human Welfare*. Deep Publ., New Delhi.

Saxena, A.P. & Vyas, K.M. 1981. Ethnobotanical records on infectious diseases from tribals of Banda district. *J. Econ. Taxon. Bot.* **2**: 191-194.

Saxena, A.P. & Vyas, K.M. 1983. Ethnobotany of Dhasan Valley. *J. Econ. Taxon. Bot.* **4**: 121-128.

Saxena, H.O. 1986. Observations on the ethnobotany of Madhya Pradesh. *Bull. Bot. Surv. India* **28**: 149-156.

Saxena, H.O., Brahmam, M. & Dutta, P.K. 1981. Ethnobotanical studies in Orissa, 232-244. *In*: Jain, S.K. (*ed.*) *Glimpses of Indian Ethnobotany*. Oxford & IBH, Publ. Co., New Delhi.

Saxena, K.G. 1991. Biological invasion in the Indian sub-continent: review of invasion by plants, 53-73. *In:* Ramakrishnan, P.S. (*ed.*): *Ecology of Biological Invasion in the Tropics.* Intl. Sci. Publ., New Delhi.

Schultes, R.E. 1960. Tapping our heritage of ethnobotanical lore. *Econ. Bot.* **8**: 224-227.

Sebastin, M.K. & Bhandari, M.M. 1984. Some plants used as veterinary medicines by Bhils. *Intern. J. Trop. Agri.* **2**: 307-310.

Sen, S. & Batra, A. 1997. Ethno-medico-botany of household remedies of Phagli Tehsil of Jaipur district (Rajasthan). *Ethnobotany* **9**: 122-128.

Sen, S., Batra, A. & Ali, Z. D. 2002. Some herbal home remedies from Jaipur Distt. Rajasthan. *Ethnobotany* **14**: 116-122.

Sen, S.K. & Behera, L.M. 2003. Ethnomedicinal plants used against skin diseases of Bargarh district in Orissa. *Ethnobotany* **15**: 90-96.

Sensarma, P. 1989. *Plants in the Indian Puranas. An Ethnobotanical Investigation.* Naya Prakash, Calcutta.

Sensarma, P. 1991. Herbal veterinary medicines in an ancient-Sanskrit work -The Garuda Purana. *Ethnobotany* **3**: 83-87.

Sethi, G.S. 2003. Potential application biotechnology in medicinal plants of Himachal Pradesh. *In: Proc. Natl. Sem. Emerging Trends in Res. & Business Management of Medicinal & Aromatic Plants,* March 11-13, 2003. CCSHAH, Hissar, Haryana.

Sethiya, N.K., Patel, M.B. & Mishra, S.H. 2010. Phytopharmacologic aspects of *Canscora decussata* Roem and Schult. *Phcog Rev.* **4**: 49-57.

Shabnam, S.R. 1964. Medicinal plants of Chamba. *Indian For.* **90:** 50-63.

Shah, G.L. 1984. Some economically important plants of Salsette Island near Bombay. *J. Econ. Taxon. Bot.* 5: 753-765.

Shah, G.L., Menon, A.R. & Gopal, G.V. 1981. An account of the ethnobotany of Saurashtra in Gujarat State. *J. Econ. Taxon. Bot.* **2:** 173-182.

Shah, M.A.U. 2005. Ethnomedicinal plants of Bhakkar tehsil in Punjab, Pakistan. *Ethnobotany* **17**: 171-175.

Shah, N.C. 1987. *Ethnobotany in the Mountainous Region of Kumaon Himalayas.* Thesis. Kumaon University, Nainital. Ph.D. (Bot.), 1-255.

Shah, N.C. 1997. Ethnobotany of *Cannabis sativa* in Kumaon region, India: *Ethnobotany* **9**: 117-121.

Shah, N.C. 2008. Prospects of ethnobotany in India. *Ethnobotany* **20:** 16-24.

Shah, N.C. & Joshi, M.C. 1971. An ethnobotanical study of Kumaon region of India. *Econ. Bot.* **25**: 414-422.

Shah, N.C. & Singh, S.C. 1990. Hitherto unreported phytotherapeutical uses from tribal pockets of Madhya Pradesh (India). *Ethnobotany* **2**: 91-95.

Sharma, B.D. 1984. *Exotic Flora of Allahabad.* B.S.I. Dehradun.

Sharma, B.D. 2003. *Medicinal Plants of India – An Encyclopaedia.* Daya Publ. House, Delhi – 110035.

Sharma, B.D. 2009. *Life-Sustaining Plants of the Himalayas.* Indus Publ. Co., New Delhi.

Sharma, B.D. & Chandel, K.P.S. 1996. Occurrence, distribution and diversity of soft fruits in N.W. and W. Himalayas and prospects of their conservation and utilization. *Indian J. Pl. Gen. Res.* **9**(2): 237-240.

Sharma, B.D. & Rana, J.C. 1999. Traditional medicinal uses of plants of Himachal hills. *J. Econ. Taxon. Bot.* **23**(1): 173-176.

Sharma, B.D. & Rana, J.C. 2005. *Plant Genetic Resources of Western Himalayas: Status and Prospects.* B.S.M.P. S., Dehradun, India.

Sharma, B.D. & Vyas M.S., 1985. Ethnobotanical studies on the ferns and fern-allies of Rajasthan, *Bull. Bot. Surv. India* **27** (1-4), 90-91.

Sharma, H.M., Devi, A.R. & Sharma, B.N. 2005. Vegetable dyes used by the Meiti community of Manipur. *Indian J. Trad. Knowledge* **4**(1): 39-46.

Sharma, H.R. & Chauhan, G.S. 2000. Physical, sensory and chemical characteristics of wheat breads supplemented with fenugreek. *Journal Food Sci.* **37:** 91-94.

Sharma, K.R. & Sood, M. 1997. *Important Medicinal Plants of Himachal Pradesh.* Vol-I & II. Dr. Y.S. Parmar Uni. Horti. For., Nauni, Solan, H.P.

Sharma, M.C. 2009. *Use of Plant Based Medicaments in Treatment of Skin Diseases.* Indian Vet. Res. Inst., Izatnagar, U.P., India.

Sharma, P.K., Chauhan, N.S. & Lal, B. 2004. Observation on the traditional phytotherapy among the inhabitants of Parvati Valley in Western Himalaya, India. *J. Ethnopharmacol.* **92**: 167-176.

Sharma, P.P. & Singh, N.P. 2001. Ethnomedicinal uses of some edible plants in Dadra, Nagar Haveli and daman (U.T.). *Ethnobotany* **13**: 121-125.

Sharma, S., Gautam, A.K. & Bhadauria, R. 2009. Some important supplementary food plants and wild edible fungi of upper hilly region of district Shimla (Himachal Pradesh), India. *Ethnobot. Leaflets* **13**: 1020-1028.

Sharma, S.C. 1996. A medicobotanical study in relation to veterinary medicines of Shahjahanpur district (Uttar Pradesh). *J. Econ. Taxon. Bot.* Addl Series **12**: 123–127.

Sharma, S.K., Bhattacharya, D.K. & Devi B. 2001. Medicinal plants used by Bodo tribe of Nalbari distt. in Assam. *Ethnobotany* **13**: 135-139.

Sharma, S.K., Bhattacharjya, D.K. & Devi, B. 2002. Traditional use of herbal medicines by Madai tribe of Nalbari district of Assam. *Ethnobotany* **14**: 103-111.

Sharma, T.R. & Borthakur, S.K. 2003. Inquire into old lead-II: ethnoveterinary medicine for treatment of horse in Assam, 69-85. *In*: *Vistas in Ethnobotany*, Vol. I, *Indian J. Pure and Appld. Biol.*

Shekhawat, G.S. & Anand, S. 1984. An ethnobotanical profile of Indian Desert. *J. Econ. Taxon. Bot.* **5**: 591-598.

Shome, U., Rawat, A.K.S. & Mehrotra, S., 1996. Time tested household remedies, 96-100. *In*: Jain, S.K. (*ed.*) *Ethnobiology in Human Welfare*, Deep Publ., New Delhi.

Shrestha, P. 1985. Research note: contribution to the ethnobotany of the Palpa area. *Contr. Nepalese Stud.* **12**(2): 63-74.

Shukla, G. & Verma, B.K. 1996. Roots -a vital plant part to cure body ailments among tribals/rural folklore of Western Bihar. *J. Econ. Taxon. Bot.* Addl. Series **12**: 392-394.

Sikarwar, R.L.S. 1996. Ethnoveterinary herbal medicines in Morena district of Madhya Pradesh, India, 194-196. *In*: Jain, S.K. (*ed.*) *Ethnobiology in Human Welfare,* Deep Publ., New Delhi.

Sikarwar, R.L.S. 2002. Ethnogynaecological uses of plants new to India. *Ethnobotany* **14**: 112-115.

Sikarwar, R.L.S., Bajpai, A.K. & Painuli, R.M. 1994. Plants used as veterinary medicines by aboriginals of Madhya Pradesh, *India. Int. J. Phamacog*. **32**(3): 251-255.

Sikarwar, R.L.S. & Kaushik, J.P. 1992. Some less known medicinal uses of trees among the Sahariyas of Morena district, M.P., India. *Ethnobotany* **4**: 71-74.

Silori, C. S. & Rana, A. R. 2000. Indigenous knowledge of medicinal plants and their uses in Narayan Sarover Sanctuary Kakhch. *Ethnobotany* **12**: 1-7.

Singh, A.K., Singh, R.N. & Singh, S. K. 1987. Some ethnobotanical plants of Terai region of Gorakhpur district I. *J. Econ. Taxon. Bot.* **9:** 407-410.

Singh, B., Chaurasia, O.P. & Jadhav, K.L. 1996a. An ethnobotanical study of Indus valley (Ladakh). *J. Econ. Taxon. Bot.* Addl. Series **12**: 92-101.

Singh, G., Singh, H.B. & Mukherjee, T.K. 2003b (*ed.*). *Ethnomedicine of North-East India.* NISCIS, New Delhi.

Singh, G.R. 1998. Traditional veterinary practices in India: review of Salhotra Sangrah, 89-92. *In:* Diwedi, S.K. *(ed.) Techniques for Scientific Validation and Evaluation of Ethnoveterinary Practices.* IVRI, Izatnagar.

Singh, G. S. 1999a. Ethnobotanical study of useful plants of Kullu district in North Western Himalaya, India. *J. Econ. Taxon. Bot.* **23** (1): 185-198.

Singh, G.S. 2003. Ethnobotanical inventory of Sariska national park in the Aravalli hills of eastern Rajasthan, India, 181-195. In: Singh, V.S. & Jain, A.P. (*eds) Ethnobotany and Medicinal Plants of India and Nepal.* Sci. Publ., Jodhpur.

Singh, G.S. 2004. Indigenous knowledge and conservation practices, tribal society of Western Himalayas: a case study of Sangla valley. *Stud. Trbes. Tribals* **2**(**1**): 29-35.

Singh, G. S., Rao, K.S. & Saxena, K.G. 1998a. Eco-cultural analysis of sacred species and ecosystem in Chhakinal Watershed, Himachal Pradesh: 301-313. *In*: Ramakrishnan, P.S., Saxena, K.G. & Chandrashekara, V.M. (*eds*): *Conserving the Sacred for Biodiversities Management.* UNESCO. Oxford & IBH Publ. Co. Pvt. Ltd., New Delhi.

Singh, G.S., Rao, K.S. & Saxena, K.G. 1998b. Environment, ecological and socio-economic impact of introduced crops in Western Himalaya: A case study of Kullu valley. *J. Human Ecol.* **9**: 63-72.

Singh, H., 1988. Ethnobiological treatment of piles by Bhoxas of Uttar Pradesh. *Anc. Sci. Life.* **8:** 167-170.

Singh, H., Kumar, B., Hyniewta, T.M. & Bora, P.J. 1997. Ethno-medico-botanical studies in Tripura, India. *Ethnobotany* **9**: 56-58.

Singh, H. & Sharma, M. 2006. *Flora of Chamba District (Himachal Pradesh)*. BSMPS, Dehradun.

Singh, H.B. & Arora, R.K. 1978. *Wild Edible Plants of India*. (1st *ed.*). ICAR Publ. New Delhi.

Singh, H.B., Singh, R.S. & Sandhu, J.S. (*eds.*) 2003a. *Herbal Medicine of Manipur: A Colour Encyclopaedia*. Daya Publ. House, New Delhi.

Singh, J., Bhuyan, T.C. & Ahmed, A. 1996b. Ethnobotanical studies on the Mishing tribes of Assam with special reference to food and medicinal plants–I. *J. Econ. Taxon. Bot.* Addl. Series **12**: 350-356.

Singh, K.K. & Kumar, K. 2000a. *Ethnobotanical Wisdom of Gaddi Tribe of Western Himalayas*. BSMPS, Dehradun.

Singh, K.K. & Kumar, K. 2000b. Observation on ethnoveterinary medicine among the Gaddi tribe of Kangra Valley, H.P. *Ethnobotany* **12**: 145-152.

Singh, K.K. & Maheshwari, J.K. 1983. Traditional phytotherapy amongst the tribals of Varanasi district, U.P. *J. Econ. Taxon. Bot.* **4:** 829-838.

Singh, K.K. & Maheshwari, J. K. 1992. Some less known medicinal uses of trees among the Sahariyas of Morena district, M.P., India. *Ethnobotany* **4**: 71-74.

Singh, K.K., Palvi, S.K. & Singh, H.B. 1980. Survey of some medicinal plants of Dharchula block-I Pithoragarh district of U.P. *Bull. Medico. Ethnobot. Res.* **1**: 1-7.

Singh, K.K. & Prakash, A. 1994. Indigenous phytotherapy among the Gond tribe of Uttar Pradesh, India. *Ethnobotany* **6:** 37-41.

Singh, K.K. & Prakash, A. 1996. Observations on ethnobotany of the Kol tribe of Varanasi district, U.P. *J. Econ. Taxon. Bot.* Addl. Series **12**: 133-137.

Singh, K.K. & Rao, R.R. 2003. Studies on some potential ethnomedicine among the tribals of Uttar Pradesh, India, 42-57. *In*: *Vistas in Ethnobotany* Vol. I, *Indian J. Appld. & Pure Biol.*

Singh, P.B. 1993. Medicinal plants of ayurvedic importance from Mandi district of Himachal Pradesh. *Bull. Medico. Ethnobot. Res.* **14** (3-4): 126-136.

Singh, P.B. 1996. Wild edible plants of Mandi district in Northwest Himalaya, 22-25. *In:* Jain, S.K. (*ed.*) *Ethnobiology in Human Welfare.* Deep Publ., New Delhi.

Singh, P.B. 1999b. *Illustrated Field Guide to Commercially Important Medicinal and Aromatic Plants of Himachal Pradesh* (with special reference to Mandi district). Herbal Med. Soc. Himalayan Biodiversity, Mandi, Himachal Pradesh.

Singh, P.B. & Aswal, B.S. 1992. Medicinal plants of Himachal Pradesh used in Indian pharmaceutical industry. *Bull. Medico-ethnobot. Res.* **13**: 172-208.

Singh, R.V. 1982. *Fodder Trees of India.* Oxford & IBH Publ. Co., New Delhi.

Singh, V. & Chauhan, N.S. 2005. Traditional practices of herbal medicines in the Lahaul valleys, Himachal Himalayas. *Indian J. Trad. Knowledge* **4** (2): 208-220.

Singh, V. & Pandey, R.P. 1996. Ethno-medicinal plants used for venereal and gynaecological diseases in Rajasthan (India). *J. Econ. Taxon. Bot.* Addl. Series **12**: 154-165

Singh, V. & Pandey, R.P. 1998. *Ethnobotany of Rajasthan, India.* Sci. Publ., Jodhpur.

Singh, V.K. & Zaheer, Anwar Ali 1996. Folk medicinal plants used for family planning in India, 184-186. *In*: Jain, S.K. (*ed.*). *Ethhnobiology in Human Welfare.* Deep Publ., New Delhi.

Sinha, S.C. 1986. *Ethnobotanical Study of Manipur*, Ph.D. Thesis, Manipur Univ., Imphal.

Sinha, S.C. 1987. Ethnobotany-medicinal Plants. *Frontier Bot.***1:** 123-152.

Sinha, S.C. 1996. Wild edible plants of Manipur, India, 42-47. *In*: Jain, S.K. (*ed.*) *Ethnobiology in Human Welfare.* Deep Publ., New Delhi.

Siwakoti, M. & Verma, S.K. 1996. Medicinal plants of the Terai of Eastern Nepal. *J. Econ. Taxon. Bot.* Addl. Series **12**: 423-438.

Sood, S.K., Bhardwaj, R. & Lakhanpal, T. N. 2005a. *Ethnic Indian Plants in Cure of Diabetes.* Sci. Publ., Jodhpur, India.

Sood, S.K. & Kaushal, B. 2008. *Fibre and Fodder Resources of India.* Satish Serial Publ. House, Azadpur, Delhi.

Sood, S.K., Koundal, V. & Dogra, K. 2011b. *Ethnoveterinary Important Plants of Bamsan, Distt. Hamirpur (HP), India.* Lap Lambert Acad. Publ., Germany.

Sood, S.K., Kumar, S., Bassi, S.K. & Rana, J.C. 2012b. *Ethnobotany of the Heritage Region of Shiwalik Himalaya.* Anamika Publ. & Dist. (Pvt.) Ltd., New Delhi.

Sood, S.K., Nath, R., & Kalia, D.C. 2001. *Ethnobotany of Cold Desert Tribes of Lahaul-Spiti (N.W. Himalaya).* Deep Publ., New Delhi.

Sood, S.K., Parmar, S. & Lakhanpal, T. N. 2005b. *Ethnic Plants of India Used in Cancer Cure -A Compendium.* BSMPS, Dehradun.

Sood, S.K., Rawat, D., Kumar, S. & Rawat, S. 2012a. *Handbook of Wild Edible Plants.* Pointer Publ., Jaipur, Rajasthan, India.

Sood, S.K., Rawat, S. & Rawat, D. 2009a. *Dye, Masticatory and Fumitory Resources of India.* Int. Book District., Dehradun.

Sood, S.K., Seema, S., Lakhanpal, T. N. & Suresh, K. 2011a. *Indian Herbs for Perfumery and Beauty Care.* Daya Publ. House, New Delhi.

Sood, S.K., Sharma, D., Kumar, S. & Lakhanpal, T. N. 2010. *Healing Herbs: Traditional Medications For Wounds, Sores and Bones.* Pointer Publ., Jaipur, Rajasthan, India.

Sood, S.K., Sharma, S.K., Kumar N., & Kumar, H. 2009b. *Ethnobotanical Studies on Trees, Shrubs and Climbers of Himalaya.* Satish Serial Publ. House, Delhi, India.

Sood, S.K. & Sudershna. 2008. *Root Drug Plants of Ethnic India.* Satish Serial Publ. House, Azadpur, Delhi.

Sood, S.K. & Thakur, S. 2004. *Ethnobotany of Rewalsar Himalaya (Dist. Mandi, Himachal Pradesh, India).* Deep Publ., New Delhi.

Spoerke, David G. & Smolinske, Susan C. 1990. *Toxicity of House Plants.* CRC Press, New York.

Srivastava, S. & Jain, S.K. 2000. Cumulative index to ethnobotany, Vol. 1-12, 1989-2000, with analysis of region and themes covered. *Ethnobotany* **12**: 139-161.

Srivastava, S.K. & Chandra Sekar, K. 2004. Ethnomedicine of the Pin Valley National Park. Himachal Pradesh: plant used in treatment of dysentery. *Ethnobotany* **16**: 62-63.

Srivastava, S.K., Mehrotra, S., Srivastava, G. K. & Chauhan, D. K. 2003. Folk uses of herbs from Mahabaleshwar region of Sahyadri hills. *J. Econ. Taxon. Bot.* **27**: 857-863.

Stainton, A. 1988. *Flowers of Himalaya, A Supplement.* Oxford Univ. Press, Delhi.

Subramani, S.P., Jishtu, Vaneet & Rawat, R.S. 2005. Plant Diversity Conservation Needs. *The Botanica* **55:** 45-59.

Sudarsanam, G., Reddy, M.B. & Nagaraju, N. 1995. Veterinary crude drugs in Rayalaseema, Andhra Pradesh, India. *Intl. J. Pharmacog.* **33**(1): 52-60.

Sundaravelu, S., Selvaraju, M. & Sundaramoorthy, P. 2003. Medicinal weed flora of South Arcot, Cuddalore district in Tamil Nadu, 87-94. *In*: *Vistas in Ethnobotany* Vol. 1, *Indian J. Pure & Appld. Biol.*

Sur, P.R. & Halder, A.C. 2002. Observation on the ethnobotany of Purulia district, West Bengal, India, 146-148. *In*: Trivedi, P. C. (*ed.*) *Ethnobotany*. Avishkar Publ. & Distri., Jaipur (Raj.) India.

Swaminathan, M. 1995. Aspects of Urban Poverty in Bombay. *Environment and Urbanisation* **7**(1): 133-143.

Swamy, A. & Gupta, B.K. 1996. A note on some commonly occurring medicinal weeds of Udhampur district (J&K). *J. Econ. Taxon. Bot.* Addl. Series **12**: 89-91.

Tarafder, C.R. 1983. Traditional medicinal plant used by the tribal of Ranchi and Hazaribagh district, Bihar: plants used in stomach trouble. *J. Econ. Taxon. Bot.* **4:** 891-896.

Tarafder, C.R. & Rai Chaudhuri, H.N. 1981. Less known medicinal uses of plants among the tribals of Hazaribagh district of Bihar-Plant, 208-217. *In:* Jain, S.K. *(ed.) Glimpses of Indian Ethnobotany.* Oxford and IBH. Publ. Co., New Delhi.

Thakur, R. S., Puri, H.S. & Hussain, A. 1989. *Major Medicinal Plants of India.* CI-MAP, Lucknow, India.

Tirkey, A. 2004. Some ethnomedicinal plant species of Chattisgarh state. *Ethnobotany* **16**: 118-124.

Tirkey, A., Khan, F. & Khan, S.S. 2003a. Medicinal plants used in the treatment of hair in Raijarh District of Madhya Pradesh (India), 104-110. *In*: *Vistas in Ethnobotany*, Vol.-I, *Indian J. Pure & Appld. Biol.*

Tirkey, A., Khan, F. & Khan, S.S. 2003b. Medicinal plants used by the tribals of Raijarh Disteict (M.P.) for Jaundice, 163-170. *In*: *Vistas in Ethnobotany*, Vol. I, *Indian Journal of Pure and Applied Biology*.

Tiwari, K.C. & Tiwari, V.P. 1996. Some important medicinal plants of the tropical, subtropical and temperate region of Siang, Subansiri and Tirap district of Arunachal Pradesh. *J. Econ. Taxon. Bot.* Addl. Series **12**: 359-363.

Tiwari, L. & Pande, P.C. 2006. Ethnoveterinary plants of Uttarkashi District, Uttaranchal, India. *Ethnobotany* **18:** 139-144.

Trivedi, P.C. 2004. *Medicinal Plants Conservation and Utilization.* Avishkar Publ., Jaipur.

Trivedi, P.C. 2009. *Medicinal Plants Conservation and Utilization* (2nd *ed.*) Avishkar Publ., Jaipur.

Trivedi, P.C. 2010. *Ethnic Tribes and Medicinal Plants.* Pointer Publ., Jaipur.

Tosh, J. 1996. Ethnobotanical study of Western Maharastra. *J. Econ. Taxon. Bot.* Addl. Series **12**: 169-174.

Trease, G.E. & Evans W.C. 1983. *Text Book of Pharmaconosy. ed.* 12, Balliere Tinad, London.

Uniyal, B. 2003. Utilization of medicinal plants by the rural women of Kullu, Himachal Pradesh. *Indian J. Trad. Knowledge* **2** (4): 366-370.

Uniyal, M.R. 1968. Medicinal plants of Bhagirathi valley lying in Uttarkashi forest division. *Indian For.* **94**: 407-420.

Uniyal, M.R., Bhat, A.V. & Chaturvedi, P.N. 1973. Preliminary observations on medicinal plants of Lahaul Spiti division in Himachal Pradesh. *Bull. Medico. Ethnobot. Res.* **3**(1): 1-26.

Uniyal, M.R. & Chauhan, N.S. 1971. Medicinal plants of Uhal valley in Kangra Forest Division. H.P. *Jour. Res. Ind. Med.* **6**(3): 287-299.

Uniyal, M.R. & Chauhan, N.S. 1973a. Traditionally important medicinal plants of Kullu forest Division, H.P. *Jour. Res. Ind. Med.* **7**(1): 76-85.

Uniyal, M.R. & N.S. Chauhan, 1973b. Traditionally important medicinal plants of Kangra Valley in Dharamsala Forest Circle, Himachal Pradesh. *J. Res. Indian Med.* **8**(1): 76-85.

Uniyal, S.K., Singh, K.N., Jamwal, P. & Lal, B. 2006. Traditional uses of medicinal plants among the tribal communities of Chhota Bhangal, Western Himalaya. *J. Ethnobiol. & Ethnomedicine* **2**: 2-11.

Upadhyay, R. & Singh, J. 2005. Ethno-medicinal uses of plants in Tikri forest of Gonda Distt (U.P.). *Ethnobotany* **17**: 167-170.

Upadhye, A. & Kumbhojkar, M.S. 1996. Ethnobotany of genus *Sterculia* L. from *India. J. Econ. Taxon. Bot.* Addl. Series **12** : 40-42.

Upadhye, A. S. & Kulkarni, D.K. 2004. Traditional phytotherapy for insect bites among the tribals in Western Maharashtra. *Ethnobotany* **16**: 59-61.

Upadhye, A. S., Vartak, V.D. & Kumbhojkar, M.S. 1994. Ethno-medico-botanical studies in Western Maharastra, India. *Ethnobotany* **6**: 25-31.

Uphof, J.C. Th. 1968. *Dictionary of Economic Plants.* H.R. Eugelmann (J. Cramer), Weinheim.

Varghese, E. 1996. *Applied Ethnobotany -A Case Study Among the Kharia of C. India.* Deep Publ. Delhi.

Varma, S.K. 1997. Comparative studies on folk drugs of tribals of Chotanagpur and Santhal Pargana of Bihar, India. *Ethnobotany* **9**: 70-76.

Varma, S.K., Srivastava, D.K. & Pandey, A.K, 1999. *Ethnobotany of Santhal Pargana,* Narendra Publ. House, Delhi.

Varshney, J.P. 1998. Traditional therapeutic resources for management of gastrointestinal disorders and their clinical evaluation, 47-51. In: Dwevedi, *(ed.) Techniques for Scientific Validation and Evaluation of Ethnoveterinary Practices.* 1VRI, Izatnagar.

Ved, D.K., Kinchal, G.A., Ravikumar, K., Pravharan, V., Uttakarsh, G., Vijaya Sankar, R. & Indresha, J.H. 2003. *Conservation Assessment and Management Prioritisation for the Medicinal Plants of Jammu and Kashmir, Himachal Pradesh and Uttranchal.* FRLHT, Banglore, India.

Venkataratnam, K.R. & Venkata Raju, R.R. 2004. Folk medicine from Gundlabrahmeswaram Wild-life Sanctuary. *Ethnobotany* **16**: 33-39.

Verma, P., Khan, A.A. & Singh, K.K. 1995. Traditional phytotherapy among the Baiga tribe of Shahdol district of Madhya Pradesh, India. *Ethnobotany* **7**: 69-73.

Verma, R. 1998. Traditional veterinary medicine used in the management of bacterial diseases of animals and their laboratory evaluation, 97-100. *In*: Dwivedi S.K. *(ed.) Techniques for Scientific Validation and Evaluation of Ethnoveterinary Practices.* IVRI, Izatnagar.

Verma, S. & Chauhan, N.S. 2006. Studies on ethno-medico-botany of Kunihar Forest Division Distt. Solan H.P. *Ethnobotany* **18**: 160-165.

Verma, S.K. & Chauhan, N.S. 2007. Indigenous medicinal plants knowledge of Kunihar forest division. *Indian J. Trad. Knowledge* **6** (3): 498-507.

Verma, S.K., Sriwastava, D.K. & Pandey, A.K. 1999. *Ethnobotany of Santhal Pargana,* Narendra Publ. House, Delhi.

Vihari, V. 1995. Ethnobotany of cosmetics of Indo-Nepal border. *Ethnobotany* **7**: 89-94.

Vir jee, G. H., Dhar, V., Kachroo, P. & Buth, G. M. 1984. Taxa-ethnobotanical studies of the rural areas in district Rajouri (Jammu). *J. Econ. Taxon. Bot.* **5**: 831-834.

Vishwanathan, M.B. 1997. Ethnobotany of the Malyalis in North Arcot District, Tamil Nadu, India. *Ethnobotany* **9**: 77-79.

Viswanathan, M.B., Kumar, P.E.H. & Ramesh, N. 2001. Ethnomedicines of Kanis in Kalakkad-Mundanthurai Tiger Reserve, Tamil Nadu. *Ethnobotany* **13**: 60-66.

Viswanathan, M.V., Kashyapa, K. & Ramachandran, K. 1981. Less known fibre-yielding plants of peninsular India. *Bull. Bot. Surv. India* **23**: 240-242.

Vishwanathan, M.V. & Singh, H.B. 1996. Plants used as household remedies In India, 105-107. *In*: Jain, S.K. (*ed.*) *Ethnobiology in Human Welfare*. Deep Publ., New Delhi.

Vyas M.S., 1987. *Ferns and Fern-allies of Rajasthan-Ethnobotany and Biochemical Analyses,* Ph.D. Thesis, J.N.V. Univ., Jodhpur.

Watt, G. 1889-1896. *A Dictionary of the Economic Products of India.* Vol. 1-V1. Periodical Experts, New Delhi.

Watt, G. 1908. *The Commercial Products of India.* John Murray, London. 249.

Watt, G. 1972. *A Dictionary of the Economic Products of India,* Vol. I -VI. Periodical Experts, Delhi (India).

Wickens, G.E. 1990. What is economic botany? *Econ. Bot.* **44**: 12-28.

Wielgorskaya, T. 1995. *Dictionary of Generic Names of Seed Plants.* BSMPS, Dehra Dun (India).

Willcox, M.L., Graz, B. & Falquet, J. 2007. *Argemone mexicana* decoction for the treatment of uncomplicated falciparum malaria. *Trans. R Soc. Trop. Med Hyg.* **101** (12): 1190-1198.

Wunderlin, R. & Hansen, B. 2008. *Synonyms of Annona squamosa. Atlas of Florida Vascular Plants.* Inst. Syst. Bot., Univ. of Florida, Florida.

Yadav, M.P. & Gupta, S. 1998. Scope of ethnoveterinary practices in equine medicine, 16-21. *In:* Dwivedi, S.K. (*ed.*) *Techniques for Scientific Validation and Evaluation of Ethnoveterinary Practices.* IVRI, Izatnagar.

Zeven, A.C. & Zhukovsky, P.M. 1975. *Dictionary of Cultivated Plants and Their Centres of Diversity.* Wageningen, Netherland.

Appendices

Appendix I: Index to Ethnobotanically Important Families

Name of the Family	Total Number of Spp. in the Genera	Total Number of Genera and Spp.
Acanthaceae	2/2	*Adhatoda (1), Andrographis (1)*
Actinidiaceae	1/1	*Actinidia (1)*
Adiantaceae	1/2	*Adiantum (2)*
Agavaceae	1/1	*Agave (1)*
Amaranthaceae	4/7	*Achyranthes (2), Aerva (1), Amaranthus (3), Celosia (1)*
Anacardiaceae	3/3	*Lannea (1), Mangifera (1), Pistacia (1)*
Annonaceae	2/2	*Annona (1), Polyalthia (1)*
Apiaceae	6/6	*Anethum (1), Carum (1), Centella (1), Coriandrum (1), Daucas (1), Foeniculum (1)*
Apocynaceae	8/9	*Alstonia (1), Carissa (2), Catharanthus (1), Holarrhena (1), Nerium (1), Rauwolfia (1), Tabernaemontana (1), Thevetia (1)*
Araceae	4/4	*Acorus (1), Amorphophallus (1), Arisaema (1), Colocasia (1)*
Arecaceae	1/1	*Phoenix (1)*
Asclepiadaceae	3/3	*Asclepias (1), Calotropis (1), Tylophora (1)*
Asteraceae	16/17	*Ageratum (1), Artemisia (2), Carduus (1), Carthamus (1), Centipeda (1), Eclipta (1), Eupatorium (1), Helianthus (1), Parthenium (1), Sonchus (1), Spilanthes (1), Tagetes (1), Taraxacum (1), Tridax (1), Vernonia (1), Xanthium (1)*
Balsaminaceae	1/1	*Impatiens (1)*
Berberidaceae	1/1	*Berberis (1)*
Bignoniaceae	3/4	*Jacaranda (1), Oroxylum (1), Tecoma (2)*

Contd...

Contd...

Name of the Family	Total Number of Spp. in the Genera	Total Number of Genera and Spp.
Boraginaceae	2/2	*Trichodesma (1), Bombax (1)*
Brassicaceae	5/7	*Brassica (3), Eruca (1), Lepidium (1), Nasturtium (1), Raphanus (1)*
Cactaceae	1/1	*Opuntia (1)*
Cannaceae	1/1	*Canna (1)*
Caricaceae	1/1	*Carica (1)*
Celastraceae	1/1	*Celastrus (1)*
Chenopodiaceae	2/3	*Chenopodium (2), Spinacia (1)*
Clusiaceae	1/1	*Mesua (1)*
Combretaceae	1/3	*Terminalia (3)*
Convolvulaceae	1/2	*Ipomoea (2)*
Cordiaceae	1/1	*Cordia (1)*
Crassulaceae	1/1	*Bryophyllum (1)*
Cucurbitaceae	6/7	*Cucumis(1), Cucurbita (2), Luffa (1), Lagenaria (1), Momordica (1), Sechium (1)*
Cupressaceae	1/1	*Thuja (1)*
Cuscutaceae	1/1	*Cuscuta (1)*
Cycadaceae	1/1	*Cycas (1)*
Dioscoreaceae	1/1	*Dioscorea (1)*
Elaeocarpaceae	1/1	*Elaeocarpus (1)*
Equisetaceae	1/1	*Equisetum (1)*
Euphorbiaceae	8/9	*Emblica (1), Euphorbia (2), Jatropha (1), Mallotus (1), Pedilanthus (1), Putranjiva (1), Ricinus (1), Phyllanthus (1)*
Fabaceae	21/26	*Abrus (1), Acacia (3), Albizia (2), Arachis (1), Bauhinia (2), Butea (1), Cassia (2), Cicer (1), Dalbergia (1), Glycine (1), Glycyrrhiza (1), Lens (1), Leucaena (1), Millettia (1), Mimosa (1), Pongamia (1), Saraca (1), Tamarindus (1), Trifolium (1), Trigonella (1), Vicia (1),*
Flacourtiaceae	1/1	*Flacourtia (1)*
Fumariaceae	1/1	*Fumaria (1)*
Gentianaceae	1/1	*Canscora (1)*
Geraniaceae	1/1	*Pelargonium (1)*
Hypericaceae	1/1	*Hypericum (1)*
Juglandaceae	1/1	*Juglans (1)*

Contd...

Contd...

Name of the Family	Total Number of Spp. in the Genera	Total Number of Genera and Spp.
Lamiaceae	9/11	*Colebrookia (1), Lavendula (1), Melissa (1), Mentha (2), Ocimum (2), Pogostemon (1), Rosmarinus (1), Roylea (1), Salvia (1)*
Lauraceae	1/2	*Cinnamomum (2)*
Liliaceae	4/7	*Allium (2), Aloe (2), Asparagus (2), Yucca (1)*
Linaceae	2/2	*Linum (1), Reinwardtia (1)*
Lythraceae	3/3	*Lagerstroemia (1), Lawsonia (1), Woodfordia (1)*
Malvaceae	4/5	*Abelmoschus (2), Gossypium (1), Hibiscus (1), Malvaviscus (1)*
Martyniaceae	1/1	*Martynia (1)*
Meliaceae	4/4	*Azadirachta (1), Cedrela (1), Soymida (1), Melia (1)*
Menispermaceae	3/3	*Cissampelos (1), Stephania (1), Tinospora (1)*
Moraceae	5/12	*Artocarpus (2), Cannabis (1), Ficus (7), Morus (1), Populus (1)*
Moringaceae	1/1	*Moringa (1)*
Musaceae	1/1	*Musa (1)*
Myrtaceae	3/4	*Callistemon (1), Eucalyptus (1), Eugenia (2)*
Nyctaginaceae	3/3	*Boerhavia (1), Bougainvillea (1),Mirabilis (1)*
Oleaceae	1/3	*Jasminum (3)*
Orchidaceae	3/3	*Habenaria (1), Rhynchostylis (1), Vanda (1)*
Oxalidaceae	1/1	*Oxalis (1)*
Papaveraceae	2/2	*Argemone (1), Papaver (1)*
Pedaliaceae	1/1	*Sesamum (1)*
Pinaceae	1/1	*Pinus (1)*
Piperaceae	1/1	*Piper (1)*
Plumbaginaceae	1/1	*Plumbago (1)*
Poaceae	14/18	*Avena (1), Bambusa (2), Cymbopogon (1), Cynodon (1), Dendrocalamus (1), Hordeum (1), Oryza (1), Paspalum (1), Pennisetum (2), Phragmites (1), Saccharum (3), Setaria (1), Triticum (1), Zea (1)*
Polygonaceae	1/1	*Rumex (1)*
Proteaceae	1/1	*Grevillea (1)*
Punicaceae	1/1	*Punica (1)*
Ranunculaceae	2/2	*Clematis (1), Ranunculus (1)*
Rhamnaceae	1/1	*Zizyphus (1)*

Contd...

Contd...

Name of the Family	Total Number of Spp. in the Genera	Total Number of Genera and Spp.
Rosaceae	7/10	*Eriobotrya* (1), *Fragaria* (1), *Prunus* (3), *Psidium* (1), *Pyrus* (2), *Rosa* (1), *Rubus* (1)
Rubiaceae	2/2	*Randia* (1), *Wendlandia* (1)
Rutaceae	5/10	*Aegle* (1), *Citrus* (5), *Murraya* (2), *Ruta* (1), *Zanthoxylum* (1)
Santalaceae	1/1	*Santalum* (1)
Sapindaceae	4/4	*Cardiospermum* (1), *Dodonaea* (1)*Litchi* (1), *Sapindus* (1)
Sapotaceae	2/2	*Achras* (1), *Madhuca* (1)
Scrophulariaceae	2/2	*Bacopa* (1), *Verbascum* (1)
Solanaceae	8/13	*Capsicum* (1), *Cestrum* (1), *Datura* (2), *Lycopersicon* (1), *Nicotiana* (1), *Physalis* (1), *Solanum* (5), *Withania* (1)
Sterculiaceae	2/2	*Abroma* (1), *Helicteres* (1)
Tiliaceae	1/1	*Grewia* (1)
Urticaceae	1/1	*Urtica* (1)
Verbenaceae	4/4	*Duranta* (1), *Lantana* (1), *Premna* (1), *Vitex* (1)
Violaceae	1/1	*Viola* (1)
Vitaceae	1/2	*Vitis* (2)
Zingiberaceae	3/3	*Curcuma* (1), *Elettaria* (1), *Zingiber* (1)

Appendix II: Index to Vernacular Names

Vernacular Names	Name of the Plant	Family	Page No.
Adrak	*Zingiber officinale* Rosc.	Zingiberaceae	326
Airand, Arind, Erand	*Ricinus communis* L.	Euphorbiaceae	268
Ajwain	*Carum copticum* Hiern.	Apiaceae	106
Ak, Madar	*Calotropis procera* (Ait.) R. Br.	Asclepiadaceae	96
Akashbel, Amarvela, Halwan	*Cuscuta reflexa* Roxb.	Cuscutaceae	139
Akharot	*Juglans regia* L	Juglandaceae	193
Akhe	*Rubus ellipticus* Sm.	Rosaceae	272
Allia, Aahali, Kanyar	*Cassia fistula* L.	Fabaceae	107
Aloo	*Solanum tuberosum* L.	Solanaceae	288
Aloo Bukhara, Plum	*Prunus domestica* L.	Rosaceae	256
Alseenia	*Leucaena leucocephala* (Lam.) de Wit.	Fabaceae	202
Alsi	*Linum usitatissimum* L.	Linaceae	203
Amb, Aam	*Mangifera indica* L.	Anacardiaceae	210
Ambla, Amla	*Emblica officinalis* Gaertn.	Euphorbiaceae	154
Amrood	*Psidium guajava* L.	Rosaceae	258
Angoor	*Vitis vinifera* L.	Vitaceae	319
Anjir	*Ficus carica* L.	Moraceae	166
Arjun tree	*Terminalia arjuna* (Roxb.) Wt. & Arn.	Combretaceae	298
Aru, Adu	*Prunus persica* Benth. & Hook. f.	Rosaceae	257
Asgandh, Ashwagandha	*Withania somnifera* (L.) Dunal	Solanaceae	320
Ashok	*Polyalthia longifolia* (Sonn.) Thw.	Annonaceae	252
Ashoka	*Saraca asoca* (Roxb.) De Wild.	Fabaceae	280

Contd...

Contd...

Vernacular Names	Name of the Plant	Family	Page No.
Bada Kaer, Roothi	*Vicia pallida* Facquem.	Fabaceae	315
Badam	*Prunus amygdalus* Baill.	Rosaceae	255
Bada-relu, Alliuna	*Cassia occidentalis* L.	Fabaceae	109
Badi-basuti	*Ipomoea carnea* Jacq.	Convolvulaceae	189
Baingan	*Solanum melongena* L.	Solanaceae	284
Baingh, Maggar	*Dendrocalamus hamiltonii* Nees & Arn. *ex* Munro	Poaceae	148
Bainjh	*Bambusa arundinaceae* (Retz.) Roxb.	Poaceae	80
Bajra, Barley	*Pennisetum americanum* L.	Poaceae	241
Balm	*Melissa officinalis* L.	Lamiaceae	214
Ban tambakhoo	*Verbascum thapsus* L.	Scrophulariaceae	313
Banfsha, Banaksha	*Viola canescens* Wall.	Violaceae	316
Banna	*Vitex negundo* L.	Verbenaceae	316
Bans	*Bambusa nutans* Wall.	Poaceae	81
Bar, Bad, Barh	*Ficus benghalensis* L.	Moraceae	164
Barseem	*Trifolium alexandrinum* L.	Fabaceae	308
Baryan	*Acorus calamus* L.	Araceae	43
Basant, Piyan	*Reinwardtia indica* Dum.	Linaceae	267
Basuti	*Adhatoda vasica* Nees	Acanthaceae	45
Behra, Bahera	*Terminalia bellirica* (Gaertn.) Roxb.	Combretaceae	299
Ber	*Zizyphus jujuba* (Lam.) Gaertn. non Mill.	Rhamnaceae	328
Beul, Bhuel	*Grewia oppositifolia* Roxb.	Tiliaceae	180
Bhabhri	*Ocimum basilicum* L.	Lamiaceae	230
Bhadra	*Aerva lanata* (L.) Juss. *ex* Schult.	Amaranthaceae	51
Bhakhadda	*Xanthium strumarium* L.	Asteraceae	323
Bhang	*Cannabis sativa* L.	Moraceae	98
Bhangru	*Rhynchostylis retusa* Blume	Orchidaceae	268
Bhangru	*Vanda roxburghii* R. Br.	Orchidaceae	313
Bhankar	*Premna latifolia* Roxb.	Verbenaceae	254
Bhindi	*Abelmoschus esculentus* (L.) Moench	Malvaceae	31
Bhringraj	*Eclipta prostrata* Roxb.	Asteraceae	152
Bhumi-Amla	*Phyllanthus urinaria* L.	Euphorbiaceae	245

Contd...

Contd...

Vernacular Names	Name of the Plant	Family	Page No.
Bhursalae	*Sonchus asper* Hill	Asteraceae	289
Bichhubuti, Choti kogsi	*Urtica dioica* L.	Urticaceae	312
Bil	*Aegle marmelos* (L.) Correa *ex* Roxb.	Rutaceae	49
Bis-Khappar	*Stephania glabra* (Roxb.) Miers	Menispermaceae	292
Bitter Gourd	*Lagenaria siceraria* (Molina) Standl.	Cucurbitaceae	195
Bottle brush	*Callistemon lanceolatus* DC.	Myrtaceae	95
Brahmi	*Centella asiatica* (L.) Urb.	Apiaceae	115
Brahmjiri	*Vernonia anthelmintica* Willd.	Asteraceae	314
Buttercup	*Ranunculus muricatus* L.	Ranunculaceae	264
Chakotra-khatta	*Citrus grandis* (L.) Osbeck.	Rutaceae	126
Chamber	*Artemisia japonica* Thunb.	Asteraceae	71
Chameli	*Jasminum humile* L.	Oleaceae	190
Chandan	*Santalum album* L.	Santalaceae	278
Chari	*Pennisetum purpureum* Schum.	Poaceae	242
Charmad	*Artemisia vulgaris* L.	Asteraceae	72
Chhota Kalpa, Rukhabi	*Trichodesma indicum* R. Br.	Boraginaceae	307
Chhota-garna	*Carissa spinarum* Hook. f.	Apocynaceae	105
Chhui-muui, Lajwanti	*Mimosa pudica* L.	Fabaceae	218
Chikku	*Achras zapota* L.	Sapotaceae	40
Chil	*Pinus roxburghii* Sarg.	Pinaceae	246
China-Rose, Shoe flower	*Hibiscus rosa-sinensis* L.	Malvaceae	183
Chitrak, Chitra	*Plumbago zeylanica* L.	Plumbaginaceae	250
Cholaee	*Amaranthus spinosus* L.	Amaranthaceae	63
Cholaee	*Amaranthus viridis* L.	Amaranthaceae	64
Cholai	*Amaranthus paniculatus* L.	Amaranthaceae	62
Chollae	*Cicer arietinum* L.	Fabaceae	119
Choti-Kandayali	*Solanum surattense* Burm. f.	Solanaceae	286
Chu, Chui	*Euphorbia royleana* Boiss.	Euphorbiaceae	163
Chu, Thor	*Opuntia dillenii* (Ker-Gawl.) Haw.	Cactaceae	233
Chuch	*Nasturtium officinale* R. Br.	Brassicaceae	227
Congress grass	*Parthenium hysterophorus* L.	Asteraceae	238

Contd...

Contd...

Vernacular Names	Name of the Plant	Family	Page No.
Daaru, Daadu	*Punica granatum* L.	Punicaceae	259
Damabuti	*Tylophora indica* (Burm.f.) Merr.	Asclepiadaceae	311
Datura	*Datura metel* auct. non L.	Solanaceae	144
Datura, Dhatura	*Datura stramonium* L.	Solanaceae	146
Devryan	*Ficus hispida* L.f.	Moraceae	168
Dhagali bel	*Clematis grata* Wall.	Ranunculaceae	129
Dhai, Dhavae, Dhaun	*Woodfordia floribunda* Salisb.	Lythraceae	321
Dhan, Jari	*Oryza sativa* L.	Poaceae	235
Dhania	*Coriandrum sativum* L.	Apiaceae	133
Dheun	*Artocarpus lakoocha* Roxb.	Moraceae	73
Doob, Dhruv, Doorva	*Cynodon dactylon* (L.) Pers.	Poaceae	142
Draek	*Melia azedarach* L.	Meliaceae	212
Dudhli, Dulal	*Taraxacum officinale* Weber *ex* Wigg.	Asteraceae	296
Dushane, Dus	*Colebrookia oppositifolia* Sm.	Lamiaceae	130
Elephant grass	*Phragmites karka* (Retz.) Trin. *ex* Steud.	Poaceae	244
Ellachi, Elaichi	*Elettaria cardamomum* (L.) Maton	Zingiberaceae	153
Gajjar	*Daucus carota* L.	Apiaceae	147
Gandhla	*Murraya paniculata* (L.) Jack.	Rutaceae	225
Gandhla, Curry patta	*Murraya koenigii* (L.) Spreng.	Rutaceae	224
Ganna	*Saccharum officinarum* L.	Poaceae	275
Garden glory	*Bougainvillea spectabilis* Willd.	Nyctaginaceae	89
Geedar dhakh	*Vitis jacquemontii* Parker	Vitaceae	318
Gehun	*Triticum aestivum* L.	Poaceae	310
Geranium	*Pelargonium graveolens* L.Hert.	Geraniaceae	241
Ghanera	*Asclepias curassavica* L.	Asclepiadaceae	74
Ghatak	*Tagetes erecta* L.	Asteraceae	294
Ghi-kawer	*Aloe barbedensis* Mill.	Liliaceae	58
Ghi-kawer	*Aloe vera* Webb. & Benth.	Liliaceae	59
Ghiya-tori	*Luffa aegyptiaca* Mill.	Cucurbitaceae	205
Ghornali	*Equisetum debile* Roxb. *ex* Vauch.	Equisetaceae	156
Gudhal	*Malvaviscus conzatii* Greenm.	Malvaceae	209
Gulab	*Rosa damascena* Mill.	Rosaceae	270

Contd...

Contd...

Vernacular Names	Name of the Plant	Family	Page No.
Gulaja, Giloy, Galon	*Tinospora cordifolia* (Willd.) Miers.	Menispermaceae	305
Gulal, 4 o'clock	*Mirabilis jalapa* L.	Nyctaginaceae	219
Gulla	*Papaver rhoeas* L.	Papaveraceae	238
Haldi, Haldan	*Curcuma domestica* (Medik.) Valeton	Zingiberaceae	137
Hallon	*Lepidium sativum* L.	Brassicaceae	201
Hansraj	*Adiantum capillus-veneris* L.	Adiantaceae	47
Hansraj	*Adiantum lunulatum* Burm.	Adiantaceae	48
Harar	*Terminalia chebula* Retz.	Combretaceae	301
Har-shingar	*Lagerstroemia indica* L.	Lythraceae	196
Hathi grass	*Setaria viridis* Beauv.	Poaceae	283
Imali	*Tamarindus indica* L.	Fabaceae	295
Itsat, Utsata	*Boerhavia diffusa* L.	Nyctaginaceae	86
Jal Brahmi	*Bacopa monnieri* (L.) Penn.	Scrophulariaceae	79
Jamun	*Eugenia jambolana* Lam.	Myrtaceae	160
Jangali Khajoor	*Phoenix sylvestris* Roxb.	Arecaceae	243
Jau, Satu	*Hordeum vulgare* L.	Poaceae	186
Jhambiri, Jambirdi, Kimb	*Citrus jambhiri* Lush.	Rutaceae	127
Joi, Javi	*Avena sativa* L.	Poaceae	77
Kachalu, Geju	*Colocasia esculenta* (L.) Schott	Araceae	131
Kachnar, Karale, Karayalae	*Bauhinia variegata* L.	Fabaceae	83
Kaddu	*Cucurbita pepo* L.	Cucurbitaceae	136
Kah-ajwain	*Chenopodium ambrosioides* L.	Chenopodiaceae	118
Kainth, Kaenth	*Pyrus pashia* Ham.	Rosaceae	262
Kakarsinghi, Kakad-singhi	*Pistacia integerrima* Stewart *ex* Brandis	Anacardiaceae	248
Kali basuti	*Eupatorium adenophorum* Spreng.	Asteraceae	162
Kali basuti, Dusyan	*Pogostemon benghalensis* Burm. f.	Lamiaceae	251
Kalmegh	*Andrographis paniculata* Wall. *ex* Nees	Acanthaceae	66
Kambhal, Kambal	*Mallotus philippinensis* (Lam.) Muell.-Arg.	Euphorbiaceae	208

Contd...

Contd...

Vernacular Names	Name of the Plant	Family	Page No.
Kaner	*Nerium indicum* Mill.	Apocynaceae	228
Kaner	*Thevetia neriifolia* Juss. *ex* Steud.	Apocynaceae	303
Kangu	*Flacourtia ramontchi* L. Herit.	Flacourtiaceae	172
Kans	*Saccharum spontaneum* L.	Poaceae	277
Kapah, Kapas	*Gossypium arboreum* L.	Malvaceae	178
Kapur, Kapoor	*Cinnamomum camphora* (L.) Nees & Eberm.	Lauraceae	121
Karela	*Momordica charantia* L.	Cucurbitaceae	220
Karkara	*Spilanthes acmella* (L.) Murr.	Asteraceae	290
Karonda, Bada-garna	*Carissa carandas* L. var. *congesta* (Wt.) Bedd.	Apocynaceae	104
Kasmal	*Berberis asiatica* Roxb. *ex* DC.	Berberidaceae	85
Kasturi bhindi	*Abelmoschus moschatus* Medic.	Malvaceae	32
Kathal	*Artocarpus integrifolia* L. f.	Moraceae	73
Kathman	*Eugenia cerasoides* Roxb.	Myrtaceae	159
Kaudi saunf	*Anetham sowa* Roxb. *ex* Flem.	Apiaceae	67
Kaun	*Martynia annua* L.	Martyniaceae	212
Kehmble	*Lannea coromandelica* (Houtt.) Merr.	Anacardiaceae	196
Kela	*Musa sapientum* L.	Musaceae	226
Khair	*Acacia catechu* Willd.	Fabaceae	36
Khatta, Galgal	*Citrus limon* (L.) Burm. f.	Rutaceae	127
Kheera, Kakri	*Cucumis sativus* L.	Cucurbitaceae	134
Kikker, Shami	*Acacia nilotica* L.	Fabaceae	39
Kirre-ri-chhhali	*Arisaema tortuosum* (Wall.) Schott	Araceae	70
Kiwi	*Actinidia chinensis* var. *deliciosa* A. Chev.	Actinidiaceae	45
Kodri grass	*Paspalum scrobiculatum* L.	Poaceae	239
Korru	*Roylea cinerea* (D. Don) Baill.	Lamiaceae	272
Kukkar-phul	*Celosia argentea* L.	Amaranthaceae	114
Kunnah, Bathu	*Chenopodium album* L.	Chenopodiaceae	117
Kurchi, Keur, Inderjaun	*Holarrhena pubescens* Wall. *ex* G. Don	Apocynaceae	184
Lahsun	*Allium sativum* L., Roxb.	Liliaceae	57

Contd...

Contd...

Vernacular Names	Name of the Plant	Family	Page No.
Lalpatta	*Euphorbia pulcherrima* Willd.	Euphorbiaceae	163
Lamb	*Tridax procumbens* L.	Asteraceae	307
Lasura	*Cordia dichotoma* Forst. f.	Cordiaceae	132
Lavender	*Lavandula angustifolia* Mill.	Lamiaceae	199
Lea	*Argemone mexicana* L.	Papaveraceae	69
Lea	*Carduus nutans* L.	Asteraceae	102
Lemongrass	*Cymbopogon citratus* (DC.) Stapf	Poaceae	141
Lichi	*Litchi chinensis* (Gaertn.) Sonn.	Sapindaceae	204
Lonku	*Sechium edule* Sw.	Cucurbitaceae	281
Lukat, Laukhat	*Eriobotrya japonica* Lindl.	Rosaceae	157
Mahua	*Madhuca indica* J. F. Gmel.	Sapotaceae	207
Makki, Bhuta	*Zea mays* L.	Poaceae	326
Makko	*Solanum nigrum* L.	Solanaceae	285
Mallathi	*Glycyrrhiza glabra* L.	Fabaceae	177
Malti	*Jasminum multiflorum* Roth.	Oleaceae	191
Maroli	*Oxalis corniculata* L.	Oxalidaceae	237
Maroli	*Rumex hastatus* D. Don	Polygonaceae	273
Marorphali	*Helicteres isora* L.	Sterculiaceae	182
Masar	*Lens culinaris* Medic.	Fabaceae	200
Mehandru	*Dodonaea viscosa* L. Jacq.	Sapindaceae	150
Mehendi	*Lawsonia alba* Lam.	Lythraceae	199
Methe, Methi	*Trigonella foenum- graecum* L.	Fabaceae	309
Mirch	*Capsicum annuum* L.	Solanaceae	100
Mooli	*Raphanus sativus* L.	Brassicaceae	265
Moongphali	*Arachis hypogaea* L.	Fabaceae	68
Motia	*Jasminum sambac* (L.) Ait.	Oleaceae	192
Mugh	*Piper longum* L.	Piperaceae	248
Nagbel	*Pedilanthus tithymaloides* (L.) Poir.	Euphorbiaceae	240
Nagkesar	*Mesua ferrea* L.	Clusiaceae	216
Nakchikdu	*Centipeda minima* (L.) A. Br. & Asch.	Asteraceae	116
Nakh, Nashpati	*Pyrus communis* L.	Rosaceae	261
Neela fulunu	*Ageratum conyzoides* L.	Asteraceae	53

Contd...

Contd...

Vernacular Names	Name of the Plant	Family	Page No.
Neelkanthi, Neel kanta	*Duranta repens* L.	Verbenaceae	151
Neem	*Azadirachta indica* A. Juss.	Meliaceae	78
Nili-gulmohar	*Jacaranda acutifolia* Humb. & Bonpl.	Bignoniaceae	189
Nimbu	*Citrus medica* L.	Rutaceae	128
Oayee	*Albizia stipulata* Boiv.	Fabaceae	55
Orchid	*Habenaria pectinata* Don	Orchidaceae	181
Palah, Palash, Plah	*Butea monosperma* (Lam.) Kuntze	Fabaceae	93
Palak, Sag	*Spinacia oleracea* L.	Chenopodiaceae	291
Panseera	*Wendlandia heynei* (Roem. & Schult.) Sant. & Merch.	Rubiaceae	319
Papita	*Carica papaya* L.	Caricaceae	102
Patharchat, Gillar pattar	*Bryophyllum calycinum* Salisb.	Crassulaceae	92
Patindu, Bhatindu	*Cissampelos pareira* L.	Menispermaceae	123
Peepal	*Ficus religiosa* L.	Moraceae	170
Petha, Dhuda petha	*Cucurbita hispida* Thunb.	Cucurbitaceae	135
Phalgu, Anjhir	*Ficus palmata* Forsk.	Moraceae	169
Phulnu	*Lantana camara* L.	Verbenaceae	197
Phultulsi	*Salvia coccinea* L.	Lamiaceae	277
Piaz	*Allium cepa* L.	Liliaceae	56
Pinli	*Hypericum cernuum* Roxb.	Hypericaceae	187
Pipermint	*Mentha piperata* L.	Lamiaceae	214
Pitpapara, Papara, Pitpapada	*Fumaria indica* (Haussk.) Pugsley.	Fumariaceae	175
Pohli	*Carthamus oxycantha* Bieb.	Asteraceae	106
Poplar	*Populus nigra* L.	Moraceae	253
Pudina	*Mentha spicata* L.	Lamiaceae	215
Puthkanda	*Achyranthes aspera* L.	Amaranthaceae	41
Puthkanda	*Achyranthes bidentata* Blume	Amaranthaceae	42
Putranjan	*Putranjiva roxburghii* Wall.	Euphorbiaceae	261
Rada	*Randia dumetorum* (Retz). Lam.	Rubiaceae	263
Rada	*Solanum erianthum* D. Don	Solanaceae	283
Rai	*Brassica juncea* (L.) Czern. & Coss.	Brassicaceae	90

Contd...

Contd...

Vernacular Names	Name of the Plant	Family	Page No.
Rai, Ohari	*Brassica nigra* (L.) Koch	Brassicaceae	91
Railway creeper	*Ipomoea cairica* (L.) Sweet	Convolvulaceae	188
Ramban	*Agave americana* L.	Agavaceae	52
Ramban	*Yucca aloifolia* L.	Liliaceae	324
Ram-Tulsi, Baramasi	*Ocimum sanctum* L.	Lamiaceae	232
Rasbhari	*Physalis peruviana* L.	Solanaceae	246
Rat-ki-rani	*Cestrum nocturnum* L.	Solanaceae	117
Rattanjot, Jamnotta	*Jatropha curcus* L.	Euphorbiaceae	192
Ratti, Raktgunja	*Abrus precatorius* L.	Fabaceae	
Reetha, Doda, Doddae	*Sapindus mukorossi* Gaertn.	Sapindaceae	279
Rosemary	*Rosmarinus officinalis* L.	Lamiaceae	271
Rudraksha	*Elaeocarpus ganitrus* Roxb. *ex* G. Don	Elaeocarpaceae	152
Ruhan, Rehana	*Soymida febrifuga* A. Juss.	Meliaceae	289
Sadabahar	*Catharanthus roseus* (L.) G. Don	Apocynaceae	110
Safeda	*Eucalyptus umbellata* Domin	Myrtaceae	159
Sago palm	*Cycas revoluta* L.	Cycadaceae	140
Salangaya	*Millettia extensa* Benth. *ex* Baker	Fabaceae	217
Sankhiryan, Sankheere	*Celastrus paniculatus* Willd.	Celastraceae	113
Sanspai	*Asparagus adscendens* Roxb.	Liliaceae	75
Santara, Santra	*Citrus aurantium* L.	Rutaceae	124
Sarpgandha	*Rauwolfia serpentina* Benth. *ex* Kurz.	Apocynaceae	266
Sarphar	*Saccharum arundinaceum* Retz.	Poaceae	274
Sarson	*Brassica campestris* L.	Brassicaceae	89
Saru , Morphankh	*Thuja orientalis* L.	Cupressaceae	304
Satavari	*Asparagus racemosus* Willd.	Liliaceae	76
Saunf, Soay	*Foeniculum vulgare* Mill.	Apiaceae	173
Shaitan-ka-per	*Alstonia scholaris* R.Br.	Apocynaceae	61
Shankhpushpi	*Canscora diffusa* (Vahl) R. Br.	Gentianaceae	99
Shehnai phool	*Tecoma grandiflora* Loisel.	Bignoniaceae	297
Shika- kai	*Acacia concinna* (Willd.) DC.	Fabaceae	38

Contd...

Contd...

Vernacular Names	Name of the Plant	Family	Page No.
Silver oak	*Grevillea robusta* A. Cunn. *ex* R. Br.	Proteaceae	179
Simbal, Simal	*Bombax ceiba* L.	Boraginaceae	87
Sirin	*Albizia lebbeck* Benth.	Fabaceae	54
Sitaphal	*Annona squamosa* L.	Annonaceae	67
Sooraj-mukhi	*Helianthus annuus* L.	Asteraceae	181
Soyabean	*Glycine max* Merr.	Fabaceae	176
Strawberry	*Fragaria chiloensis* (L.) Duchesne	Rosaceae	175
Sudab	*Ruta graveolens* L.	Rutaceae	274
Sudarshan, Hardarshan	*Canna indica* L.	Cannaceae	97
Suhanjan, Sungna	*Moringa oleifera* Lam.	Moringaceae	221
Sukh-chain	*Pongamia pinnata* Pierre.	Fabaceae	253
Tagar	*Tabernaemontana* divaricata (L.) R. Br.	Apocynaceae	293
Tahali, Shisham	*Dalbergia sissoo* Roxb. *ex* DC.	Fabaceae	143
Tamatar	*Lycopersicum esculentum* Mill.	Solanaceae	206
Tambaku, Tammaku	*Nicotiana tabacum* L.	Solanaceae	230
Taramira	*Eruca sativa* Lam.	Brassicaceae	158
Tardi	*Dioscorea bulbifera* L.	Dioscoreaceae	149
Tat-palanga	*Oroxylum indicum* Vent.	Bignoniaceae	234
Tecoma, Shehnai phool	*Tecoma stans* (L.) H.B. &K.	Bignoniaceae	297
Tejpatra	*Cinnamomum tamala* Nees & Eberm.	Lauraceae	122
Teur, Balsam	*Impatiens balsamina* L.	Balsaminaceae	187
Tikku-malu	*Cardiospermum halicacabum* L.	Sapindaceae	101
Til	*Sesamum indicum* DC.	Pedaliaceae	281
Tirmer	*Zanthoxylum armatum* DC.	Rutaceae	324
Tooni	*Cedrela toona* Roxb. *ex* Rottl.	Meliaceae	111
Toot	*Morus alba* L.	Moraceae	223
Tor	*Bauhinia vahlii* Wight & Arn.	Fabaceae	82
Triamble, Trayambalu	*Ficus roxburghii* Wall. *ex* Miq.	Moraceae	172
Ulat kamal	*Abroma augusta* Jacq.	Sterculiaceae	34
Umreyan, Umara	*Ficus glomerata* Roxb.	Moraceae	167
Zimmikand	*Amorphophallus campanulatus* (Roxb.) Blume *ex* Decne.	Araceae	65

Appendix III: Index to Glossary of Words Used by the Local Populace

English	Local Dialect
Ailments and State of the body	
Asthma	Dama
Belching	Dkar
Bladder stone	Patthri
Blind	Anna
Bodly pains	Pira, Darda
Boil	Phora
Cold	Najla, Jkhwam
Colic	Sul
Constipation	Kabji
Cough	Khang
Deaf	Tona
Diarrhoea	Atisar
Disk slip	Cas
Dumb	Gunga
Epilepsy	Mirgi
Fever	Tap, Bkhwar
Fit	Dora
Jaundice	Pusar rog, Peolia
Leprosy	Kor
Loose motion	Jlab, Dast
Measles	Trasal
Mumps	Knedde
Nose bleeding	Naksir
Odema, swelling	Phara, Sojas
One eyed	Kana
Paralysis	Trang
Phlegm	Kapph, Balgum
Phobia	Alkh
Pneumonia	Nomunia
Pus	Pak
Reddishness	Lalgi
Sciatica	Ringan
Sick	Bmar, Rogi

Contd...

Contd...

English	Local Dialect
Smallpox	Mata
Sneez	Chik
Stool	Cara
Sweating	Psinna
Throat irritation	Khriri
Tonsilitis	Kandi
Toothache	Klwanda
Tuberculosis	Chairog
Twisting Ankle	Mos
Urine	Psab, Muttar
Vomit	Ulti
Body	Parts
Arm	Ban
Ankle	Gitta
Armpit	Kacch
Back	Pith
Belly	Pet
Cheek	Khakh
Chest	Chatti
Chin	Chunni, Thodi
Calf	Pilli
Ear	Kan
Eye	Akh
Eye ball	Della
Face	Munh
Finger	Ongli, Gutthi
Foot	Per
Fore head	Kaudi
Gum	Masude
Hair	Kes
Head	Sir
Hand	Atth
Heart	Kalja
Heel	Addi
Knee	Godda
Liver	Jiggar
Lip	Labbar
Lung	Phephra
Head	Mattha
Mole	Tilla
Mouth	Muhn
Nail	Na
Neck	Kyari
Nose	Nakk
Nostril	Nasan
Shoulder	Munda
Skull	Khopri
Soft head	Talu

Contd...

Contd...

English	Local Dialect
Sole	Toe
Testicle	And
Thigh	Pat
Throat	Kandu
Thumb	Guttha
Tongue	Jib
Tooth	Dand
Waist	Lakk
Wrist	Maunna
Kin-ship (Relation)	
Aunt	Cacci
Brother	Pau, Pai
Child	Bacca
Daughter	Di
Daughter' husband	Jwai
Daughter-in-law	Nu
Father	Bab, Bappu, or Babbul
Father's elder brother	Taya
Grandfather	Dadda, Daddu
Great-grandfather	Pardadda
Grandson	Potru
Husband	Lara
Mother	Ma, Amma
Sister	Ben, Bebbe
Son	Puttar
Son's wife	Nu
Uncle	Cacca
Wife	Jnana
Wife's brother	Sala
Clothes	
Cap	Topu
Clothes	Kapre
Duppata	Cadru
Gloves	Dastane
Lengha	Ghagru
Pocket	Khisa
Shirt	Jaggu
Socks	Jurab
Thread	Taga
Trouser	Suthunu
Turban	Safa
Wool	Un
Fruits, Vegetable, Cereals, Drinks	
Apple	Seb, Seaun
Butter	Nauni
Curd	Dehu
Food	Khana
Grass	Ka

Contd...

Contd...

English	Local Dialect
Liquor	Gurangi
Meat	Skaar
Milk	Dud
Maize	Chali
Potato	Alu
Rice	Chaul
Sugar	Khand
Tea	Cha
Wheat	Kanak
Parts of Tree	
Branch	Tehani
Flower	Phul
Leaf	Patta
Root	Jarh
Thorn	Kanda
Tree	Rukh, Butta
Wood	Lakri
Dwelling Units	
Courtyard	Baranda
Door	Dwar
House	Kar, Makan
Stable	Tabella
Ground floor common room	Owan
Room	Kamra
Room on first floor	Bor
Roof top	Ladda
Sitting room	Bethak
Wall	Kand
Window	Dwari
Animals and related terms	
Ass	Khota
Bull	Bald
Calf	Bacchu
Cat	Billie
Crow	Ko
Dog	Kutta
Egg	Anda
Fish	Machi
Goat	Bakri
Hare	Serru
Hen	Kukri
Jackel	Gidder
Lion	Ser
Lice	Gun
Monkey	Bander
Mule	Kachar
Nest	Alana
Rat	Chua

Contd...

Contd...

English	Local Dialect
Tail	Poonch
Domestic Articles	
Almirah	Almari
Balance	Trakri
Blow pipe	Painpla
Bowl	Katori
Box	Dabba
Broom	Caru
Comb	Kangi
Cot	Manja
Key	Chabbi
Lid	Dakan
Lock	Janda
Mirror	Sisa
Needle	Sui
Plate	Thalu
Spoon	Chamcha
Table	Mej
Thread	Dhaga
Tools	
Axe	Kulhari
Digger	Chabal
Hammer	Thora
Plough	Al
Sickle	Drat, Drati
Stick	Sothi
Week Days	
Sunday	Twar
Monday	Swar
Tuesday	Mangal
Wednesday	Bud
Thursday	Bir bar
Friday	Sukkar
Saturday	Saniccar
Meals/ Dishes	
Breakfast	Datyallu
Butter	Noni
Chapati	Rotti
Clarified butter	Kiu
Cooked rice	Bhat, Col
Curd	Dau, Dai
Fried fermented sweet wheat chapati	Babru
Halwa	Kra
Milk pudding	Khir
Rice pan cake	Enklu
Snack	Tukru
Sour dish	Mani, Khatta
Sweet dish	Mittha

Contd...

Contd...

English	Local Dialect
Tea	Ca
Butter milk dish	Reru, Kari, Karu
Wheat fermented chapati	Baturu
Heavenly bodies	
Air	Oa
Clouds	Baddal
Earth	Tart
Earthquake	Hillan
Hailing	Aan
Lighting	Bijli
Moon	Chandarma
Moonlight	Chanani
Rain	Barkha
Sky	Ambar
Stars	Tare
Sun	Suraj
Sunshine	Tuppa
Customs, Rites, Ceremonies	
Dowry	Daj
Engagement	Karmai
Fast	Barat
Marriage	Bya
Marriage party	Brat
Married women	Pyoki
Parents house	
Marriage procession	Jnet
Marriage party member	Jneti
Day of death	Tith
10th day after death	Kapre
13th day after death	Sog, Tera, Kirya
30th day after death	Mkwan
Ist death anniversary	Barkhi
4th death anniversary	Cabarkh
Numerals	
One	Ik
Two	Do
Three	Tin
Four	Char
Five	Panj
Six	Che
Seven	Sat
Eight	Atth
Nine	Naun
Ten	Das
Eleven	Gyara
Twelve	Bara
Thirteen	Tera
Fourteen	Choda

Contd...

Contd...

English	Local Dialect
Sixteen	Sola
Seventeen	Stara
Eighteen	Athara
Ninteen	Unni
Twenty	Bi
Thirty	Ti
Forty	Chali
Fifty	Panja
Sixty	Sath
Seventy	Sattar
Eighty	Assi
Ninty	Nabbe
Hundred	So
Thousand	Zahar

Plates

Pl. 01: A. *Abelmoschus esculentus* (L.) Moench. **B**. *Abelmoschus moschatus* Medic. **C**. *Abroma augusta* Jacq. **D**. *Abrus precatorius* L.

Pl. 02: A. *Acacia catechu* Willd. **B**. *Acacia concinna* (Willd.) DC. **C**. *Acacia nilotica* L. **D**. *Achras zapota* L.

Pl. 03: A. *Achyranthes aspera* L. **B.** *Achyranthes bidentata* Blume. **C.** *Acorus calamus* L. **D.** *Actinidia chinensis* var. *deliciosa* A. Chev.

Pl. 04: A. *Adhatoda vasica* Nees **B.** *Adiantum capillus-veneris* L. **C.** *Adiantum lunulatum* Burm. **D.** *Aegle marmelos* (L.) Correa ex Roxb.

Pl. 05: A. *Aerva lanata* (L.) Juss. *ex* Schult. **B**. *Agave americana* L. **C**. *Ageratum conyzoides* L. **D**. *Albizia lebbeck* Benth.

Pl. 06: A. *Albizia stipulata* Boiv. **B**. *Allium cepa* L. **C**. *Allium sativum* L. **D**. *Aloe barbedensis* Mill.

Pl. 07: A. *Aloe vera* (L.) Webb & Berth. **B**. *Alstonia scholaris* R.Br. **C**. *Amaranthus paniculatus* L. **D**. *Amaranthus spinosus* L.

Pl. 08: A. *Amaranthus viridis* L. **B**. *Amorphophallus campanulatus* (Roxb.) Blume ex Decne. **C**. *Andrographis paniculata* Wall. ex Nees. **D**. *Anetham sowa* Roxb.

Pl. 09: **A.** *Annona squamosa* L. **B.** *Arachis hypogaea* L. **C.** *Argemone mexicana* L. **D.** *Arisaema tortuosum* Schott.

Pl. 10: **A**. *Artemisia japonica* Thunb. **B.** *Artemisia vulgaris* L. **C**. *Artocarpus integrifolia* L. **D.** *Artocarpus lakoocha* Roxb.

Pl. 11: **A**. *Asclepias curassavica* L. **B**. *Asparagus adscendens* Roxb.
C. *Asparagus racemosus* Willd. **D**. *Avena sativa* L.

Pl. 12: **A**. *Azadirachata indica* A. Juss. **B**. *Bacopa monnieri* (L.) Penn.
C. *Bambusa arundinacca* (Retz.) Roxb. **D**. *Bambusa nutans* Wall.

Pl. 13: **A.** *Bauhinia vahlii* Wight & Arn. **B.** *Bauhinia variegata* L. **C.** *Berberis asiatica* Roxb. ex DC. **D.** *Boerhavia diffusa* L.

Pl. 14: **A**. *Bombax ceiba* L. **B.** *Bougainvillea spectabilis* Willd. **C**. *Brassica campestris* L. **D**. *Brassica juncea* (L.) Czern. & Coss.

Pl. 15: **A.** *Brassica nigra* (L.) Koch. **B.** *Bryophyllum calycinum* Salisb. **C.** *Butea monosperma* (Lam.) Kuntze **D.** *Callistemon lanceolatus* DC.

Pl. 16: **A.** *Calotropis procera* (Ait.) R.Br. **B**. *Canna indica* L. **C**. *Cannabis sativa* L. **D**. *Canscora diffusa* (Vahl) R.Br.

Pl. 17: **A.** *Capsicum annuum* L. **B.** *Cardiospermum halicacabum* L. **C.** *Carduus nutans* L. **D.** *Carica papaya* L.

Pl. 18: **A.** *Carissa carandas* L. var. *congesta* (Wt.) Bedd. **B.** *Carissa spinarum* Hook. f. **C.** *Carthamus oxycantha* Bieb. **D.** *Carum copticum* Hiern.

Pl. 19: **A**. *Cassia fistula* L. **B**. *Cassia occidentalis* L. **C.** *Catharanthus roseus* G. Don **D**. *Cedrela toona* Roxb.

Pl. 20: **A.** *Celastrus paniculatus* Willd. **B.** *Celosia argentea* L. **C.** *Centella asiatica* (L.) Urb. **D.** *Centipeda minima* (L.) A. Br. & Asch.

Pl. 21: **A.** *Cestrum nocturnum* L. **B.** *Chenopodium album* L. **C.** *Chenopodium ambrosioides* L. **D.** *Cicer arietinum* L.

Pl. 22: **A.** *Cinnamomum camphora* (L.) Nees & Eberm **B.** *Cinnamomum tamala* Nees & Eberm **C.** *Cissampelos pareira* L. **D.** *Citrus aurantium* L.

Pl. 23: A. *Citrus grandis* (L.) Osbeck. **B.** *Citrus jambhiri* Lush. **C.** *Citrus limon* (L.) Burm. f. **D.** *Citrus medica* L.

Pl. 24: A. *Clematis grata* Wall. **B.** *Colebrookia oppositifolia* Sm. **C.** *Colocasia esculenta* (L.) Schott **D.** *Cordia dichotoma* Forst. f.

Pl. 25: **A.** *Coriandrum sativum* L. **B.** *Cucumis sativus* L. **C.** *Cucurbita hispida* Thunb. **D.** *Cucurbita pepo* L.

Pl. 26: **A.** *Curcuma domestica* (Medik.) Valeton **B.** *Cuscuta reflexa* Roxb. **C.** *Cycas revoluta* L. **D.** *Cymbopogon citratus* (DC.) stapf.

Pl. 27: **A.** *Cynodon dactylon* (L.) Pers. **B.** *Dalbergia sissoo* Roxb. **C.** *Datura metel* auct. non L. **D.** *Datura stramonium* L.

Pl. 28: **A.** *Daucus carota* L. **B.** *Dendrocalamus hamiltonii* Nees & Arn. **C.** *Dioscorea bulbifera* L. **D.** *Dodonaea viscosa* L. Jacq.

Pl. 29: **A.** *Duranta repens* L. **B.** *Eclipta prostrata* Roxb. **C.** *Elaeocarpus ganitrus* Roxb. *ex* G. Don **D.** *Elettaria cardamomum* (L.) Maton.

Pl. 30: **A.** *Emblica officinalis* Gaertn. **B.** *Equisetum debile* Roxb. ex Vauch. **C.** *Eriobotrya japonica* Lindl. **D.** *Eruca sativa* Lam.

Pl. 31: A. *Eucalyptus umbellata* Domin **B.** *Eugenia cerasoides* Roxb. **C.** *Eugenia jambolana* Lam. **D.** *Eupatorium adenophorum* Spreng.

Pl. 32: A. *Euphorbia pulcherrima* Willd. **B.** *Euphorbia royleana* Boiss. **C.** *Ficus benghalensis* L. **D.** *Ficus carica* L.

Pl. 33: A. *Ficus glomerata* Roxb. **B.** *Ficus hispida* L. f. **C.** *Ficus palmata* Forsk. **D.** *Ficus religiosa* L.

Pl. 34: A. *Ficus roxburghii* Wall. *ex* Miq. **B.** *Flacourtia ramontchi* L. Herit. **C.** *Foeniculum vulgare* Mill. **D.** *Fragaria chiloensis* (L.) Duchesne.

Pl. 35: A. *Fumaria indica* (Haussk.) Pugsley. **B.** *Glycine max* Merr. **C.** *Glycyrrhiza glabra* L. **D.** *Gossypium arboreum* L.

Pl. 36: A. *Grevillea robusta* A. Cunn. ex R. Br. **B.** *Grewia oppositifolia* Roxb. **C.** *Habenaria pectinata* Don **D.** *Helianthus annuus* L.

Pl. 37: A. *Helicteres isora* L. **B.** *Hibiscus rosa-sinensis* L. **C.** *Holarrhena pubescens* Wall. *ex* Don **D.** *Hordeum vulgare* L.

Pl. 38: A. *Hypericum cernum* Roxb. **B.** *Impatiens balsamina* L. **C.** *Ipomoea cairica* (L.) Sweet **D.** *Ipomoea carnea* Jacq.

Pl. 39: A. *Jackranda acutifolia* Humb. & Bonpl. **B.** *Jasminum humile* L. **C.** *Jasminum multiflorum* Roth. **D.** *Jasminum sambac* (L.) Ait.

Pl. 40: A. *Jatropha curcus* L. **B.** *Juglans regia* L. **C.** *Lagenaria siceraria* (Molina) Standl. **D.** *Lagerstroemia indica* L.

Pl. 41: **A**. *Lannea coromandelica* (Houtt.) Merr. **B.** *Lantana camara* L. **C.** *Lavandula angustifolia* Mill. **D.** *Lawsonia alba* Lam.

Pl. 42: **A**. *Lens culinaris* Medic. **B.** *Lepidium sativum* L. **C.** *Leucaena leucocephala* (Lam.) de Wit. **D.** *Linum usitatissimum* L.

Pl. 43: A. *Litchi chinensis* Sonn. B. *Luffa aegyptica* Mill. C. *Lycopersicum esculentum* Mill. D. *Madhuca indica* J. F. Gmel.

Pl. 44: A. *Mallotus philippinensis* (Lam.) Muell-Arg. B. *Malvaviscus conzatii* Greenm. C. *Mangifera indica* L. D. *Martynia annua* L.

Pl. 45: A. *Melia azedarach* L. **B.** *Melissa officinalis* L. **C.** *Mentha piperata* L. **D.** *Mentha spicata* L.

Pl. 46: A. *Mesua ferrea* L. **B.** *Millettia extensa* Benth. ex Baker **C.** *Mimosa pudica* L. **D.** *Mirabilis jalapa* L.

Pl. 47: **A**. *Momordica charantia* L. **B.** *Moringa oleifera* Lam. **C.** *Morus alba* L. **D.** *Murraya koenigii* Spreng.

Pl. 48: **A**. *Murraya paniculata* (L.) Jack. **B.** *Musa sapientum* L. **C.** *Nasturtium officinale* R. Br. **D.** *Nerium indicum* Mill.

Pl. 49: A. *Nicotiana tobaccum* L. **B.** *Ocimum basilicum* L. **C.** *Ocimum sanctum* L
D. *Opuntia dillenii* Haw.

Pl. 50: A. *Oroxylum indicum* Vent. **B.** *Oryza sativa* L. **C.** *Oxalis corniculata* L.
D. *Papaver rhoeas* L.

Pl. 51: A. *Parthenium hysterophorus* L. **B.** *Paspalum scrobiculatum* L.
C. *Pedilanthus tithymaloides* (L.) Poir. **D.** *Pelargonium graveolens* L'Hert.

Pl. 52: A. *Pennisetum americanum* L. **B.** *Pennisetum purpureum* Schum.
C. *Phoenix sylvestris* Roxb. **D.** *Phragmites karka* Trin. ex Steud.

Pl. 53: **A.** *Phyllanthus urinaria* L. **B.** *Physalis peruviana* L. **C.** *Pinus roxburghii* Sarg. **D.** *Piper longum* L.

Pl. 54: **A.** *Pistacia integerrima* Stewart ex Brand. **B.** *Plumbago zeylanica* L. **C.** *Pogostemon benghalensis* Burn. f. **D.** *Polyalthia longifolia* (Sonn.) Thw.

Pl. 55: **A.** *Pongamia pinnata* Pierre. **B.** *Populus nigra* L. **C.** *Premna latifolia* Roxb. **D.** *Prunus amygdalus* Baill.

Pl. 56: **A.** *Prunus domestica* L. **B.** *Prunus persica* Benth. & Hook. f. **C.** *Psidium guajava* L. **D.** *Punica granatum* L.

Pl. 57: **A.** *Putranjiva roxburghii* Wall. **B.** *Pyrus communis* L. **C.** *Pyrus pashia* Ham. **D.** *Randia dumetorum* (Retz.) Lam.

Pl. 58: **A.** *Ranunculus muricatus* L. **B.** *Raphanus sativus* L. **C.** *Rauwolfia serpentina* Benth. ex Kurz. **D.** *Reinwardtia indica* Dumort.

Pl. 59: A. *Rhynchostylis retusa* Blume **B.** *Ricinus communis* L. **C.** *Rosa damascena* Mill. **D.** *Rosmarinus officinalis* L.

Pl. 60: A. *Roylea cinerea* (D. Don) Baill. **B.** *Rubus ellipticus* Sm. **C.** *Rumex hastatus* Don **D.** *Ruta graveolens* L.

Pl. 61: A. *Saccharum arundinaceum* Retz. **B.** *Saccharum officinarum* L. **C**. *Saccharum spontaneum* L. **D.** *Salvia coccinea* L.

Pl. 62: A. *Santalum album* L. **B.** *Sapindus mukorossi* Gaertn. **C.** *Saraca asoca* (Roxb.) De Wild. **D.** *Sechium edule* Sw.

Pl. 63: **A.** *Sesamum indicum* DC. **B.** *Setaria viridis* Beauv. **C.** *Solanum erianthum* Don **D.** *Solanum melongena* L.

Pl. 64: **A.** *Solanum nigrum* L. **B.** *Solanum surattense* Burm. f. **C.** *Solanum tuberosum* L. **D.** *Sonchus asper* Hill.

Pl. 65: A. *Soymida febrifuga* A. Juss. **B.** *Spilanthes acmella* Murr. **C.** *Spinacia oleracea* L. **D.** *Stephania glabra* (Roxb.) Miers.

Pl. 66: A. *Tabernaemontana divaricata* R. Br. **B.** *Tagetes erecta* L. **C.** *Tamarindus indica* L. **D.** *Taraxacum officinale* Wigg.

Pl. 67: **A.** *Tecoma grandiflora* Loisel. **B.** *Tecoma stans* (L.) H.B. & K. **C.** *Terminalia arjuna* Wt. & Arn. **D.** *Terminalia bellirica* (Gaertn.) Roxb.

Pl. 68: **A.** *Terminalia chebula* Retz. **B.** *Thevetia neriifolia* Juss. ex Steud. **C.** *Thuja orientalis* L. **D.** *Tinospora cordifolia* (Willd.) Miers.

Pl. 69: **A.** *Trichodesma indicum* R. Br. **B.** *Tridax procumbens* L. **C.** *Trifolium alexandrinum* L. **D.** *Trigonella foenum-graecum* L.

Pl. 70: **A.** *Triticum aestivum* L. **B.** *Tylophora indica* (Burm. f.) Merr. **C.** *Urtica dioica* L. **D.** *Vanda roxburghii* R. Br.

Pl. 71: A. *Verbascum thapsus* L. **B.** *Vernonia anthelmintica* Willd. **C.** *Vicia pallida* Facquem. **D.** *Viola canescens* Wall.

Pl. 72: A. *Vitex negundo* L. **B.** *Vitis jacquemontii* Parker **C.** *Vitis vinifera* L. **D.** *Wendlandia heynei* (Roem. & Schult.) Sant & Merch.

Pl. 73: **A.** *Withania somnifera* (L.) Dunal **B.** *Woodfordia floribunda* Salisb. **C.** *Xanthium strumarium* L. **D.** *Yucca aloifolia* L.

Pl. 74: **A.** *Zanthoxylum armatum* DC. **B**. *Zea mays* L. **C**. *Zingiber officinale* Rosc. **D.** *Zizyphus jujuba* Mill.